All Rights Reserved
Copyright © 1999
INTERNATIONAL ASTRONOMICAL UNION
98bis, bd Arago – 75014 Paris – France
Tel: +33 1 4325 8358; Fax: +33 1 4325 2616;
E-mail: iau@iap.fr; Web Site: www.iau.org

No part of the material protected by this copyright notice may be reproduced or utilized in any form or by any means, electronic or mechanical including photocopying, recording or by any information storage and retrieval system, without written permission from the IAU.

Published on behalf of the
INTERNATIONAL ASTRONOMICAL UNION

by
Astronomical Society of the Pacific
First published 1999

Managing Editor, D. H. McNamara
Production Manager, Enid Livingston

EDITORIAL/PUBLISHING OFFICE:		CATALOG/BOOK ORDERS:	
Managing Editor		IAU Publications	
PO Box 24463		390 Ashton Avenue	
211 KMB Brigham Young University		San Francisco, CA 94112-1722	
Provo, UT 84602-4463		USA	
USA			
		(415) 337-1100	Phone
(801) 378-2298	Phone	(415) 337-5205	FAX
(801) 378-2265	FAX	catalog@aspsky.org	E-Mail
pasp@astro.byu.edu	E-Mail	www.aspsky.org	Web Site

Printed by Sheridan Books, Inc.
Chelsea, Michigan

Library of Congress Catalog Card Number: 99-64824
ISBN: 1-886733-008-0

INTERNATIONAL ASTRONOMICAL UNION
UNION ASTRONOMIQUE INTERNATIONALE

ACTIVE GALACTIC NUCLEI AND RELATED PHENOMENA

Proceedings of the 194th Symposium
of the International Astronomical Union
held in Byurakan, Armenia
17-22 August 1998

Edited by:

YERVANT TERZIAN
Department of Astronomy, Cornell University
Ithaca, New York, USA

DANIEL WEEDMAN
Department of Astronomy & Astrophysics, Pennsylvania State University
University Park, Pennsylvania, USA

and

EDWARD KHACHIKIAN
Byurakan Astrophysical Observatory, Armenian National Academy of Sciences
Byurakan, Armenia

Publisher

Information on other IAU Symposium proceedings is given at the back of this volume

Table of Contents

Dedication xi
Preface xv
SOC/LOC xvii
List of Participants xviii
Conference Photo.. xxi

I. OBSERVATIONAL PROPERTIES OF AGN

An Overview of AGN
    Ian Robson 3
HST Observations of Active Galactic Nuclei
    F. Duccio Macchetto 12
Infrared Emission from AGN
    David B. Sanders 25
Radio Properties of Quasars and AGN
    Kenneth I. Kellermann 39
Megamaser Activity in Active Galaxies
    Willem A. Baan 46
The Mid-Infrared Spectral Properties of Starburst Galaxies and Active Galactic Nuclei
    Olivier Laurent and I.F. Mirabel 54
VLBI Imaging of Luminous Infrared Galaxies: Starbursts & AGN
    Harding E. Smith et al. 60
HST Observations of the Jet of 3C120
    Gerard Lelièvre et al. 66
Are LINERs Starbursts or Mini-quasars? A Comparative Study of their Ultraviolet Spectra
    Anuradha Koratkar 71
Near-IR Polarimetry of the Obscured Nucleus in the Circinus Galaxy
    David M. Alexander et al. 75
Multiwavelength Properties of Narrow-line Seyfert 1's: Studying One Extreme of the AGN Primary Eigenvector
    Paul Hirst et al. 77
Kinematic Mapping of the Narrow Line Region of NGC4151
    Mary E. Kaiser et al. 79
On the Nature of Radio Emission of AGNs: Spectra, Milli-arcsecond Structure and Polarization
    Yu.A. Kovalev and Y.Y. Kovalev 82
Surface Photometry of Barred AGN Arakelian 564
    G. Petrov et al. 84
Spectrophotometry of Selected AGN: Seyfert Galaxy AKN 564
    L.S. Slavcheva 87

Global Oscillations of Masing Disks in Megamasers
  Atsuo T. Okazaki 90

II. OBSERVATIONAL PROPERTIES OF AGN AND RELATED OBJECTS

Quasars
  Patrick S. Osmer 97
The Evolution and Luminosity of Quasars
  Vahé Petrosian 105
Blazars: Clues to Jet Physics
  Rita M. Sambruna 113
Markarian and Seyfert Galaxies
  Edward Ye. Khachikian 123
Feeding the Central Engine in Giant Radio Galaxies
  I. Felix Mirabel and O. Laurent 133
On the QSO Content of the First Byurakan Survey and the Completeness of the Palomar Green Survey
  Philippe Véron et al. 140
$2.5 - 11.6\mu$ Spectrophotometry and Imaging of AGNs
  Jean Clavel et al. 145
The Second Byurakan Survey: Faint Markarian Galaxies, AGN and QSOs
  Jivan A. Stepanian 151
Investigation of a New Sample of IRAS Galaxies
  Areg M. Mickaelian 156
Galaxies in Selected Fields of the Second Byurakan Survey
  Susanna A. Hakopian and S.K. Balayan 162
Identification of Faint Markarian Galaxies with IRAS Sources
  V.T. Ayvazyan 168
The Study Of Extended Radio Galaxies
  M.A. Hovhannisyan 171
The Variability of Optically Selected QSO Sample
  Keliang Huang and H. Zhou 172
Photometric Monitoring of Selected Quasars: The Highly Luminous Quasar HS 1946+7658
  B.M. Mihov et al. 175
Monitoring of 1-22 GHz Instantaneous Spectra of 550 Compact Extragalactic Objects in 1997-1998
  Y.Y. Kovalev et al. 177
The Continuum Spectra of the Core of Hotspots of Cygnus-A in the Millimetre and Submillimetre
  L.L. Leeuw and E.I. Robson 179
Hidden Quasars in Ultraluminous Infrared Galaxies
  H.D. Tran et al. 181

The Variability of QSOs in the Optical Band
  Dario Trèvese et al. 184

III. AGN THEORY AND MODELS

Energy Sources and Physical Processes in Active Galaxies
  Daniel W. Weedman 191
What Physics Drives the Unified Model?
  Michael A. Dopita 199
Magnetohydrodynamic Origin of Jets from Accretion Disks
  Richard V.E. Lovelace et al. 208
Active Galaxies and Cosmic Evolution
  Malcolm S. Longair 219
The Jets of Quasar 3C 345 and 1803+784
  Leonid I. Matveenko and A.I. Witzel 229
Are Gamma-Ray Bursts Signals of Supermassive Black Hole Formation?
  Kevork Abazajian et al. 235
The Emission Line Properties of the 3CR Radio Galaxies at Redshift
  One: Shocks, Evolution, and the Alignment Effect
  Philip Best et al. 241
Accretion Disks and Star Formation
  Suzy Collin and J.P. Zahn 246
Origin of Blazar Activity
  Marina M. Romanova 256
Velocity Fields in Spiral Galaxies
  Alexei M. Fridman and O.V. Khoruzhii 269
Nuclear kpc-sized Disks of Spiral Galaxies
  Anatoly V. Zasov and A.V. Moiseev 279
Broad Emission Lines in AGN: Phenomenology and Models
  Jack W. Sulentic et al. 285
Quantized Redshifts - New Physics or Old Muddle?
  William M. Napier 290
The Nature of the UV-optical Continuum in Seyfert 2 Galaxies
  Thaisa Storchi-Bergmann et al. 295
An Asymmetric Relativistic Model for FRII Radio Sources
  Tigran G. Arshakian and M.S. Longair 301
Luminosity Correlation of the X-ray Selected Radio-Loud AGNs
  Qirong Yuan et al. 306
Black Hole Masses and Unification of Seyferts
  Rumen Bachev et al. 311
Bardeen-Petterson Effect and Broad HI Profiles of Seyferts
  Rumen Bachev 313
Estimating the Number of Broad-Line Region Clouds
  Matthias Dietrich 317

Diffusion Mechanism of Radiation of a Charged Particle on the Randomly Spaced Dust Grains in the X-Rays
    Z.S. Gevorkian et al. 319
General Discussion of Accretion Disks
    Vahagn G. Gurzadyan 321
Compact Nuclei of Galaxies and Sources of Their Energy
    L.S. Grigoryan and G.S. Sahakian 323
Magnetic Field Strengths of GPS Radio Sources
    V.G. Panajyan 324
Quark-Hadron Hybrid Stars as Remnants of Supernovae Explosions
    Gevorg S. Poghosyan 327
A New Numerical Method for Investigation of Evolution of Stars and Active Galactic Nuclei
    Avetis A. Sadoyan et al. 330
Can the Sonic Radius in ADAF Be Large?
    Feng Yuan et al. 332
M87 Solves the Problem of AGN
    S.G. Iskudarian 334

IV. AGN RELATED PHENOMENA

Activity in Interacting and Binary Galaxies
    Yervant Terzian 341
Redshifts of New Galaxies
    Halton Arp 347
AGN Variability
    Toshihiro Kawaguchi and S. Mineshige 356
Supernova Types, Star Formation and AGN
    Massimo Turatto et al. 364
The Gravitational Lens System B1030+074. Discovery and Follow-up
    Emily Xanthopoulos et al. 373
An Analysis of 900 Optical Rotation Curves
    David F. Roscoe 379
Variable Sources in Active Galactic Nuclei
    Vladimir A. Hagen-Thorn et al. 384
A UV Flare at the Center of the Elliptical Galaxy NGC 4552
    Lucio M. Buson et al. 389
Emission-Line Galaxies in Shahbazian Compact Groups
    H. Tiersch et al. 394
Morphology and Photometry of the UV-Excess Galaxy Pair NGC 7770/7771
    Gamal B. Ali and A.M.I. Osman 399
The Orientation of Extragalactic Radiosources Relative to the Optical Axes of their Host Galaxies
    R. Andreasyan and H. Sol 404
Rapid Variations in the Broad $H\beta$ Profile of the Radio Galaxy 3C 390.3
    Norair S. Asatrian et al. 406

Rapid Profile Variations in the Broad Hα Line of the Seyfert Galaxy
  Markarian 6
    Norair S. Asatrian et al. 409
Multiwave Monitoring of the Nuclei of Seyfert Galaxies and Quasars
  at Several Telescopes of FSU and European Countries in the Frame
  of an INTAS Program
    N.G. Bochkarev et al. 411
Variability of the Broad Emission H$_\beta$ Profile in 3C390.3
    A.N. Burenkov et al. 414
Flare in the Hydrogen Line Region of the NGC 3227 Nucleus on
  January 12-15, 1977
    Iraida Pronik et al. 416
Strong Radio Outbursts in Six Active Galactic Nuclei in 1997-1998
    Y.Y. Kovalev 418
Long Term X-ray Variability of Galactic Nuclei
    John Cunniffe and E.J.A. Meurs 420
Searching for Low-Mass Supermassive Black Holes
    Michele Cappellari et al. 422
Seyfert Activity in Ring Galaxies Due to Galactic Interactions
    Tapan K. Chatterjee 424
The Role of Neutral Hydrogen in the Evolution of Spiral and
  Irregular Galaxies
    Victor A. Ambartsumian and A.L. Gyulbudaghian 426
Are There Quasars at Non-Cosmological Distances?
    Haik A. Harutyunian 428
First Ranked Galaxy Morphology and Morphological Content in
  Groups of Galaxies
    A.P. Mahtessian and V.H. Movsessian 432
Activity Phenomena Observed at Radio Frequencies in Spiral Galaxies
    V.H. Malumian 434
Simultaneous UBVRI Light Curves of the Seyfert Galaxy NGC 4151
  During the Extraordinary Brightening in 1989-1996
    N. Merkulova et al. 436
The Anisotropy in the Distribution of the Major Axes of Elongated
  Shakhbazian CGCG
    R.A. Vardanyan 438
NGC 3077: H$_\alpha$ Features and the Young Population
    H. Abdel-Hamid and P. Notni 439
A New Sample of Candidates GPS Radio Sources with Intermediate
  Flux Densities
    Gabriel A. Ohanian 442
Determination of the OB Stellar Population of IZw18 on the Basis
  of its H$_\gamma$ and H$_\delta$ Absorption Lines
    A. Sinanyan et al. 444

V. CONFERENCE SUMMARY

Future Directions in AGN Research
    Julian H. Krolik 453

VI. REFLECTIONS OF V.A. AMBARTSUMIAN

Influence of V. A. Ambartsumian on the Development of Astronomy
    Alexander Boyarchuk 467
Ambartsumian's Greatest Insight - The Origin of Galaxies
    Halton Arp 473

We dedicate this volume to the memory
of the eminent astrophysicist

VICTOR AMAZASP AMBARTSUMIAN

VICTOR AMAZASP AMBARTSUMIAN
1908-1996

VICTOR AMAZASP AMBARTSUMIAN (1908-1996)

In a productive career lasting more than 70 years, Victor Amazasp Ambartsumian contributed extensively to the progress of science. His creative research ideas stimulated physics and astrophysics, while his leadership in scientific policy developed and maintained widespread scientific activities in postwar Armenia and abroad.

Ambartsumian was born on September 18, 1908 in Tibilisi, Georgia. He was educated at the University of Leningrad and graduated in 1928. In 1932 he became the head of astrophysics at the same university. He moved to Armenia in 1943 and was one of the founding members of the Armenian Academy of Sciences. In 1946 he founded the Byurakan Astrophysical Observatory and was its director until 1988. In 1947 he became the president of the Armenian Academy of Sciences, a post he retired from in 1993. His scientific works include the physics of gaseous nebulae and radiation transfer theory, the physics of multiple light scattering, the invariance principle, stellar evolution, stellar associations, protostars, and the evolution of galaxies and in particular the study of active galactic nuclei. His work and leadership made the Byurakan Observatory world famous.

His career was rewarded by many honors internationally and within the Soviet Union. He served as President of the IAU from 1961 to 1964, and in 1968 and again in 1970 he was twice elected President of the International Council of Scientific Unions. He was a member of many foreign academies including the U. S. National Academy of Sciences, the Royal Society, and the Academie Franaise. In the Soviet Union among other honors he received five times the Order of Lenin, and the Hammer and Sickle Gold Medal. In 1994 he was proclaimed National Hero of the Republic of Armenia, by then an independent country.

At the age of 80, Ambartsumian went on a protest fast in Moscow to attract world attention to the conflict in Nagorno-Karabakh. He was a scientist, a leader and a humanitarian. He died in Byurakan on August 12, 1996.

Byurakan and the 2.6 Telescope

PREFACE

Since its founding in 1946, the Byurakan Astrophysical Observatory has been a significant astronomical institution. Victor Ambartsumian's insights regarding the importance of astronomical activities in the Armenian SSR were distinctive examples of genius, both in scientific research and scientific administration. Theoretical astrophysics flourished while various observational undertakings, especially the Markarian survey for galaxies with ultraviolet continua, provided frontier data eagerly pursued by astronomers in the rest of the world.

For decades, international astronomers were awed at how the small Armenian Republic could maintain such a major research facility. By the end of the 1970s, progress for the future was being set in place with construction of the 2.6-m telescope. By the conclusion of the 1980s, Byurakan Observatory had a versatile, world-class telescope to accompany the famous and unique 1-meter Schmidt photographic telescope. A new, large scientific office and laboratory building had also been constructed.

Beginning in 1990, astronomical progress was disrupted by political events accompanying the dissolution of the USSR. Interruption of society's infrastructure had widespread and unpredictable negative consequences. Astronomers throughout the world felt the tragedy of events in Armenia as we saw the deprivations of our professional colleagues. We were humbled and inspired by the determination of the Armenian astronomers to maintain their Observatory and their astronomy in the face of overwhelming difficulties. Attending Symposium 194 of the International Astronomical Union in 1998 gave the international community of astronomers the opportunity to acknowledge their Armenian colleagues for their successful efforts in continuing research at Byurakan.

The "Active Galaxies" theme of the Symposium reflected major past contributions from Byurakan, and stimulating presentations from throughout the world heralded invigorating new research, both in the topic and at the Observatory. This Symposium summarized the exciting progress being made toward understanding the most energetic and the most distant objects in the universe, these objects being Active Galaxies. Invited reviews and contributed papers, which follow in this Volume, utilize observational techniques from gamma ray to radio wavelengths. Starburst galaxies, blazars, Seyfert galaxies, and radio galaxies are discussed in terms of how various observed characteristics relate to hypothesized massive black holes in galactic nuclei, and to the extensive star formation taking place in disks surrounding these nuclei.

Unification schemes are summarized that accomodate diverse observational properties within similar models for all forms of active galactic nuclei. Theoretical papers summarize issues of jet formation, mass flow from disk into nucleus, and triggering mechanisms for starbursts. New observations in infrared and submillimeter wavelengths demonstrate how the high redshift universe is dominated by ultraluminous infrared galaxies. New results are presented utilizing the First and Second Byurakan Sky Surveys (the First being the original Markarian survey). A few papers also utilize anomalies of redshift distributions or quasar arrangements to argue that a need remains for new physics to explain some observations.

Preface

On behalf of the International Astronomical Union, the Editors of this Volume thank many people and organizations who contributed to the success of Symposium 194 and to the wide-ranging astrophysics which was included. The greatest effort was expended by the Local Organizing Committee, who worked months in advance to prepare all logistical arrangements and who worked tirelessly during the time of the Symposium to accommodate all needs of the participants. Chairman of the LOC was H. Haratyunian and Secretary was A. Mickaelian, both assisted by E. Balayan, S. Balayan, K. Gigoyan, A. Gyulbudaghian, V. Hambaryan, T. Magakian, A. Mahtessian, M. Melikian, T. Movsessian, V. Movsessian, E. Nikoghossian, and A. Yeghiazarian.

Topics and papers for the Symposium were selected by the Scientific Organizing Committee, co-Chaired by E. Khachikian and Y. Terzian. Other members were H. Arp, F. Bertola, A. Boyarchuk, G. Burbidge, D. Kunth, M. Longair, J. Narlikar, V. Trimble, D. Weedman, L. Woltjer, and A. Zasov.

Financial support for the Symposium was provided by the IAU, the Armenian Academy of Sciences, the Byurakan Astrophysical Observatory, The Armenia Fund, Cornell University, INTAS (International Association for the promotion of cooperation with scientists from the new independent states of the former Soviet Union), and the Associazione Italia-Armenia.

Finally, the Editors express great appreciation to Craig Sheppard of Cornell University for his excellent and extensive work in assembling, formating, and editing the contributions to this Volume.

Y. Terzian, Ithaca, New York, USA
D. Weedman, State College, Pennsylvania, USA
E. Khachikian, Byurakan, Republic of Armenia

April, 1999

SCIENTIFIC ORGANIZING COMMITTEE

Edward Khachikian (Co-Chair)	Armenia
Yervant Terzian (Co-Chair)	USA
Halton Arp	Germany
Franesco Bertola	Italy
Alexander Boyarchuk	Russia
Geoffrey Burbidge	USA
Daniel Kunth	France
Malcolm Longair	UK
Jayant Narlikar	India
Virginia Trimble	USA
Daniel Weedman	USA
Lodewijk Woltjer	France
Anatoly Zasov	Russia

LOCAL ORGANIZING COMMITTEE

Haik Harutyunian	Chairman
Areg Mickaelian	Secretary

E. Balayan, S. Balayan, K. Gigoyan, A. Gyulbudaghian, V. Hambaryan, T. Magakian, A. Mahtessian, N. Melikian, T. Movsessian, V. Movsessian, E. Nikoghossian, A. Yeghiazarian

LIST OF PARTICIPANTS OF IAU SYMPOSIUM 194

Abazajian, Kevork N., University of California, San Diego, USA
Abrahamian, Martin G., Yerevan State Pedagogical Institute, Armenia
Alexander, David M., University of Hertfordshire, UK
Ali, Gamal Bakr, Ntl Research Inst. of Astron. & Geophysics, Egypt
Amirkhanian, Arthur S., Byurakan Astrophysical Observatory, Armenia
Andreasyan, Ruben R., Byurakan Astrophysical Observatory, Armenia
Arp, Halton C., Max-Planck-Inst. fur Astrophysik, Garching, Germany
Arshakian, Tigran G., Univ. of Cambridge, UK / Byurakan Obs., Armenia
Asatrian, Norair S., Byurakan Astrophysical Observatory, Armenia
Baan, Willem A., NFRA - Westerbork Observatory, Netherlands
Balayan, Smbat K., Byurakan Astrophysical Observatory, Armenia
Bergmann, Thaisa S., Instituto de Fisica - UFRGS, Brasil
Bertola, Francesco, Dipartimento di Astronomia, Padova Univ., Italy
Best, Philip N., Sterrewacht Leiden, Huygens Laboratory, Netherlands
Bochkarev, Nikolai G., Sternberg Astronomical Institute, Russia
Boyarchuk, Alexandr A., Institute of Astronomy, Moscow, Russia
Burenkov, Alexandr N., Special Astrophysical Observatory, Russia
Buson, Lucio Maria, Osservatorio di Capodimonte, Napoli, Italy
Cappellari, Michele, Dipartimento di Astronomia, Padova Univ., Italy
Chatterjee, Tapan K., FCFM, Universidad Autonoma de Puebla, Mexico
Clavel, Jean, European Space Agency, Astrophys. Division, Spain
Collin-Zahn, Suzy, Observatoire de Paris-Meudon, France
Courvoisier, Thierry, INTEGRAL Sci Data Centre, Obs. de Geneve, Switzerland
Cunniffe, John, Dunsink Observatory, Dublin, Ireland
Denissyuk, Eduard K., Astrophysical Institute, Alma-Ata, Kazakhstan
Dietrich, Matthias, Landessternwarte Heidelberg, Germany
Dodonov, Serguei N., Special Astrophysical Observatory, Russia
Dopita, Michael A., Mt.Stromlo and Siding Springs Observatory, Australia
Egikian, Anahit G., Byurakan Astrophysical Observatory, Armenia
Fridman, Alexei M., Institute of Astronomy, Moscow, Russia
Grigoryan, Levon Sh., Inst. of Applied Problems in Physics, Armenia
Gurzadyan, Vahagn G., Yerevan Physics Institute, Armenia
Gyulbudaghian, Armen L., Byurakan Astrophysical Observatory, Armenia
Gyulzadyan, Marietta V., Byurakan Astrophysical Observatory, Armenia
Hagen-Thorn, Vladimir, Sankt-Petersburg State University, Russia
Hakopian, Susanna A., Byurakan Astrophysical Observatory, Armenia
Harutyunian, Haik A., Byurakan Astrophysical Observatory, Armenia
Hirst, Paul, Dept. Physics & Astronomy, Univ. Leicester, UK
Hovhannessian, Martin, Byurakan Astrophysical Observatory, Armenia
Huang, Keliang, Physics Department, Nanjing Normal University, China
Iskudarian, Sofik G., Byurakan Astrophysical Observatory, Armenia
Israelian, Garik L., Instituto de Astrofisica de Canarias, Spain
Kaiser, Mary E., Johns Hopkins University, Baltimore, USA
Kalloghlian, Arsen T., Byurakan Astrophysical Observatory, Armenia
Kandalyan, Rafik A., Byurakan Astrophysical Observatory, Armenia

List of Participants

Kawaguchi, Toshihiro, Dept of Astronomy, Kyoto University, Japan
Kellermann, Kenneth I., National Radio Astronomical Observatory, USA
Khachikian, Edward Ye., Byurakan Astrophysical Observatory, Armenia
Koratkar, Anuradha P., Space Telescope Science Institute, USA
Kovalev, Yuri Yu., Astronomical Space Centre, Phys. Inst., Moscow, Russia
Krolik, Julian H., Johns Hopkins University, Baltimore, USA
Laurent, Olivier, Service d'Astrophysique, Saclay, France
Leeuw, Lerothodi L., Joint Astron.Center & Univ.of Central Lanchashire, USA
Lelievre, Gerard H., Observatoire de Paris, DASGAL, France
Lominadze, Jumber G., Abastoumani Astrophysical Observatory, Georgia
Longair, Malcolm S., Univ. of Cambridge, Cavendish Laboratory, UK
Lovelace, Richard V.E., Cornell University, Astronomy Dept, USA
Macchetto, F. Duccio, Space Telescope Science Institute, USA
Mahtessian, Abraham P., Byurakan Astrophysical Observatory, Armenia
Malumian, Vigen H., Byurakan Astrophysical Observatory, Armenia
Matveenko, Leonid I., Space Research Institute, RAN, Russia
Melikian, Norair D., Byurakan Astrophysical Observatory, Armenia
Mickaelian, Areg M., Byurakan Astrophysical Observatory, Armenia
Mirabel, I. Felix, Service d'Astrophysique, Saclay, France
Movsessian, Vardan H., Byurakan Astrophysical Observatory, Armenia
Napier, William M., Armagh Observatory, Northern Ireland, UK
Navasardyan, Hripsime, Byurakan Astrophysical Observatory, Armenia
Nikoghossian, Elena H., Byurakan Astrophysical Observatory, Armenia
Notni, Peter, Astrophysikalisches Institut Potsdam, Germany
Ohanian, Gabriel A., Byurakan Astrophysical Observatory, Armenia
Okazaki, Atsuo T., Hokkai-Gakuen University, Japan
Osmer, Patrick S., Ohio State University, Dept of Astronomy, USA
Otterbein, Kai, Landessternwarte Heidelberg, Germany
Panajian, Vazgen G., Byurakan Astrophysical Observatory, Armenia
Petrosian, Artashes R., Byurakan Astrophysical Observatory, Armenia
Petrosian, Vahe, Stanford Univ., Ctr. Sp. Sci & Astrophysics, USA
Petrov, Georgy T., Institute of Astronomy, Sofia, Bulgaria
Pronik, Iraida I., Crimean Astrophysical Observatory, Ukraine
Robson, E. Ian, Joint Astronomy Centre, Hawaii, USA
Romanova, Marina M., Cornell University, Astronomy Dept, USA
Roscoe, David, Univ. of Sheffield, Dept of Applied Maths, UK
Sadoyan, Avetis H., Yerevan State University, Armenia
Sahakian, Gurgen S., Dept of Theor. Physics, Yerevan State Univ., Armenia
Sambruna, Rita M., Dept of Astronomy, Pennsylvania State Univ., USA
Sanders, David B., Institute for Astronomy, Univ. of Hawaii, USA
Sinanian, Armen A., Byurakan Astrophysical Observatory, Armenia
Smith, H. Eugene, University of California, San Diego, USA
Stepanian, Jivan A., Special Astrophysical Observatory, Russia
Sulentic, Jack W., Univ. of Alabama, Dept of Physics & Astron., USA
Terzian, Yervant, Cornell University, Dept of Astronomy, USA
Tran, Hien D., Inst. of Geophys. and Planetary Physics, USA
Trevese, Dario, Istituto Astronomico, Universita di Roma, Italy
Turatto, Massimo, Osservatorio Astronomico di Padova, Italy

Vardanyan, Rafik A., Byurakan Astrophysical Observatory, Armenia
Vardapetian, Ruben, INTAS, Brussels, Belgium
Veron, Marie-Paule, Observatoire de Haute-Provence, France
Veron, Philippe, Observatoire de Haute-Provence, France
Weedman, Daniel W., Pennsylvania State Univ, Dept of Astron Astrophys, USA
Xanthopoulos, Emily, NRAL Jodrell Bank, UK
Zahn, Jean-Paul, Observatoire de Paris-Meudon, France
Zasov, Anatoly V., Sternberg Astronomical Institute, Moscow, Russia

CONFERENCE PHOTO

Group picture at Byurakan

I. OBSERVATIONAL PROPERTIES OF AGN

Active Galactic Nuclei and Related Phenomena
IAU Symposium, Vol. 194, 1999
Yervant Terzian, Daniel Weedman, Edward Khachikian, eds.

An Overview of AGN

Ian Robson

Joint Astronomy Centre, 660 N. A'ohoku Place, Hilo, HI 96720, USA

Abstract.
This paper gives an overview of the global properties of Active Galactic Nuclei, introducing and putting into context those areas that will be more fully described later in this volume. For the new entrant into AGN research, a semi-historical description of how the field developed is also given; showing how far we have come in a relatively short time. The pioneering work is now done, the big picture of how AGNs work is understood, we now embark on the search for the precise details of the emission mechanisms and their interrelations.

1. Introduction

The study of active galactic nuclei is just over fifty years old and in that time tremendous strides have been made in understanding what they are and how they work. Active galactic nuclei are the most luminous objects in the Universe and present a unique opportunity to study the physics of supermassive black holes (SBH). They bring together a stimulating interplay between observation and theory and that is one of the great appeals. Another is that AGN research requires observations over the entire electromagnetic spectrum, from the radio to gamma regimes, including imaging, photometry, spectroscopy, and polarimetry. Flux limitations continually drive for ever larger telescopes, and it is clear that the requirement for the highest spatial resolutions is nowhere near being satisfied. This lack of spatial resolution has meant that variability studies are extremely important as a means of mapping out the inner regions around the central engine, the broad- and narrow-line regions, while also determining the response of other parts of the AGN through delays to other (and usually longer) wavelengths.

So what are AGNs? In a nutshell they are galaxies with specific peculiar properties that *may* include the following: extremely bright and point-like nuclei; variability on timescales ranging from minutes upwards; highly ionized gas moving at high velocities (up to a few thousand km s^{-1}); X and gamma emission; extended radio lobes and jets. However, not all these features are seen in all objects, either because of observational selection effects (such as observer orientation) or intrinsic differences in the sources. Explaining these differences in terms of a unifying model of physical emission processes has been one of the major challenges of the last twenty years.

2. A historical perspective

2.1. The discovery epoch

This overview will be a personal synopsis of the subject and I will begin, primarily for the benefit of the younger astronomers, with a very brief history of the field, selecting those events that for me made major breakthroughs in understanding. The story began in the 1940's when Carl Seyfert found that a handful of spiral galaxies possessed extremely bright nuclei compared to normal galaxies and that their optical spectra showed the presence of broad emission lines of highly ionized species (Seyfert, 1943). This was clearly an unusual and very rare phenomenon, but to determine exactly how rare and how they fitted into the overall picture of galaxies required optical surveys, such as those undertaken by Markarian and colleagues at Byurakan.

Within ten years, the newly emerging field of radio astronomy discovered what were believed to be radio-stars, but which turned out to be something far more interesting. The early radio surveys showed that there were two types of sources; those that were extended and corresponded to supernova remnants in our Galaxy, and compact, unresolved sources that had no known origin. Eventually, good radio positional determination led to the discovery of the optical counterpart of a small number of the latter, and they were point-like, hence the suggestion of radio-stars. With the eventual interpretation of the optical spectrum of one of them, 3C273, quasars burst on the scene in 1963. For a more extensive description on the discovery of quasars see Robson (1996).

Briefly, 3C273 was a very bright, compact, radio source with a jet, and a bright (V=13) optical point-source also with a jet. It showed strong and broad emission lines, and with a redshift of z=0.158, it was obviously not a star in our Galaxy. Assuming the cosmological interpretation of the redshift, which most people did, led to the luminosity being an incredible 5×10^{12} $L_\odot$. As more objects were discovered and past photographic plates were re-examined, it was found that these objects showed variability on timescales of less than a year, hence the emission regions had to be very compact, and the source of energy became an immediate problem. Early observational clues all pointed to a central phenomenon in a galaxy; the point-like nuclei, radio-lobes and the jets. The energy supply solution was soon agreed by most to be the gravitational field of a supermassive black hole. Subsequent observational evidence during the last thirty years has served to underline the correctness of this basic picture, a tribute to the theoretical wisdom of the time.

2.2. The classification epoch

We now enter the realm of what I call astronomical botany, which was absolutely essential, but has also caused confusion for students ever since. The classification of AGNs according to various observational criteria led to: Seyfert galaxies; Markarian galaxies; N-galaxies; radio-loud and radio-quiet quasars; OVV quasars; BL Lac objects; radio galaxies (which were eventually split into FRI and FRII sub-categories); LINERs and Starburst galaxies. The big question is how these objects are related and it is important, especially for new students entering the field to remember the big picture and the big questions; it is only too easy to get sucked into a very narrow channel and lose sight of the overall goals.

2.3. First steps in unification

Very soon, there were observational hints at distinct differences in the categories, and also important similarities. I will give three examples. A major advance was the tentative sub-classification of Seyferts into type 1 and 2 by Weedman in 1970. This was soon confirmed and turned into an entire industry, which eventually led to the classification of 'intermediate' type 1.5, 1.8 and 1.9 objects. This to me clearly pointed the way to a unification scenario rather than two different and distinct types of intrinsic phenomena.

Also, radio-loud BL Lacs when fading to very low levels occasionally showed a type I spectrum (i.e. broad optical emission lines) and this, as well as a number of other pointers, led to the classification of OVV quasars and BL Lacs as Blazars in 1978. (I should point out that while this was helpful at the time, we now believe that the blazar characteristics are the manifestation of a jet-driven phenomenon, and the OVV quasars and BL Lacs are different populations of objects (see Sambruna this volume).

A further key series of observations leading to our modern view of unification were IUE studies of the classical Seyfert 1 galaxy NGC4151. Over a number of years, with the decline in the continuum emission it changed into a Seyfert 2 (Ulrich et al 1984). This was another boost for unification as clearly this showed that the type 1 and 2 objects were not intrinsically different, rather the observational differences were an internal aspect of the object such as energy supply and/or environment.

2.4. The obscuring torus

One of the most important observations for unification was that of the polarized emission from the Seyfert 2 galaxy NGC1068 by Antonucci and Miller in 1985. Unlike the continuum type 2 spectrum, the very weak polarized spectrum was found to show a type 1 form. This provided the first direct evidence for unification by orientation, a suspicion that had been gaining ground for some years. The interpretation was that the polarized emission was revealing a broad-line region that was hidden from our direct line-of-sight, and was only detected due to the reflection from some form of reflecting screen, or 'mirror', such as dust or an electron cloud. The obscuring medium preventing us from seeing the nucleus and BLR directly was attributed to an extended gas and dust torus of dimensions from tens to hundreds of pc. This unification by orientation is now at the corner-stone of current assumptions and has had a high degree of success. (Further examples of NGC1068 will be given by Macchetto (this volume).) In this global picture, if we can see the central engine directly, then it is a type 1 object, with strong X-ray emission and a strong broad-line region. If our line-of-sight intercepts the torus, then we do not see the central engine directly, and it is a type 2 object and spectroscopically characterised by the narrow-line region. Figure 1 shows a cartoon of this scheme.

However, while the obscuring torus and orientation is now the accepted paradigm for type 1 and 2 differences, very recently some doubt has been cast over this simple view with HST snapshots of over 250 Seyfert galaxies (Malkan et

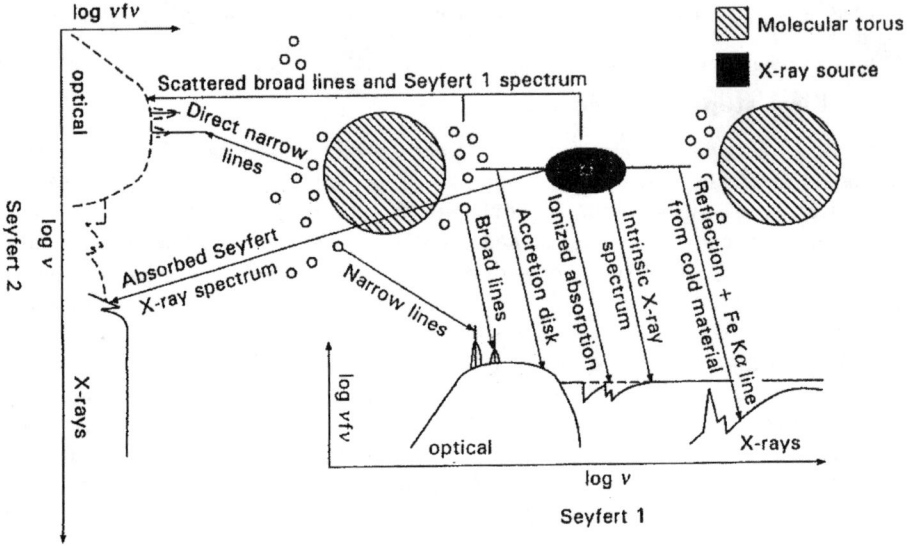

Figure 1. Cartoon of the inner parsec of a radio-quiet AGN showing the main components and possible interaction processes. (Courtesy of Mushotsky, Done and Pounds, Ann.Rev.Astron.Astrophys., 31, 717, 1993.)

al 1998). Their conclusions show that all the galaxies with point-like nuclei are, as expected, Seyfert 1s, but 34 Seyfert 1s have resolved nuclei that are slightly dimmer–both suggesting the presence of dust. Also, both S1 and S2s have central surface brightness similar to bulges of normal spiral galaxies and that both types of galaxy show the presence of dust features in the nuclear regions. So far so good, however, they then go on to conclude that the S1s reside in earlier type spirals than S2s, and if this is the case, then the difference between S1 and S2 is not solely due to orientation, but to something else, something that is host galaxy dependent. The authors conclude that instead of an absorbing molecular torus providing extinction, perhaps galactic clouds of dust may be the answer, being more prevalent in later-type spirals. However, this cannot explain all the type 1 to 2 differences as the absorbing column densities in a number of S2s (for example NGC1068) are known to exceed 10^{24} cm^{-2} and this cannot be produced by aggregates of diffuse clouds along the line of sight. While not being convinced about the host galaxy differences from the data, the presence of intervening dust sheets in the galaxy is a clear clue that even if a torus is present, the picture is probably more complex. Furthermore, while we always picture 'perfect' tori (and there are some excellent examples such as NGC4261), in reality, ragged and leaky tori must also be a possibility.

3. The continuum spectral energy distribution

Turning to the continuum emission from AGNs one striking fact is that for the same IR-optical-X-ray luminosity, two quasars (for example) can have radio powers differing by up to three orders of magnitude (see later). We know the radio-loud emission is due to synchrotron radiation from the core-jet-lobes, which can now be separated out using VLBI imaging. But why only some have these powerful jets creating this huge difference in power remains to be determined.

If extensive dust is present in the parent AGN, this is manifest by a peak of emission (in flux terms) in the submillimetre to far-infrared (see the contribution by Sanders– this volume). The near-infrared is generally the overlap region where the old stellar population of the galaxy shines through (apart from the blazars where the beamed synchrotron overpowers the galaxy contribution). Moving to the optical-UV, we find (for the more luminous examples) evidence for a hot ($\sim$30,000K) blackbody, the so-called big blue bump, usually attributed to an accretion disk (but see the contribution by Courvoisier in this volume).

The peak of the luminosity for many AGNs lies in the extended UV regime, one that has been poorly served by observations due to lack of telescopes and the fundamental limitation that neutral hydrogen extinction plays havoc for extragalactic observations. The X-ray regime is very complex, showing a range of properties, and the lack of a baseline to longer wavelengths (lowest energy) hampers the precise interpretation of what the X-ray continuum represents. At the longest wavelength we probably see the tail of the big blue bump plus additional thermal emission from hotter plasma surrounding the central black hole. At higher energies there is evidence of an underlying power-law that may be the initial source of photo-ionization from the black hole. As might be expected, being the region that is probably the primary photon generation territory and deeply buried within the central zone of the accretion process, depending on the object (and orientation) the X-ray spectrum can be made up of direct photon emission, reflected components, and absorbed/re-radiated components. All of this makes disentangling the components of X-ray observations exceedingly difficult and we await the next generation of high sensitivity and high spatial resolution X-ray facilities (AXAF and XMM) with great anticipation.

The gamma-ray region appears to be important only for blazars, where it can easily dominate the bolometric luminosity of the source (although there is bound to be beaming present to distort this picture). The gamma-emission is assumed to be produced by interactions of photons with the radio jet, but the precise mechanism is still not pinned down (see Sambruna this volume). This latter point brings us to the importance of variability. For essentially no change in radio through far infrared luminosity, the gamma-rays can increase by over two orders of magnitude, accompanied by changes elsewhere such as the X-ray and IR. It is only through variability studies that the exact emission mechanism (internal or external self-Compton) can be determined.

4. The zone of the central engine

Major efforts were invested in the 1980's to interpret the wealth of high quality spectroscopic optical-UV data from AGNs. The outcome was that two regions were identified, the broad-line region and the narrow-line region. These were two distinct regions of ionization and hence location. The current picture is still unclear about the exact make-up of these regions, especially the BLR which is now itself split into high and low ionisation zones (see Collin-Zahn, & Sulentic this volume), as the interaction of the primary radiation with the surrounding gaseous envelope and the putative accretion disk is becoming ever more complex. Whatever the precise picture, it is clear that to fuel the central engine, matter must be accreted directly to the central zone. A thin accretion disk is the favourite mechanism (the puffed up thick torus now seems to have been abandoned), although how the inner mass transport works remains unclear. Indeed, firm evidence for an accretion disk is still not totally accepted (see Courvoisier this volume). Nevertheless, disk fuelling and ensuing radiation by the central engine, subsequent photon interactions and re-radiations with the various components of the BLR and the disk give a fairly acceptable, if not precise picture of what the central region probably looks like.

5. Observational evidence for supermassive black holes

Although supermassive black holes are widely assumed to be the driving force, we must look for observational evidence to support this concept. For myself I find three convincing strands. First there is the pioneering studies of Kormendy and others on stellar dynamics. They find convincing evidence for very large, non-radiative objects in the centres of a number of local (and non-active) galaxies (e.g. Kormendy & Richstone 1995). Because alternative explanations are even more complex and contrived, I shall assume this object is a supermassive black hole. The second piece of evidence is the water vapour maser distributions that have now been seen in a number of AGNs. The first was in NGC4258 (Miyoshi et al 1995 and Greenhill et al 1995) and the masers trace out a very thin ring of gas orbiting in a gravitational potential, about 0.1pc from the radio continuum source. The simplest interpretation is that this is a disk of gas orbiting a supermassive black hole of mass $\sim 3 \times 10^7$ $M_\odot$. The third piece of evidence is the ASCA observation of the extreme red-broadened wing of the X-ray Kα iron fluorescence line in the Seyfert 1 galaxy MCG-6-30-15 (Tanaka et al 1995). This has a velocity width of 100,000 km s^{-1} and is interpreted as high velocity gas in the gravitational field of a supermassive black hole.

6. The Central engine and evolution

In terms of the SBH we now generally believe that the mass and the accretion rate (and perhaps spin) are critical factors in terms of the energy production rate, which can be of order L_{Edd}. It should be stressed that although a SBH is necessary for an AGN, the mere presence of a SBH does not lead to an AGN, because unless fuel can be fed to the hole, then it will be dormant and invisible to photon detection processes apart from its influence on the surrounding material through its gravitational attraction. And as we saw above, we know this must be true due to the presence of SBHs in our own Galaxy and local non-active galaxies (Richstone et al 1998).

This leads us into evolution. We know that AGNs are not active all the time, their energy supply is limited by the available gas in the galaxy and the most luminous quasars have lifetimes of order 10^8y. Indeed, we see a quasar-epoch in the luminosity function, peaking between redshifts of 2 and 3 and then showing a steep decline to higher redshifts. Imaging is now showing that many young quasars show disturbed morphologies and signs of interactions and mergers. This is a key factor. Interactions and merging are clear ways of feeding gas to the SBH and it is well known that interactions and mergers were far more prevalent in the past than at the present epoch apart (see Mirabel this volume for some spectacular 'local' merging systems).

Two observations have opened up new areas of work in the evolution of AGNs: the Hubble Deep Field (HDF) and the discovery of the submillimetre-far-IR background radiation. Observations of the HDF and other regions by the new submillimetre camera SCUBA on the James Clerk Maxwell Telescope have shown that in all probability, this background radiation is produced by point-sources, in this case ultra-luminous IR galaxies (ULIRGs), the super-starbursts (Hughes et al 1998, Barger et al 1998). Furthermore, for the highest luminosity ULIRGs there is compelling evidence that they probably harbour buried AGNs. This evidence also extends to the handful of these objects that have been discovered at redshifts exceeding 2 and 3. The SCUBA data show that there is a population of ULIRGs that have previously been missed by optical-UV studies and that if they also contain AGNs, then the AGN population is also under sampled. However, probably even more important is the effect these galaxies have on the production of metals in the Universe. The submillimetre data show rates of star formation and dust masses at redshifts exceeding 3 of up to five times that claimed from UV studies (see Longair this volume).

Let me now return to what for me remains one of the key unanswered questions about AGNs. What is it that makes a galaxy radio-loud? By radio-loud I mean specifically it has a powerful radio jet. Given the two facts that Seyfert galaxies are (virtually) all spiral galaxies and that radio galaxies exist only in elliptical systems, there has long been the temptation to assume that that it was the host galaxy morphology that governed the production of the jet (either by black hole properties or a dense interstellar medium quenching the jet). Ergo, radio-quiet quasars resided in spiral galaxies. Work that I was involved with a few years ago tested this paradigm through highly accurate IR imaging from UKIRT (Taylor et al 1996). While the sample was not large and was restricted to the top-end of the luminosity function of radio galaxies, radio-loud and radio-quiet quasars, the results were clear. Radio-loud galaxies and radio-loud quasars were housed exclusively in ellipticals while radio-quiet quasars have both elliptical and spiral hosts (although there was a suspicion that as the luminosity increased all the hosts tended to become elliptical). Furthermore, in contrast to data from the HST at the time, quasars did not reside in galaxies of luminosity $<L^*$. Given a few assumptions it was tantalising to note that a good fraction ($\sim 100 L > 2L^*$) might have been quasars in the past.

This work has now been extended using the HST and has produced powerful data (McLure et al 1998). These support all the earlier conclusions and go further by clearly showing that the most luminous of the radio-quiet quasars also reside in elliptical systems. This demonstrates that host galaxy morphology is not the sole answer to the radio-quiet question. (It should also be noted that high sensitivity radio observations have shown small radio jets in a number of nearby AGNs.) Furthermore, the elliptical hosts are identical in almost all global properties to normal 'local' giant elliptical galaxies, except that nearly all show tidal tails or signs of interactions, again suggesting a possible trigger mechanism for activity. The results are also in excellent agreement with the relationship between the mass of the spheroidal bulge component of the galaxy and the mass of the black hole (Magorrian et al. 1998). Using the observed luminosity of the spheroidal bulge of the quasars and the normal elliptical M-L ratio gives the M_{sph} and hence the - *expected* M_{SBH}. Using this to infer L_{Edd} shows L $\sim$ 0.1-1.0 L_{Edd}. This strongly suggests that for RQQs and RLQs the optical luminosity arises from a similar process of accretion onto a SBH. Furthermore, in this picture RLQs and RQQs of similar R-band magnitude have black holes of comparable mass (M $> 3 \times 10^9$ $M_\odot$), i.e. SBHs can only be housed in galaxies with a spheroidal bulge mass of M $> 5 \times 10^{11}$ $M_\odot$.

The results also agree well with the radio-luminosity to black-hole mass correlations of Franceschini, Vercellone & Fabian (1998). The RQQs also show a correlation, albeit better with radio-core luminosity, which tends to give lower M_{SBH} (by a factor of $\sim$2) than using the above technique of M_{sph}. If this conclusion is correct, it implies that *no RQQ* in the sample has a $M_{SBH} > 10^{10}$ $M_\odot$ which in turn suggests the RQQs have lower mass black-holes ($\simeq 10^9$ $M_\odot$) radiating close to the Eddington limit. So, we have that (a) if radio luminosity (5 GHz P_{core}) is the better predictor of the black-hole mass, then the FRII radio sources simply require black holes of M $> 10^{10}$ $M_\odot$, or, (b) if M_{sph} is the better predictor, then some of the RQQs have $M_{SBH} > 10^{10}$ $M_\odot$ and so some other interpretation is required such as SBH angular momentum, for example to explain why black holes of comparable mass can have radio powers differing by two orders of magnitude for a similar optical luminosity.

7. Conclusions

I hope the above has given a flavour of where we are and how we got here. While we have a relatively good overall picture of what makes an AGN and how it works, there are still some major questions remaining unanswered, which means that the field of AGN research will continue to push observational and theoretical ingenuity for many years to come. For me these questions are: the conditions required for the formation of SBHs; the cosmological evolution of galaxies and AGNs; the key parameter(s) that determine what 'type' of AGN we see, in particular the radio-loud, radio-quiet split; how are powerful radio jets produced.

References

Antonucci,R & Miller,J., 1985. ApJ, 297, 621
Barger,A. et al., 1998. Nature, 394, 248
Franceschini, Vercellone & Fabian,A., 1998. MNRAS, 297, 817
Greenhill et al., 1995. ApJ, 440, 619

Hughes,D.H. et al., 1998. Nature, 394, 241
Kormendy,J & Richstone,D., 1995. ARA&A, 33, 581
Magorrian et al., 1998. AJ, 115, 2285
Malkan,M.A., Gorjian,V., & Tam,R., 1998. ApJS, 117, 25
McLure,R.J. et al., 1998. Astro-ph 9809030
Miyoshi,M. et al., 1995. Nature, 373, 127
Richstone,D. et al., 1998. Nature, 395, A14
Robson,I. 1996. *Active Galactic Nuclei*, Wiley-Praxis Press
Seyfert,C.K. 1943. ApJ, 97, 28
Tanaka,Y. et al., 1995. Nature, 375, 659
Taylor,G.L., Dunlop,J.S., Hughes,D.H. & Robson,E.I. 1996. MNRAS, 283, 930
Ulrich,M.H. et al., 1984. MNRAS, 206, 221
Weedman,D.W. 1970. ApJ, 159, 405

HST Observations of Active Galactic Nuclei

F. Duccio Macchetto

Space Telescope Science Institute, 3700 San Martin Drive, Baltimore, MD 21218, USA; E-mail: macchetto@stsci.edu

On assignment from the Space Science Department of ESA

Abstract. The *HST* has made many contributions to all areas of research in the field of AGN, and I have selected three topics where major progress in our understanding has been made over the last two years. The study of the NLR is key to understanding to what extent the unified model for AGN is applicable. In particular, understanding how the NLR is ionized and how its morphology is defined makes an important contribution to clarify the differences between Seyfert 1 and Seyfert 2. Work by Macchetto et al., Capetti et al., Axon et al., and Wilson et al. has helped clarify the picture. Our work on NGC 1068, Mrk 3, Mrk 7, Mrk 348, Mrk 6, and Mrk 573 has shown the following important properties: a) all AGN with a linear radio jet show emission-line morphology ([O III], [O II], Hα) which is aligned along the jet in a surrounding cocoon; b) those AGN with radio lobes show emission-line morphology which is filamentary and coincident with the position of the lobes; and c) in NGC 1068 and in Mrk 3 we have measured transverse velocities to the radio-jet as large as 1700 km s^{-1}. These velocities measured at different positions across the radio-jet show an almost perfect velocity ellipsoid, indicating that the cocoon around the jet is expanding, compresses the ISM and shocks and ionizes the region.

The main conclusion in this field is that the radio-jet is responsible for defining the morphology, both through the expanding cocoon and at the working surface of the radio lobe, and is largely responsible for the observed ionization. The ionization parameter Q in these sources is either constant or actually increases with distance from the nucleus; therefore, nuclear ionization alone cannot explain the observations, whereas local ionization by shocks is fully capable of providing the required flux of ionizing photons.

Many *HST* observations (Ford et al., Ferrarese et al., van der Marel et al.) have shown the presence of extended accretion disks ($\sim$ 200–300 pc). For the first time however, we have been able to show that in the case of M87 we are dealing with a true Keplerian disk. We carried out long-slit spectroscopy with the FOC and measured in particular the [O II] emission line, developed a sophisticated model which took into account the impact parameter, the shape of the PSF, the unknown SED of the nucleus in the 0.06″ inner region, and built a set of models which best fitted the data and were fully self-consistent. We derived an inclination of the disk to our line-of-sight of 51° and a black-hole mass of 3($\pm$0.5) $\times$

10^9 $M_\odot$. We showed that this mass is concentrated within the inner 3 pc and cannot be uniformly distributed; therefore, it must be a black hole.

We have investigated the optical counterparts of all the 3C radio jets and discovered a number of new optical jets. (Sparks et al.). By comparing the radio and optical data we conclude that beaming is responsible for the optical visibility of these objects. We have also measured the proper motions of the jet in M87 over a 3-year period, and we find apparent velocities of features which vary from $1.5c$ to $6c$ for different knots. This is a very important observation which shows among other things that the bulk and pattern velocities differ.

1. Introduction

The investigation of the physical properties of the nuclear regions of active galaxies has been the subject of many *HST* observing programs. In the now standard paradigm for AGN, the basic differences between the different classes of objects are simply explained as a result of different orientations to the line-of-sight. In the unified picture, a central energy source, generally assumed to be a massive black-hole, is surrounded by two spatially and kinematically distinct regions. The Broad Line Region (BLR) has scales of the order of a parsec in diameter and emits broad permitted emission lines with widths up to 10000 km s^{-1}. The Narrow Line Region (NLR) emits narrow (a few hundred km s^{-1}) permitted and forbidden lines and has sizes of up to a kiloparsec in diameter. The orientation effects and, therefore, the classes of AGN, are determined by an optically thick torus composed of dense molecular clouds, whose inner diameter is comparable to the size of the BLR and can extend for tens of parsecs. The symmetry axis of the torus determines the direction of the kiloparsec scale radio jet and is independent of the orientation of the galactic disc. In the standard model if the torus is seen face on, we can see the continuum source and BLR directly and the galaxy is classified as a Seyfert 1. If, instead, the torus obscures the central regions, the galaxy is classifed as a Seyfert 2.

While *HST* observations to date seem to confirm the broad validity of this picture, high spatial resolution observations as well as the important new imaging polarimetry results have shed new light on the fundamental physical phenomona that are at play in the nuclear regions of active galaxies.

Another important field of research has been the study of the optical counterparts to the radio jets. We know that these jets play a fundamental role in transporting energy from the central source to the extended radio lobes. Observations at optical and ultraviolet wavelengths with the *HST* are essential to obtain spatial resolutions similar to, or better than, those achieved in the radio band and, thus, provide the possibility of directly comparing the sites and mechanisms responsible for the emission at these different wavelengths.

In all cases to date, the emission has been attributed to the synchrontron mechanism, and since the electron lifetime is a strong function of the observed frequency, observations at optical and ultraviolet wavelengths offer the possibility to determine the precise location where particle acceleration occurs. Comparison of the radio and optical morphologies further allows the study of the confinement mechanisms and diffusion processes within the jet. A number of important discoveries and observations that place the theoretical models on firmer observational grounds have been published based on the *HST* data.

In this review, I will discuss in detail examples for the two main categories of Seyfert galaxies. I will show the significant progress made in our understanding of the NLR, and I will highlight *HST* discoveries related to the optical counterparts to radio jets.

2. The Seyfert 1.5 Prototype: NGC 4151

NGC 4151 is the nearest (13.3 Mpc; $0.1'' = 6.4$ pc) example of a (sometime) type 1 Seyfert galaxy although it is rather a low-luminosity example of the class. The broad emission component of Hβ, which is a characteristic of Seyfert 1 galaxies, varies dramatically and, in a low state, can almost disappear. This led to NGC 4151 being reclassified as Seyfert 1.5. Both the permitted and the continuum emission show variations on time scales as short as days. FOC observations in [O III] and nearby continuum have shown that the NLR is resolved into a number of emission-line clouds (sizes ~ 10 pc) with elongated morphology. These are distributed in a biconical structure with apices coincident with the central point source and a cone opening-angle projected on the sky of $\sim 75° \pm 10°$. The cone position angle of $60°/240° \pm 5°$ is aligned with the extension of the nuclear VLBI radio source, suggesting that the same mechanism may align both the optical ionizing radiation field and the parsec scale radio structure.

Long-slit spectroscopic observations were carried out with the (uncorrected) FOC (Boksenberg et al. 1995). They found a high and a low radial velocity component within the narrow emission lines and identify the low-velocity component with the bright, extended, knotty structure within the cones, and the high velocity component with more confined diffuse emission. Also present are strong continuum emission and broad Balmer emission-line components, which are attributed to the extended point spread function arising from the intense nuclear emission.

Winge et al. (1997) have carried out Faint Object Camera long-slit spectroscopy of the inner $8''$ of the Narrow Line Region of NGC 4151 at a spatial resolution of $0.029''$. The emission gas is characterized by an underlying general orderly behaviour consistent with galactic rotation over which are superposed kinematically distinct and strongly localized emission structures. High velocity components shifted up to ~ 1500 km s^{-1} from the systemic velocity are seen, associated with individual clouds located preferentially along the edges of the radio knots. Off-nuclear blue continuum emission is also observed associated with the brightest emission-line clouds. Emission line ratios of key diagnostic lines vary substantially between individual clouds. The high spatial resolution long-slit spectroscopy of the nuclear region of NGC 4151 by Winge et al. (1997), shows that the popular picture of anisotropic illumination by the nuclear source is overly simplistic to explain the complex morphology and kinematics observed in the NLR. Comparison of the radio and optical data shows that the line emission is enhanced along the edges of the radio knots, as would be expected from the interaction between the jet and the surrounding medium. This clear association implies that the interaction with the radio jet plasma, and not illumination effects, is the dominant mechanism in determining the morphology, as well as the physical conditions in the individual clouds.

3. The Seyfert 2 Prototype: NGC 1068

A number of observations of the prototypical Seyfert 2 galaxy NGC 1068 ($D \sim$ 22.7 Mpc, $0.1'' = 11$ pc) have been carried with the FOC and WFPC2 (Macchetto et al. 1994, Capetti et al. 1995b, 1996a, Bower et al. 1995, Capetti, Axon and Macchetto, 1997b). These include visible and UV continuum and emission-line observations, as well as FOC imaging polarimetry (F253M, F372M & F501N). These observations show that the inner morphology is very complex. In first approximation, it appears as a "bi-cone," with the axis of symmetry changing with distance from the nucleus. This may be due to the different location of the gas as it streams around the inner bar. Alternately, the ionization and scattering cone is rather narrow and tracks the radio jet more closely. The brightest feature in the continuum and line emission images is "Cloud-B" which is highly polarized (65%) and contrary to previous claims is not the location of the nucleus. The polarization data show that the true nucleus is located some $0.6''$ South of Cloud B and is obscured. This is very consistent with the unified scheme of AGNs. A strange feature, the "twin crescent" is also highly polarized (45%). The distance from the active nucleus $\sim 0.1''$ makes it unlikely that it has any physical relations to it and thus, it remains a mystery. The high-degree of polarization in the circumnuclear region implies that most, if not all, of the observed UV light is scattered light from the nucleus, with either dust or electrons providing the scattering medium.

High precision ground-based and *HST* astrometry (Capetti, Macchetto & Lattanzi, 1997) has allowed the alignment of the optical and radio data to be carried-out. The maps have been registered with a precision of better than $\pm 0.030''$. With this determination the position of the obscured nucleus, as derived from the polarization measurements, falls within the S1 and S2 peaks of the radio data, but is not coincident with any of these two peaks, whereas it is within $1\ \sigma$ of the position of the H_2O maser source.

Capetti et al. (1997b) have used *HST*/WFPC2 imaging to investigate the emission-line structure of the NLR of NGC 1068 and its relationship to the extended radio emission. Previous works showed that the brightest knots of the NLR lie along the opposite edges of the radio-jet. These WFPC2 images indicate that a similar association also holds on a larger scale. In NGC 1068 both radio-jets and lobes are present and the NLR also displays a quasi-linear emission feature associated with the radio-jet and emission-line filaments around the radio lobe. Outside the radio lobe, the emission-line surface brightness drops dramatically. This NLR structure is, therefore, strongly reminiscent of that which has been seen in most Seyfert galaxies studied to date with *HST*. We conclude (Axon et al. 1998) that in NGC 1068, the morphology of the NLR is determined by the presence of a radio-outflow which is sweeping and compressing the surrounding interstellar gas causing the line-emission to be highly enhanced in the region where this interaction occurs.

Furthermore, it appears that the role of the radio jet is not just limited to determining the morphology of the NLR but is physically involved in its ionization structure. The brightest knots of the NLR, located along the radio jet, show a much higher ionization state than the surrounding gas. The density measurements of individual knots derived from archival FOS/*HST* spectra and compared with ground based data indicate that they are high density condensations within the NLR. Their higher ionization and their higher density imply that these knots are illuminated by an incident ionizing flux greater by at least one order of magnitude than the rest of the NLR. Shocks can produce large ionizing flux as soon as their velocities exceed a few hundreds km s^{-1}. This source of ionizing photons can prevail locally over the nuclear radiation field and dominates the ionization conditions in these regions of the NLR. While there are hints of such complex NLR ionization structure changes in the *HST* observations of other more distant Seyferts, only in NGC 1068 can we really resolve it.

At a distance of $\sim 4''$ (~ 300 pc) from the nucleus, the ionization structure of the NLR shows a sharp and well-defined boundary between an inner low-ionization zone and an outer higher zone in correspondence to the transition in the radio structure from jet-like to lobe-like. This ionization change can best be explained with a density drop where the jet enters the lobe which we interpret as evidence for backflowing jet material.

4. Radio Outflow and NLR Structure

For many years, it was generally accepted that in the NLR of Seyfert galaxies, the gas is photoionized by nuclear radiation. The discovery of the NLR with "conductive" morphology on initial *HST* observations of some nearby Seyfert galaxies (NGC 1068, Evans et al. 1991; NGC 4151, Evans et al. 1993; NGC 5728, Wilson et al. 1993) seemed to give further support to that view. However, ground-based studies have shown that the NLR is cospatial with the radio emission and its kinematics clearly shows signs of the effect of interactions with the ejected radio-plasma. This association, now clearly confirmed by the observations of NGC 4151 and NGC 1068 described in the previous sections has been given even stronger support by the observations of several other Seyfert galaxies with *HST*. (Bower et al. 1994, 1995, Capetti et al. 1995c, 1996a,b).

These and other *HST* observations of the nuclear regions of Seyfert 2 have been discussed in a seminal paper by Capetti et al. (1996b). They show that in all cases the physical structure of the NLR in Seyfort 2s is closely related to the radio-emission. The optical morphology is dominated by the interaction with the radio ejecta. The NLR appears to take a different form depending on the structure of the radio emission. Where there are radio lobes, there are shell-like emission-line structures. For Mrk 573, Mrk 78 and Mrk 348, the emission-line structures are bow-shocks. Where a collimated radio jet is present, the morphology is different. In Mrk 3, the NLR follows the jet morphology. Capetti et al. (1996b) show that this dichotomy implies that bow-shock emission-line structures are produced by the sweeping-up of gas at the advancing working surface of the ejected radio plasma. The corrugated structure indicates that instabilities have developed in the compressed gas. Where a jet is apparent, it is surrounded by a halo of hot gas which expands radially from the jet axis and the emission line region forms a cylindrical cocoon on the outer cooling surface. In all cases, the ionization conditions, as determined for example from the emission-line ratios, are such that the ionization parameter increases with distance from the nucleus. This requires a source of ionization in addition to the nuclear ionizing flux.

Capetti et al. (1996b) show that shock mechanisms are the best candidates to produce the relatively small amount of locally produced excess radiation. They are fully consistent with both the measured ionization conditions and the observed filamentary morphology.

5. Black Hole Masses

The presence of massive black holes at the center of galaxies is widely believed to be the common origin of the AGN phenomena. The black hole model is very appealing because it provides an efficient mechanism that converts gravitational energy, via accretion, into radiation within a very small volume as required by the rapid variability of the large energy output of AGNs (e.g., Blanford 1991).

The standard model comprises a central black hole with mass in the range $\simeq 10^6$–10^9 $M_\odot$ surrounded by an accretion disk that releases gravitational energy. The radiation is emitted thermally at the local black body temperature and is identified with the "blue bump," which accounts for the majority of the bolometric luminosity in the AGNs. The disk posesses an active corona, where infrared synchrotron radiation is emitted along with thermal bremsstrahlung X-rays. The host galaxy supplies this disk with gas at a rate that reflects its star formation history and, possibly, its overall mass (Magorrian et al. 1996), thereby accounting for the observed luminosity evolution. Broad emission lines originate homogeneously in small gas clouds of density $\geq 10^9$ cm^{-3} and size $\simeq 1$ AU in random virial orbits about the central continuumn source. Plasma jets are emitted perpendicular to the disk. At large radii, the material forms an obscuring torus of cold molecular gas. Orientation effects of this torus to the line of sight naturally account for the differences between some of the different classes of AGNs (see Antonucci 1993). While this picture has been supported and improved by a number of observations, direct evidence for the existence of accretion disks around supermassive black holes is sparse and detailed measurements of their physical characteristics are conspicuous by their absence.

Ground-based observations of the giant elliptical galaxy M87 first revealed the presence of a cusplike region in its radial light profile accompanied by a rapid rise in the stellar velocity dispersion and led to the suggestion that it contained a massive black hole (Young et al. 1978; Sargent et al. 1978). Stellar dynamical models of elliptical galaxies showed, however, that these velocity dispersion rises did not necessarily imply the presence of a black hole but could instead be a consequence of an anisotropic velocity disperion tensor in the central 100 pc of a triaxial elliptical potential.

Considerable controversy has surrounded this and numerous other attempts to verify the existence of the black hole in M87 and other nearby giant ellipticals using ground-based stellar dynamical studies (e.g., Dressler & Richstone 1990; van der Marel 1994). To date, the best available data remain ambiguous largely because of the difficulty of detecting the high-velocity wings on the absorption lines that are the hallmark of the black hole.

One of the major goals of *HST* has been to establish or refute the existence of black holes in active galaxies by probing the dynamics of AGNs at much smaller radii than can be achieved from the ground.

HST emission line imagery (Crane et al. 1993b; Ford et al. 1994) of M87 has led to the discovery of a small-scale disk of ionized gas surrounding its nucleus which is oriented approximately perpendicularly to the synchrotron jet. This disk is also observed in both the optical and UV continuum (Macchetto 1996a, 1996b). Similar gaseous disks have also been found in the nuclei of a number of other massive galaxies (Ferrarese et al. 1994; Jaffe et al. 1993).

Because of surface brightness limitations on stellar dynamical studies at *HST* resolutions, the kinematics of such disks are in practice likely to be the only way to determine if a central black hole exists in all the very nearest galaxies. In the case of M87 FOS observations at two locations on opposite sides of the nucleus separated by 0.5″ showed a velocity difference of $\simeq 1000$ km s^{-1}, a clear indication of rapid motions close to the nucleus (Harms et al. 1994). By *assuming* that the gas kinematics determined at these and two additional locations arise in a thin rotating Keplerian disk, Ford et al. (1996) estimated the central mass of M87 to be $\simeq 2 \times 20^9$ M$_\odot$ with a range of variation between 1 and 3.5×10^9 M$_\odot$.

HST FOC f/48 high spatial resolution long-slit spectroscopy of the ionized circumnuclear gas disk of M87, at three spatially separated locations 0.2″ apart, was carried out by Macchetto et al. 1997, who analyzed the emission lines and derived rotation curves that extend to a distance of $\sim 1''$ from the nucleus. Within the uncertainties, these data are insensitive to density variations over a broad range of values that are larger than the constraints on density derived from the FOS archive data (Ford et al. 1994). The rotation curve is compatible with that obtained from the archival FOS archive data, given their substantially larger intrinsic errors.

To analyze the data Macchetto et al. first constructed a simple analytical model for a thin Keplerian disk around a central mass condensation, and fitted the model function to the observed rotation curve. Since the number of free parameters is large they carried out trial minimization of the residual errors by using different estimates for the values of the key parameters. Using this simple model they derived two extreme sets of self-consistent solutions that provide good fits to the observational data.

They then conducted a more realistic analysis incorporating the finite slit width, the spatial PSF and the intrinsic luminosity distribution of the gas. This analysis showed that the thin Keplerian disk with a central hole in the luminosity function provides an excellent match to the data, and the resulting parameters of the disk are $i = 51°$, $\theta = -9°$, $V_{sys} = 1290$ km s^{-1} and a corresponding mass of $(3.2 \pm 0.9) \times 10^9$ M$_\odot$, where the error in the mass allows for the uncertainty of each parameter. They showed that this mass must be concentrated within a sphere of less than 3.5 pc and concluded that the most likely explanation is a supermassive black hole.

Another major result has come from high-spatial resolution long-slit observations of the nuclear region of NGC 4151 (Winge et al. 1999). They carried out a detailed study of the kinematics of the gas in both the extended and inner NLR of NGC 4151 using ground-based and *HST* data with high spatial resolution that allow them to separate the underlying velocity field of the emission gas in the NLR from the effects of the radio jet, and to probe its connection with the large scale rotation of the ENLR in the galactic disk.

They decomposed the [O III] λ5007 line profile in multiple Gaussian components and traced the main kinematic component of the ENLR across the nuclear region, connecting smoothly the emission gas system with the large scale rotation defined by H I observations. The individual clouds in the NLR ($R < 4''$) are kinematically disturbed by the interaction with the radio jet, but underlying these perturbations the cloud system is moving in a pattern best described by disk rotation. High velocity components (up to ± 1000 km s^{-1}, relative to systemic) and broad (FWHM up to 1800 km s^{-1}) bases are detected in the [O III]λ5007 profile of the brightest clouds. Such regions are invariably at the edge of the radio knots, and this association, together with the overall morphology of the velocity field, show that the main kinematic system in the inner region of NGC 4151 is still rotation in the plane of the disk, disturbed but not defined by the interaction with the radio jet and the AGN emission.

They fitted the data with a planar rotation and showed that the ENLR gas ($R > 4''$) has a kinematic behaviour well represented by rotation in the galactic disk, with characteristics similar to other normal spiral systems. They obtain $i = 21°$, and $\Psi_o = 34°$–$43°$ for the inclination to the line of sight and position angle of the line of nodes of the disk, respectively. The velocity field of external knots at $R \sim 6''$ and $20''$ transverse to the radial direction presents evidence of non-planar or non-circular movements, probably associated with gas turbulence and streaming motions along the bar. The NLR emission component believed to represent the continuation of the disk velocity field was also found to be consistent with planar rotation, although disturbed by the jet, as expected. However, while the velocity field of the extended ENLR gas is dominated by the potential of the galactic bulge, they find that the behaviour of the gas in the inner NLR is best represented by a Keplerian-like potential, with the kinematics of the gas up to $4''$ dominated by the $\sim 10^9$ M$_\odot$ mass concentration located within the 0.54 turn-over radius of the rotation curve.

Thus in NGC 4151 they were able to directly measure, for the first time, the mass of the central black hole.

6. Optical Counterparts to Radio Jets

To date, thirteen optical synchrotron jets have been identified and most of these have been discovered with the *HST*. The jets are located in the radio galaxies M87 (Curtis 1918), 3C15 (Martel et al. 1998), 3C66B (Fraix-Burnet et al. 1989), 3C78 (Sparks et al. 1995), 3C120, 3C200 (de Koff 1996), 3C264 (Crane et al. 1993b), 3C273 (Bahcall et al. 1995), 3C346 (Sparks et al. 1995), 3C371 (Nilsson et al. 1997), PKS 0521-365, 3C212, and 3C245.

In the standard model of jet formation, material accretes on a supermassive black-hole leading to the ejection of a collimated, relativistic hot plasma. The jet becomes visible in the optical if the Doppler boosted synchrotron radiation is "beamed" towards the observer. High-resolution optical imaging with *HST* has revealed common properties among the jets and their nuclei, consistent with this hypothesis: the jets originate from a bright, compact, unresolved nucleus, characteristic of an AGN, no counter jet is observed, the jets are curved and possess a distorted morphology, and they are smaller and narrower in appearance than their radio counterparts (Sparks et al. 1995). Presently, the only known exception to this relativistic beaming picture is 3C 66B whose counterjet has been imaged in both the radio (Hardcastle et al. 1996) and the optical (Fraix-Burnet 1997), suggesting that the observed synchrotron optical emission is due to environmental effects. On the other hand, despite a superficial consistency with the relativistic beaming picture, Sparks et al. (1995) showed that qualitatively, the statistics of optical jet sizes and power favor, instead, an intrinsic difference between the optically emitting jets and those radio jets which show no optical emission.

Other important issues regarding the formation, collimation, acceleration, evolution and lifetime of optical jets still need to be addressed comprehensively. For example, the acceleration mechanism responsible for the emission of optical synchrotron radiation is still a matter of debate, especially when the radio and optical structures are broadly similar (Meisenheimer 1996). Either the particle acceleration is *in situ* to the jet through processes such as shocks or magnetic reconnection or the electrons are accelerated in the nucleus and are transported along the jet in a channel having low magnetic field and consequently low radiation losses as suggested for M 87 (Owen, Hardee, & Cornwell 1989). Eilek & Arendt (1996) have shown that the observed synchrotron spectrum can be attributed not only to power-law particle distribution functions, but also to power-law magnetic field distributions, instead of the uniform magnetic field assumed in the standard model. This has important ramifications for spectral aging studies since observed synchrotron sources can be significantly older than predicted by the standard model. Ultimately, the properties of optical jets need to be understood in the context of their host galaxies and their environment. Since so few optical jets have been identified, it is important to carefully study the few that are known to understand their physical properties and improve the theoretical models.

Although few in number, there are several common features shared by the optical jet radio sources: (i) all have relatively prominent nuclei in both radio and optical domains; (ii) the nuclei all have flat radio spectra; (iii) the jets are small compared to typical radio jet dimensions, with the arguable exception of 3C 273; (iv) there are no two-sided optical jets—they are all one-sided with a large jet to counterjet lower limit; (v) there is noticeable jet curvature; (vi) in addition, where measured, the optical emission is more localized than the radio and the optical jet is narrower than the radio jet.

An obvious candidate explanation for most of these characteristics is relativistic 'beaming', in which the jet becomes visible in the optical only when pointing towards the observer. In the beaming picture, the jet appears brightened, foreshortened and with a prominent active nucleus. Beaming is also expected to blueshift the synchrotron 'break' frequency.

As an alternative to relativistic beaming, environmental effects may be considered. If the pressure is higher in the vicinity of the optically emitting sources, then the additional confinement may act to enhance their radiative luminosity while suppressing the growth of the source, thereby giving rise to the correlation between size and power. This does not immediately suggest an explanation for the core dominance and one-sidedness; however, there may be instabilities which cause jet disruption and optical emission that are sufficiently rapid that only one side is visible at a given time. 'Age' may provide yet a third alternative, with the optical jet sources being young, in the process of forcing their way out through the interstellar medium, and ram pressure playing a similar role in enhancing the visibility.

There are statistical uncertainties at present, however an extensive analysis of many more optical jets should provide results that will be essential in elucidating the nature of extragalactic synchrotron jets.

7. M87: The Best Studied Optical Jet

The giant elliptical galaxy M87 contains the closest extragalactic jet, which makes it a prime target for studies of jet structure and kinematics. Optical and ultraviolet observations of M87 have been carried out with *HST* and have been extensively reported (Macchetto 1991, Boksenberg et al. 1992; Macchetto, Biretta & Sparks 1992).

While the radio and optical images present a remarkable degree of similaity, there are nevertheless significant differences. The optical/UV images show intrinsically higher contrast than the radio, with compact regions of emission localized within the knots. The jet is narrower in the optical/UV, and more concentrated to the jet center in the optical/UV than in the radio. The radio-to-optical spectral index of the inter-knot regions is steeper than that of the knots themselves. There are also differences in the detailed knot structure of the optical emission compared to the radio, and there is a weak overall spectral steepening with distance from the nucleus beyond knot A.

Capetti et al. (1997c) have analyzed polarization observation of the M87 jet taken with the FOC in the ultraviolet and with the WFPC1 in the visual. The degree of polarization is typically 30% over most of the jet. At the edges of the jet the polarization is as high as 60%, requiring a highly ordered magnetic field. In the center of the jet the small scale structure of the magnetic field produces significant cancellation reducing the polarization to $\sim 10\%$. The degree of polarization and the polarization pattern are very similar at radio and optical wavelengths. No significant depolarization or Faraday rotation have been detected, in agreement with previous radio determinations. However, the morphology of knot D is considerably different in the VLA observations by Owen et al. (1989) and these *HST* observations. Knot D1 appears to be relatively brighter and closer to the nucleus in the optical than in the radio images. Capetti et al. (1997c) conclude that this component is associated with a shock front. At the location of a shock front acceleration of relativistic electrons occurs, enhancing the synchrotron emission at shorter wavelengths, and the transverse component of the magnetic field is amplified by the compression produced by the shock.

As part of a long-term ongoing study, Biretta, Sparks and Macchetto (1999) have measured proper motions for twelve features within the first $6''$ (500 pc) of the M87 jet. Of these, ten appear to be superluminal with eight having apparent speeds in the range $4c$ to $6c$. The two sub-luminal features, knot L and HST-1 EAST, have speeds of $0.63 \pm 0.27c$ and $0.84c \pm 0.11c$, respectively, and coincidentally, are the two features nearest the nucleus.

The most natural explanation for the observed superluminal speeds is that they are due to bulk relativistic flow in the context of the relativistic jet model (Blandford and Königl 1979a). The predominance of speeds in the range $4c$ to $6c$ suggests these are closely linked to an underlying bulk flow. Several regions have much lower speeds, but they attribute this to obstruction or standing shocks within the jet (e.g., Blandford and Königl 1979b). Alternative models which invoke phase effects to produce superluminal motion seem unlikely to work in M87. The consistency of the large speeds in different regions of the M87 jet, as well as the lack of large negative speeds, argues against phase effects as the source of motion.

The observed speeds provide strong constraints on the bulk flow speed and line-of-sight angle for the jet; the strongest constraints result from the largest apparent speeds, of $6c$. Assuming that the jet's velocity is parallel to the jet axis, the relativistic jet model requires a bulk flow with Lorentz factor $\gamma \geq 6$ and a jet orientation within $\theta \sim 19°$ of the line-of-sight.

All solutions imply very large jet to counter-jet brightness ratios owing to strong dimming of the receding counter-jet. Ratios are 4×10^4 or larger, which are entirely consistent with the lack of a detected counter-jet.

These results strongly confirm "unified models" which propose that FR-I radio galaxies like M87 are the parent population of BL Lac objects (Browne 1983). Urry, Padovani, and Stickel (1991) predict that FR-I radio galaxies should have jets with bulk flow speeds in the range $\gamma \sim 5$ to ~ 35, with most near $\gamma \sim 7$, which is in good agreement with speeds $\gamma \geq 6$ implied by the observed $6c$. Furthermore, they derive a critical angle $\theta_{crit} \sim 10°$ for the FR-I/BL Lac division, which is consistent with the angle $\theta \sim 10°$ to $19°$ found for M87. These speeds of $6c$ are in fact the fastest yet seen for an FR-I radio source.

References

Antonucci, R. 1993, Ann. Rev. Astron. Astrophys., 31, 473

Axon, D. J., Marconi, A., Capetti, A., Macchetto, F. D., Schreier, E., & Robinson, A. 1998, ApJL, 496, L78

Bahcall, J. N., Kirhakos, S., Schneider, D. P., Davis, R. J., Muxlow, T. W. B., Garrington, S. T., Conway, R. G., & Unwin, S. C., 1995, ApJL, 452, L91

Biretta, J. A., Sparks, W. B., & Macchetto, F. D. 1999, ApJ, in press

Blandford, R. D. 1991, Proc. Heidelberg Conf., Physics of AGN, ed. W. J. Duschl & S. J. Wagner (Berlin: Springer), 3

Blandford, R. D. and Königl, A. 1979a, ApJ, 232, 34

Blandford, R. D. and Königl, A. 1979b, ApJL, 20, 15

Boksenberg, A., et al. 1992, A&A, 261, 393

Bower, G. A., Wilson, A. S., Morse, J. A., Gelderman, R., Whittle, M., & Mulchaey, J. 1995, ApJ, 454, 106

Bower, G. A., Wilson, A. S., Mulchaey, J. S., Miley, G. K., Heckman, T. M. & Krolik, J. H., 1994, AJ, 107, 1686

Browne, I. W. A. 1983, MNRAS, 204, p23

Capetti, A., Axon, D. J., Kukula, M., Macchetto, F., Pedlar, A., Sparks, W. B., & Boksenberg, A. 1995a, ApJ, 454, l85
Capetti, A., Axon, D. J., Macchetto, F., Sparks, W. B., & Boksenberg, A. 1995b, ApJ, 446, 155
Capetti, A., Macchetto, F., Sparks, W. B., & Boksenberg, A. 1995c, ApJ, 448, 600
Capetti, A., Macchetto, F., Axon, D. J., Sparks, W. B., & Boksenberg, A. 1996a, ApJ, 466, 169
Capetti, A., Axon, D. J., Macchetto, F., Sparks, W. B., & Boksenberg, A. 1996b, ApJ, 469, 554
Capetti, A., Macchetto, F., & Lattanzi, M. G. 1997a, ApJ, 476, 67
Capetti, A., Axon, D. J., & Macchetto, F. D., 1997b, ApJ, 487, 560
Capetti, A., Macchetto, F., Sparks, W. B., & Biretta, J. A. 1997c, A&A, 317, 637
Crane, P., et al. 1993, ApJL, 402, L37
Crane, P., et al. 1993, AJ, 106, 1371
Curtis, H. D. 1918, Publ. Lick Obs., 13, 11
de Koff, S., Baum, S. A., Sparks, W. B., et al. 1996, ApJS, 107, 621
Dressler, A. & Richstone, D. O. 1990, ApJ, 348, 120
Eilek, J. A., & Arendt, P. N., 1996, ApJ, 457, 150
Evans, I. N., Ford, H. C., Kinney, A. L., Antonucci, R. R. J., Armus, L. & Caganott, S, 1991, ApJL, 369, L27
Evans, I. N., Tsvetanov, Z., Kriss, G. A., Ford, H. C., Caganott, S., & Koratkar, A. P., 1993, ApJ, 417, 82
Ferrarese, L., van den Bosch, F. C., Ford, H. C., Jaffe, W., & O'Connell, R. W. 1994, AJ, 108, 1598
Ford, H. C., et al. 1994, ApJ, 435, L27
Ford, H. C., Tsvetanov, Z. I., Hartig, G. F., Kriss, G. A., Harms, R. J., Dressel, L. L. 1996, in Science with the *HST* II, ed. P. Benvenuti, F. Macchetto, & E. Schreier (Washington: GPO), 192
Fraix-Burnet, D., 1997, MNRAS, 284,911
Fraix-Burnet,D., Nieto, J. L., Leliévre, G., Macchetto, F., Perryman, M. A. C., & di Serego Alighieri, S. 1989, ApJ, 336, 121
Hardcastle, M. J., Alexander, P., Pooley, G. G., & Riley, J. M., 1996, MNRAS, 278, 273
Harms, R. J., et al. 1994, ApJ, 435, L35
Jaffe, W., Ford, H. C., Ferrarese, L., van den Bosch, F. C., & O'Connell, R. W. 1993, Nature, 364, 213
Macchetto, F. 1991, Proceedings, Physics of AGN, ed. S. J. Wagner, W. J. Duschl, Springer-Verlag: Berlin, 325
Macchetto, F. 1996a, in IAU Symp. 175, Extragalactic Radio Sources, ed. R. Ekers et al. (Dordrecht: Reidel), 195
Macchetto, F. 1996b, in Science with the *HST* II, ed. P. Benvenuti, F. Macchetto, & E. Schreier (Washington: GPO), 394

Macchetto, F., Biretta, J. A., & Sparks, W. B. 1992, Proceedings 182nd AAS Meeting, 24, 1183

Macchetto, F., Capetti, A., Sparks, W. B., Axon, D. J. & Boksenberg, A. 1994, ApJL, 435, L15

Macchetto, F. D., Marconi, A., Axon, D. J., Capetti, A., Sparks, W. B., & Crane, P., 1997, ApJ, 489, 579

Magorrian, J., Tremaine, S., Gebhardt, K., Richstone, D., & Faber, S. 1996, BAAS, 189,111

Marconi, A., Axon, D. J., Macchetto, F. D., Capetti, A., Sparks, W. B., & Crane, P., 1997, MNRAS, 2891, 21M

Martel, A. R., Sparks, W. B., Macchetto, F. D., Biretta, J. B., Baum, S. A., Golombek, D., McCarthy, P. J., de Koff, S., & Miley, G. K., 1998, ApJ, 496, 203

Meisenheimer, K., 1996, in Jets from Stars and Galactic Nuclei, ed. W. Kundt (New York: Springer)

Nelson, C. H., Mackenty, J. W., Simkin, S. A., & Griffiths, R. E. 1996, ApJ, 466, 713

Nilsson, K., Keidt, J., Pursimo, T., Sillanpää, A., Takalo, L. O., & Jaäger, A. 1997, ApJ, 484, L107

Owen, F. N., Hardee, P. E. & Cornwell, T. J. 1989, ApJ, 340, 698

Sargent, W. L. W., Young, P. J., Boksenberg, A., Shortridge, K., Lynds, C. R., & Hartwick, F. D. A. 1978, ApJ, 221, 731

Sparks, W. B., Golombek, D., Baum, S. A., Biretta, J., de Koff, S., Macchetto, F. D., McCarthy, P., & Miley, G. K., 1995, ApJ, 450, L95

Urry, C. M., Padovani, P., & Stickel, M. 1991, ApJ, 382, 501

van der Marel, R. P. 1994, MNRAS, 270, 271

Wilson, A. S., Braatz, J. A., Heckman, T. M., Krolik, J. H., & Miley, G. K., 1993, ApJ, 419, L61

Winge, C., Axon, D. J., Macchetto, F. D. & Capetti, A., 1997, ApJL, 487, L121

Winge, C., Axon, D. J., Macchetto, F. D., Capetti, A., & Marconi, A., 1999, ApJ, in press

Young, P. J., Westphal, J. A., Kristian, J., Wilson, C. P., & Landauer, F. T. 1978, ApJ, 221, 721

Infrared Emission from AGN

D. B. Sanders

Institute for Astronomy, University of Hawaii, 2680 Woodlawn Drive, Honolulu, HI 96822

Abstract. Infrared observations of complete samples of active galactic nuclei (AGN) have shown that a substantial fraction of their bolometric luminosity is emitted at wavelengths $\sim$8–1000μm. In radio-loud and Blazar-like objects much of this emission appears to be direct non-thermal synchrotron radiation. However, in the much larger numbers of radio-quiet AGN it is now clear that thermal dust emission is responsible for the bulk of radiation from the near-infrared through submillimeter wavelengths. Luminous infrared-selected AGN are often surrounded by powerful nuclear starbursts, both of which appear to be fueled by enormous supplies of molecular gas and dust funneled into the nuclear region during the strong interaction/merger of gas rich disks. All-sky surveys in the infrared show that luminous infrared AGN are at least as numerous as optically-selected AGN of comparable bolometric luminosity, suggesting that AGN may spend a substantial fraction of their lifetime in a dust-enshrouded phase. The space density of luminous infrared AGN at high redshift may be sufficient to account for much of the X-Ray background, and for a substantial fraction of the far-infrared background as well. These objects plausibly represent a major epoch in the formation of spheroids and massive black holes (MBH).

1. Introduction

Infrared observations[1] of AGN historically have lagged behind observations at shorter and longer wavelengths, therefore it is not surprising that much of the literature is still biased toward studies of radio and optical/X-Ray selected AGN. While radio-loud objects and the highly variable optical/X-Ray sources (e.g. Blazars, OVVs) provide an opportunity for studying the physics of AGN, they draw attention away from the much larger fraction of radio-quiet and dust-enshrouded sources. These dusty AGN appear to hold important clues for understanding the origin and evolution of all AGN, and the relation of AGN to other classes of extragalactic objects.

---

[1] "Infrared" is used here to include rest-frame emission over the broad wavelength range $\sim$8–1000μm (i.e. the mid-infrared, far-infrared and submillimeter, but not the near-infrared). Definitions for observed quantities such as L_{ir}, L_B, etc. are taken from Table 1 of the review by Sanders & Mirabel (1996). $H_o = 75$ and $q_o = 0$ is assumed throughout this article.

This review focuses on the relatively large body of infrared continuum data[2] that is now available for complete samples of nearby AGN, and stresses new results that have been published since the previous IAU Symposium on AGN in Geneva, 1993 ("Multiwavelength Continuum Observations of AGN", S159). A major highlight of the most recent work is the clear identification of a large population of dust-enshrouded AGN that may be more numerous than optically selected AGN in the Universe.

In reviewing the infrared continuum properties of AGN, it is instructive to trace the highlights of infrared studies from the first mid- and far-infrared measurements of selected nearby targets in the late 60's and 70's, through to the latest spacecraft results that bear directly on the AGN population in the more distant Universe. Infrared observations took a great step forward following the all-sky surveys carried out by the Infrared Astronomical Satellite (*IRAS*), and it seems natural to divide our initial discussion accordingly, and then to show how infrared studies of AGN in the local Universe can help in understanding the AGN population at high redshift.

2. Pre-*IRAS*: A Brief Historical Review, 1968-83

The pioneering infrared observations of Low & Kleinmann (1968), and Kleinmann & Low (1970a,b), followed by more accurate photometry by Rieke & Low (1972), made it clear that strong infrared emission could dominate the spectral energy distributions (SEDs) of Seyfert galaxies, and even singled out a class of objects with "ultra-high" infrared luminosities that rivaled the bolometric luminosity of QSOs. The first clear evidence that the mid-infrared emission from most Seyferts might not be direct synchrotron radiation was provided by new $10\mu m$ data for NGC1068 which showed lack of variability (Stein et al. 1974), plus an extended source (Becklin et al. 1973). The infrared spectrum appeared to be better explained by models of thermal reradiation from dust (e.g. Rees et al. 1969; Burbidge & Stein 1970).

Mid-infrared ground-based photometry of large samples of Markarian Seyferts and starbursts (Rieke & Low 1975; Neugebauer et al. 1976), Seyfert galaxies (Rieke 1978), plus mid- and far-infrared observations of a few nearby Seyferts with the Kuiper Airborne Observatory (Harper & Low 1973; Telesco & Harper 1980) proved that "infrared-excess" was a common property of Seyferts as well as starburst galaxies, and with the possible exception of radio-loud objects and QSOs, that this emission could indeed be understood in terms of thermal emission from dust. Rieke (1978) perhaps summarized it best (from a study of 50+ Markarian Seyferts) by stating that "strong infrared excess is a virtually universal characteristic of these sources. ...the infrared continuum of a number of type 1s and most type 2s is dominated by thermal reradiation by dust", while also pointing out that the strength of the infrared excess was correlated with the

---

[2]Mid- and far-infrared spectroscopy of AGN is not covered in this review. Until very recently, such data were available for only a small number of optically selected targets. However, data for a larger number of nearby AGN are now available from the *Infrared Space Observatory* (*ISO*), and the reader is referred to articles by O. Laurent and J. Clavel at this conference, and to the excellent paper by Genzel et al. (1998).

strength of the reddening (as measured by Hα/Hβ), and that the proportionality between the 10μm and 21cm fluxes noted earlier (e.g. Rieke & Low 1972) was confirmed.

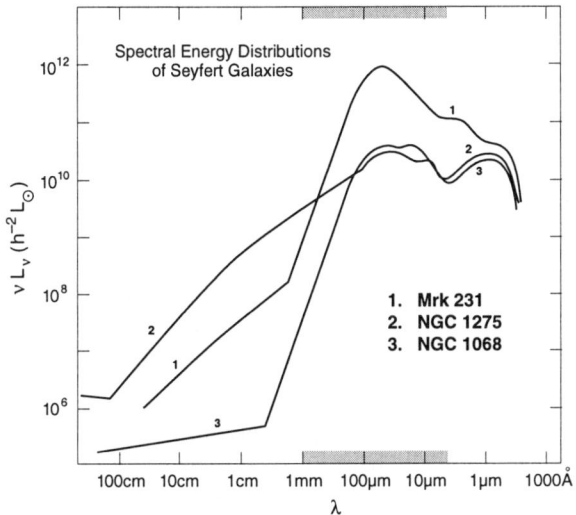

Figure 1. Spectral energy distributions from UV to radio wavelengths for three well-known AGN: the Sy1 galaxy Mrk231, the Sy2 galaxy NGC1068, and the Sy2 powerful radio galaxy NGC1275 (Perseus A).

Figure 1 shows SEDs using the most recent radio-to-UV continuum data for three "classic" Seyferts, which were among the first AGN observed in the mid- and far-infrared. The steep submillimeter spectral index, $\alpha > 2.5$ (where $f_\nu \propto \nu^\alpha$), for radio-quiet objects confirms earlier suggestions that the "infrared bump" at wavelengths $\sim 5-500\mu m$ is due to thermal emission from dust. Only in the radio-loud sources such as NGC1275 (Perseus A) is there any reason to believe that a substantial portion of the infrared emission is simply the short wavelength extension of the non-thermal emission seen at radio-to-millimeter wavelengths (e.g. Edelson & Malkan 1986).

3. Post-*IRAS*: Infrared Properties of Optical Samples of AGN

IRAS was the first telescope with sufficient sensitivity to detect large numbers of extragalactic sources at mid- and far-infrared wavelengths (Neugebauer et al. 1984). *IRAS* surveys of optically selected Seyfert galaxies (Miley et al. 1985) and QSOs (Neugebauer et al. 1985, 1986) confirmed that active galaxies could be strong infrared emitters; most optically selected AGNs had ratios L_{ir}/L_B in the range 0.2 to 1.0 with higher values in only a small number of objects.

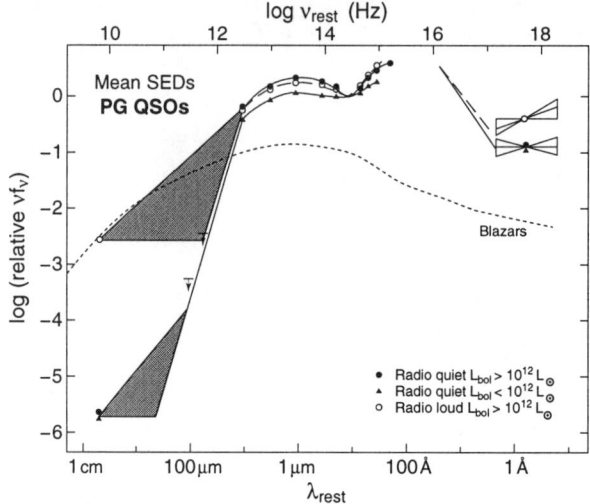

Figure 2. Mean spectral energy distributions from X-ray to radio wavelengths for optically selected radio-loud and radio-quiet QSOs (Sanders et al. 1989) from the Palomar-Green Bright QSO Survey (Schmidt & Green 1983), and for Blazars (Impey & Neugebauer 1988).

3.1. QSOs

A more complete accounting of the infrared properties of optically selected Sy1s and QSOs is given by the data in Figure 2. The interpretation by Sanders et al. (1989) was that the gross shape of the SEDs between 3000Å and 300μm is remarkably similar for all QSOs (except the flat-spectrum radio-loud quasars like 3C273) and that this can broadly be interpreted by two broad components of thermal emission; the "big blue bump" representing 10^5–10^6 K thermal emission from an accretion disk, and an "infrared bump" made up of reradiation from dust in a distorted disk extending from $\sim$0.1 pc to more than 1 kpc.

Only for flat-spectrum radio-loud QSOs is there good evidence that much of the infrared emission is probably direct non-thermal emission from the central AGN. These objects also tend to exhibit variability on relatively short time-scales (hours to weeks). In the highly variable Blazars there is strong evidence that infrared emission is truly just part of a single non-thermal spectral component from millimeter to optical wavelengths (see Figure 2). The range of variability in both the spectral index and flux density for Blazars increases with decreasing wavelength, with the variability in the far-infrared being less than half that observed in the optical. However, a substantial fraction of Blazars, when in their minimum variable state, also show evidence for underlying emission lines and thermal infrared components, suggesting that these objects may still contain substantial amounts of dust.

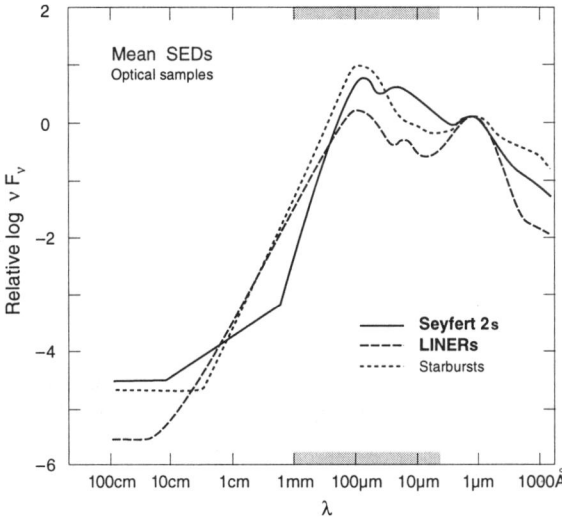

Figure 3. Mean spectral energy distributions from X-ray to radio wavelengths (normalized to log $\nu F_\nu(7700\text{Å}) \equiv 0$) for optically selected NLAGN (Sy2s and LINERs) compared with luminous starbursts (adapted from Schmitt et al. 1997).

3.2. Narrow-line AGN

A more complete accounting of the infrared properties of optically selected narrow-line AGN (NLAGN) is given by the data in Figure 3, which suggest a larger mean infrared excess ($\equiv L_{\rm ir}/L_{\rm B}$) than previously assumed; the mean value is ~ 1 for LINERs and ~ 5 for Sy2s. Both Seyferts and LINERs have a far-infrared peak at ~ 60–200μm characteristic of relatively "cool" dust ($T_{\rm dust} \sim 25$–50K), with the strength of the infrared emission in Sy2s actually being somewhat larger than that observed in starbursts (!), primarily due to added emission from a second "warm" ($T_{\rm dust} \sim 100$–500K) dust peak in Sy2s at wavelengths ~ 5–50μm.

4. Infrared Selected AGN

The true extent of infrared excess exhibited by all types of extragalactic objects was only begun to be realized following extensive ground-based follow-up studies of complete infrared selected samples. deGrijp et al. (1985) found that searches based on "warm" infrared colors ($f_{25}/f_{60} \gtrsim 0.3$) could be useful for discovering *new* infrared-luminous AGN, a technique that appeared to be motivated by the shape of the infrared spectrum of the Sy2 galaxy NGC1068 (Telesco & Harper 1980), and the discovery of a similar "warm" component in the broad-line, infrared-luminous radio galaxy 3C290.3 (Miley et al. 1984). These infrared-selected AGN had a mean ratio of $L_{\rm ir}/L_{\rm B} \sim 10$ and included a few objects with ratios as large as ~ 30–50. deGrijp et al. (1985) suggested that the true space density of AGNs could be a factor of two larger than previously assumed (with

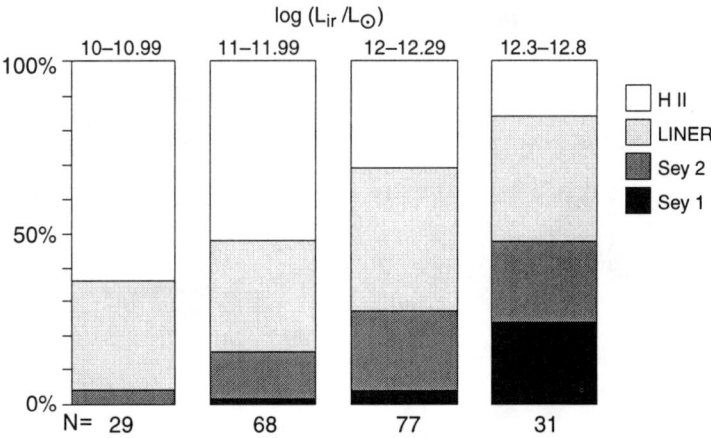

Figure 4. The optical spectral classification of flux-limited (60μm) samples of infrared selected galaxies versus infrared luminosity (Veilleux et al. 1999).

the majority of the new infrared-selected objects being a mixture of Sy2s and LINERs).

The first infrared-selected "bonifide" QSOs [3] were also found during color-selected searches of the *IRAS* database at the faintest flux levels (Beichman et al. 1986; Vader & Simon 1987; Low et al. 1988; Sanders et al. 1988b). Additionally, the most luminous infrared sources (with $\bar{L}_{ir} \sim 10^{13} L_\odot$), all of which were found to be dusty Sy2s in direct emission (e.g. Kleinmann & Keel 1987; Hill et al. 1987; Frogel et al. 1989; Cutri et al. 1994) were subsequently shown to be obscured infrared QSOs (i.e. Sy1s) in polarized light (e.g. Hines et al. 1995). Whether all infrared selected Sy2s harbor obscured Sy1s is not yet clear. However, it seems plausible that "unified models" invoked to understand the polarization properties of optically selected AGN (e.g. Antonucci & Miller 1985), which suggest that the observed spectral type (Sy1 vs. Sy2) depends largely on the orientation of a circumnuclear dust torus to the line of sight (see Antonucci 1993 for a more complete review) could play a major role in infrared-selected AGN. Models with even larger dust shrouds (>100 pc) where the obscuring material covers most of the sky as seen from the central source (e.g. Fabian et al. 1998) may also need to be invoked to account for the objects with the largest infrared excess.

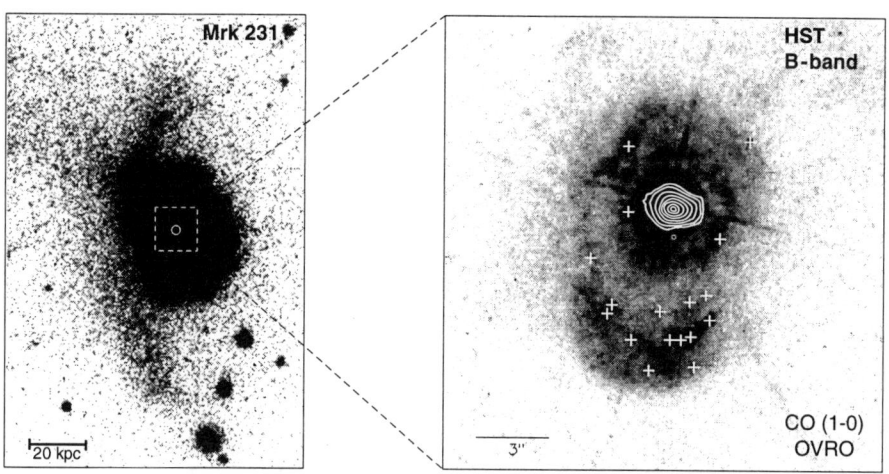

Figure 5. The advanced merger/ULIG/QSO Mrk 231 – Left panel: optical image (Sanders et al. 1987) and CO contour (Scoville et al. 1989). Right panel: HST B-band image and identified stellar clusters ('+') from Surace et al. (1998). The high resolution CO contours are from Bryant & Scoville (1997).

4.1. AGN versus Infrared Luminosity

Perhaps the most important spectroscopic result from studies of flux-limited samples of extragalactic infrared objects is the increasing fraction of AGN among the most luminous sources. It is clear from Figure 4 that the fraction of Seyferts increases systematically with increasing $L_{\rm ir}$, to where *Seyferts account for nearly half of all objects at the highest infrared luminosities*. The ratio of Sy1s to Sy2s also increases to the point where both are ~25% of the total number of objects with $L_{\rm ir} > 10^{12.3} L_\odot$. Whereas from optical surveys alone it had been thought that Sy2s were very rare at high bolometric luminosities, it now seems clear that high luminosity Sy2s were simply hiding as a subset of the most luminous infrared selected galaxies. Likewise for luminous H II galaxies, although the fraction of H II galaxies diminishes to $\lesssim 25\%$ for ultraluminous infrared galaxies (ULIGs) at $L_{\rm ir} > 10^{12} L_\odot$. LINERs appear to remain constant at ~1/3 of the sample at all $L_{\rm ir}$, and although LINERs have sometimes been lumped together with NLAGN, recent evidence suggests that the emission lines in many of these infrared-selected LINERs may be powered primarily by shocks and superwinds from massive stars (e.g. Veilleux et al. 1999).

5. The Starburst-AGN Connection

There is increasing direct evidence that powerful circumnuclear starbursts and AGN may be intimately related. An excellent review of this subject can be found in the Taipei Workshop - Relationships between AGN and Starburst Galaxies (1992). For the current discussion, the case of Mrk231 (Figure 5) is instructive. Mrk231, like other ULIGs, contains a large population of relatively unobscured luminous star clusters at galactocentric radii $\sim$0.5–3kpc (Surace et al. 1998). The inner 1kpc region still contains an enormous supply of gas and dust, much of which may be arranged in a subkiloparsec disk (Carilli et al. 1998). A partial face-on orientation for this disk may be the explanation for why we can see the redenned Sy1 nucleus.

It has been suggested that both the intense circumnuclear starburst and the AGN currently contribute approximately equally to the bolometric infrared luminosity in Mrk231 (D. Weedman and H.E. Smith, this conference). However, the nature of the dominant power source for ULIGs continues to be the subject of great debate (e.g. Ultraluminous Galaxies: Monsters or Babies, Ringberg Workshop, 1998) with arguments favoring both circumnuclear starbursts and AGN as well as nearly equal mixtures of the two phenomena. Extensive multiwavelength spectroscopic studies are currently underway in an attempt to resolve the issue (e.g. Genzel et al. 1998), but due to heavy dust obscuration it is not yet clear which process dominates in all objects.

5.1. The Origin and Evolution of Infrared-luminous AGN

Extensive ground-based observations of complete samples of infrared-selected galaxies now clearly show that strong interactions/mergers of gas-rich spirals play a dominant role in triggering the most luminous infrared systems (see the review by Sanders & Mirabel 1996). The fraction increases from $\sim$30% at $L_{\rm ir} = 10^{11} L_\odot$ to $\gtrsim$95% for ULIGs at $L_{\rm ir} > 10^{12} L_\odot$ (e.g. Sanders et al. 1988a; Mirabel et al. 1990; Kim 1995; Murphy et al. 1996; Clements et al. 1996). Prominent tidal tails, similar to what is seen for Mrk231 (Figure 5), are indeed found in all nearby ULIGs, suggesting that the mergers involve two relatively large gas-rich spirals. Figure 6 illustrates the ubiquitous tidal tails, and in some cases double nuclei, that can still be detected in deep images of even more distant luminous infrared objects, in this case three optically selected AGN.

Strong interactions/mergers of gas-rich spirals appear to be extremely efficient at funneling large amounts of gas into the merger nuclei (e.g. Barnes & Hernquist 1992; Mihos & Hernquist 1994). Kormendy & Sanders (1992) have summarized the evidence that these objects are elliptical galaxies in formation. Intense starbursts are clearly involved in producing the bulk of the infrared luminosity throughout much of the initial stages of this process, and may continue to do so through the most intense infrared phase, although it seems reasonable to assume that during the peak infrared phase (which is close in time to when the two nuclei merge and also corresponds to the most compact concentration of gas in the nuclear region), conditions would be most optimum for building and

---

[3](i.e. Sy1s with $L_{\rm ir} > 10^{12} L_\odot$; equivalent to the bolometric luminosity of optically selected QSOs with $M_{\rm B} < -23$)

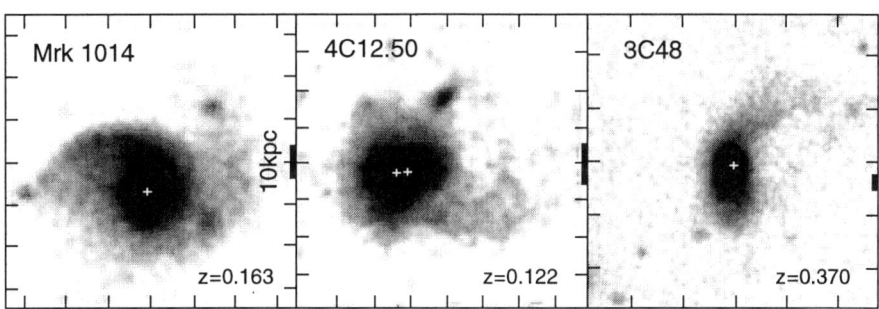

Figure 6. Optical images of infrared-excess, optically selected QSOs, powerful radio galaxies, and infrared selected QSOs (MacKenty & Stockton 1984; Kim 1995; Stockton & Ridgway 1991). The '+' sign indicates the position of putative optical nuclei. Tick marks are at 5″ intervals and the scale bar represents 10 kpc. All three objects exhibit strong nuclear concentrations of molecular gas, with typically $\sim 10^{10} M_\odot$ concentrated at galactocentric radii $\lesssim 1$ kpc (Sanders et al. 1988c; Mirabel et al. 1989; Scoville et al. 1989).

fueling an AGN, and that the AGN may rival if not dominate the luminosity output of the system. Mrk231 and most other ULIGs would seem to be at this stage.

Eventually, powerful superwinds (e.g. Armus et al. 1989), that are indeed observed in ULIGs, may clear away much of the surrounding nuclear dust shroud. It seems reasonable that this housecleaning process could terminate much of the circumnuclear starburst (as it already may have in the 1-3 kpc annulus of Mrk231), but that the fueling of the AGN may continue due to the strong self-gravitation of the inner accretion disk surrounding the AGN. This later period may still be marked by infrared excess determined by conditions on much smaller scales such as the thickness, orientation, and opening angle of the dust and gas torus thought to surround the AGN. Eventually the "big blue bump" normally associated with optically selected QSOs should emerge, as has happened already in the case of Mrk1014 and 3C48 (Figure 6). This scenario would seem to provide a plausible explanation for a correlation between black hole mass and bulge mass as recently found in nearby ellipticals (Kormendy & Richstone 1995) since the mass of both the black hole and spheroid may show similar dependence on the total mass of gas funneled into the merger nucleus.

6. Deep Infrared and X-Ray Surveys: AGN and the Distant Universe

There is now good evidence that both ULIGs and NLAGN may be sufficiently numerous at high redshift ($z >1$) to account for the infrared background and the X-Ray background respectively. There is also increasing circumstantial evidence that suggests NLAGN may be a substantial subset of these high-redshift ULIGs, which if true, would suggest that a major epoch in the formation of spheroids and MBH may have finally been discovered.

The evidence from the X-Ray side is as follows. Deep X-Ray surveys with ROSAT are consistent with a space density that evolves as steeply as $(1 + z)^5$ for high-luminosity AGN out to at least $z \sim 2$ with a constant value at higher redshift (e.g. Hassinger et al. 1998). These sources are sufficiently numerous to account for nearly all of the observed X-Ray background; however, the shape of the X-Ray background spectrum implies that most of these AGN are heavily absorbed (e.g. Fabian & Barcons 1992; Boyle et al. 1995; Almaini et al. 1998). These "narrow-line X-Ray galaxies" (NLXGs: Hassinger 1996) have recently been characterized by Maiolino et al. (1998) as having "extremely heavy obscuration along the line of sight, $(N_H > 10^{25} \text{cm}^{-2})$ in most cases".

The evidence from the infrared side is as follows. Evidence for space density evolution as steep as $(1 + z)^5$ is consistent with a series of studies at varying far-infrared flux levels and wavelength bands. In the mid- and far-infrared, the deepest surveys carried out by *IRAS* (e.g. Hacking & Houck 1987; Lonsdale & Hacking 1989; Gregorich et al. 1995; Kim & Sanders 1998), and more recently the surveys with *ISO* (e.g. Taniguchi et al. 1997; Kawara et al. 1998; Aussel et al. 1998; Puget et al. 1998) are consistent with evolution at least as steep as $(1 + z)^5$ out to $z \sim 1$. Within the past year, submillimeter surveys with the Submillimeter Common User Bolometer (SCUBA) on the James Clerk Maxwell Telescope (Smail et al. 1997; Hughes et al. 1998; Barger et al. 1998; Eales et al. 1998) have revealed a substantial population of ULIGs, consistent with steep evolution out to at least $z \sim 2$–3, and with constant space density at higher redshift. These high redshift ULIGs, which are almost certainly the high-z extension of the sources detected by *IRAS* and *ISO*, are sufficiently numerous to account for *all* of the far-infrared/submillimeter background (e.g. Barger et al. 1999).

To test the relationship between ULIGs and NLXGs at high redshift requires sufficiently sensitive X-Ray and far-infrared/submillimeter surveys in overlapping regions of the sky. These data should be available within the next few years from new X-Ray satellites and submillimeter interferometers. For now we can only note the interesting result that all of the ULIGs uncovered by *IRAS* at $z > 0.4$ (all of which have $L_{ir} \sim 10^{13} L_\odot$) show direct evidence for powerful AGN (e.g. Kleinmann & Keel 1987; Rowan-Robinson et al. 1991; Cutri et al. 1994), as does the first identified SCUBA source (Ivison et al. 1998). It would thus appear that the relationship shown in Figure 4 indeed continues to higher luminosities and to higher redshift.

7. Conclusions

Infrared observations have shown clearly that thermal emission plays an important, and often dominant, role in the total luminosity output of most AGN. Although in some objects – most notably the flat-spectrum, radio-loud AGN and the highly variable Blazars – a substantial fraction of the observed infrared emission appears to be direct non-thermal synchrotron radiation, the much larger number of radio-quiet Sy1s and NLAGN appear to have infrared SEDs dominated by thermal emission from dust.

The discovery by *IRAS* of a substantial population of infrared-selected AGN suggests that a substantial fraction of the energy produced by accretion may be absorbed by dust. Ground-based observations of these dusty objects show that powerful circumnuclear starbursts ($r \lesssim 1$kpc) are often closely linked with the building and fueling of AGN, and that the most luminous sources (ULIGs) are often associated with strongly interacting/merger galaxies.

There is now strong evidence that the space density of ULIGs was much larger in the past ($z > 1$), and that they may account for a large fraction, if not all, of the far-infrared/submillimeter background. If the trend of increasing AGN fraction versus increasing infrared luminosity observed for local ULIGs continues to high redshift, then dust-enshrouded AGN may account for much of the X-Ray background as well. This large population of high-redshift ULIGs may represent an important stage in the formation of both spheroids and massive black holes.

Acknowledgments. I am grateful to Karen Teramura for assistance in preparing the figures, and to JPL contract no. 961566 for partial financial support.

References

Almani, O., et al. 1998, Astr.Nachr., 319, 55

Antonucci, R. 1993, ARA&A, 31, 473

Antonucci, R. & Miller, J.S. 1985, ApJ, 297, 621

Armus, L., Heckman, T.M., & Miley, G.K. 1989, ApJ, 347, 727

Aussel, H., Cesarsky, C.J., Elbaz, D., & Starck, J.L. 1998, A&A, in press

Barger, A.J., Cowie, L.L., & Sanders, D.B. 1999, ApJ, submitted

Barger, A.J., et al. 1998, Nature, 394, 248

Barnes, J.E., & Hernquist, L. 1992, ARA&A, 30, 705

Becklin, E.E., Matthews, K., Neugebauer, G., & Wynn-Williams, G.C. 1973, ApJ, 186, L69

Beichman, C.A., et. al. 1986, ApJ, 308, L1

Bryant, P.M., & Scoville, N.Z. 1996, ApJ, 457, 678

Burbidge, G.R., & Stein, W.A. 1970, ApJ, 160, 573

Carilli, C.L., Wrobel, J.M., & Ulvestad, J.S. 1998, A&A, 115, 928

Clements, D.L., Sutherland, W.J., McMahon, R.G., & Saunders, W. 1996, MN-RAS, 279, 477

Cutri, R.M., et al. 1994, ApJ, 424, L65

de Grijp, Miley, G.K., Lub, J., & de Jong, T. 1985, Nature, 314, 240

Eales, S., et al. 1998, ApJ, in press

Edelson, R.A., & Malkan, M.A. 1986, ApJ, 308, 59

Fabian, A.C., & Barcons, X. 1992, ARA&A, 30, 429

Fabian, A.C., & Barcons, X., Almaini, O., & Iwasawa, I. 1998, MNRAS, 297, L11

Frogel, J.A., Gillett, F.C., Tendrup, D.M., & Vader, J.P. 1989, ApJ, 343, 672

Genzel, R., et al. 1998, ApJ, 498, 579

Gregorich, D.T., Neugebauer, G., Soifer, B.T., Gunn, J.E., & Herter, T.L. 1995, AJ, 110, 259
Hacking, P.B., & Houck, J.R. 1987, ApJS, 63, 311
Harper, D.A., & Low, F.J. 1973, ApJ, 182, L89
Hassinger, G. 1996, A&AS, 120, 607
Hassinger, G., et al. 1998, A&A, 329, 482
Heckman, T.M. 1980, A&A, 87, 152
Hill, G.J., Wynn-Williams, C.G., & Becklin, E.E. 1987, ApJ, 316, L11
Hines, D.C., et al. 1995, ApJ, 450, L1
Hughes, D.H., et al. 1998, Nature, 394, 241
Impey, C., & Neugebauer, G. 1988, AJ, 95, 307
Ivison, R., et al. 1998, MNRAS, 298, 583
Kawara, K., et al. 1998, A&A, 336, L9
Kim, D.-C., et al. 1995, PhD Thesis, University of Hawaii
Kim, D.-C., & Sanders, D.B. 1998, ApJS, 119, 41
Kleinmann, D.E., & Low, F.J. 1970a, ApJ, 159, L165
Kleinmann, D.E., & Low, F.J. 1970b, ApJ, 161, L203
Kleinmann, S.G., & Keel, W.C. 1987, in Star Formation in Galaxies, ed. C.J. Lonsdale-Persson, Wash. DC: US GPO, 559
Kormendy, J., & Richstone, D. 1995, ARA&A, 33, 581
Kormendy, J., & Sanders, D.B. 1992, ApJ, 390, L53
Lonsdale, C.J., & Hacking, P. 1989, ApJ, 339, 712
Low, F.J., Huchra, J.P., Kleinmann, S.G., Cutri, R.M. 1988, ApJ, 327, L41
Low, F.J., & Kleinmann, D.E. 1968, AJ, 73, 868
MacKenty, J.W., & Stockton, A. 1984, ApJ, 283, 64
Maiolino, R., et al. 1998, A&A, 338, 781
Mihos, J.C., & Hernquist, L. 1994, ApJ, 431, L9
Miley, G.K., Neugebauer, G., & Soifer, B.T. 1985, ApJ, 293, L11
Miley, G.K., et al. 1984, ApJ, 278, L79
Mirabel, I.F., Booth, R.S., Garay, G., , L.E.B., & Sanders, D.B. 1990, A&A, 236, 327
Mirabel, I.F., Sanders, D.B., & Kazès, I. 1989, ApJ, 340, L9
Multiwavelength Continuum Emission of AGN 1994, eds. T.J.-L. Courvoisier, A. Blecha (Dordrecht: Kluwer)
Murphy, T.W., et al. 1996, AJ, 111, 1025
Neugebauer, G., Becklin, E.E., Oke, J.B., & Searle, L. 1976, ApJ, 205, 29
Neugebauer, G., Soifer, B.T., & Miley, G.K. 1985, ApJ, 295, L27
Neugebauer, G., Soifer, B.T., & Miley, G.K., & Clegg, P.E. 1986, ApJ, 308, 815
Neugebauer, G., et al. 1984, ApJ, 278, L1
Puget, J.-L., et al. 1998, A&A, in press
Rees, M.J., Silk, J.I., Werner, M.W., & Wickramasinghe, N.C. 1969, Nature, 223, 37

Relationships Between AGN and Starburst Galaxies 1992, ed. A.V. Filippenko (San Francisco: PASP Conf Ser, v.31)

Rieke, G.H. 1978, ApJ, 226, 550

Rieke, G.H., & Lebofsky, M.J. 1979, ARA&A, 17, 477

Rieke, G.H., & Low, F.J. 1972, ApJ, 176, L95

Rieke, G.H., & Low, F.J. 1975, ApJ, 200, L67

Rowan-Robinson, M., et al. 1991, Nature, 351, 719

Sanders, D.B., & Mirabel, I.F. 1996, ARA&A, 34, 749

Sanders, D.B., et al. 1989, ApJ, 347, 29

Sanders, D.B., Scoville, N.Z., & Soifer, B.T. 1988c, ApJ, 335, L1

Sanders, D.B., et al. 1988b, ApJ, 328, L35

Sanders, D.B., et al. 1987, ApJ, 312, L5

Sanders, D.B., et al. 1988a, ApJ, 325, 74

Schmidt, M., & Green, R.F. 1983, ApJ, 269, 352

Schmitt, H.R., Kinney, A.L., Calzetti, D., & Storchi-Bergmann, T. 1997, AJ, 114, 592

Scoville, N.Z., et al. 1989, ApJ, 345, L25

Smail, I., Ivison, R.J., & Blain, A.W. 1997, ApJ, 490, L5

Stein, W.A., Gillett, F.C., Merrill, K.M. 1974, ApJ, 187, 213

Stockton, A., & Ridgway, S.E. 1991, AJ, 102, 488

Surace, J.A., et al. 1998, ApJ, 492, 116

Taniguchi, Y., et al. 1997, A&A, 328, L9

Telesco, C.M., & Harper, D.A. 1980, ApJ, 235, 392

Ultraluminous Galaxies: Monsters or Babies 1998, Ap&SS, in press

Vader, J.P., & Simon, M. 1987, Nature, 327, 304

Veilleux, S., & Osterbrock, D.E. 1987, ApJS, 63, 295

Veilleux, S., Kim, D.-C., & Sanders, D.B. 1999, ApJ, in press

Gerard Leliévre, Dave Sanders and Duccio Macchetto

Radio Properties of Quasars and AGN

K. I. Kellermann

NRAO, 520 Edgemont Road, Charlottesville VA 22903

Abstract. We discuss the compact radio sources associated with quasars and AGN with particular emphasis on the implications of their structure and kinematics for relativistic beaming models.

1. Introduction

About ten to fifteen percent of optically selected quasars and AGN are radio loud defined by $P_6 > 10^{25}$ W/Hz or $R > 10$, where P_6 is the radio flux density at $\lambda = 6$ cm and R is the ratio of radio to optical flux density measured at 6 cm and 4400 A respectively (Kellermann, et al. 1989; Miller et al. 1990). The radio loud quasars and AGN are conveniently divided into two categories, the lobe-dominated sources primarily associated with radio galaxies and the core-dominated sources which are mostly identified with quasars and AGN. For both the lobe-dominated and core-dominated sources, the source of energy appears to lie in a compact central engine generally assumed to be due to accretion from a surrounding material onto a massive black hole. Radio emission from the extended lobes arises from optically thin steep spectrum synchrotron radiation which may extend up to hundreds of kiloparsecs from the parent galaxy. The core dominated sources, on the other hand, are dominated by very compact self-absorbed, often variable, synchrotron sources. High resolution radio observations give a unique opportunity to study in detail the processes occurring close to the central engine.

Much less is known about the nature of the radio quiet quasars and AGN. At least some of the low luminosity radio emission appears to be the result of intense star-forming activity, generally coincident with the visible optical disk and showing little or no detailed radio structure (e.g., Richards et al. 1998). However, some radio quiet quasars also have more extended radio lobes as well as compact radio components more characteristic of the activity of a central engine. It seems likely that the low luminosity radio emission observed from radio quiet objects is a combination of star-forming activity and ejection from a compact central engine.

At the time of the IAU Symposium No. 29 on *Non Stable Phenomena in Galaxies* which was held in Byurukan in 1966, it was realized that the rapid time variability observed in many quasars and AGN implied such small angular dimensions that corresponding physical conditions appeared to be unrealistic (Kellermann and Pauliny-Toth 1968). It is now appreciated that the observed radio emission from quasars and AGN can be understood in terms of the twin relativistic jet model described by Blandford and Konigl (1979) based on ideas first discussed by Shklovsky (1963) and later extended by Rees (1966) and Woltjer (1966). We note that only at radio wavelengths is it possible to see through the gas and dust close to the AGN where relativistic plasma is focused into a narrow beam or jet. Also, only at radio wavelengths, is it possible to directly observe the relativistic flow from the central engine.

There are three observational consequences of relativistic motion. First, of course, owing to the Doppler shift, the emission from the approaching beam is shifted toward shorter wavelengths, and the receding component toward longer wavelengths by the Doppler factor $\delta = 1/\gamma_f (1 - \beta \cos \theta)$. Second, as a result of aberration, the observed flux density from a relativistically moving source is boosted by a factor $\delta^{n+\alpha}$. Finally, because a relativistically moving source nearly catches up with its own radiation, the time scale is compressed for an observer located close to the direction of motion. The corresponding apparent transverse velocity, $\beta_{app} = \beta_p \sin \theta / (1 - \beta_p \cos \theta)$, and can exceed the speed of light when $2\beta_p / (1 + \beta_p^2) > \cos \theta$. In these expressions, $\gamma = (1 - \beta^2)^{-1/2}$ is the usual Lorentz factor, $\beta = v/c$; θ is the angle between the motion and the line of sight; the subscripts f and p refer respectively to the relativistic flow velocity of the radiating source and the velocity of the moving pattern; and n = 2 or 3 depending on whether the motion is in the form of a continuous jet or discrete components.

The effect of differential Doppler boosting between the approaching and receding sources was the basis of the original unified models which attempt to describe the different appearance of lobe-dominated and core-dominated sources as the result of the different orientation of intrinsically similar sources containing both extended unbeamed radio lobes as well as compact relativistic jets (Orr & Browne 1982). Later Antonucci and Miller (1985), and Barthel (1989) included the effect of observation of the nucleus by a dusty torus surrounding the central engine to extend the unified picture to optical wavelengths.

In the simplest relativistic beaming models, $\gamma_f = \gamma_p$ and there are simple relations between the observed component velocities and the degree of Doppler boosting which allow quantitative tests of the beaming models and the intrinsic velocity and orientation to be uniquely determined from observation of the apparent component motion and flux density ratio of the approaching and receding components. In this paper we briefly describe the high resolution observations of quasars and AGN and their implication to the nature of the compact central engine and the formation and propagation of relativistic jets. We assume throughout $H_0 = 65$/km/sec/Mpc and qo=1/2. Other recent relevant reviews can be found in Cohen & Kellermann (1995), Zensus (1997), Zensus et al. (1998), and Urry and Padovannini (1995).

2. Morphology

High resolution radio observations using the Very Long Baseline Array (VLBA), the European Very Long Baseline Network (EVN), and the Southern Hemisphere (SHEVE) give images of the radio emission near the central engine with an angular resolution exceeding 0.001 arcseconds (1 milliarcsec), or orders of magnitude better than images made with any other instrument operating in any wavelength band, on the ground or in space. The corresponding linear scale is only a few tens of parsecs even at cosmological distances ($z \sim 1$) and only a few tenths of a parsec for nearby ($z \sim 0.01$) galaxies. For the more powerful quasars and AGN with $P_6 > 10^{25}$ W/Hz the structure is invariably one-sided, with the parsec scale jet lying on the same side as the much larger kiloparsec scale jets which feed the extended radio lobes. It is assumed that the kiloparsec scale jets are intrinsically two-sided as suggested by the symmetric appearance of the radio lobes. Thus, if the one-sided appearance of the kiloparsec jet is due to differential Doppler boosting, the kiloparsec scale jets must also be relativistic, and must be focused on scales of a few parsecs or less, yet remain collimated out to hundreds of kiloparsecs.

Earlier VLBI observations suggested a radio morphology which consisted of a compact core with multiple secondary components lying along a jet-like structure. The new VLBA (e.g., Kellermann et al. 1998a) and EVN observations, which have greater sensitivity to low surface brightness structure, often show a complex continuous jet with condensations possibly the result of shock fronts. The compact core is self absorbed, sometimes even at short centimeter or millimeter wavelengths. The jets are sometimes very narrow and appear unresolved in the transverse direction with linear dimensions of the order of a few parsecs or less. Often, the jets have multiple bends or twists, sometimes up to 90 degrees or more, especially close to the core. Some jets show an oscillating appearance characteristic of a rotating nozzle or the effect of Kelvin-Helmholtz instabilities. These one-sided compact structures typically have a flat radio spectrum or one with multiple maxima and minima which are thought to be the superposition of multiple components each with a different self absorption cutoff frequency.

Some AGN as well as a few quasars show symmetric structure up to hundreds of parsecs in extent. These compact symmetric objects (CSOs) frequently contain a very compact central component which we identify with the center of activity (Readhead et al. 1996). CSOs appear to be related to the class of source with GHz peaked spectra (GPS) with a spectral peak near one GHz. It seems likely that the spectral peak is due to synchrotron self absorption, but we can not rule out the presence of free-free absorption by an intervening ionized medium. Possible interpretations of CSOs and GPS sources which have been discussed include a) gravitational "minilensing" from black holes of the order of 10^6 M$_\odot$ located in galactic halos (Lacey & Ostriker 1985); b) young classical doubles (e.g., Phillips and Mutel 1980); c) "frustrated" or confined doubles (O'Dea et al. 1991); d) reborn doubles (Baum et al. 1990); e) smothered doubles (Baum et al 1990); or f) intrinsically confined doubles (Gopal-Krishna & Wiita (1991).

Often, the low luminosity AGN show a bilateral or two-sided jet symmetry at short wavelengths but at longer wavelengths, the structure may appear asymmetric due to free-free absorption in the nuclear region, apparently by ionized circumnuclear gas contained in a torus or other structure oriented perpendicular to the jet (Levinson, et al. 1995; Walker, et al.1998, Kellermann et al. 1998a). In these sources, the jets appear to propagate through a rich and complex medium, possibly including the broad-line region and containing both molecular (e.g., Claussen et al. 1998) and atomic gas (Vermeulen et al. 1999). The AGN found in NGC 5128 (Jones, et al. 1996), NGC 1052 (Kellermann, et al. 1998b), NGC 1275 (Dhawan, et al. 1998), and Cygnus A (Krichbaum, et al. 1998) all show signs of asymmetric absorption from an intervening medium.

High resolution polarimetry to determine the distribution of linear polarization gives further insight into the strength and orientation of the magnetic field in quasars and AGN as well as other information about jet physics (e.g., Wardle et al. 1994). In general the orientation of the magnetic field is aligned with the jet direction for quasars but tends to be orthogonal to the jet in weak lined sources such as those associated with BL Lac objects (Cawthorne et al. 1993).

Until recently, it has been unclear whether relativistic jets are composed of a "normal" electron-proton plasma or electron-positron pairs. Wardle et al. (1998), have reported the detection of circular polarization in several quasars and AGN which they argue is evidence that the relativistic jets must be composed primarily of electrons and positrons. Future observations of circular polarization combined with measurements of Faraday rotation of linear polarization will give new insight on the energy and mass content of relativistic jets and help to understand how they are created.

3. Kinematics

The observed peak radio brightness temperatures of quasars and AGN are typically in the range of 10^{11-12} K. The maximum brightness temperature for a static incoherent synchrotron source is limited by inverse Compton cooling to about 3×10^{11} K (Kellermann & Pauliny-Toth 1969) and somewhat less if there is equilibrium between the relativistic particles and the magnetic field (Readhead 1994). But the observed values can be larger if there is a coherent mechanism involved, if the source is not stationary (Slish 1992), or if Doppler boosting is significant. Comparison of inverse Compton scattered x-ray emission with observed peak brightness temperatures can be used to estimate the amount of Doppler boosting. Ghisellini et al. 1993 find good agreement with inverse Compton Doppler factors and observed superluminal velocities, but Guerra et al. (1997) find better agreement with equipartition Doppler factors and observed motions. Recently, Bower & Backer (1998) have used space VLBI observations to directly measure a peak brightness temperature of about 3×10^{12} K in the quasar NRAO 530 significantly in excess of inverse Compton or equilibrium values expected from a static source.

When examined over a period of time, the flux density of most quasars and AGN show variability on time scales ranging from a few hours to a few years. The most rapid so-called "intraday" variables (Heeschen et al. 1987, Wagner & Witzel 1996, Wagner 1998) present a challenge to standard models. If the variations are intrinsic as suggested by the similarity of observed variations over a wide range of wavelength (Wagner et al. 1996), light travel time arguments suggest extremely small dimensions corresponding to brightness temperatures as great as 10^{21} K. Assuming synchrotron radiation, such large brightness temperatures require unrealistic large values of the Doppler factor up to 10^3 (Kedziora-Chudczer et al 1998), non stationary conditions in the relativistic plasma (Slish 1992), or coherent emission mechanisms. It has been argued that the observed rapid time variations may not be intrinsic but may be due to refractive scintillations in the interstellar medium. However, in order that such rapid scintillations occur, the angular size is still limited to about 5 as for the most rapid sources and the corresponding brightness temperatures are still an uncomfortably high value of 10^{14} K (Cordes 1998).

Corresponding to the observed changes in total flux density, individual features appear to be ejected from the central engine with apparent linear velocities mostly in the range $0 < v/c < 10$ with a median value about 3c. Radio sources associated with strong gamma-ray sources have somewhat faster moving components suggesting either a larger intrinsic velocity or motion more closely aligned to the line of sight. In every case with a well determined velocity, the motion is outward along the jet. It is perhaps surprising that there is no direct evidence from the radio data of the postulated infall into a massive black hole; only outflows are observed.

Compact radio sources at the center of symmetric double lobed quasars typically have slower velocities than the core-dominated asymmetric sources (e.g., Hough 1996) as expected from unified models that place the lobe dominated sources near the plane of the sky. Motions have been observed in only a few FR II radio galaxies. In Cygnus A the motion is surprisingly subluminal (Krichbaum et al. 1998), while in 3C 111 and 3C 390.3 it is superluminal (Kellermann et al. 1998b) with observed velocities comparable to typical core dominated quasars.

The low luminosity components in AGN appear to have significantly slower moving components. For a given luminosity, there appears to be no difference in component velocities found in BL Lacs, AGN, and quasars. This is perhaps surprising, since the relativistic beaming models consider BL Lacs, AGN, and quasars to be increasingly aligned along the line of sight so that the observed kinematics should differ among these populations. Moreover, in the two-sided low luminosity jets, the observed transverse velocities are often much less than the speed of light. AGN in the galaxies NGC 1052, M87, Cygnus A, and NGC 5128 show velocities in the range 0.1c to 0.5c (Kellermann, et al. 1998b; Tingay et al. 1998; Krichbaum, et al. 1998).

Relativistic beaming models all require a common intrinsic velocity very close to the speed of light. From Section 1, we see that according to the twin-jet model, the apparent transverse velocity cannot be less than the speed of light for any orientation. It is therefore difficult to understand how these two-sided compact sources or the asymmetric cores of the symmetric lobe-dominated FR II galaxies fit into the simple unified beaming models which consider the low radio luminosity AGN and the lobe dominated quasars and FR II radio galaxies to differ from the high luminosity quasars only by orientation. Rather, it would appear that there are intrinsic differences as well.

Individual sources often show apparent accelerations along the jet, e.g., M87, NGC 1275, Cygnus A, 3C 345, but it is not clear if these represent real changes in velocity, or are just the effect of motion along a curved trajectory with a changing inclination to the line of sight (Dhawan, et al. 1998; Krichbaum, et al. 1998; Unwin et al. 1997).

In a randomly oriented sample, most sources should lie close to the plane of the sky so the observed transverse velocity of the jet components should be close to c (or 2c if measured between oppositely moving jets. On the other hand, in a flux density limited sample, differential Doppler boosting results in a bias toward sources moving close to the line of sight. In this case, the observed distribution of apparent velocities depends in detail on the intrinsic value of the Lorentz factor, γ, the spectral index, α, and the slope of the radio source count. Vermeulen (1995) has shown that for a wide range of parameters the observed distribution of apparent velocity should be closely peaked near a value of γc. The observed distribution of apparent linear velocity is not consistent with any simple ballistic model and appears to require either a spread in intrinsic velocity or a difference between flow and pattern velocities (Vermeulen 1995, Kellermann et al. 1998b). Lister & Marscher (1997) discuss the observed statistics of superluminal velocities in flux limited samples of radio sources.

Acknowledgments. Some of the results described here are based on unpublished work done in collaboration with R. Vermeulen, J. A. Zensus, and M. Cohen. The National Radio Astronomy Observatory is a facility of the National Science Foundation, operated under a cooperative agreement by Associated Universities, Inc.

References

Antonucci & Miller 1985. ApJ, 297, 621.

Barthel, P. 1989. ApJ, 336, 606.

Baum, S.A., 1990, A&A, 232, 19.

Blandford, R.D. & Konigl, A. 1979. ApJ, 232, 34.

Bower, G.C. & Backer, D.C. 1998, ApJ, 507, L117.

Cawthorne, T.V. et al. 1993. ApJ, 416, 519.

Cohen, M.H. & Kellermann, K.I. 1995. *Proc.of the Nat. Academy of Sciences*, 92, 11339.

Claussen, et al. 1998, ApJ, 500, L129.

Cordes, J. 1998. in *Radio Emission from Compact Galactic and Extragalactic Radio Sources*, eds. A. Zensus et al., ASP., San Francisco, p. 329.

Dhawan, V. et al. 1998, ApJ, 498, L111.

Ghisellini, G. et al. 1993, ApJ 407, 65.

Gopal-Krishna & Wiita, P. J. 1991, ApJ, 373, 325.

Guerra, E. et al. 1997. ApJ, 491, 483.

Hough, D. et al. 1996. ApJ, 459, 64.

Heeschen, D.S., et al. 1987. AJ, 94, 1493.

Jones, D. et al. 1996, ApJ, 466, L63.

Kedziora-Chudczer, L.L. et al. 1998. in *Radio Emission from Compact Galactic and Extragalactic Radio Sources*, eds. A. Zensus et al., ASP., San Francisco, p. 267

Kellermann, K.I. & Pauliny-Toth, I.I.K. 1968, in *Non-Stable Phenomena in Galaxies*, Academy of Sciences of Armenian SSR, Yerevan, p. 245.

Kellermann, K.I., et al. 1989, AJ, 98, 1185.

Kellermann, K.I. et al. 1998a. AJ, 115, 1295.

Kellermann, K.I., et al. 1998b, in *Proc. of EVN/JIVE Symp. No. 4, New Astron. Rev., in press.*
Krichbaum, T.P., et al. 1998, A&A, 329, 873.
Lacey, C. G. & Ostriker, J.P. 1985, ApJ, 299, 649.
Levinson, A., et al. 1995, ApJ, 448, 589.
Lister, M.L. & Marscher, A.P. 1997. ApJ, 476, 572.
Miller, L. et al. 1990. MNRAS, 244, 207.
O'Dea, C.P. et al.1991, ApJ, 380, 66.
Orr, M.J.W. & Browne, I.W.A. 1982. MNRAS, 200, 1067.
Phillips, R.B. & Mutel, R.L. 1980, ApJ, 236, 89.
Readhead, A.C.S. 1994. ApJ, 426, 51.
Readhead, A.C.S. et al. 1996. ApJ, 460, 612.
Rees, M. 1966. Nature, 211, 468.
Richards, E.R., et al., 1998. AJ, 116, 1039.
Shklovsky, J. 1963, Sov.-Astron., 7, 748.
Slish, V.I. 1992. ApJ, 391, 453.
Tingay, S. et al. 1998, AJ, 115, 960.
Unwin et al. 1997. ApJ, 480, 596.
Urry, M. & Padovannini, 1995. PASP, 107, 803.
Vermeulen, R.C. 1995. *Proceedings of the National Academy of Sciences*, 92, 11385
Vermeulen, R.C. 1999, in preparation.
Wagner, S.J. et al. 1996. AJ, 111, 2187.
Wagner, S.J. 1998, in *Radio Emission from Compact Galactic and Extragalactic Radio Sources*, eds. A. Zensus et al., ASP., San Francisco, p. 257
Wagner, S.J. & Witzel, A. 1995, ARA&A, 33, 163.
Walker, R.C. et al. 1998, in *Radio Emission from Compact Galactic and Extragalactic Radio Sources*, eds. A. Zensus et al., AS., San Francisco, p. 133
Wardle, J.F.C. et al. 1994. ApJ, 437, 122.
Wardle, J. 1998. *Nature*, 395, 457.
Wotjer, L. 1966, ApJ, 146, 597.
Zensus, J.A. 1997. ARA&A. 33, 607.
Zensus et al. 1998, *Proceedings of IAU Colloquium No. 164, Radio Emission from Compact Galactic and Extragalactic Radio Sources*, Astron. Soc. Pac., San Francisco.

Megamaser Activity in Active Galaxies

Willem A. Baan
Netherlands Foundation for Research in Astronomy
Westerbork Observatory, 7990 AA Dwingeloo, The Netherlands

Abstract. Megamaser emission provides an unprecedented view of the inner nuclear regions of active galaxies. The characteristics of OH, H_2CO, and H_2O sources identify different physical environments in the nuclei of megamaser host galaxies. Various classification schemes suggest a dominant starburst nature for the OH and H_2CO sources providing the NIR/FIR radiation field as a pump. The molecular emission originates in a 50-100 parsec disk/torus, which is not yet completely observed. For H_2O masers in relatively dust-free LINER/Seyfert 2 sources, the AGN provides the X-ray radiation field for pumping in a compact sub-parsec thin disk.

1. General Characteristics of Megamasers

Studies of megamaser emission allow a close-up look at molecular material inside the active nuclei at unprecedented small scales and provide diagnostic tools of their physical environment and dynamics. The different excitation conditions for the masering molecules OH, H_2CO, CH, and H_2O ensure that significantly different nuclear environments are observed.

The initial scenario for OH megamasers involving relatively low gain and mostly unsaturated amplification of the radio continuum still applies for OH and H_2CO sources but less so for H_2O sources (see Baan et al. 1982; Baan 1997). Optical depths with a significant range between zero and a few (or many) would be expected because of the varying optical depths resulting from imperfect superposition of radio continuum and molecular clouds.

The connection between megamaser activity and the nature of the galactic nuclei has now clearly been established (see Baan 1985 for early evaluation). OH megamasers are clearly correlated with (mostly) edge-on luminous FIR sources and are pumped by the FIR radiation field. The formaldehyde masers follow the same FIR population but with less stringent FIR temperature criteria. The H_2O megamasers occur in LINER/Seyfert 2 nuclei.

The relatively low finding rates for OH and H_2O megamasers among selected samples result in part from the non-isotropic emission pattern due to the spatial distribution of the amplifying molecular gas. If this rate is solely due to an axisymmetric molecular torus distribution, the detection rates would tell us about the opening angle and size of the molecular structure. The opening angles for such disk geometries have been presented in Figure 1 for OH and H_2O sources (Baan 1997). OH and H_2CO sources have thick molecular structures and the

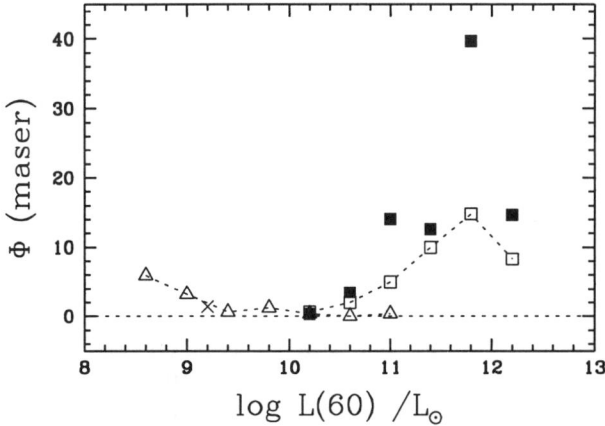

Figure 1. The effective opening angle of the emission pattern for OH and H_2O megamaser galaxies. The open (connected) and filled (unconnected) squares for OH follow from the probability of detecting megamasers among FIR sources using two different luminosity functions. The open triangles for H_2O follow from the probability of finding masers among LINER/Seyfert 2 sources (Baan 1997).

radial distance of the amplifying material is estimated to be 50 - 100 pc (Henkel et al. 1987). The H_2O disks are very thin as has been verified for NGC 4258 (Myoshi et al. 1994).

In the present paper, we will address some megamaser issues that have direct bearing on the nature of the nuclear activity and the spatial characteristics of the molecular emission components. General properties of megamasers have been discussed elswhere (Baan 1991, 1997; Henkel, Baan & Mauersberger 1993).

2. Compact Disks of H_2O Megamasers

The emission regions in a few powerful and well-studied H_2O megamaser sources are confined to Keplerian disks (see Haschick et al 1994; Myoshi et al. 1995). The radial distance of the masering cloud within the thin compact disk in NGC 4258 is approximately 0.13 pc, while for NGC 1068 the emitting section has an inner radius of 1.3 pc (Gallimore et al. 1996). The likely pumping agent for the H_2O masers is the X-ray radiation field from the nucleus (Neufeld et al. 1994). For the limited sample of H_2O sources, the line luminosity is found to be correlated with the product of the X-ray and the radio continuum luminosities suggesting that both radiation fields are of interest (Baan 1997; Braatz et al. 1997). The water-vapor masers provide a high-resolution probe of the nuclear environment of relatively quiescent and low luminosity but "naked" (without much dust) AGNs.

3. Optical and Radio Classification of OH Megamasers

OH megamasers are a dominant sub-class of the (super-) luminous FIR galaxies, which at high luminosity show mostly optical AGN characteristics and are in part inter-acting systems. At lower FIR luminosities the sample is predominantly composed of starburst (SBN) nuclei. *Optical classification* studies of the OH megamaser sample support this general trend with 45 % showing AGN and with 22.5 % Composite (AGN+SBN) characteristics; the other 32.5 % shows pure starburst characteristics (Baan, Salzer & LeWinter 1998). On the other hand, the *radio classification* of the same sample suggests that a majority of 74.5 % shows no or only weak AGN characteristics and resembles radio SBNs (Baan et al 1999). A total of 51 % of the sample show no sign at all of radio AGN activity, which is in agreement with earlier radio surveys of luminous FIR galaxies (Condon et al. 1991). The discrepancy between the optical and radio classifications of OH/H_2CO megamasers suggests a uniqueness of luminous FIR sources regard to several characteristics.

(1) In comparison the radio characteristics of the nuclei of OH and H_2O sources are found to be quite different. The OH sources have mostly radio-quiet nuclei while only three sources have outstanding radio-loud sources. Earlier studies suggest that such radio-loud sources are powered by AGNs, while these radio-quiet sources are powered by starbursts (see Helou et al. 1985). In comparison, the majority of the H_2O nuclei are more radio-loud but they also have lower radio and FIR luminosities than the OH sources.

(2) A principal compromise picture for luminous FIR sources is a combination of a circumnuclear starburst and a central AGN. The relative activity level of these two components would then determine its outward appearance. However, the locus of LINER/Seyfert 2 sources in the [O III]/Hβ versus [S II]/Hα diagnostic diagrams for the OH megamaser sample (Baan et al. 1998) do not represent a simple mixing curve of two luminosity (AGN + SBN) components. In this alternative picture, the radio and FIR luminosities would be dominated by the (more extended) starburst, while the nuclear optical emission is still dominated by the more obscured AGN component. Several lines of argument could support this picture but it still leaves some unanswered questions.

(3) Large concentrations of molecular gas and dust will affect the optical emissions and diagnostic line ratios, because different line emissions originate at different radial locations and at different obscuration depths within the galaxy. For instance, the majority of the OH sources have (buried) edge-on dust lanes, giving high values of A_V for the emission from the nucleus itself but lower A_V for emissions from the outlying regions above and below the plane.

(4) A mixture of starburst and AGN activity has also been proposed on the basis of recent ISO data. An infrared "diagnostic diagram" based on the ratio of high- and low-excitation mid-IR emissions lines versus the strength of the 7.7 μm PAH feature shows that the ultra-luminous FIR galaxies form a distinct AGN - SBN "mixing sequence". (Genzel et al. 1998). About 70% - 80% of the luminous FIR galaxies are predominantly powered by starbursts rather than by AGNs, which is very consistent with the radio classification for megamaser galaxies (Baan et al. 1999). For example, several prominent OH megamasers IR17208, IR12112, and Arp 220 (see section below) which have LINER/Seyfert 2 optical spectra fall in this group in the radio and according to the infrared

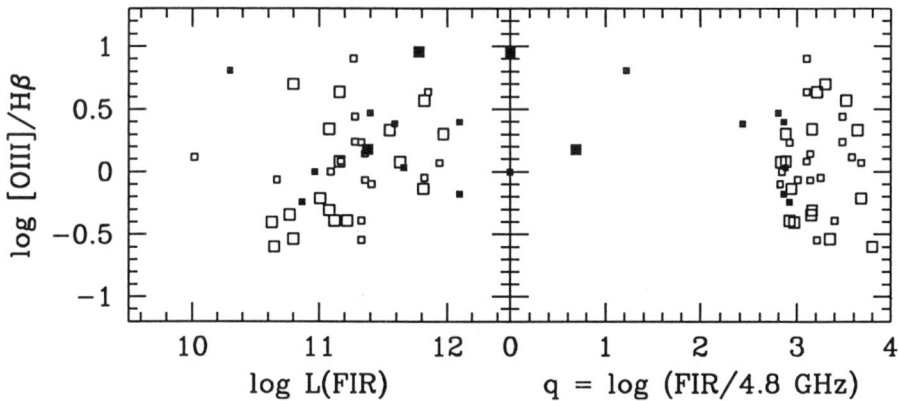

Figure 2. The relation between the optical diagnostic [O III]/Hβ ratio for OH megamasers and (a) their 60 μm FIR luminosity and (b) their "q-ratio" of the FIR luminosity and 4.8 GHz radio continuum luminosity. The galaxies have been marked to indicate their radio characteristics as follows: (1) large open squares show no evidence of AGN activity, (2) small open square show minimal AGN activity, (3) small filled squares show moderate AGN activity, and (4) large filled squares show strong AGN characteristics (Baan et al. 1998; 1999).

"mixing sequence" (Genzel et al. 1998). We find agreement between the nuclear classification in the radio and the infrared but we also find that the optical classifications can be discrepant from the other two.

(5) The combination of a modified mass function for the nuclear starburst with a higher mass cutoff and of superwinds emanating from the region could somewhat harden the ionizing radiation field and consequently change the optical emission line ratios. Nuclear superwinds (Heckman et al. 1990) and OH molecular outflows showing outflow velocities up to 1000 km s^{-1} (Baan et al. 1987) provide a consistent picture of a general blow-out in galaxies undergoing massive central starbursts.

(6) A critical parameter in the optical classification is the [O III]/Hβ ratio, which increases significantly when moving from starburst nuclei to LINERs and to Seyferts. In Figure 2 the optical line ratio has been plotted against the 60 μm FIR luminosity and against the "q-ratio" of FIR luminosity over the 6 cm radio luminosity (see Baan et al. 1998; 1999). Figure 2(a) shows a roughly linear increase of [O III]/Hβ with the FIR luminosity, which could suggest that the generation of optical emission and FIR emission is due to the same process independent of its optical classification. Frame 2(b) shows that only two AGN galaxies stand out clearly with a low q-value characteristic of AGNs; most of the galaxies have a q-ratio in the range characteristic of SBNs.

The above information suggests that contrary to this optical line characteristics, the majority of megamaser galaxies are indeed predominantly powered by nuclear starbursts. Due to a variety of circumstances the intense starburst region in a (super-)luminous FIR/OH galaxy appears to mimic radio-quiet AGNs in their optical appearance. If AGNs were present in these nuclei, they must be radio-quiet but optical-loud. This result implies that the standard optical classification scheme for these high activity super-luminous galaxies may not be accurate enough as a diagnostic.

4. OH Emission Structure of III Zw35

The OH megamaser III Zw35 has been studied with MERLIN (Montgomery & Cohen 1992) and more recently with the VLBA (Trotter et al. 1997). The galaxy exhibits several extended emission regions with a total extent of 50 pc. At high resolution the more extended emission regions breaks into compact maser features, which together account for about 50 percent of the flux in the double peaked OH emission profile. The distribution of the OH emission and the fact there is missing flux support the idea that the maser features originate in a ring/torus centered on the nucleus of III Zw35. One could argue that the two main emission regions at 48 parsec separation and 100 km s^{-1} velocity separation signify the tangential and frontal regions of a Keplerian torus. If this were true, the nuclear mass needs to be 1.4×10^7 $M_\odot$. The rest of the missing flux should also come from this torus. The inferred radial distance of the masering clouds in the torus is also consistent with FIR pumping models (Henkel et al. 1987; Kylafis & Pavlakis 1998).

5. OH and Radio Continuum Structure of Arp 220

The prototype megamaser galaxy Arp 220 (IC 4553) has its most prominent OH and H_2CO emission at the western nucleus and less prominent emission at the eastern nucleus (Baan & Haschick 1995). Spectral line VLBI studies with the EVN reveal three high brightness OH components: two components south of the eastern radio nucieus separated by 40 pc and one north of the western radio nucleus (Diamond et al. 1990; see Baan & Haschick 1995). These components appear not to coincide with the radio nuclei themselves as predicted for any compact AGN - nuclear torus scenario. Very recent VLBI results reveal a cluster of compact radio supernovae at both nuclei of Arp 220 (Smith et al. 1998). Two extended OH emission regions in the western nucleus straddle the cluster of supernova remnants and could together form a north-south torus structure with radius of 45 pc (Lonsdale et al. 1998). At the eastern nucleus two components also straddle the cluster but lie slightly south of the radio peak. However, the emission found (to date) at the eastern nucleus is not at the nuclear velocity and thus unrelated to the nucleus itself (see paragraph below). The VLBI features found at the nuclei account for about 40 % of the single dish flux. More diffuse emission must be present in the intervening regions, particularly at the western nucleus.

Also the H_2CO emission also comes predominantly from the western nucleus with only weak emission from a double north - south structure southeast from the eastern nucleus and from some more intervening regions between the two nuclei (Baan & Haschick 1995). Recent NIR images of Arp 220 made with NICMOS (Scoville et al. 1998) show an emission structure almost identical to that of the H_2CO emission. Such coincidences suggest that (1) obscuring dust plays havoc in this region and that large fractions of the nuclear regions are heavily obscured, and that (2) the formaldehyde and OH masers each appear to be pumped by the FIR/NIR radiation fields.

A simple orbital model has been proposed for the dynamics of the two nuclei based on molecular line characteristics (Baan & Haschick 1995; Scoville et al. 1996). Both nuclei show OH emission closely to 5300 km s^{-1}, which may be due to foreground material at the eastern nucleus. The OH features at the systemic velocity of the eastern nucleus are to be found in the second spectral feature, which is in fact a mixture of the 1667 MHz lines from the eastern nucleus and the 1665 MHz lines from the western nucleus (see Baan & Haschick 1987).

6. High Redshift OH Sources

Megamaser searches of galaxies at high radial velocities have in the past been severely limited by receiver capabilities and spectral coverage. Only for OH it has been possible to explore higher redshift regimes. The formaldehyde sources found to date are too weak to be detectable much beyond the local universe. On the other hand, the H_2O megamasers are very powerful at lower redshifts but their detection rate has been very low. Again the H_2O searches at higher redshifts will be severely limited by the receiver capabilities at the larger telescopes.

Powerful OH megamasers have now been found to redshifts of about z = 0.26 or v = 80,000 km s^{-1}, which is similar to those of luminous (thermal) CO sources. If OH megamaser activity is rather common locally, OH emission may be found in the very red objects in the range z = 1 - 2, where the population of quasars peaks. With a total velocity width of 2400 km s^{-1}, the currently highest redshift OH maser IRAS 14070+0525 suggests that the dynamic structure of such sources may be quite different from those observed at lower redshifts (Baan et al. 1992). Although there may be a violent starburst or merging activity in this source, the radio and optical images do not yet reveal any interaction or merger nature.

A quadratic luminosity relation observed for the OH sources may render OH megamasers observable to very high redshifts. For current observing systems, a megamaser is required to have $L_{OH} = 10^{3.8}$ $L_\odot$ to be observable at z = 2; already two nearby masers have been detected with such luminosities. Calculations of the number of detectable OH sources suggest that the observable universe increases faster than the decrease in the spatial density of OH galaxies, and that at redshifts z = 1 - 2 large numbers of detectable sources are expected (Baan 1991; Briggs 1998). The OH detection rate at high redshifts will depend on the actual L(FIR) - L(OH) relation and on the galaxy merger rate, that plays a significant role in making the population of luminous FIR sources (see Briggs 1998). The OH detection rate will also depend strongly on the sensitivity of the observing systems favoring telescopes such as the upgraded Arecibo and the to-be-built Square Kilometer Array (SKA). Probing the higher redshift ranges for OH activity will provide independent tests of cosmological models and the nuclear evolution of galaxies.

7. Conclusions

Radio and molecular line studies of megamasers provide independent probes of the nuclear environment and the dynamics of active galaxies. The compact disks in H_2O sources tell us about the physics of an AGN and the nuclear radiation environment. The torus/ring structures in OH and H_2CO sources provide an excellent probe of the critical circumnuclear environment that shapes the (super-)luminous FIR galaxies. The missing flux in both III Zw35 and Arp 220 likely come from lower brightness (maser) emission regions distributed along the dusty torus.

The radio and NIR classifications of luminous OH/FIR sources largely agree on a dominant starburst powering these galaxies rather than an AGN, although the optical classification may still suggest an AGN nature. At high redshift no clear connection is yet available between OH/H_2CO activity and a particular part of the galaxy population. Possibly the very red objects found at higher redshift or even type 2 quasars are the megamaser host galaxies. If so, continued molecular studies will provide valuable new insights into the nature of such peculiar objects.

References

Baan, W.A. 1985, Nature, 315, 26
Baan, W.A. 1989, Ap.J., 338, 804
Baan, W.A. 1991, in Skylines, Third Haystack Observatory Conference, A.D. Haschick and P.T. Ho, P.A.S.P. Conf. Series, 16, 45
Baan, W.A. 1997, in High Sensitivity Radio Astronomy, R.J. Davis & N.J. Jackson, Cambridge University, 73
Baan, W.A. & Haschick, A.D. 1987, Ap.J., 318, 139
Baan, W.A. & Haschick, A.D. 1995, Ap.J., 454, 745
Baan, W.A., Haschick, A.D. & Uglesich, R. 1993, Ap.J., 415, 140
Baan, W.A., Henkel, C. & Haschick, A.D. 1987, Ap.J., 320, 154
Baan, W.A. et al. 1992, Ap.J., 396, L99
Baan, W.A., Kloeckner, H.-R., Haschick, A.D., & Besenfelder, E. 1999, Ap.J., submitted
Baan, W.A., Salzer J.J., & LeWinter, R.D. 1998, Ap.J., December 15
Baan, W.A., Wood, P.A.D. & Haschick, A.D. 1982, Ap.J., 260, L49
Braatz, J., Wilson, A.S., & Henkel, C. 1997, Ap.J. Suppl., 110, 321
Briggs, F.H 1998, Astron. Ap.,
Condon, J.J, Huang, Z.-P., Yin, Q.F., & Thuan, T.X. 1991, Ap.J., 378, 65
Diamond, P., Norris, R. Baan, W.A., & Booth, R. 1990, Ap.J., 340, L49
Gallimore, J., et al. 1998, Ap.J., 462, 740
Genzel, R. et al. 1998, Ap.J., 498, 579
Haschick, A.D., Baan, W.A., & Peng, E. 1994, Ap.J., 437, L35
Heckman, T.M., Armus, L., & Miley. G.K. 1990, Ap.J. Suppl., 74, 833

Henkel, C., Baan, W.A., & Mauersberger, R. 1993, Astron. Ap. Rev., 3, 47
Henkel, C., Güsten, R. & Baan, W.A. 1987, Astron. Ap., 185, 14
Helou, G., Soifer, B.T., & Row, R.M. 1985, Ap.J., 298, L11
Kylafis, N.D. & Pavlakis, K.G. 1998, IAU Highlights of Astronomy, 23, in press
Lonsdale, C.J., Lonsdale, C.J., Diamond, P.J., & Smith, H.E. 1998, Ap.J., 493, L13
Montgomery, A.S. & Cohen, R.J. 1992, M.N.R.A.S., 254, 23p
Myoshi, M. et al. 1995, Nature, 373, 127
Neufeld, D., Maloney, P., & Conger, S. 1994, Ap.J., 436, L127
Scoville. N.Z., Yun, M & Bryant, P. 1997, Ap.J., 484, 702
Scoville, N.Z. et al. 1998, Ap.J., 492, L107
Smith, H.A, Lonsdale, C.J., Lonsdale, C.J., & Diamond, P.J. 1998, Ap.J., 493, L17
Trotter, A.S. et al. 1997, Ap.J., 485, L79

Active Galactic Nuclei and Related Phenomena
IAU Symposium, Vol. 194, 1999
Yervant Terzian, Daniel Weedman, Edward Khachikian, eds.

The Mid-infrared Spectral Properties of Starburst Galaxies and Active Galactic Nuclei

O. Laurent and I.F. Mirabel

CEA/DSM/DAPNIA. Service d'Astrophysique F-91191 Gif-sur-Yvette, France

Abstract.

We present a new diagnostic tool based on mid-infrared spectra (5-16μm) for distinguishing the emission triggered by the active galactic nucleus (AGN) and by the star formation activity. We show that the AGN spectra, contrary to the starburst spectra, present an important continuum below 9μm and an absence of Unidentified Infrared Bands (UIBs).

1. Introduction

Nuclear regions harboring starbursts or AGNs are usually embedded in a large concentration of dust producing heavy obscuration especially for interacting galaxies which contain a large amount of molecular gas in their inner regions (see Sanders & Mirabel 1996 for a review). Since large amounts of molecular gas are needed for fueling nuclear starbursts as well as AGNs, the absorption makes the distinction between starburst and AGN difficult. Mid-IR wavelengths, being less affected by the absorption ($A_V/A_{8\mu m} \sim 50$), can probe obscured regions usually not visible in optical and near-IR wavelengths as well as the re-radiated emission from dust (Mirabel et al. 1998a, 1998b). We present a statistical approach to classify the spectral diversity of a large sample of galaxies observed in mid-IR wavelengths.

2. Observations

Based on the spectral-imaging capabilities of the Infrared Space Observatory (ISO)[4], a sample of 25 active/interacting galaxies was observed with the mid-infrared camera on-board ISO (Cesarsky et al. 1996a). Standard data reduction techniques were applied using the CAM Interactive Analysis software (CIA)[5].

---

[4]Based on observations with ISO, an ESA project with instruments funded by ESA Member States (especially the PI countries : France, Germany, the Netherlands and the United Kingdom) and with participation of ISAS and NASA.

[5]CIA is a joint development by the ESA astrophysics division and the ISOCAM consortium led by the ISOCAM PI, C. Cesarsky, Direction des Sciences de la Matière, C.E.A. France.

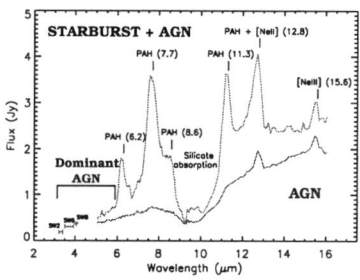 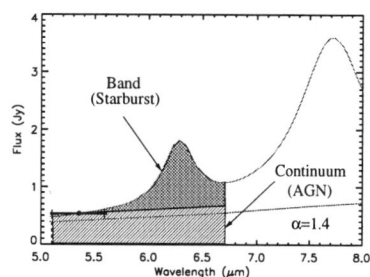

Figure 1. Left panel: Mid-IR spectra of NGC 5128 with the lower curve for the nuclear region (100pc size) and the upper curve for a larger integrated region (800pc size). The AGN spectrum represents 45% of the energy between 5μm and 16μm and it highly dominates at short wavelengths below the UIB at 6.2μm for which the emission associated with starburst activity is not detected with the three narrow band filters below 5μm. Right panel: the spectrum between 5.1μm and 6.7μm is divided into two different parts. The flux above the thick line (power law with a spectral index $\alpha \sim 1.4$ fitting the AGN spectrum continuum and rescaled to the average flux between 5.1μm and 5.6μm) estimates the flux arising from the UIB at 6.2μm and traces the starburst activity. The flux below the line selects the continuum arising mainly from the AGN. The band/continuum indicator varies similarly to the starburst/AGN ratio.

3. Mid-IR features of AGN and Starburst spectra

We observed nearby prototypical active galaxies (NGC 5128, NGC 1068) in order to separate with the highest spatial resolution (~100-400 pc with 5″) the distinct mid-IR emission produced by the AGN from that produced by the circumnuclear starburst. Using observations of nearby star forming galactic regions (NGC 7023, M17, Cesarsky et al. 1996b, 1996c), we identified a typical mid-IR spectrum of a pure ionized region (HII) heated by young stars and a spectrum characterizing photo-dissociation regions (PDRs) surrounding HII regions. Those two spectra were used to separate into two components the emission from larger unresolved star forming regions which can be interpreted as a mixing of several PDRs and HII regions. Based on these results, we are able to estimate the relative contribution of each component to the mid-IR emission of unresolved regions observed in galaxies (see also Tran 1998).

3.1. Diagnostics for distinguishing AGN, HII and PDR emission

The UIBs, the carbonaceous species that have been called polycyclic aromatic hydrocarbons (PAHs) seen at 6.2, 7.7, 8.6, 11.3 and 12.8μm, are detected mainly in PDRs, whereas a noticeable continuum at short wavelengths[5-9μm] without UIBs is present in our sample of AGN spectra. Presumably, the UIBs are destroyed in AGNs by the intense radiation field. We estimate the PDRs/AGN fraction with the UIB(6.2μm)/continuum[5-6.7μm] ratio described in Figure 1 (see also Genzel et al. 1998).

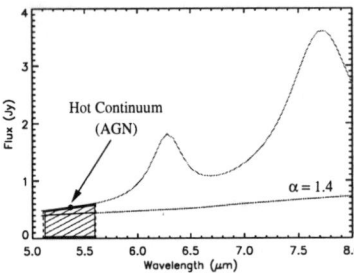 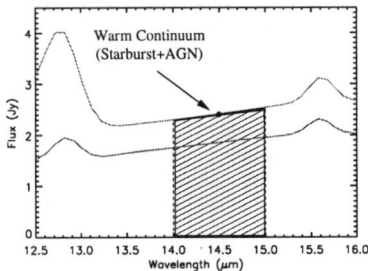

Figure 2. Left panel: The hot continuum[5.1-5.6μm] originates mainly from the underlying emission of the AGN. Right panel: Both star forming regions and the AGN contribute to the warm continuum[14-15μm]. The ratio of the warm over hot continuum can be used to differentiate between strong starburst spectra with faint UIBs and AGN spectra.

Since the HII region spectrum has very faint thermal emission at short wavelengths compared to that of an AGN, the warm continuum[14-15μm]/hot continuum[5-5.6μm] ratio can be used as an estimator of the HII/AGN fraction (see Figure 2).

3.2. Mid-IR diagram

In Figure 3, we present the three mid-IR spectra used as templates for identifying different components associated with AGNs or starbursts. The HII regions have absent or faint UIBs and a strong rising continuum associated with strong emission lines of [ArIII] (9μm), [SIV](10.5μm), [NeII](12.8μm) and [NeIII](15.5μm). The regions dominated by PDRs, observed in galactic discs, present strong UIBs largely dominating the total emission and a faint continuum. Finally, the AGN mid-IR emission is mainly characterized by strong continuum at 5-9μm without UIBs. Assuming that the spectra integrated over the whole galaxy are composed of the mixing of spectra originating from HII regions, PDRs and an AGN, we have built composite spectra from the three mid-IR templates. This diagram can be decomposed in three different regions where we select preferentially the AGN (bottom left), the HII region (top) and the PDR spectra (bottom right).

In Figure 4, we have built the same diagram including 34 spectra of galaxies containing star forming regions and/or AGNs. The different galaxies are placed in the diagram. The extreme starburst galaxies like Arp 220 and Mrk 171(A) are classified as galaxies dominated by strong star forming activity presenting a large fraction of pure HII regions compared to PDRs. The AGNs detected for nearby objects are located in the bottom left part of the diagram. The star forming regions observed in galactic discs are situated near the region dominated by PDRs.

Using this diagram, we conclude that the well-known ultraluminous galaxies like Arp 220, IRAS 23128-5919 and NGC 6240 present mid-IR spectra dominated by strong starburst activity. The AGN contribution is only detected in nearby objects for which we are able to separate the nuclear regions and the galactic discs.

Mid-IR Spectral Properties of Starburst Galaxies and AGN

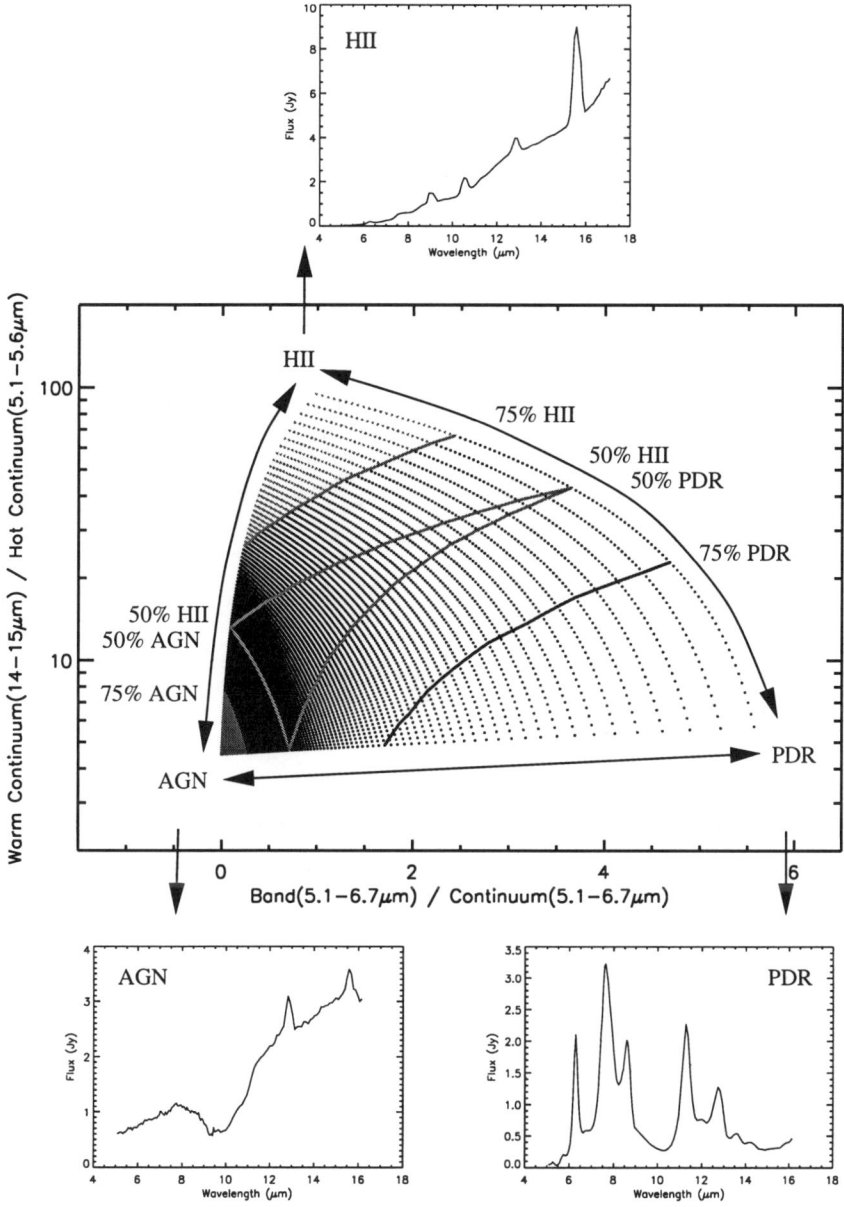

Figure 3. Based on synthetic spectra containing variable fractions of three mid-IR templates observed in a pure HII region (M17), a PDR region (NGC 7023) and an AGN (NGC 5128), the diagnostic diagram defines three distinct areas. In the upper part, we select spectra dominated by HII regions (strong starbursts). On the left, the AGN spectra are dominant in a very small region, and finally, the PDR spectra fall in the right part. The percentages indicate the fraction of each mid-IR contribution used in the composite spectra.

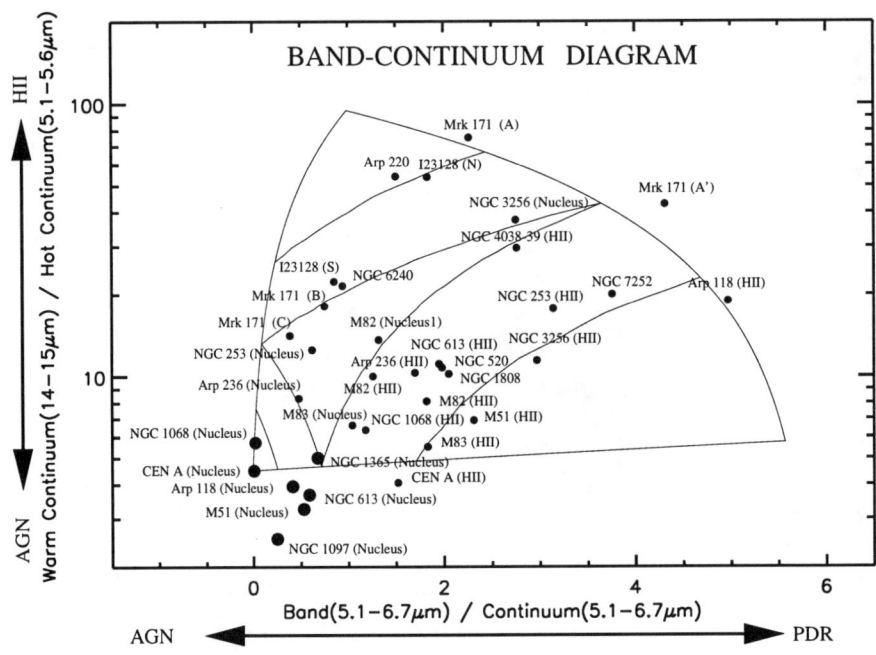

Figure 4. Diagnostic diagram including 34 spectra covering a large sample of mid-IR emission (nulcear regions, disc starburst regions called HII and unresolved galaxies). The galaxies hosting an intense starburst are in the upper part. The galactic central regions (size of several kpc) hosting an AGN (big circles) are all located in the bottom left part of the diagram. The spectra of galactic discs lie near the PDR region. The curves demarcate the AGN, PDR and HII regions according to figure 3.

4. Conclusion

We have obtained a new AGN/Starburst diagnostic using the UIB at 6.2μm and the mid-IR continuum based on two independent points:
1) The presence of a non-negligible continuum below 9μm in AGN spectra which we attribute to hot dust emission directly heated by the central engine.
2) The absence in "pure" AGN spectra of UIBs whose carriers are presumably destroyed by the strong UV-X radiation field.

Acknowledgments. We thank V. Charmandaris, H. Roussel and D. Tran for providing data in advance of publication as well as P. Gallais and S. Madden for their help on different aspects of this work.

References

Cesarsky C., Abergel A., Agnese P. et al. 1996a, A&A, 315, L32
Cesarsky D., Lequeux J., Abergel A. et al. 1996b, A&A, 315, L305
Cesarsky D., Lequeux J., Abergel A. et al. 1996c, A&A, 315, L309
Genzel R., Lutz D., Sturm E. et al. 1998, Ap.J., , 498, 579
Mirabel I.F., Vigroux L., Charmandaris V. et al. 1998a, A&A, 333, L1
Mirabel I.F., Laurent O., Sanders D.B. et al. 1998b, A&A, in press
Sanders D.B., & Mirabel I.F. 1996, ARA&A, 34, 749
Tran D. 1998, PhD thesis, University of Paris XI

VLBI Imaging of Luminous Infrared Galaxies: Starbursts & AGN

Harding E. Smith[1]

Center for Astrophysics & Space Sciences and Department of Physics, University of California, San Diego, La Jolla, CA 92093-0424, USA

Carol J. Lonsdale

Infrared Processing and Analysis Center, California Institute of Technology, Pasadena, CA 91125, USA

Colin J. Lonsdale

Haystack Observatory, Massachusetts Institute of Technology, Westford, MA 01886, USA

Philip J. Diamond

National Radio Astronomy Observatory, Socorro, NM 87801, USA

Abstract.
Luminous Infrared Galaxies (LIGs) are locally more numerous than normal galaxies, AGN, and QSOs above $L \sim 10^{11} L_\odot$ and may be the evolutionary precursors of classical radio-quiet quasars. VLBI observations of a complete sample show that high-T_b radio cores are common, perhaps universal among LIGs. VLBI imaging shows that these radio cores may be produced by intense starbursts which generate luminous radio supernovae, as in the case of Arp 220 (Smith *et al.* 1998), or by a classical AGN core, as in the case of Mrk 231, which we interpret as a newly formed QSO emerging from a starburst. Compact OH 1667MHz maser emission appears to be common in LIGs and may be related to AGN activity. These results lend further support to the scenario suggested by Sanders *et al* (1988) in which mergers of gas-rich galaxies lead first to luminous starbursts which evolve into radio-quiet quasars.

1. Luminous Infrared Galaxies

The most luminous galaxies in the Local Universe are Luminous Infrared Galaxies (LIGs) which emit the vast majority of their radiant power in the far-infrared between about 40–120μm. These are gas-rich systems which are in the late stages of collisions or mergers and extrapolation from the properties of lower luminosity starburst galaxies suggests that the LIGs should be active star-forming systems

---

[1]also, Infrared Processing and Analysis Center, Caltech/JPL, Pasadena, CA 91125

(see Sanders & Mirabel 1996 for a review). The LIGs also show many characterisatics of AGN and their luminosities reach values comparable to those of luminous QSOs. Much effort has been focused on whether LIGs are powered principally by starburst or AGN activity, although both types of activity are almost certainly present. The discussion has been framed around a scenario proposed by Sanders et al. (1988) in which a merger of gas-rich disk galaxies stimulates a massive nuclear starburst which in turn feeds a coalescing AGN core in the galaxy nucleus. As the AGN turns on, radiation pressure drives out the shroud of dust, revealing a nascent quasar. The goal must be not only to understand the dominant source of energy in LIGs, but to understand the relationship between starburst and AGN activity and other galaxy characteristics, and to place them into an evolutionary context. We have approached this question by studying a complete sample of LIGs defined by Condon et al. (1991; CHYT); this work has concentrated on VLBI observations which offer unique AGN/starburst diagnostics in one of the few wavelength regimes where the optical depths to the active regions may fall below unity.

In an 18-cm VLBI survey of Luminous Infrared Galaxies for compact, high-T_b emission commonly associated with AGN activity, Lonsdale, Smith, and Lonsdale (1993; Paper I) showed that milli-arcsecond scale emission, $T_b >> 10^7 K$, is common, perhaps universal in LIGs. Furthermore, the LIGs follow a common relationship between core radio power and bolometric luminosity with radio-quiet QSOs (Lonsdale, Smith & Lonsdale 1995). This lends support to the interpretation of LIGs as dust-enshrouded AGN. On the other hand, a recent detailed analysis of our VLBI survey data, (Smith, Lonsdale & Lonsdale 1998; Paper II) investigated a starburst origin for LIGs in which the compact, high-T_b emission is produced by luminous radio supernovae (RSN). This analysis indicates that most, but not all, LIG VLBI-scale emission may be modelled with starburst-generated RSN, provided the RSN are *extremely luminous* and, in most cases, clustered. These predictions were confirmed with the detection of luminous RSN in the nuclei of Arp 220, consistent with a starburst origin for the infrared luminosity (Smith et al. 1998), described in §2. The AGN view is shown in §3 where we present our images of Mrk 231 which clearly show an AGN core which ignited within the last $\sim 10^6$ years in the center of a starburst disk.

2. Arp 220: An Intense Starburst

Arp 220 (= IC 4553/4 = UGC 9913 = IRAS 15327+2340) is the archetype LIG with $L_{fir} \approx 10^{12} L_\odot$ at a distance of 76 Mpc ($H_0 = 75$ km s^{-1}Mpc^{-1}). In previous 18cm radio studies we showed that about 3% of the 18cm radio power originates from compact, high-T_b regions associated with its merging nuclei (Lonsdale, Smith & Lonsdale 1993). First epoch images (1994 November) of Arp 220 at 3 × 8 mas angular resolution (Smith et al. 1998), showed over a dozen unresolved sources, $S_{18cm} = 0.2 - 1.2$ mJy, within a $0.2 \times 0.4''$ (75 × 150 pc) region centered on the NW nucleus and at least two additional sources in the SE nucleus. We interpreted these compact sources as luminous radio supernovae and presented a simple starburst model for Arp 220 which has a star-formation rate of $50 - 100\, M_\odot yr^{-1}$ and a luminous radio supernova rate, $\nu_{LRSN} \sim 2\, yr^{-1}$.

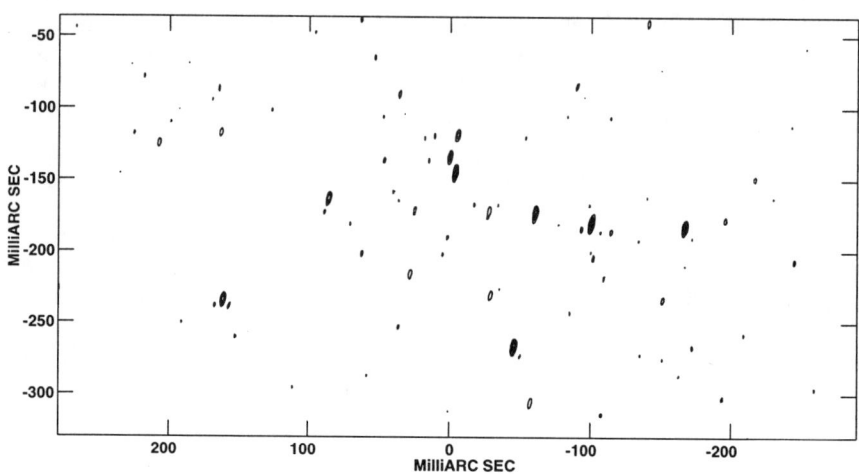

Figure 1. 18 cm VLBI image of the W Nucleus of Arp 220 from Smith, Lonsdale, Lonsdale & Diamond (1998). Over a dozen unresolved sources are interpreted as luminous radio supernovae in an intense starburst.

Our second epoch observations, 3 years following the first epoch, show that most, but not all, of the compact radio sources have declined in brightness, but no new compact sources have appeared. Two of the compact sources are detected at 6cm; these sources have inverted 1.67–5.0GHz radio spectra, suggesting large optical depths, $\tau_{ff}(18cm) \gtrsim 1.0$–$1.5$ and possibly young ages. These observations are consistent with the luminous RSN interpretation, but require a slower decline in the 18cm radio light-curve with a concommittantly lower luminous supernova rate, $\nu_{LRSN} \lesssim 0.3\,yr^{-1}$, consistent with the lack of new RSN. If Arp 220 is powered by an enormous starburst, then only a fraction of its supernovae are luminous RSN, either because luminous RSN only occur in the most extreme environments, or because only very massive stars ($M \gtrsim 20 M_\odot$) produce luminous RSN. We believe that it is likely that we are seeing only the upper part of the RSN luminosity function, formed in the densest molecular regions of the compact nuclear starburst.

2.1. Compact OH Maser Emission

VLBI spectroscopic studies have shown that the 1667MHz OH maser emission from Arp 220 is concentrated in four pc-scale sources, two in each of the nuclei (Lonsdale et al. 1998). Other LIGs (*e.g.* IIIZw35 & IRAS17208-0014, Diamond et al. 1998) show similar compact maser emission. These compact maser sources exhibit *only* the OH 1667MHz feature, but may account for up to 2/3 of the 1667MHz luminosity. The maser characteristics may be summarized as follows:

1. Compact maser dimensions are of order $R \sim$ a few pc.

2. The masers are high gain, with amplification factors of order $10^2 - 10^3$, and are almost certainly highly saturated.

3. The compact masers do not subtend sufficient solid angle for the infrared radiation to produce the pumping; they must be collisionally pumped, perhaps by shocks.

4. The compact OH masers exhibit substantial velocity gradients, with $\frac{\Delta v}{\Delta r} \sim 30 km\ s^{-1} pc^{-1}$. If interpreted as rotation these gradients imply mass concentrations in excess of $10^6 M_\odot$.

These characteristcis are very different from the classical extended megamasers described by Baan (this volume). We speculate that these compact masers may be related to the onset of AGN activity.

3. Mrk 231: A Nascent Quasar Emerging from A Starburst

Mrk 231 (=UGC 5058; $D = 173$ Mpc) has long been recognized as a remarkable galaxy: it is the most luminous galaxy in the local ($z \lesssim 0.1$) Universe with properties that place it among classical AGN and also infrared galaxies. The remarkable properties of Mrk 231 are summarized by Weedman (this volume).

3.1. Mrk 231 as a QSO

Mrk 231 has long been considered an infrared quasar — the luminosity of Mrk 231, $L_{bol} \gtrsim 10^{46} erg\ s^{-1}$, is comparable to other low-redshift radio-quiet QSOs and it falls in the midst of the radio power–luminosity relation for QSOs (and LIGs) constructed by Lonsdale, Smith & Lonsdale (1995). Mrk 231 shows a strong, broad-emission-line spectrum with strong FeII emission and highly reddened line and continuum ($A_V \approx 2$; Boksenberg et al. 1977). Mrk 231 is reported to be variable at both optical (Hamilton & Keel 1987) and radio wavelengths (McCutcheon & Gregory 1978). Mrk 231 has three broad-absorption-line systems: $v_{ej} \approx 4700\ km/s$, $v_{ej} \approx 6000\ km/s$, $v_{ej} \approx 8000\ km/s$ (Adams & Weedman 1972, Rudy, Foltz & Stocke 1985). The $v_{ej} \approx 8000\ km/s$ absorption-line system is variable on timescales of order 2–3 yrs (Boroson et al. 1991). Mrk 231 is underluminous in x-rays (0.1–4.5kev; Rush et al. 1996) and soft γ–rays (Dermer et al. 1997) a characteristic it shares with other BAL objects, but it shows an AGN-like hard x-ray spectrum (Turner 1998).

3.2. Mrk 231 as a Luminous Infrared Galaxy

With an infrared luminosity, $log\ L_{FIR} = 12.35(L_\odot)$, Mrk 231 is the most luminous infrared galaxy in the complete sample of CHYT. Like other well-studied LIGs, the system is rich in molecular gas, $log\ M_{H_2} \approx 10.2(M_\odot)$ (Solomon et al. 1997) and it is one of the original OH megamaser systems (Baan 1985). Furthermore, Mrk 231 shows tidal tails and other evidence for merger or disturbance in the host galaxy, which is itself a luminous system with evidence for rapid star formation (Hamilton & Keel 1987, Sanders et al. 1988).

Bryant & Scoville (1996) have interpreted the CO distribution as a molecular disk containing an estimated $M_{H_2} \approx 3 \times 10^9 M_\odot$ within 1″(=840pc) with

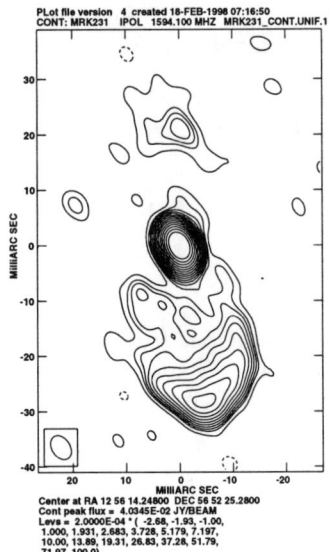

Figure 2. 18 cm VLBI image of the central 100pc of Mrk 231.

a scale height, $h \approx 30pc$. Carilli, Wrobel and Ulvestad (1998) have detected HI *absorption* against the inner disk ($r \lesssim 160pc$). The HI observations suggest $N_{HI} \approx 6 \times 10^{22}(T_s/1000K)cm^{-2}$, $<n_{HI}> \approx 250(T_s/1000K)\,cm^{-3}$, and $M_{HI} \approx 2 \times 10^7(T_s/1000)M_\odot$. Our OH maser observations show only extended 1667MHz emission within the central 100pc. The radio continuum emission from the disk suggests that there remains significant star-formation activity, $\dot{m} \sim 60\text{--}100 M_\odot\,\mathrm{yr}^{-1}$.

3.3. The Compact AGN Source at the Center

Our image of the center of Mrk 231 shows a compact asymmetric source classified as a Compact Symmetric Object (CSO; Readhead *et al.* 1996b). The CSOs are believed to be young objects (age $\lesssim 10^4\mathrm{yr}$) which will evolve into FRII radio sources. Our analysis supports this view for Mrk 231. Simple optically-thin synchrotron modelling of the dominant (southern) radio lobe, which is almost certainly the working surface of a nuclear sub-relativistic jet upon the ambient medium, produces an estimate of the pressure in particles and field, $P_{rel} \approx 7 \times 10^{-7} dyn\,cm^{-2}$. Equating this pressure to the ram pressure, $P_{ram} \sim \rho v^2$, provides an estimate of the advance speed of the south lobe, provided we have an estimate of the ambient density. The mean molecular density in the star-forming disk is estimated to be of order $n \sim 250\,cm^{-3}$. A stronger limit on the density of the medium surrounding the south lobe comes from our map of the spectral index distribution, and by inference the free-free optical depth, around the working surface, $\rho \lesssim 2 \times 10^{-19} g\,cm^{-3}$. This results in a *lower limit*, $v_{adv} \gtrsim 10^{-4}c$, with a resulting upper limit on the timescale for the onset of activity, $\tau_{AGN} \lesssim 10^6 yr$. Mrk 231 is a young quasar; the estimate of its age is substantially less than the inferred age of the starburst, $\tau_{*B} \sim 10^8 yr$.

4. Summary

The VLBI characteristics of many Luminous Infrared Galaxies, like Arp 220, may be explained by starburst-generated luminous RSN, but some LIGs, like Mrk 231, must harbor AGN cores. Compact OH megamaser emission is frequently present in LIGs and may be related to AGN activity. The scenario of Sanders et al. (1988) from merger — luminous starburst — QSO remains an attractive picture for the interpretation of LIGs.

References

Adams, T., & Weedman, D. 1972, Ap.J., , 173, L109.
Baan, W. 1985, Nature, 315, 26. Ap.J., , 419, 553.
Boksenberg, A., et al. 1977, M.N.R.A.S., , 178, 451.
Boroson, T., Meyers, K., Morris, S., & Persson, S. E. 1991, Ap.J., , 370, L19.
Bryant, P. & Scoville, N. 1996, Ap.J., , 457, 678.
Carilli, C., Wrobel, J., & Ulvestad, J. 1998, Astron. J., , 115, 928.
Condon, J., Huang, Z.-P., Yin, Q., & Thuan, T. 1991, Ap.J., , 378, 65.
Dermer, C., et al. 1997, Ap.J., , 484, L121.
Diamond, P., Lonsdale, C., Lonsdale, C. & Smith, H. E. 1998, Ap.J., , in press.
Hamilton, D. & Keel, W. 1987, Ap.J., , 321, 211.
Lonsdale, C., Diamond, P., Lonsdale, C. & Smith, H. E. 1998, Ap.J., , 493, L13.
Lonsdale, C., Smith, H. E., & Lonsdale, C. 1995, Ap.J., , 438, 632.
Lonsdale, C., Smith, H. E., & Lonsdale, C. 1993, Ap.J., , 405, L9 (Paper I).
McCutcheon, W. & Gregory, P. 1978, Astron. J., , 83, 566.
Readhead, A., Taylor, G., Pearson, T., & Wilkinson, P. 1996, Ap.J., , 460, 634.
Rudy, R., Foltz, C. & Stocke, J. 1985, Ap.J., , 288, 531.
Rush, B., Malkan, M., Fink, H., & Voges, W. 1996, Ap.J., , 471, 190.
Sanders, D. B. & Mirabel, I. F. 1996, ARA&A, 34, 749.
Sanders, D., et al. 1988, Ap.J., , 325, 74.
Smith, H. E., Lonsdale, C., & Lonsdale, C. 1998, Ap.J., , 492, 137 (Paper II).
Smith, H. E., Lonsdale, C., Lonsdale, C. & Diamond, P. 1998, Ap.J., , 493, L17.
Solomon, P., Downes, D., Radford, S., & Barrett, J. 1997, Ap.J., , 478, 144.
Turner, T. J. 1998, Ap.J., , in press.

HST Observations of the Jet of 3C120

G. Lelièvre

Observatoire de Paris, DASGAL, URA 335, 75014 Paris CEDEX, France

A. Bijaoui

Observatoire de la Côte d'Azur, Cerga, UMR6527, 06304 Nice CEDEX 04, France

G. Wlérick

Observatoire de Paris-Meudon, DASGAL, URA 335, 92195 Meudon CEDEX, France

Abstract. The WFPC2 observations made by J. Westphal, in July 1995, allow us to confirm the existence of condensations O', O1 and O2; they are located at intersections between the radio jet and filaments emitting in the continuum and in the lines. The depolarization of the radio jet at the position of knot R1 is related to the interaction of the jet with a strongly ionized region of the galaxy. We also detect additional knots closer than 2.5" from the nucleus. The high angular resolution brings precise measurements in position and flux. The optical and radio positions are in good agreement and the radio/optical spectral indices of the knots are as expected in the case of synchrotron radiation.

1. Introduction

In two cases, observations of extragalactic jets are performed with nearly similar resolutions at both optical and radio wavelengths : 3C 273 and M87. In 1994, Lelièvre et al. detected an optical jet out of the nucleus of 3C 120, using the CFH Telescope and an atmospheric compensation device providing elementary images with a resolution of 0.4" (FWHM) and 0.66" after filtering. They associated several optical features with radio knots in the range 2.5"– 7" from the nucleus. In 1995, Hjorth et al. could not find clear evidence of optical counterparts of radio knots on images with poorer resolution ($> 0.76"$). Nevertheless, they claim detection of a continuous flux between 6" and 15" from the nucleus. There was a clear need to get data with better resolution and to explore the inner part of the jet.

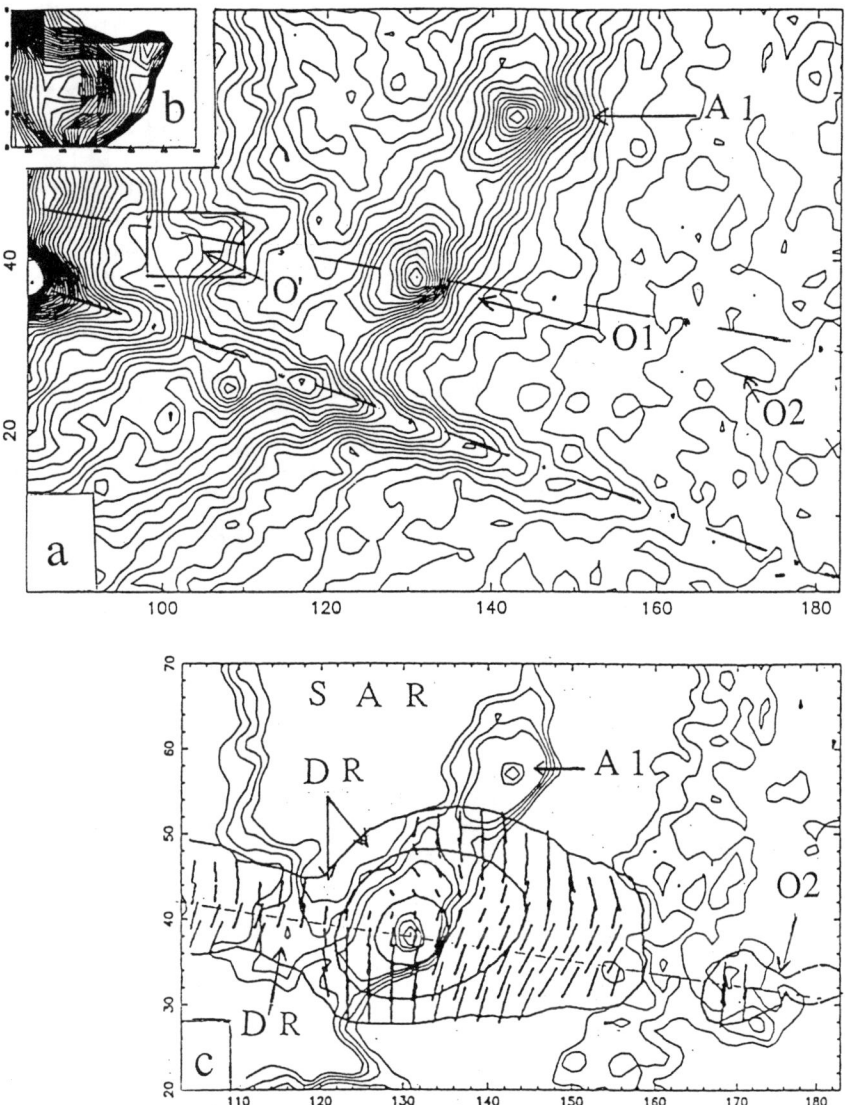

Figure 1. Main condensation O1 of the optical jet. Filter F814W. North is up and east left. Resolution is 0.15". The dotted lines show the direction NR1 of the radio jet and the axis of the diffraction spike. Condensations O' and O2 are also visible. The isophotes increase linearly from 11 to 61 in steps of one DN (data number). b) The detailed map of O' showing two maxima. c) Superposition of the map of figure 2a and the polarization map of the radio jet (Walker et al. 1987, 5 GHz, resolution 0.37"). Isophotes: 12.5, 13, 13.5, 22, 23, 24, 25, 26, 32, 33 and 34DN. The depolarized regions DR occur in strongly absorbed regions SAR of the optical image.

2. Observations and results

Four sets of data have been obtained by J. Westphal on July 25, 1995, with the WFPC2 of the HST. They consist of 3 exposures (one short and two long) through four filters at wavelengths 555 nm, 547 nm, 675nm and 814 nm. The scale is 0.0455" per pixel and the resolution is close to 0.1" (FWHM). To study the jet in the continuum, we selected data taken through the 814 nm filter that, outside of the nucleus, is only weakly polluted by the presence of one of the SIII lines.The main difficulty resides in the presence of one diffraction spike very close to the jet. For data reduction, we had to adopt different techniques according to the distance to the nucleus.

TABLE 1   3C 120 Relative distance of condensations to the nucleus

RADIO Distance (")	R"" 0.2" and 0.4"	R"' 1" to 1,75"			R" 1.8" to 2.3"	R'a 2,46"	R'b 2,80"	R1 3,76"	R2a 5,60"	R2b 6,01"	R3 7,03"
OPTICAL Distance (")	O"" ~0.3"	O"'a 1.18	O"'b 1.44	O"'c 1.76	O"a 2.15	O"b 2.30	O'a 2.53"	O'b 2.70"	O1 3.76"	O2ab 5.76"	
Index (corrected)	O""/R"" 0.83	O"'/R"' 0.78			O"/R" 0.83	O'/R' 0.68			O1/R1 0.72 (0.84)	O2/R2 0.70	

The optical features O1, O' and O2 of Lelièvre et al. (1994) are illustrated in figure 1; O1 is clearly separated from condensation A. Figure 2 shows two new condensations O" and O"' ; Additional sub-structures (O"a, etc) are also detected. Closer to the center, we use the short exposure, which is not saturated, and detect a condensation O"" that can be associated with radio knots R"" between 0.2" and 0.45".

Table 1 lists radio positions of the knots as measured on the figures of papers (Benson et al. 1988, Muxlow and Wilkinson 1991 and Walker et al. 1987). Precisions are typically 0.05". Our measured optical positions are also listed in table 1. Precisions are close to 0.1" and even better for O1. The positions of R1 and O1 are the same. Fluxes have been estimated at 5GHz (from figure 15 in Walker et al. 1987) and compared to I magnitudes in order to compute the spectral indices. In the case of the main knot O1, we were able to substract not only the contribution of the galaxy but also the flux contributed by condensation A. The corresponding corrected value is : $\alpha(R/O) = 0.84$.

3. Discussion

There is a good agreement between the positions of optical and radio features. Thus, like the radio knots, the optical condensations from O"" to O2, are not strictly aligned on the direction Nucleus-R1. The structures at 0.1" resolution show sub-structures that are not identical but similar to the radio structures and as pointed out by Walker (1997) for radio knots, condensations O1 and O2 also show a south-north asymmetry with a steep edge towards the south.

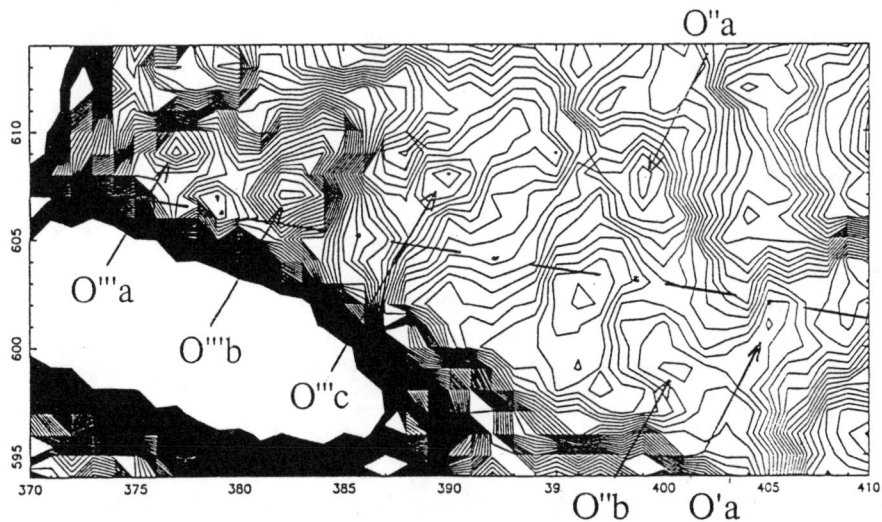

Figure 2. Region of condensations O" and O"'. Distance from the nucleus : 1" to 2.5". The mean radial gradient has been substracted. The optical jet is very close to the diffraction spike. Isophotes increase linearly in step of 0.25 DN starting at 5.

Unlike the radio jet at resolution 0.37", in optics at resolution 0.15", we do not see, between the knots, a continuous structure that can be undoubtedly attributed to the jet.

The HST data do not include polarization but Hjorth et al. (1995) found that condensation A is polarized as the radio jet. This is likely due to the main component O1 of A.

For all these reasons, the optical condensations appear as good counterparts of the synchrotron radio knots.

The galaxy is gas-rich and for O1, there are, at least, three arguments in favor of an interaction between the jet and ionized gas: a) in the region of O1, Baldwin et al. (1980) found a ratio $OIII/H_\beta$ higher than in any other part of the galaxy ; b) on a spectrum taken along filament A, Axon et al. (1989) found a maximum of OIII emission at O1 ; c) on figure 1c, depolarization of the radio jet is divided in two parts : north-east to the center of O1, depolarization is due to a plasma related to the knot itself and towards the east, a total depolarization in two regions reveals the existence of ionized and absorbing zones. Soubeyran et al. (1989) already interpreted this depolarization as due to the jet going through a layer of ionised gas.

The emission by the jet, at optical wavelengths, reinforces the idea of Grandi et al.(1997) that the emission, at 100 KeV, is due to the jet. Between the nucleus and 6", there is no significant variation of the radio/optical index ; this indicates that no important loss of energy of the jet occurs along the path. This case is similar to the one of M87 (Meisenheimer et al. 1996) and different from the behaviour of 3C 273 (see Lelièvre et al. 1984 and Bahcall et al. 1995).

These observations are compatible with the idea that the jet is driven by a precession motion and, as suggested by Soubeyran et al. (1989), their filaments A,B and D could be star formation regions produced by the successive encounters of the jet with regions of ionized gas.

References

Axon, D.J., Unger, S.W., Pedlar, A., Meurs, E.J.A., White, D.M., & Ward, M.J. 1989, Nature, 341, 631

Bahcall, J.N., Kirhakos, S., Schneider, D.P., Davis, R.J., Muxlow, T.W.B., Garrington, S.T., Conways, R.G., & Unwin, S.C. 1995, ApJ, 452, L91

Baldwin, J.A., Carswell, R.F., Wampler, E.J., Burbidge, E.M., & Boksenberg, A. 1980, ApJ, 236, 388

Benson J.M., Walker, R.C., Unwin, S.C., Muxlow, T.W.B., Wilkinson, P.N., Booth, R.S., Pilbratt, G., & Simon, R.S. 1988, ApJ, 334, 560

Grandi, P., Sambruna, R.M., Maraschi, L., Matt, G., Urry, C.M., & Mushotzky, R.F. 1997, ApJ, 487, 636

Hjorth, J., Vertergaard, M., Sörensen, A.N., & Grundhal, F. 1995, ApJ, 452, L17

Lelièvre, G., Wlérick, G., Sebag, J., & Bijaoui, A. 1994, C.R. Acad. Sci. Paris, 318, série 2, 905

Lelièvre, G.,Nieto,J.-L., Horville, D., Renard, L., & Servan, B., 1984, A & A, 138,49

Muxlow, T.W.B., & Wilkinson, P.N. 1991, MNRAS, 251, 54

Meisenheimer, K., Roser, H.-J.& Schotelburg, 1996, A & A, 307,61

Soubeyran, A., Wlérick, G., Bijaoui, A., Lelièvre, G., Bouchet, P., Horville, D., Renard, L., & Servan, B. 1989, A & A, 222, 27

Walker, R.C. 1997, ApJ, 488, 675

Walker, R.C., Benson, J.M., & Unwin, S.C. 1987, ApJ, 316, 546

Active Galactic Nuclei and Related Phenomena
IAU Symposium, Vol. 194, 1999
Yervant Terzian, Daniel Weedman, Edward Khachikian, eds.

Are LINERs Starbursts or Mini-quasars? A comparative Study of their Ultraviolet Spectra

Anuradha Koratkar

Space Telescope Science Institute

Abstract.

Low Ionization Nuclear Emission Line Regions (LINERs) are found in ~30% of all bright galaxies. The nuclear luminosities in these objects are such that they can be produced by a number of mechanisms and there have been heated debates on the nature of ionizing sources in LINERs. The variety of ionizing mechanisms suggested are low luminosity AGNs, starbursts, shocks, or any combination of these. We have studied Hubble Space Telescope (*HST*) ultraviolet (UV) spectra of seven LINERs having compact nuclear UV sources.

The picture emerging from this comparison is that the compact source observed in these LINER galaxies, at least in some cases, is a nuclear star cluster rather than a low-luminosity active galactic nucleus (AGN). In these cases, the UV luminosity is driven by tens of thousands of O-type stars, depending on the assumed extinction for these objects. The O-stars could be the high-mass end of a bound stellar population, similar to those seen in super star clusters. Our data do not exclude the possibility that a similar stellar continuum source could dominate in all the LINERs. Alternatively, there may be two types of UV-bright LINERs: those where the UV continuum is produced by a starburst, and those where it is nonstellar.

The "clearly-stellar", weak [O I] emitters, LINERs have relatively weak X-ray emission, and their stellar populations probably provide enough ionizing photons to explain the observed optical emission-line flux. The other LINERs, strong [O I] emitters, have severe ionizing photon deficits, for reasonable extrapolations of their UV spectra beyond the Lyman limit, but have an X-ray/UV power ratio that is higher by two orders of magnitudes than that of the "clearly-stellar" LINERs. A component which emits primarily in the extreme-UV may be the main photoionizing agent in these objects.

Recent results show that nuclear-starburst and quasar-like activity are often intermingled. Our results extend this result to the lower luminosities of the LINERs.

1. Why study LINERs?

Low Ionization Nuclear Emission Line Regions (LINERs) are galaxies which show low nuclear activity with narrow optical emission lines. Since the original survey of bright galaxies by Heckman (1980), several other surveys (see Ho 1996 for a review) have established that LINER activity is found in ~30% of the nearby galaxies. Thus, the LINER phenomenon is the commonest form of nuclear activity in galaxies.

The luminosities of most LINERs are unimpressive compared to "classical" AGNs. The nuclear luminosities in these objects are such that they can be produced by a number of mechanisms and there have been heated debates on the nature of ionizing source in LINERs. Several other ionizing mechanisms such as low luminosity AGNs (Ferland & Netzer 1983), hot high-metallicity O-type stars (Filippenko & Terlevich 1992; Shields 1992), radiative shocks in accretion flows or winds (Dopita & Sutherland 1995), or UV-bright post-AGB stars (Binette et al. 1994) have been invoked as the energy sources of LINERs.

If LINERs represent the low-luminosity end of the AGN phenomenon, then they are the nearest and most common examples, and their proximity can be used to get a fundamental understanding of the AGN phenomenon. Their study can also be used to understand the connection between AGNs and starbursts.

What is the nature of the energy source powering LINERs? Are LINERs the low luminosity end of AGNs? These are some of the fundamental questions still unanswered. To properly address these questions, the various proposed energy sources have to be distinguished observationally, and their energy contribution quantified. Studies involving only optical emission line ratios are not capable of differentiating among the several proposed ionizing mechanisms. Ideally multi wavelength observations are preferred, but the UV is a good starting place as it provides a number of discriminants. The x-ray and IR investigations are just starting to achieve the high spatial and spectral resolution demanded by these investigations.

2. UV spectroscopy of LINERs

When a search for UV point sources in LINERS was conducted, only ~22% of the LINERs showed a compact UV-bright source with HST (Maoz et al. 1996a; Barth et al. 1996; 1998). Since internal extinction effects play a crucial role in the detection of UV-bright sources, this fraction is a lower limit of the total fraction of LINERs with a central UV compact source . The compactness of the sources suggested that they could either be nonstellar in nature, or compact star clusters. The UV luminosity in these compact sources was consistent with the object being a low luminosity AGN, star cluster or shock. UV spectroscopy can be used to distinguish the ionizing mechanism, since all these ionizing mechanisms have UV spectral diagnostics. I present UV (1150–3200 Å) spectra for seven LINERs and compare their properties.

Figure 1 shows the 1200 – 1590 Å spectra for the seven LINERs. We immediately see that only two of these LINERs have broad emission lines similar to those seen in "classical" AGN. Except for the two LINERs with broad emission lines (M81 and NGC 4579), the LINERs have weak or no detectable UV emission lines.

The spectra of NGC 404, NGC 4569, and NGC 5055 show clear absorption-line signatures of massive stars, indicating a stellar origin for the UV continuum. The UV luminosity, in these object, is driven by tens to thousands of O-type

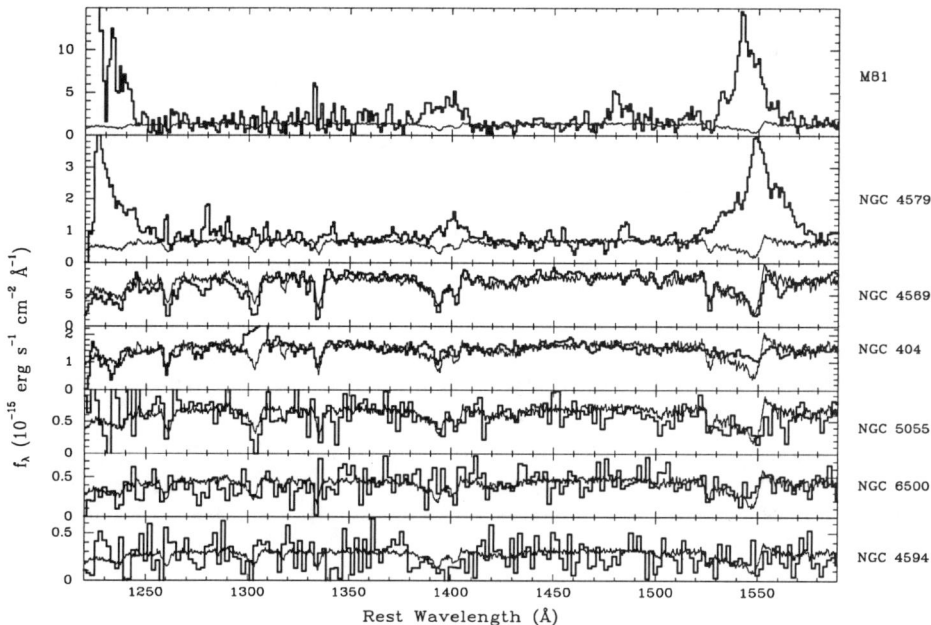

Figure 1. The UV spectra of seven LINERs (bold line), ordered with the two broad-lined objects on the top, and then in decreasing f_λ. The spectrum of the starburst NGC 1741-B is overlayed in each case. The starburst spectrum is normalized to be flat in f_λ and scaled to match the LINER continuum level. The compact central UV continuum source that is observed in these galaxies is a nuclear star cluster rather than a low-luminosity AGN, at least in some cases. The same stellar signatures *may* be present but undetectable in M81 and NGC 4579, due to superposed strong, broad emission lines.

stars, depending on the object and the extinction assumed. The O-stars could be the high-mass end of a bound stellar population, similar to those seen in super star clusters (e.g., Maoz et al. 1996b). Spectral signatures of massive stars are probably also present in NGC 6500. The same stellar signatures *may* be present but undetectable in NGC 4594, due to the low signal-to-noise ratio of the spectrum, and in M81 and NGC 4579, due to superposed strong, broad emission lines. Thus the UV continuum in all these objects could have a stellar origin. Circumnuclear starbursts within 300 pc. of the galaxy nucleus are often seen and these can contribute significantly in the UV. But is the observed stellar continuum sufficient to power the LINERs?

A comparison of the Hα line flux to the continuum flux at 1300Å indicates that the emission line strengths are not correlated to the observed UV continuum, indicating that the emission lines are not driven by the observed UV continuum. To determine if the stellar continuum was sufficient to power the LINERs we computed the ionizing photons due to young star clusters. We found that in the three LINERs whose UV emission is clearly dominated by stars (NGC 404, NGC 4569, and NGC 5055), ionization by the stellar population can provide the required power, but only if very massive stars are still present. Incidentally these objects are also weak [O I] emitters. In the other four LINERs (strong [O I] emitters) there was a ionizing photon deficit, indicating an additional energy source beyond that implied by the observable UV.

To search for an additional source of photoionizing photons we compared the UV flux with the 2-10 keV X-ray flux. We found that the strong [O I] emitters have X-ray power comparable to the UV power. The weak [O I] emitters have X-ray power which is at least an order of magnitude lower than the UV. Thus the strong [O I] emitters have an ionizing source which dominates in the X-ray and Extreme-UV. Thus, a non-stellar source may be significant or even dominate at other wavelengths. Although the three "clearly-stellar" LINERs do not obviously require the existence of such an additional ionizing source, our data do not exclude the existence of such a source.

Acknowledgments. This work was supported by *HST* GO grant GO-6112 provided by the Space Telescope Science Institute, which is operated by the Association of Universities for Research in Astronomy Inc., under NASA contract NAS 5-26555.

References

Barth, A.J., Ho, L.C., Filippenko, A.V., & Sargent, W.L.W. 1996, in "The Physics of LINERs in View of Recent Observations", eds. M. Eracleous et al. (San Francisco: ASP), p. 153

Barth, A.J., Ho, L.C., Filippenko, A.V., & Sargent, W.L.W. 1998, ApJ, 496, 133

Binette, L., Magris, G., Stasinska, G., & Bruzual, G. 1994, A&A, 292, 13

Dopita, M., & Sutherland, R. 1995, Ap.J., , 455, 468

Ferland, G. J., & Netzer, H. 1983, ApJ, 264, 105

Filippenko, A. V., & Terlevich, R. 1992, ApJ, 397, L79

Heckman, T.M. 1980, A&A, 87, 152

Ho, L.C., Filippenko, A.V., & Sargent, W.L.W. 1996, ApJ, 462, 183

Maoz, D., Filippenko, A.V., Ho, L.C., Macchetto, F.D., Rix, H.-W., & Schneider, D.P. 1996a, ApJS, 107, 215

Maoz, D., Barth, A.J., Sternberg, A., Filippenko, A.V., Ho, L.C., Macchetto, F.D., Rix, H.-W., & Schneider, D.P. 1996b, AJ, 111, 2248

Near-IR Polarimetry of the Obscured Nucleus in the Circinus Galaxy

D. M. Alexander[1,2], M. Ruiz[1], S. Young[1], C. Heisler[3], S. Lumsden[4], J. H. Hough[1], J. Bailey[4]

[1] *University of Hertfordshire, Hatfield, Herts, UK, AL10 9AB*
[2] *International School for Advanced Studies, Trieste, Italy*
[3] *Mount Stromlo & Siding Spring Observatory, Canberra, Australia*
[4] *Anglo-Australian Observatory, Epping NWS, Australia*

Abstract. We present J and K band imaging polarimetry and optical and K band spectropolarimetry of the Circinus galaxy. The imaging polarimetry shows a bipolar scattering cone in the J band and a more compact structure in the K band. The spectropolarimetry shows broad polarised hydrogen alpha in the optical, however broad lines are not detected in the K band. Analysis of the observations show that galactic and stellar processes dominate in the optical total and polarised flux whilst the nucleus dominates in the near-IR polarised flux. Modelling of the observations show that the small fraction of nuclear polarisation in the optical is due to electron scattering whilst the K band polarisation is dominated by dichroism.

1. Introduction

The Circinus galaxy is a nearby massive spiral that lies close to the plane of the Galaxy and is, optically, difficult to detect (Freeman et al, 1977). The galactic disc is inclined by 65 degrees, with a dust lane to the SE of the nucleus making the Hubble type difficult to determine. The nucleus displays both Seyfert and starburst activity and at a distance of only 4 Mpc is the closest Seyfert/starburst galaxy and therefore an excellent object to test the unified theory of AGN. It is highly polarised in the optical and a broad hydrogen alpha line has been detected in polarised flux (Oliva et al, 1998).

2. Observations and Data Reduction

The observations were taken in 1997 at the Anglo-Australian Telescope. The Royal Greenwich Observatory spectrograph and the University of Hertfordshire waveplate modulator were used for the optical observations. The IRISPOL imaging spectrometer and the University of Hertfordshire polarimeter were used for the near-IR observations (IRISPOL). The spectral slits were positioned on the nucleus at a position angle of 150 degrees.

3. Results and Discussion

The J band polarisation is 2% across the galaxy with a small nuclear enhancement. In polarised flux, prominent bipolar scattering cones are observed with the NW cone coincident with the [OIII] ionisation cone (Marconi et al, 1994). The K band polarisation is 0.5% across the galaxy, but higher (3%) at the nucleus where a compact structure dominates in polarised flux.

Polarised broad hydrogen alpha is detected with a width and flux consistent with that found by Oliva et al (1998). Strongly polarised optical stellar lines show that the galactic polarisation contribution is very high. Based on our measured broad hydrogen alpha flux, and a non-stellar R band continuum to broad hydrogen alpha line flux correlation that we've found for Seyfert 1s, we determine that just 0.1% of the measured R band polarisation is from the nucleus. In the K band, broad brackett gamma is not detected, putting an upper limit of 7.7 mags on the scattered visual extinction, assuming electron scattering and the Case B approximation.

We've modelled the spectropolarimetric results with the cone scattering model of Young et al (1995) with electron scatterers visually extincted by 5 mags and distributed in a 45 degree opening half-angle cone, consistent with the J band bipolar scattering cones, inclined at 50 degrees, giving an intrinsic polarisation of 26%. To fit to the near-IR an additional polarisation source of dichroism was required with a visual extinction to the near-IR emission region of 35 mags through the dusty torus.

The starlight fraction in the optical is greater than 99% whilst the radio and IR luminosities are lower than any other observed Seyfert 2 galaxy with polarised broad lines. The high optical polarisation of the galaxy is produced mostly through galactic processes and it is only in the near-IR that the nucleus is clearly detected. As this is the closest Seyfert 2 galaxy this suggests considerable difficulty in detecting polarised broad lines in other low powered obscured AGN.

References

Freeman, K.C., et al., 1977, A&A, 55, 445

Marconi, A., Moorwood, A.F.M., Origlia, L., Oliva, E., 1994, Messenger, 78, 20

Oliva, E., Marconi, A., Cimatti, A., di Serego Alighieri, S., 1998, A&A, 329, L21

Young, S., Hough, J.H., Axon, D.J., Bailey, J.A., Ward, M.J., 1995, MNRAS, 272, 513

Multiwavelength Properties of Narrow-line Seyfert 1's: Studying One Extreme of the AGN Primary Eigenvector

Paul Hirst, Duncan Law-Green, Martin Ward

X-ray Astronomy Group, Department of Physics & Astronomy, University of Leicester, Leicester LE1 7RH, UK

Abstract. Narrow Line Seyfert 1 galaxies (NLS1s) are an important subclass of radio quiet AGN having extreme optical and X-ray spectroscopic properties. Their relationship to other types of Seyferts remains unclear. NLS1s exhibit many characteristics of Seyfert 1 AGN, but their optical spectra show narrow permitted lines, and some high ionization species. This may result from us viewing them close to 'pole-on' orientation with respect to an inner torus geometry. We present comparisons of the radio and X-ray properties of NLS1s with those of the $12\mu m$ and CfA Seyfert samples.

1. Introduction

"Narrow-Line Seyfert 1s" (NLS1s) were first identified as a distinct subclass of Seyferts by Osterbrock & Pogge (1985). They have intriguing optical properties: while their permitted lines are narrow like Seyfert 2s (on the order of 1000km s^{-1}), they show other characteristics which are normally associated with Seyfert 1s. They have [O III]$\lambda 5007$Å/Hβ flux ratios of < 3, indicating the presence of a High-Density Region (HDR). They also exhibit blends of Fe II emission multiplets and high ionisation lines such as [Fe VII]$\lambda 6807$Å and [Fe X]$\lambda 6375$Å.

The X-ray properties of NLS1's are also unusual; their X-ray spectra are typically steep, and exhibit soft X-ray excesses. They may show large-amplitude, rapid soft X-ray variability; IRAS 13224-3809 varies by a factor ~ 60 in the ROSAT HRI band with a doubling time of ~ 900s (Boller et al. 1997). The cores of bright emission lines such as Lyα also appear to vary.

2. Observations

To date, the only systematic radio imaging of NLS1's has been the work of Ulvestad et al. (1995). We extend Ulvestad et al.'s work in three important respects: (a) we have a systematically-selected sample for comparison with normal Seyferts, (b) we use multi-frequency scaled-array observations to construct accurate spectral maps, and (c) MERLIN 6cm observations have been obtained or approved for 10 of our 15 sources. The combination of MERLIN and VLA data will give high sensitivity to both compact and extended structures.

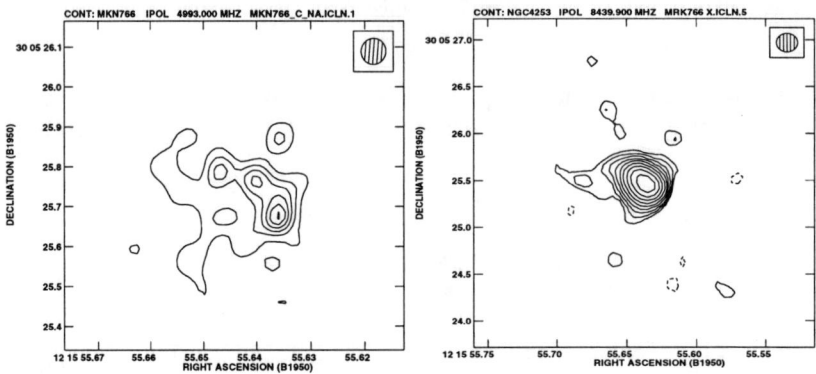

Figure 1. Left: 5GHz MERLIN image of Mkn 766 (Andy Thean, Priv. Com). Right: 8GHz VLA A array map of Mkn 766, from archive data

3. Radio Structures

If NLS1's are generally "pole-on", then we might expect them to have radio structures which are somewhat "blazar-like", containing a significant fraction of their flux in a compact, flat-spectrum core, and having foreshortened, distorted radio jets. Initial MERLIN imaging of the NLS1 Mkn 766 supports this; the jet is one-sided and curved through $\sim$ 90° within 200mas. Radio-loud quasars exhibit an inverse correlation between Hβ line width and radio core dominance (Wills & Browne 1986). The situation for radio-quiet objects is less clear, but the narrow permitted line width in NLS1's may be similarly linked with beaming.

There are currently no published studies of radio variability in NLS1's, despite their known strong variability in other wave bands. We are currently reducing multi-epoch VLA data on one target, IRAS F13349+2438, which we believe to have undergone strong variation (factor $\sim$ 10 in 5 years).

Data on about 25 objects have been retrieved from the VLA archive. Snapshot images of another 20 will be obtained in B array at 6cm under our recent successful observing proposal. Two NLS1s (Mrk 766 and I Zw 1) have been previously studied by MERLIN. Images of a further 8 NLS1s have been obtained under our successful observing proposal.

References

Boller, T., Brandt, W.N., Fabian, A.C. & Fink, H.H. 1997, M.N.R.A.S., , 289, 393
Osterbrock, D.E. & Pogge, R.W. 1985, Ap.J., , 297, 166
Ulvestad, J.S., Antonucci, R.R.J. & Goodrich, R.W. 1995, Astron. J., , 109, 81
Wills, B.J. & Browne, I.W.A. 1986, Ap.J., , 302, 56

Kinematic Mapping of the Narrow Line Region of NGC4151 [1]

M.E. Kaiser [2], L.D. Bradley II

Dept of Physics and Astronomy, Johns Hopkins University, Baltimore, MD 21218

J.B. Hutchings [2]

Dominion Astrophysical Observatory, National Research Council of Canada, 5071 W. Saanich Rd., Victoria B.C. V8X 4M6, Canada

S.B. Kraemer [2], D.M. Crenshaw, J. Ruiz

Dept of Physics, Catholic University of America, Washington DC 20064

D. Weistrop [2], C. Nelson

Dept of Physics, University of Nevada, Las Vegas, 4505 Maryland Pkwy, Las Vegas, NV 89154-4002

T.R. Gull [2]

NASA Goddard Space Flight Center, Lab for Astronomy and Solar Physics, Code 681, Greenbelt MD 20771

We present *HST* Space Telescope Imaging Spectrograph (STIS) slitless spectroscopy of the NGC4151 narrow line region (NLR) as a probe of the kinematic stucture of the extended emission-line gas emanating from the nucleus. Using slitless spectroscopy at two roll angles (with a spatial resolution of 0.051″/pixel and a point source spectral resolution of 0.55 Å) augmented with narrow band images, we have mapped the velocity field of the NLR as defined by ∼60 discrete cloud structures in [OIII]. Flux measurements of [OII], Hβ, [OIII], [OI], and [SII] were made for individual cloud structures wherever possible.

The emission line clouds have a biconical distribution which extends further SW than NE of the nucleus (Figure 1). In general, the SW clouds have approaching velocities, whereas the NE clouds are receding. Higher velocity clouds, as well as higher velocity dispersion clouds, are located within approximately ±180 pc of the nucleus. The bicone axis is roughly coincident with the radio axis (Pedlar et al., 1993, Winge et al., 1997).

---

[1] Based on observations with the NASA/ESA *Hubble Space Telescope*, obtained at the Space Telescope Science Institute, which is operated by AURA Inc under NASA contract NAS5-26555

[2] Co-investigator STIS Instrument Defininition Team, funded in response to NASA Announcement of Opportunity OSSA-4-84

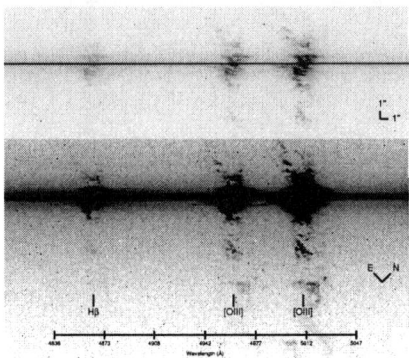

Figure 1. Slitless spectrum of NGC4151, illustrating the capability of this mode to spatially map the biconical emission line structure while simultaneously providing a kinematic map of the Hβ and [OIII] clouds.

Our velocity measurements indicate that there are two distinct kinematic components. One component is characterized by lower velocities ($|v_{rel}| < 400$ kms^{-1}). The other population of clouds is characterized by both high velocities and high velocity dispersions ($400 < |v_{rel}| < 1700$ km s^{-1}, $\sigma_V \gtrsim 130$ km s^{-1}).

We have modelled the velocity distribution by simple Keplerian rotation about a central source. At large radii the gas kinematics are consistent with normal galaxy rotation and have a low amplitude since NGC4151 is quite face-on. We find a rather poor fit to both the high and low velocity clouds closer to the nucleus, effectively eliminating rotation around a black hole as the origin for the gas motions.

For the high velocity cloud population another physical mechanism, such as wind driven outflow, must be responsible for the gas kinematics. We argue against infall based upon opacity and similarity arguments. (1) The blueshifted clouds are brighter and have a more extended distribution. Since we expect clouds on the near side of the nucleus to appear brighter, this implies outflow. (2) The CIV absorbers all have outflow velocities (Hutchings et al. 1998, Weymann et al. 1997). (3) Outflow is consistent with the radio observations (Ulvestad et al., 1998, Pedlar et al. 1993).

Several of the clouds lie near, but not coincident with, the radio axis. However, we have found no strong correlation between either velocity or velocity dispersion and proximity to the radio cores.

Flux measurements of the individual clouds are consistent with photoionization. The [OIII]/Hβ ratio declines roughly linearly within an inner $\sim 3''$ (200 pc) radius from the nucleus. This decrease in the flux ratio indicates a decreasing ionization parameter (Ferland and Netzer, 1983). Since the ionization parameter is proportional to $r^{-2}n^{-1}$, it appears that the density, n, must decrease as $\sim r^{-1}$. At distances further from the nucleus, the [OIII]/Hβ ratio is roughly constant and comparable to its initial value, indicating that either the ionization parameter is roughly constant at larger radii with the density decreasing as r^{-2} or we are subject to projection effects, local gradients, or both.

References

Ferland, G.J., and Netzer, H., Ap.J., , 264, 105
Hutchings, J.B., Crenshaw, D.M., Kaiser, M.E., Kraemer, S.B., Weistrop, D., Baum, S., Bowers, C.B., Feinberg, L. D., Green, R.F., Gull, T.R., Hartig, G.F., Hill, G., Lindler, D.J., 1998, ApJ, 492, L115
Pedlar, A., Kukula, M.J., Longley, D.P.T., Muxlow, T.,W.B., Axon, D.J., Baum, S., O'Dea, C., Unger, S.W. 1993, M.N.R.A.S., , 263, 471
Ulvestad J.S., Roy, A.L., Colbert, E.J.M., Wilson, A.S., 1998, Ap.J., , 496, 196
Weymann, R.J., Morris, S.L., Gray, M.E., Hutchings, J.B., Ap.J., , 483, 717
Winge, C., Axon, D.J., Macchetto, F.D., Capetti, A., 1997, ApJ, 487, L121

On the Nature of Radio Emission of AGNs: Spectra, Milli-arcsecond Structure and Polarization

Yu.A. Kovalev, Y.Y. Kovalev

Astro Space Center of the Lebedev Physical Institute, Profsoyuznaya 84/32, Moscow, 117810 Russia

Abstract. The main observational properties of AGNs at centimeter–decimeter wavelengths — spectra, variability on various time scales, polarization and milli-arcsecond maps — are in agreement with calculations in a jet model with a strong quasi radial magnetic field. Typical spectra and VLBI structures can be explained by a simple case of a straight line jet, unlike the degree of the linear polarization. Curved jets in the model allow decreasing the polarization to observational values as well.

1. Results

We calculate and analyze synchrotron spectra, structure and polarization of a stationary and non-stationary continuous jet, ejected in a strong radial magnetic field from an active nucleus (magnetic pressure is much higher than jet pressure). An approach for the Hedgehog model, suggested by N.S.Kardashev in 1969 and discussed in details by Kovalev & Kovalev (1997), is used.

The spectral shapes and the milli-arcsecond maps for stationary straight line jets agree with typical observations (Figure 1). Polarized structures have position angles transversal to the jet at the frequencies higher than the HF spectra maximum (similar to the Figure 1a, P), parallel to it — at the frequencies lower than the LF frequency twist, with both orientations in turn in two components — at the middle frequencies for the quasi–flat spectra region (similar to the Figure 1b, P, if the angular resolution is high enough), as expected.

Nevertheless, the linear polarization for the brightness in the model can be up to 70% in some regions — much more than it is usually observed by VLBI. This discrepancy can be explained by the following specific reasons in addition to known ordinary mechanisms of depolarization.

1. Observed polarized intensity maps strongly depend on the size, the shape and the orientation of the VLBI beam. This results in a decreasing of the polarization of *visible* intensity for a straight line stationary jet, if the angular resolution is not high along the jet.

2. A high polarized region located in the external optically thin part of a jet can be non-visible, if its brightness is much lower than that of the low polarized region. In the model it can occur in a wide frequency band.

3. Strong depolarization of visible intensity on the VLBI maps can occur, if a narrow jet is curved or its ejection is non–stationary. Moreover, total spectra can be the same as those of non-curved jets, for a special class of curvatures.

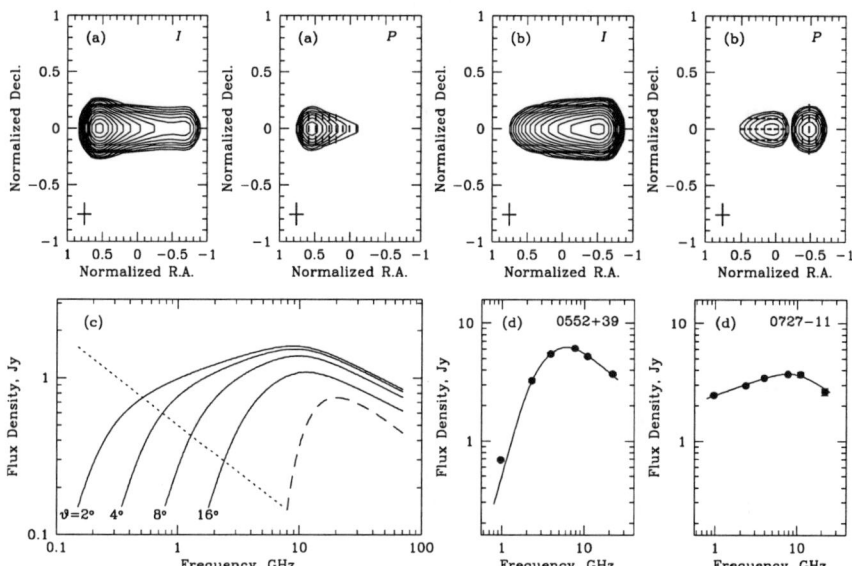

Figure 1. Some examples of the model analysis for a stationary jet: maps for total and polarized intensity at various frequencies (on top; beam used is shown by the cross) and spectra at various angles ϑ of the jet's axis to an observer line of sight (below). Maps (a) and (b) correspond to the frequencies higher and lower than the frequency of the maximum of spectra shown in (c), respectively. For a comparison we give the spectrum of a homogeneous synchrotron source (dashed lines). Total spectra (not shown) are often obtained by a sum of spectra with a spectrum of LF background component (dots). Fits for two AGNs from our 1–22 GHz observations (points) in 1998, March, at the RATAN-600, are presented in (d) by solid lines.

2. Conclusions

The jet model does not contradict the main observational properties of AGNs, including the existence and transitions between quasi-stationary and variable phases for spectra, structure and polarization (see e.g. simulation of the flare evolution in BL 0236+164 in Kovalev & Larionov, 1994).

Increasing the resolution angle (in Space VLBI or at mm–wavelengths) can increase the polarization of the visible intensity of VLBI maps along jets, if the model is valid and if the curvature of jets is not too high.

Low values of integral polarization may indicate that many compact objects in the model can have curved jets with a non-stationary ejection.

References

Kovalev, Yu. A., & Kovalev, Y. Y. 1997, Ap&SS, 252, 133
Kovalev, Y. Y., & Larionov, G. M. 1994, ALett., 20, 3

Surface Photometry of Barred AGN Arakelian 564

G. Petrov, L. Slavcheva, R. Bachev, B. Mihov

Institute for Astronomy, Bulgarian Academy of Sciences, Bulgaria

The Seyfert 1.5 galaxy Akn564 [1] is a well known X-ray active galactic nucleus (AGN) included in our list of selected barred AGN. The galaxy was observed during the August 1996 season at the 2-m RCC telescope of the Astronomical Observatory "Rozhen" of the Bulgarian Academy of Sciences. ST-6 with standard Schott V, R, I filters were used. MIDAS'96 package was used for the data reduction with teh Richter's expansion for the surface photometry [2]. Most of the basic data for the galaxy are shown in the Table 1 below.

Table 1. Observational data for Sy1.5 galaxy Arakelian 564

Coordinates (1950)	22h 40m 18s	+29d 27' 48"	Type: SBb		[1]
Dimensions	0.8 x 0.52 arcm				[2]
Photometric data:	V = 13.67	U-B = -0.42	B-V 0.69	V-R = 0.76	[2]
Redshift	0.025				[3]
Other data	[5007]/Hβ = 1.0 F12 < 0.25 Jy	S6 = 49 mJy F25 = 0.72 Jy	F60 = 3.46 y	F100 = 5.53 Jy	[2]
Table of Observations	13.08.1996 V - 2x600 sec	2 m RCC tel. R - 2x300 sec	ST-6 CCD I - 2x300 sec	0.295 "/px	

References to the Table: [1] Arakelian, M.A., 1975, Publ.Bjurak.Obs, 47,3. [2] Lipovettsky, V.A. et al., 1987, Comm. of SAO (UdSSR), 55, 5. [3] Arakelian, M.A. et al., 1976, Astrophysics, 12, 456.

Following Richter, flat-fielded, dark-subtracted CCD frames were adaptive filtered, and a multiple masking technique was applied to have "galaxy on a pure background." The real objects were modeled with ellipses according to Bender & Moelenhoff [3] methods. Cumulative magnitudes and surface brightness for V,R and I frames and for the models have been evaluated, as well as the distribution of the position angles of the major axes and the colors. The parameters of the disk in 3 - 18 arcsec radius for the objects and models in the three colors are presented in Table 2.

Here SB = Mc + (r/rc)exp(1/n). On figure 1 (only R color is shown) the twisted bar, wider and stronger in the south part, is clearly seen and some bright star formation regions are traced. The extremely bright starlike nucleus with diameters 8 - 10 arcsec includes about 60 % of the luminosity of the galaxy. The disk is clearly visible to 50 arcsec on SB = 26 mag/sqr.arcsec. The nucleus is bigger in R and I colors – see the sharp minimum in the distribution of V-R colors for the object and model. Figure 2 presents the surface brightness (SB), color index (V-R), axis ratio (b/a) and position angle (PA) as a function of the radius.

Table 2. Akn564 Fitting parameters

mod_obj	n	mc	err_mc	rc	err_rc
SB_v_mod	+0.40	21.36	0.782	+0.0029	+.00026
SB_r_mod	+0.40	21.94	0.168	+0.0018	+.00005
SB_i_mod	+0.40	+19.81	2.030	+0.0042	+.00036
SB_v_obj	+0.40	+20.37	+0.780	+0.0039	+.00029
SB_r_obj	+0.40	+20.31	+0.732	+0.0032	+.00019
SB_i_obj	+0.90	+18.92	+0.391	+0.1298	+.00718

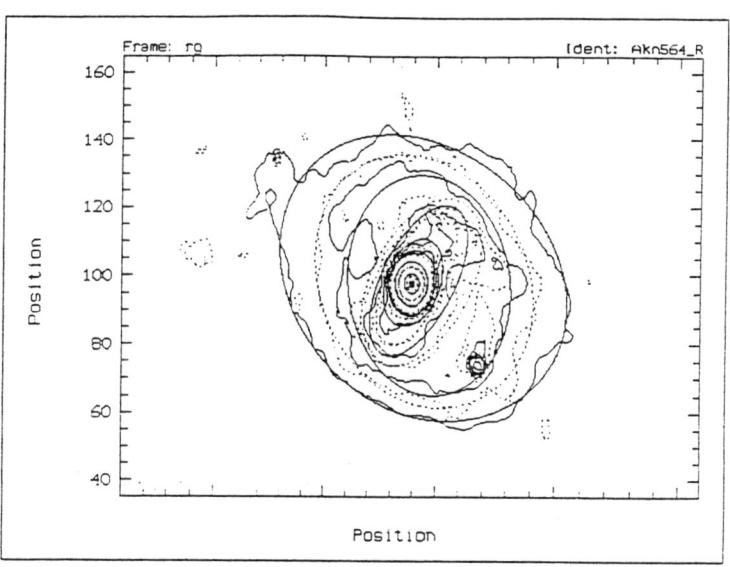

Figure 1. Fitting of the R-frame of Akn 564 with ellipses according to [4]

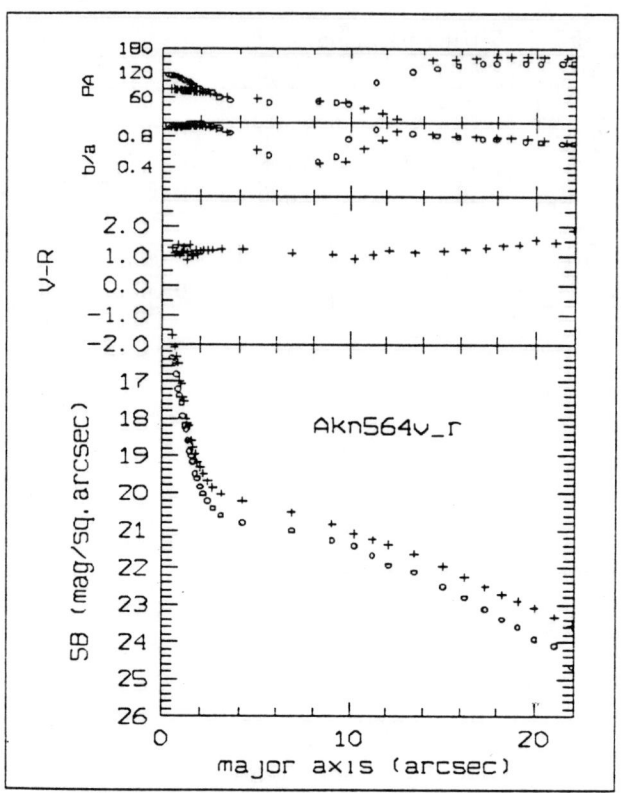

Figure 2. Surface brightness (SB), color index (V-R), axis ratio (b/a) and position angle (PA) as a function of the radius.

References

Arakelian, M.A., 1975, Publ.Bjurak.Obs., 47, 3
Vennik, J., Hopp, U., Kovachev, B., Kuhn, B., Elsasser, H., 1996, AApSuppl., 117, 261
Lorenz, H., Richter, G., Cappaccioli, M., Longo, G., 1993, A&A, 277, 321
Bender, R., Moelenhoff, C., 1987, A&A, 177, 71

Spectrophotometry of Selected AGN Seyfert Galaxy AKN 564

L.S. Slavcheva, B.M. Mihov, G.T. Petrov, R.S. Bachev

*Institute for Astronomy, Bulgarian Academy of Sciences,
lslav@astro.bas.bg*

Akn 564 ($\alpha_{1950} = 22^h40^m18.3^s$, $\delta_{1950} = 29°27'47"$) is a Sy1.5G SBb type galaxy. According to Zwicky (1966) it has a photographic magnitude $m_p = 14.4$ and a redshift of 0.025. The spectra of the galaxy were obtained at the 2.6-m telescope of the Crimean Astrophysical Observatory with a spectrograph having a dispersion of 100 A mm^{-1}. They were processed with the help of SPEC and LONG packages integrated in MIDAS. As a result of the spectrophotometry we obtain the fluxes at $\lambda\lambda$ 4363, 4959, 5007 A: I(4363), I(4959), I(5007). The spectrum of the galaxy in $\lambda\lambda$ 4000-7000 is shown in Figure 1. We use the relation of the fluxes of those narrow forbidden emission lines:

$$R = [j(\lambda 4959) + j(\lambda 5007)]/j(\lambda 4363) \tag{1}$$

$$= [8.32 exp(3.29 \times 10^4/T]/(1 + 4.5 \times 10^{-4} Ne/T^{1/2}, \tag{2}$$

sensitive at a greater extent to the electron temperature T_e than to the electron density n_e. The value of R = 74.3 we got, having a typical value of $n_e = 5 \times 10^5$ cm^{-3} for the NLR (Narrow Line Region), leads to the estimation of a typical temperature of $T_e = 10^4$ K.

We can evaluate the effective volume V_{eff} and respectively the size R_{eff}, the mass M_g and the kinetic energy E_k of the emitting gas in the NLR with $n_e = 5 \times 10^5$ cm^{-3} and $T_e = 10^4$K assumed and I(5007) measured via the equations (Dibay 1980):

$$L(H_\beta) = 4\pi R^2(1+z)^2 I(H_\beta); \tag{3}$$

$$V_{eff} = R^2 I(H_\beta)/j(H_\beta); \tag{4}$$

$$R = cz/H; \tag{5}$$

$$V_{eff} = fV; \tag{6}$$

$$R_{eff} = (3V_{eff}/4)^{1/3}; \tag{7}$$

$$M_g = n_e m_p V_{eff}/M_o; \tag{8}$$

$$E_k = 1/2 M_g v^2 = 1/4 M_g FWHM; \tag{9}$$

$$M_c = 3v_v^2 R/G, \tag{10}$$

where V is the geometrical volume of the region, f $\approx 10^{-3}$ is the filling factor and j is the emmission coefficient.

T_e and n_e in the BRL (Broad Line Region) cannot be estimated directly. We accept representative of the BLR values of $n_e = 5 \times 10^5$ cm^{-3} and $T_e =$

10^4K acquired by comparing photoionizational models with some observational parameters. As a result we evaluate V_{eff}, R_{eff}, M_g, E_k and the mass of the central object M_c, all of them given in the following table:

NLR		BLR	
n_e, [cm^{-3}]	5×10^5	n_e, [cm^{-3}]	10^9
T_e, [K]	10^4	T_e, [K]	10^4
I([OIII] λ5007), [erg.cm^{-2}.s^{-1}]	1.04×10^{-12}	I(H$_\beta$), [erg.cm^{-2}.s^{-1}]	5.85×10^{-13}
FWHM([OIII] λ5007), [cm.s^{-1}]	663×10^5	FWHM(H$_\beta$), [cm.s^{-1}]	899×10^5
L([OIII] λ5007), [erg.s^{-1}]	9.18×10^{41}	L(H$_\beta$), [erg.s^{-1}]	5.18×10^{41}
j([OIII] λ5007), [erg.cm^{-3}.s^{-1}]	1.15×10^{-19}	j(H$_\beta$), [erg.cm^{-3}.s^{-1}]	6.63×10^{-9}
V_{eff}, [cm^3]	1.6×10^{55}	V_{eff}, [cm^3]	6.19×10^{48}
R, [pc]	5	R, [pc]	0.037
M_g, [Mo]	6.68×10^3	M_g, [Mo]	5.17
E_k, [erg]	7.34×10^{51}	E_k, [erg]	1.04×10^{49}
		M_c, [Mo]	0.52×10^7

The errors of the fluxes are about 7×10^{-15} erg cm^{-2} s^{-1} and the errors of the other parameters are about 10-30 %.

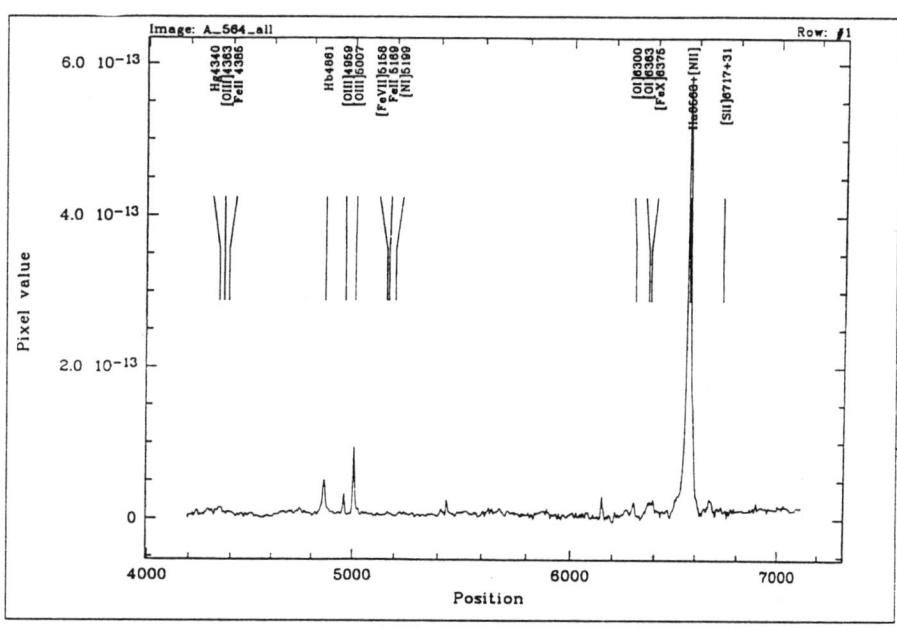

Figure 1. Energy distribution in $\lambda\lambda$ 4000-7000 A for Akn564. The data reduction was made by MIDAS 95NOV packages. The strongest forbidden and permitted lines are marked.

References

Arakelian, M., 1975, Publ.Bjurak.Obs., 47, 3
Dibay, E., 1980, Astron. J., , 57, 677
Zwicky, F., 1966, Ap.J., , 143, 192

Global Oscillations of Masing Disks in Megamasers

Atsuo T. Okazaki

Faculty of Engineering, Hokkai-Gakuen University, Toyohira-ku, Sapporo 062-8605, Japan

Abstract. We study the characteristics of global oscillation modes of masing disks in megamasers and the effect of the modes on the disk kinematics. We find that the eccentric mode is responsible for the observed sub-Keplerian velocity distribution of the maser source of NGC 1068, whereas in the masing disk of NGC 4258 the warping mode is dominant so that the angular rotation velocity remains near Keplerian.

1. Introduction

Recently, VLBI observations have revealed the detailed spatial and velocity distribution of maser emission in megamaser sources. In this paper, we studied the effect of global oscillation modes on the kinematics of masing disks in megamasers, using a simplified disk model.

2. Eccentric Mode in the Masing Disk of NGC 1068

Greenhill et al. (1996) found that the line-of-sight velocity of the redshifted maser sources in the nucleus of NGC 1068 decreases as $r^{-0.31}$. This sub-Keplerian line-of-sight velocity distribution can be attributed to the fundamental eccentric mode shown in Figure 1.

3. Warping Mode in the Masing Disk of NGC 4258

The position and velocity distribution of the water megamaser sources at the center of NGC 4258 are fitted well with a warped Keplerian disk model (Miyoshi et al. 1995). We found that the characteristics of the fundamental warping mode shown in Figure 2 agree well with the observed Keplerian velocity distribution of maser sources of NGC 4258.

References

Miyoshi, M., Moran, J., Herrnstein, J., Greenhill, L., Nakai, N., Diamond, P., and Inoue, M. 1995, Nat, 373, 127

Greenhill, L. J., Gwinn, C. R., Antonucci, R., and Barvainis, R. 1996, Ap.J., , 472, L21

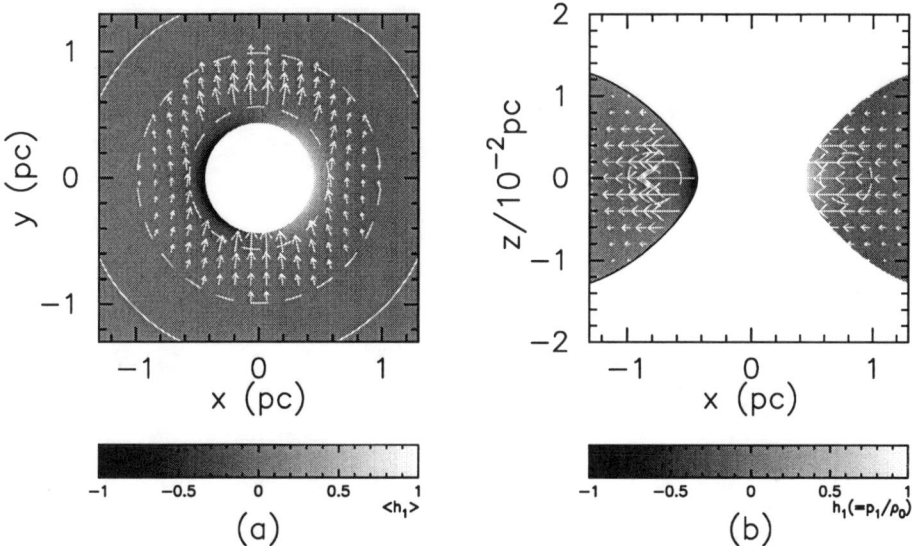

Figure 1. Fundamental eccentric mode in the masing disk of NGC 1068. (a) The (r, ϕ)-distribution of the perturbations averaged vertically over the upper half of the disk. (b) The (r, z)-distribution of the perturbations. A gray-scale representation denotes the enthalpy perturbation h_1. Arrows superposed on the gray-scale plot are the perturbed velocity vectors in the maser emission region (surrounded by dashed lines).

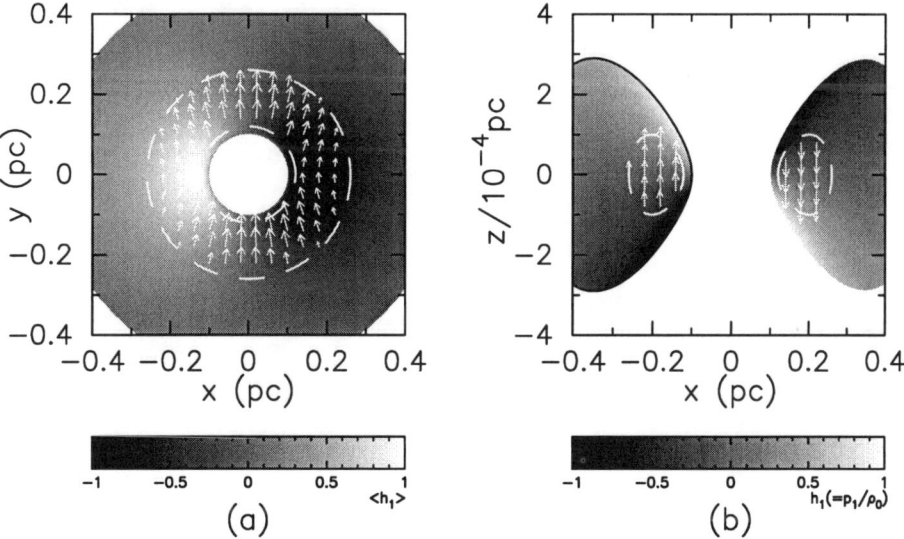

Figure 2. Fundamental warping mode in the masing disk of NGC 4258. The format of the figure is the same as that of Figure 1.

II. OBSERVATIONAL PROPERTIES OF AGN AND RELATED OBJECTS

Pat Osmer and Ken Kellermann

Jivan Stepanian and Vahé Petrosian

Active Galactic Nuclei and Related Phenomena
IAU Symposium, Vol. 194, 1999
Yervant Terzian, Daniel Weedman, Edward Khachikian, eds.

Quasars

Patrick S. Osmer

Department of Astronomy, The Ohio State University, 174 W. 18th Ave., Columbus, OH 43210, USA, posmer@astronomy.ohio-state.edu

Abstract.
I review recent results for quasars and discuss how they are related to activity in galaxies. Topics included are studies of quasar host galaxies with HST; searches for quasars in the Hubble Deep Field; evolution of the quasar luminosity function; news highlights from astro-ph; and current observational research problems and their relation to theoretical work.

1. Introduction

Our meeting occurs 35 years after the discovery of quasars, a discovery that transformed our concepts of active galactic nuclei (AGN), even though the connection between quasars and AGN was not clear at the time. We now consider quasars as the most luminous class of AGNs. Their great luminosity, which can be more than 1000 times that of an L^* galaxy, is part of their mystery on the one hand, while on the other hand it enables us to observe them at the greatest distances and earliest epochs at which they occur in the universe.

One of the great values of this symposium is that it brings together people from all fields of AGN research and provides us with an opportunity to take a fresh look at the state of the field and the key research problems.

In this talk I will cover some of the highlights of quasar history as they apply to our topic and review the main properties of quasars as we define them today. I will discuss recent results in quasar research, especially those that bear on the relation of quasars to activity in galaxies. I will also describe some current research problems and consider future opportunities for the field that will be provided by large telescopes and the large quasar surveys that are under way.

2. History

While the discovery of quasars in 1963 (Schmidt 1963) is well known[1], I would like to mention that it was preceded in 1958, forty years ago, by a key paper by Burbidge at the Paris Symposium. He pointed out that tremendous energy, 10^{60} ergs, resided in extragalactic radio sources. This was an unprecedented amount for the time, and the paper was influential in forcing people to think

---

[1]The 1st Texas Symposium on Relativistic Astrophysics (Robinson, Schild, & Schucking 1965) still makes excellent reading about that feverish first year of work.

about non-stellar sources of energy in galaxies, that is, what we now call activity in galaxies.

To continue with the topic of milestone years, we can also note that the evolution of the quasar population was discovered by Schmidt in 1968, thirty years ago, a discovery that was an essential first step to showing that the characteristic time-scale for quasar activity is quite short in cosmological terms.

3. Definitions

Schmidt's classic definition is that quasars are star-like objects of large redshift. More quantitatively, quasars are generally considered to have $z > 0.1$ and $M_B < -23$ mag ($H_0 = 50$) (see Schmidt and Green 1983). Traditionally, i.e. at resolutions of $1 - 2$ arcsec, they were considered as being star-like, a description that is intertwined with the redshift and absolute magnitude limits just given. We now know that better spatial resolution often yields evidence of a host galaxy. Other key properties of quasars are that they have broad emission lines[2] in their spectra and that they can emit continuum radiation across the electromagnetic spectrum from γ-rays to radio waves, with ultraviolet and X-ray emission usually being very prominent. Also, quasars show variability on time scales of days to years.

How well can we explain all these properties? Although it is generally believed that the picture of an accretion disk surrounding a black hole is correct, agreement between current models and observations is distressingly poor in many cases.

Operationally, it is important to be aware of the effect of the apparent size and luminosity limit in the definition of quasars on modern surveys. For example, as the angular resolution of surveys improves to 1 arcsec or better and the depth of surveys increases either because of the use of larger aperture telescopes or longer exposures, the host galaxies of quasars will be increasingly visible. In such cases, strict imposition of the "star-like" criterion for quasars will exclude bona-fide AGNs.

Similarly, we now know that high-luminosity Seyfert galaxies can overlap in absolute luminosity with low-luminosity quasars, and analyses of surveys must allow for this. For example, deep surveys for quasars with good angular resolution will find both quasars and AGNs. It is important for the determination of the evolution of the entire AGN population that surveys consider what classes of objects they are including and perhaps rejecting.

4. Recent Results

4.1. Host Galaxies

The Hubble Space Telescope (HST) has provided critical new information on the nature of the host galaxies in which quasars reside and about the nature

---

[2]For this article, BL Lac objects will be considered as a separate class, although they are members of the AGN family.

of quasar environments. The excellent image quality of the repaired telescope gives the best combination of angular resolution and light gathering power yet applied to quasars. Here I report on two papers, which of course build on previous ground-based work.

Bahcall et al. (1997) presented results with the Wide-Field Camera of HST for 20 luminous quasars with $z < 0.3$. For the host galaxies, they found that 2 were as bright as the brightest cluster galaxies, 10 were like normal elliptical galaxies, 3 were normal spirals, 3 were complex, interacting systems, and in 2 cases there was faint nebulosity surrounding the quasar. For the radio-quiet quasars, 7 occurred in elliptical galaxies and 3 in spirals. For the 6 radio-loud quasars, 3 to 5 of them were in elliptical galaxies. On average, the host galaxies were 2.2 magnitudes brighter than normal field galaxies. In 8 cases, they detected companion galaxies within a projected distance of 10 kpc from the quasar nucleus. The interactions, presence of companions, and higher density of galaxies seen around quasars suggest that interactions are important to quasar activity.

Boyce et al. (1998) used HST in a complementary study of 14 low-redshift quasars. They find that 9 occur in elliptical galaxies (all 6 of the radio-loud quasars and 3 radio-quiet objects); 2 radio-quiet quasars are in disk galaxies, and the other 3, which are radio-quiet, ultraluminous IR objects, occur in violently interacting systems. The average luminosity of the quasar host galaxies is 0.8 magnitudes brighter than L^*, while the radio-loud objects are 0.7 magnitudes brighter than the radio-quiet ones.

It is evident, as Bahcall et al. point out, that the hosts and environments of quasars are complex, and that the previous ideas about radio-quiet quasars residing in spiral galaxies and radio-loud quasars in ellipticals may not hold up. However, it is perhaps more important to realize that the HST observations provide powerful support for the concept of quasars residing at the centers of galaxies and that galaxy interactions play an important role in quasar activity.

4.2. The Hubble Deep Field

The Hubble Deep Field (HDF) has given us unprecedented new views of distant galaxies. In combination with spectroscopic observations with the Keck Telescopes, studies of galaxies are now well advanced at $z > 3$, redshifts that were unattainable previously. Consequently, we now have the opportunity to study directly the relationship of galaxies and quasars at and beyond the redshift of peak quasar activity.

Recently Conti et al. (1998) have carried out a detailed search for compact quasars and AGNs in the HDF to $V_{606} = 27$ mag to study their presence and behavior at luminosities corresponding to AGNs in the nearby universe. Although the HDF contains more than 3000 galaxies, Conti et al. found an upper limit of 20 for the number of quasar candidates. Based on spectroscopic observations to date, the actual number may be much smaller, even close to 0. However, because of the great depth of the HDF exposures and the ~ 0.1 arcsec image quality, it is possible that any AGNs in the HDF are spatially resolved, and the next step is to develop sensitive techniques to detect AGN within faint, resolved galaxies in the HDF. A complication is that many of the distant galaxies being found by HST and Keck are undergoing intense star formation, which gives them colors similar to those of many quasars. Jarvis and MacAlpine (1998) report identification of 12 resolved objects harboring candidate AGN. The crucial next step will be to confirm the nature of the candidates with follow-up spectroscopy, a very difficult task because of their faintness.

4.3. Evolution of the Luminosity Function

One of the most striking observed features of quasars is the evolution of their luminosity function. The space density of luminous quasars increases by a factor of ~ 1000 between the present epoch and redshift $2-3$ and then falls steeply toward higher redshifts (Warren, Hewett, & Osmer 1994, WHO; Schmidt, Schneider, & Gunn 1995, SSG; Kennefick, Djorgovski, & de Carvalho 1995). A straightforward explanation of this behavior is that we are seeing back to the epoch of peak quasar activity, an epoch that presumably has to do with the formation of black holes at the centers of galaxies and the time of significant fueling of the quasar activity via the infall of material to the center.

However, a persistent question about the nature of the peak is whether it is affected significantly by dust absorption along the line of sight. If so, there could be an important population of quasars at high redshift that are hidden at optical/UV wavelengths, indicating that the epoch of peak activity was even earlier. There is no doubt that some quasars are highly reddened; the basic question is how many.

One way to answer this question is to use samples of radio-selected quasars with complete optical identifications. Dust is transparent to radio radiation, and so samples with complete optical identifications provide an excellent test, as long as the ratio of radio quasars to the total number of quasars does not change significantly with epoch.

Hook, Shaver, and McMahon (1998) have carried out just such a program and find that the evolution of quasars in their sample is remarkably similar to that found by WHO and SSG. This suggests that dust is not the cause of the apparent decline in activity at $z > 3$. Similarly, Benn et al. 1998 used IR observations in the K band of radio-selected quasars and found no evidence for a large population of reddened and dust-absorbed quasars. These results are in contrast to those of Masci (1998), who does claim evidence for a population of reddened objects.

The ultra-deep ROSAT survey of Hasinger et al. (Hasinger 1998) has yielded important new X-ray results. The good positional accuracies of the survey show that most of the sources are quasars/AGNs and narrow emission-line galaxies are only a small fraction, in contrast with some previous work. Their new determination of the X-ray luminosity function is not consistent with pure luminosity evolution but can be fit by pure density evolution from $z = 0$ to $z \approx 2$. Their results suggest that black holes should be common in massive galaxies at the present epoch, as discussed in more detail below.

5. The News

The astro-ph electronic preprint archive has had a large impact on our field by greatly increasing the accessibility of preprints and making them instantly available around the world. It also provides a convenient way of tracking the latest developments. Here I mention a few highlights gleaned from postings to astro-ph in the last year and from other sources.

The Most Luminous. Irwin et al. (1998) reported the discovery of APM 0279 + 5255, a broad-absorption line quasar with $z = 3.87$ and $R = 15.2$ mag. The object is coincident with an IRAS FSC source, and the estimated luminosity is $\approx 5 \times 10^{15} L_\odot$, making it the intrinsically most luminous object known. There is evidence that the source is gravitationally lensed, which amplifies the true emitted luminosity.

The Most Distant. Weymann et al. (1998) find from Keck spectroscopy an emission line in the galaxy HDF4-473.0 that, if identified with Lyα, yields a redshift of $z = 5.60$. The galaxy is in the Hubble Deep Field and is the most distant object with slit spectroscopy that has yet been identified. It is not a quasar or AGN, and the absence of quasars with $z > 4.9$, despite continuing surveys for them, is beginning to appear significant in view of the increasing number of confirmed and candidate galaxies with $z > 5$.

The Smallest. Kedziora-Chudczer et al. (1998) observed significant radio variability on timescales less than an hour in the radio quasar PKS 0405 − 385, which would make it the smallest extragalactic source observed. They attribute the variation to interstellar scintillation of a source with an angular size smaller than 5 microarcsec. The inferred brightness temperature is well above the inverse Compton limit. If interpreted as steady relativistic beaming, the Lorentz factor would be 1000.

The first FIRST gravitational lens. Schechter et al. (1998) found that the quasar FBQ 0951+2635, with $V = 16.9$ mag and $z = 1.24$, from the FIRST radio survey, is a gravitational lens with two images separated by 1.1 arcsec.

Update to the Verón-Cetty and Verón Catalog. Verón-Cetty and Verón released the 8th edition of their catalog during the year. It contains entries for 11,358 quasars, 357 BL Lac objects, and 3334 AGNs and is available electronically at http://obshpz.obs-hp.fr/www/catalogues/veron2_8.html. Such catalogs continue to be an vital resource for the community, especially as new surveys yield so many new quasars and AGNs. Also, the electronic availability of the catalogue makes it even more accessible and valuable than it was previously.

6. Some Current Research Problems

Here I call attention to some current research problems that need further work. Their eventual solution should improve our understanding of quasars and AGNs in important ways.

The Disagreement between Observations and Predictions for Accretion Disks. Koratkar (1997) points out that observations do not confirm most predictions of accretion disk models. For example, the Zheng et al. (1997) composite spectrum for ultraviolet wavelengths does not match predictions, and soft X-ray fluxes are observed to be too flat. Fewer Lyman edges are observed than predicted. Polarization is not seen either, which seems to rule out scattering as a way of smoothing the Lyman edges. An additional theoretical question is how the radiation from the accretion disk couples with that of the hot (X-ray) corona. It is important to resolve these issues if we are to have confidence in this basic part of our concept for quasars and AGNs.

What powers Ultra-luminous IRAS galaxies? Observations by Genzel et al. (1998) indicate that massive stars predominate in 70–80% of the cases, with AGNs dominating in the others. At least half of the systems probably have both an AGN and a circumnuclear ring of starburst activity. They see no clear trend for the AGN component to dominate in the most compact and presumably most advanced mergers.

Do all galaxies have massive black holes? van der Marel (1997) notes that available data appear consistent with most galaxies having black holes, whose mass roughly correlates with the luminosity of the spheroid (cf. Magorrian et al. 1998). The black holes could have formed in or prior to a quasar phase and grown via mass accretion. Some of the implications of this work are discussed below under the theory section.

Is the broad Fe $K\alpha$ line produced directly near a black hole? How well do we understand the origin of X-ray emission in general? Observations of the broad Fe $K\alpha$ line in AGNs are widely interpreted as arising in the inner part of accretion disks around black holes and therefore providing both confirmation of the presence of black holes as well as direct information about conditions in the disks. However, Weaver and Yaqoob (1998) have raised questions about whether the emission in fact does occur so close to the centers of AGNs. More generally, intensive monitoring of NGC 7469 in X-rays and the ultraviolet by Nandra et al. (1998) provides strong constraints on quasar models. The data are not consistent with the UV emission being reprocessed by gas absorbing X-rays nor with the X-rays arising from Compton upscattering of the UV radiation.

These are just some examples of research problems in need of solution for us both to have confidence in our general picture of quasars and AGNs being powered by accretion onto massive black holes and to develop a quantitative understanding that explains the major observed features of these objects.

7. Theory

In addition to the above types of problems, considerable research is directed to basic questions such as, Do we understand how quasars form and evolve? Can we connect theories of galaxy and black hole formation with the observations of quasars at high redshift and the incidence of black holes in galaxies at low redshift? Here I mention briefly some recent theoretical work that demonstrates progress in our understanding of quasars and ties in with present and future observational work.

Haiman, Madau, and Loeb (1998) point out that the scarcity of quasars at $z > 3.5$ in the Hubble Deep Field implies that the formation of quasars in halos with circular velocities less than 50 km/s is suppressed (on the assumption that black holes form with constant efficiency in cold dark matter halos). They note that the Next Generation Space Telescope should be able to detect the epoch of formation of the earliest quasars.

Cavaliere and Vittorini (1998) note that the observed form for the evolution of the space density of quasars can be understood at early times when cosmology and the processes of structure formation provide material for accretion onto central black holes as galaxies assemble. Quasars then turn off at later times because interaction with companions cause the accretion to diminish.

Haehnelt, Natarajan, and Rees (1998) show that the peak of quasar activity occurs at the same time as the first deep potential wells form. The Press-Schechter approach provides a way to estimate the space density of dark matter halos. But the space density of $z = 3$ quasars is less than 1% that of star-forming galaxies, which implies the quasar lifetime is much less than a Hubble time. For an assumed relation between quasar luminosity and timescale and the Eddington limit, it is possible to connect the observed quasar luminosity density with dark matter halos and the numbers of black holes in nearby galaxies. The apparently large number of local galaxies with black holes implies that accretion processes for quasars are inefficient in producing blue light.

8. Future Directions and Possibilities

The research problems and theoretical ideas described in this article are already open to observational study and testing with 8-10m class telescopes and the Hubble Space Telescope, as we have discussed in the case of studies of quasar host galaxies, high-redshift galaxies, and black holes in galaxies. As the capabilities of the large ground-based telescopes improve (via infrared optimization and adaptive optics, for example), and when the Next Generation Space Telescope is completed, we will be able to study directly the relation of AGNs and galaxies over virtually the entire range of their evolutionary history. Similarly, the X-ray observatories AXAF and XMM will offer very significant new capabilities for the study of both the nature of quasars and AGNs and their evolution.

In the meantime, large-area, ground-based surveys such as the Sloan Digital Sky Survey[3] and the 2dF[4] survey will increase the number of known quasars by more than an order of magnitude. We may expect that the combination of the new samples, the new observatories, and continued theoretical advances will answer many of the questions raised here.

Acknowledgments. I thank Brad Peterson and David Weinberg for comments and suggestions on a first draft of this article. I am grateful to the Organizing Committee and the National Science Foundation (via grant AST-9529324) for financial support.

References

Bahcall, J. N., Kirhakos, S., Saxe, D. H., & Schneider, D. P. 1997, ApJ, 479, 642

Benn, C. R., Vigotti, M., Carballo, R., Gonzalez-Serranon, J. I., & Sanchez, S. F. 1998, MNRAS, 295, 451

Boyce, P. J. et al. MNRAS, 298, 121

---

[3] www.sdss.org

[4] msowww.anu.edu.au/~rsmith/QSO_Survey/qso_surv.html

Burbidge, G. 1958, in Paris Symposium on Radio Astronomy, ed. R. Bracewell, (Stanford: Stanford Univ. Press), p. 541

Cavaliere, A., & Vittorini, V. 1998, in The Young Universe: Galaxy Formation and Evolution at Intermediate and High Redshift, ed. S. D'Odorico, A. Fontana, & E. Giallongo, ASP Conf. Series, 146, 26

Conti, A., Kennefick, J. D., Martini, P., & Osmer, P. 1998, AJ, in press (astro-ph/9808020)

Genzel, R. et al. 1998, ApJ, 498, 579

Haehnelt, M. G., Natarajan, P., & Rees, M. J. 1998, MNRAS, 300, 817

Haiman, Z., Madau, P., & Loeb, A. 1998, ApJ, submitted (astro-ph/980528)

Hasinger, G. 1998, Astron. Nach., 319, 37

Hook, I. M., Shaver, P. A., & McMahon, R. G. 1998, in The Young Universe: Galaxy Formation and Evolution at Intermediate and High Redshift, ed. S. D'Odorico, A. Fontana, & E. Giallongo, ASP Conf. Series, 146, 17

Irwin, M. J., Ibata, R. A., Lewis, G. F., & Totten, E. J. 1998, ApJ, 505, 529

Jarvis, R. M., & MacAlpine, G. M. 1998, AJ, in press (astro-ph/9810491)

Kedziora-Chudczer, L., Jauncey, D. L., Wieringa, M. H., Walker, M. A., Nicolson, G. D., Reynolds, J. E., & Tzioumis, A. K. 1998, ApJ, 490, L9

Kennefick, J. D., Djorgovski, S. G., & de Carvalho, R. R. 1995, AJ, 110, 2553

Koratkar, A. 1997, in Accretion Processes in Astrophysical Systems: Some Like it Hot, Maryland conf., to be published

Magorrian, J. et al. 1998, AJ, 115, 2285

Masci, F. I. 1998, preprint (astro-ph/9801181)

Nandra, K., Clavel, J., Edelson, R. A., George, I. M., Malkan, M. A., Mushotzky, R. F., Peterson, B. M., & Turner, T. J. 1998, ApJ, 505, 594

Robinson, I., Schild, A., & Schucking, E. L. 1965, Quasi- stellar sources and gravitational collapse, proc. of the 1st Texas Symposium on Relativistic Astrophysics, (Chicago: Univ. of Chicago)

Schechter, P. L., Gregg, M. D., Becker, R. H., Helfand, D. J., & White, R. L. 1998, AJ, 115, 1371

Schmidt, M. 1963, Nature, 197, 1040

Schmidt, M. 1968, ApJ, 151, 393

Schmidt, M., & Green, R. F. 1983, ApJ, 269, 352

Schmidt, M., Schneider, D. P., & Gunn, J. E. 1995, AJ, 110, 68

van der Marel, R. 1997, in IAU Symp. 186, Galaxy Interactions at Low and High Redshift, 102

Warren, S. J., Hewett, P. C., & Osmer, P. S. 1994, ApJ, 421, 412

Weaver, K. A., & Yaqoob, T. 1998, ApJ, 502, L139

Weymann, R. J., Stern, D., Bunker, A., Spinrad, H., Chaffee, F. H., Thompson, R. I., & Storrie-Lombardi, L. J. 1998, ApJ, in press (astro-ph/9807208)

Zheng, W., Kriss, G. A., Telfer, R. C., Grimes, J. P., & Davidsen, A. F. 1997, ApJ, 475, 469

Active Galactic Nuclei and Related Phenomena
IAU Symposium, Vol. 194, 1999
Yervant Terzian, Daniel Weedman, Edward Khachikian, eds.

The Evolution and Luminosity Function of Quasars

Vahé Petrosian

Center for Space Science and Astrophysics, Varian 302c, Stanford University, Stanford, CA, 94305-4060

Abstract. I report results from analysis of data from several quasar samples (Durham/AAT, LBQS, HBQS and EQS) on the density and the luminosity evolution of quasars. We have used new statistical methods whereby we combine these different samples with varying selection criteria and multiple truncations. With these methods the luminosity evolution can be found through an investigation of the correlation of the bivariate distribution of luminosities and redshifts. Of the two most commonly used models for luminosity evolution, $L = e^{kt(z)}$ and $L = (1+z)^{k'}$, we find that the second form, with $k' = 2.58$ (one σ range $[2.14, 2.91]$), gives a better description of the data at all luminosities. Using this form of luminosity evolution we determine a global luminosity function and the evolution of the co-moving density for the two classes of cosmological models. We find a gradual increase of the co-moving density up to $z \sim 2$, at which point the density peaks and begins to decrease rapidly. This is in agreement with results from high redshift surveys and in disagreement with the pure luminosity evolution (i.e. constant co-moving density) model. We find that the local luminosity function exhibits the usual double power law behavior. The luminosity density is found to increase rapidly at low redshift and to reach a peak at around $z \approx 2$. This result is compared with those from high redshift surveys and with the evolution of the star formation rate.

1. Introduction

This work is an outcome of collaborations with Professor Bradly Efron of Department of Statistics and Alexander Maloney, a student at Department of Physics at Stanford University. A more complete description of this work can be found in Efron & Petrosian (1999) and Maloney & Petrosian (1999).

The first aim of this work is to determine the so called **statistical evolution** of quasars and other active galactic nuclei (AGNs) as described by the luminosity function and its variation with cosmic time t or redshift z; $\Psi(L, z)$. This evolution is different than what one may call the **physical evolution** (see, e.g. Lynds & Petrosian 1972) such as the rate of formation or birth of sources as a function of cosmological epoch, $S(L, z)$, and the rate of the variation of the luminosities, $L(z)$ or $\dot{L} = dL/dz$. These two types of evolutions are connected via the continuity equation (see, e.g. Cavaliere & Padovani 1988; Caditz & Petrosian 1990):

$$\partial\Psi(L,z)/\partial z + \partial(\dot{L}\Psi(L,Z))/\partial L = S(L,z). \tag{1}$$

The ultimate goal is to relate the physical evolution functions to the models for production of the luminosity (see, e.g. Caditz, Petrosian & Wandel 1991) and other cosmic evolutionary processes.

Investigations of the statistical evolution of the quasar have played a major role in the development of our ideas about these sources. From the very beginning (Schmidt 1968, and Lynds and Wills 1972) it has been evident that quasars have undergone rapid evolution which was then described by the pure density evolution (PDE) models; $\Psi(L,z) = \psi(L)\rho(z)$, with $\rho(z)$ describing the co-moving density evolution. However, both the source counts and the redshift distribution of optically selected samples of quasars (see e.g. Marshall 1985) clearly showed that PDE cannot be correct and favored the pure luminosity evolution (PLE) model, with $\Psi(L,z) = \psi(L/g(z))/g(z)$. The function $g(z)$ (with $g(0) = 1$) describes the luminosity evolution of the population with $L_o = L/g(z)$ as the luminosity adjusted to its present epoch value. As we shall see below, this model also appears to be inadequate.

Without loss of generality, we can write the luminosity function as

$$\Psi(L,z) = \rho(z)\psi(L/g(z), \alpha_i)/g(z), \tag{2}$$

where $\psi(L_o, \alpha_i)$ gives the local luminosity function. Here we explicitly include the shape parameters α_i, which could also vary with redshift. A surprising result has been the absence of evidence for strong shape variation so in this paper we ignore these variations.

In the next section we give a brief description of the new statistical methods we have developed for accurate determination of $\Psi(L,z)$. In §3 we list the characteristics of the data used for this purpose and in §4 we present the results.

2. The Statistical Methods

The presence of the function $g(z)$ implies that the variables L and z may be correlated. The statistical problem at hand is to first determine the degree of this **correlation** and then determine the univariate **distributions** $\rho(z)$ and $\psi(L_o)$ from an observed bivariate distribution which suffers from selection biases and is subject to multiple truncations. The left panel of Figure 1 shows some generic truncations. The distribution may be truncated parallel to the axis (dotted lines) which can be referred to as untruncated because there is no bias within the observed ranges. More interesting cases are when the truncation is not parallel to the axis. The data may suffer a one-sided truncation from below (solid curve) or above (dashed curve), truncated both from above **and** below or in a more complex way. The most general truncation is when each data point, say $[L_i, z_i]$, has its individual upper or lower limits, $L_i^- < L_i < L_i^+$ and $z_i^- < z_i < z_i^+$, as shown by the large cross for one point. In several papers (Petrosian 1992, Efron & Petrosian 1992 and 1998) we have developed new methods for dealing with all of these situations. These are essentially non-parametric methods which avoid the usual arbitrary binning and the consequent loss of data.

Briefly, the determination of the correlation (i.e. the luminosity evolution) function $g(z)$ is based on the rank order R_i of each source among its *comparable* or *eligible* set $J_i = \{j : y_j > y_i, y_j \in (y_i^-, y_i^+)\}$, where y stands for either variable. One then defines the test statistic $\tau = \sum_i (R_i - E_i)/\sqrt{\sum_i V_i}$ with $E_i = \frac{1}{2}(N_i + 1)$ and $V_i = \frac{1}{12}(N_i^2 - 1)$, where N_i is the number of points in J_i. This statistic is equivalent to Kendall's τ test and for independent variables its distribution should be a gaussian with mean of zero and dispersion of unity. Thus, the luminosity and redshift will be uncorrelated or stochastically independent if $|\tau| < 1$, in which case one may assume that there is no luminosity evolution ($g(z) = 1$) and proceed with the determination of the univariate distributions $\psi(L)$ and $\rho(z)$ using the methods mentioned below. However, if $|\tau| \geq 1$ then L and z cannot be considered independent and one may assume that the most likely explanation is the presence of luminosity evolution ($g(z) \neq$ constant). One can then determine the function $g(z)$ parametrically as follows.

Given a parametric form for the luminosity evolution $g_k(z)$ one can transform the luminosities into $L_o(k) = L/g_k(z)$ and proceed with the determination of the test statistic $\tau(k)$ for the new variables L_o and z as a function of k. The most likely value of k is that with $\tau(k) = 0$ and the range of k for 1 σ confidence level is $\{k : |\tau(k)| < 1\}$. The right panel of Figure 1 shows an example of our results using this procedure.

The last step is the determination of the univariate distributions $\psi(L_o)$ and $\rho(z)$ with our methods which are generalizations of Lynden-Bell's (1971) C^- method.

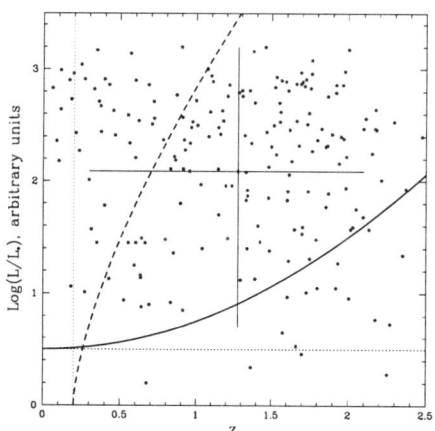
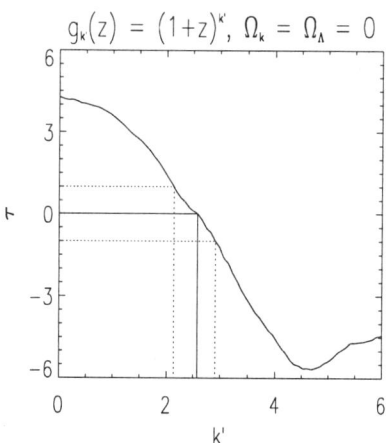

Figure 1. Left Panel: Demonstration of various types of data truncations: Parallel to axis (dotted lines), from below (the solid curve), from above (the dashed curve), and a general truncation when each data point has its specific observable range (shown by the cross for only one of the points). Right Panel: A determination of the luminosity evolution parameter for the parametric form $g_{k'}(z) = (1+z)^{k'}$. The correlation statistic τ is shown as a function of k' for the combined data set for the Einstein - de Sitter cosmological model. The solid line at $\tau = 0$ gives the optimal value $k' = 2.58$ and the dotted lines at $|\tau| = 1$ demonstrate the 1 σ range [2.14, 2.91].

3. The Data

The redshifts and B magnitudes of the data used here are shown in Figure 2. It includes the Large Bright QSO survey (LBQS) containing 1055 QSOs in the magnitude range $16.0 < B_J < 18.85$ and redshift range $0.2 < z < 3.4$ (Hewett et al. 1995) and the Homogeneous Bright QSO Survey (HBQS) (Cristiani et al. 1995) with 285 QSOs in the range $15.5 < B_J < (18.25 \text{ to } 18.85)$ and $0.3 < z < 2.2$. The lower cluster of dots in Figure 2 show the above two data sets in the $0.3 < z < 2.2$ range. The upper cluster of points show the Durham/AAT survey containing 419 QSOs in the magnitude range $17.0 < b < 21.27$ (Boyle et al. 1990) and redshift range $0.3 < z < 2.2$, giving information about the QSO luminosity function in a different regime than the previous two samples. Finally we have added the crosses, a subsample of the Edinburgh QSO Survey (EQS) consisting of 8 QSOs brighter than $B = 16.5$ that fall in the redshift range $0.3 < z < 2.2$ (Goldschmidt et al. in 1992), and give information about the luminosity function at the bright end.

There have been several previous analyses of the above surveys (Boyle et al. 1992, La Franca & Cristiani 1996, Hatziminaoglou et al. 1998) using binning and PLE or PDE models. Our approach is unique in that we combine all of these data and determine both the luminosity and density evolution laws independently.

Before presenting our results we briefly discuss the QSO Hubble diagram and the question of the non-cosmological origin of their redshifts.

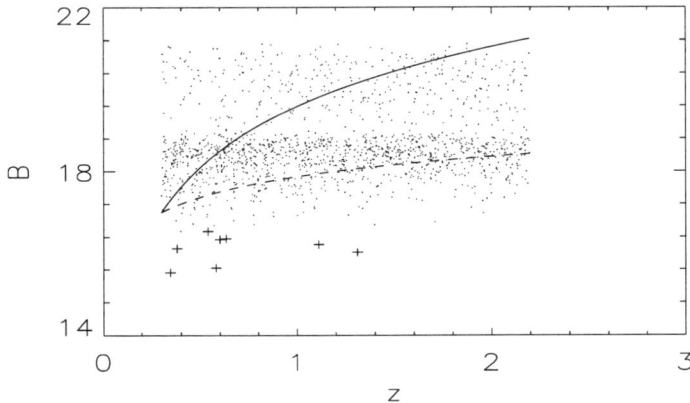

Figure 2. The B magnitude - redshift data for the four samples LBQS and HBQS (lower cluster), Durham/AAT (upper cluster), and EQS (crosses) for $0.3 < z < 2.2$. The data are fitted to the parametric form $B(z) = B_o - \beta \log(d_L^2(z) K(z))+$ constant. The best fit ($\beta = 0.84$) is shown by the dashed line. The solid line shows the expected relation between B and z for standard candles ($\beta = 2.5$). We have used the Einstein-de Sitter model.

3.1. The Quasar Hubble Diagram

At first glance there does not seem to be clear evidence for a Hubble type relation in Figure 2. This well known result has been used as an argument against the cosmological origin of quasar redshifts (see, e.g. Burbidge & O'Dell 1973). However, this is not the only possible interpretation. One would not expect a simple Hubble diagram for sources with a broad luminosity function (nonstandard candles, see e.g. Petrosian 1974). The absence of an obvious Hubble relation can also arise from approximate cancellation between cosmological dimming and luminosity evolution. Exact cancellation of these two effects is highly implausible and could bring into question the basic assumptions about the distribution of quasars. To clarify this situation we have applied the correlation tests described in §2 to the combined data and find a correlation with $\tau = 3.63$. This result rejects the hypothesis of independence between B and z at the 99.97% confidence level. In addition, given a cosmological model with parameters Ω_i, we may test the parametric fit $B(z) = B_o - \beta \log(d_L^2(z, \Omega_i) K(z)) +$ constant, where d_L is the luminosity distance and $K(z)$ is the K-correction term. We find that for the Einstein - de Sitter model $B(z)$ and z are uncorrelated, i.e. $\tau(\beta) = 0$, for $\beta = 0.84$, which is shown by the dashed line in Figure 2. Here we also show the expected relation for standard candles ($\beta = 2.5$, the solid line). This results shows that β, while clearly less than 2.5, differs significantly from the value of zero expected in case of the complete absence of a Hubble relation or the exact cancellation described above.

We now turn to the determination of the evolution of the luminosity function. We have used two classes of cosmological models; models with zero cosmological constant Λ or $\Omega_\Lambda = \Lambda/3H_o^2 = 0$, and flat or inflationary models with non-zero cosmological constant or $\Omega_k \equiv 1 - \Omega_M - \Omega_\Lambda = 0$. Here Ω_M is the matter density parameter and $H_o = 70$ km/(s Mpc) is our assumed value of the Hubble constant, although most results are independent of this assumption. Here we present our results for the Einstein-de Sitter model ($\Omega_\Lambda = \Omega_k = 0$). The other models give qualitatively similar results.

4. The Results

The results we have obtained can be summarized as follows:
- We find a strong **correlation** between luminosity and redshift, indicating presence of a rapid luminosity evolution.
- For determination of the **Luminosity evolution** we examine the two commonly used forms: the power law $g(z) = (1+z)^{k'}$ and the exponential $g(z) = e^{kt(z)}$, where $t(z)$ is the fractional lookback time. We find that the first form provides a better description of the data than the second. We find the value of $k' = 3.53$ for the Durham/AAT sample which is similar to the values found by others (3.45, Boyle 1992; 3.2, Caditz & Petrosian 1990). However, the value 2.58 that we obtain for the combined data is smaller than the previous estimates (e.g. 3.26, La Franca & Cristiani 1996) most of which assume a pure luminosity evolution, i.e. a constant $\rho(z)$. The exponent k' of the luminosity evolution is somewhat coupled to the strength of the density evolution (see also below). An incorrect assumption about the latter will result in an incorrect value for k'. In

our method we obtain k' without any assumptions about the form or strength of the density evolution.

In order to better model this evolution future analyses of quasar evolution could consider parametric forms, with more than one free parameter. More complex forms of luminosity evolution have sometimes been expressed in terms of a luminosity dependent luminosity (or density) evolution. These forms can be turned into a simple luminosity *independent* luminosity evolution form with more than one parameter. For example, the form with the exponent $k' = k_1 - k_2(z)\ln(L/L_*)$ for $L > L_*$ used by La Franca & Cristiani (1996) is the same as the simpler luminosity independent luminosity evolution with $k' = k_1/[1 + k_2\ln((1+z)/(1+z_*))]$ for z greater than some z_*.

Given the form of the luminosity evolution we make the simple transformation of all luminosities to their hypothetical present epoch values, $L_0 = L/g(z)$, so that L_0 and z are uncorrelated. This allows us to use our methods to determine the **univariate distributions** of z and L_0.

• We find that the **co-moving density evolution** depends somewhat on the cosmological model. As shown in the left panel of Figure 3 our results show a relatively slow increase ($\rho \sim (1+z)^{2.5}$) for low redshifts and a rapid decline ($\rho \sim (1+z)^{-5}$) for $z > 2$, which is similar to the high redshift results from Schmidt et al. (1995) and Warren et al. (1994). These exponents are very approximate because, as evident in this figure, power laws are not good representations of the results.

As mentioned above the values of these exponents are coupled to the parameters describing the luminosity evolution. A higher value of k' would give a slower density evolution for $z < 2$. For example, Caditz & Petrosian (1990) use a higher value of k' and find a constant or slowly decreasing co-moving density. Similarly, previous analyses (e.g. Schmidt 1968) assuming no luminosity evolution ($k' = 0$) have consistently given a large (> 5) value for the exponent of the pure density evolution power-law. Miyaji, Hasinger & Schmidt (1998) make the same assumption and find similar results for the soft X-ray luminosity function. On the other hand, we find that the decline of the co-moving density for $z > 2$ is steeper than the previous results for both the optical luminosity function (Schmidt et al. 1995 and Warren et al. 1994) and the soft X-ray luminosity function (the constant $\rho(z)$ results of Miyaji et al. 1998). This is partly due to the assumption of pure density evolution in all three of these previous works and partly due to the incompleteness at high redshift of the data used in our analysis.

• We find that the cumulative local **luminosity function**
$$\Phi(L_o) = \int_{L_o}^{\infty} \psi(x)dx$$
can be described by a double power law form found previously. Our values for the low and high luminosity power law indices, $k_1 = 1.05$ and $k_2 = 3.17$, are consistent with values of 1.35 to 1.50 and 3.6 to 3.9 obtained previously (Caditz & Petrosian 1990, La Franca & Cristiani 1996). There appears to be little variation with redshift of the shape of the cumulative and differential luminosity functions, thus the $\alpha_i =$ constant prescription [see eq. (2)] seems adequate.

• The above results allow us to determine the **luminosity density evolution** $\mathcal{L}(z) \propto \rho(z)g(z)$. As shown in the right panel of Figure 3, this measure increases rapidly with z at low redshift, peaks around $z \approx 2$ and then decreases.

Here we have included all of the LBQS data extending to $z = 3.3$ and the high redshift survey quasars (Schmidt et al. 1995, Warren et al. 1994). It has been claimed (Cavaliere & Vittorini 1998, Shaver et al. 1998) that this rise and fall of $\mathcal{L}(z)$ with redshift is similar to the behavior of the star formation rate (SFR), which has recently been extended to high redshifts (see, e.g. Madau 1997, and Hughes et al. 1998, from SCUBA, presented also in these proceedings). We have shown these rates in the above figure as well. Although the general trends of rise and fall of the SFR and $\mathcal{L}(z)$ are similar, there is considerable difference in the detailed variations. (Our quasar results seem to be in better agreement with the SCUBA than the Madau results.) The similarity may indicate some relation between the SFR and the feeding of the central engine of the quasars (e.g. both are affected by mergers). However, considering the many differences between the star formation process and the generation of energy in quasars the above differences are not surprising.

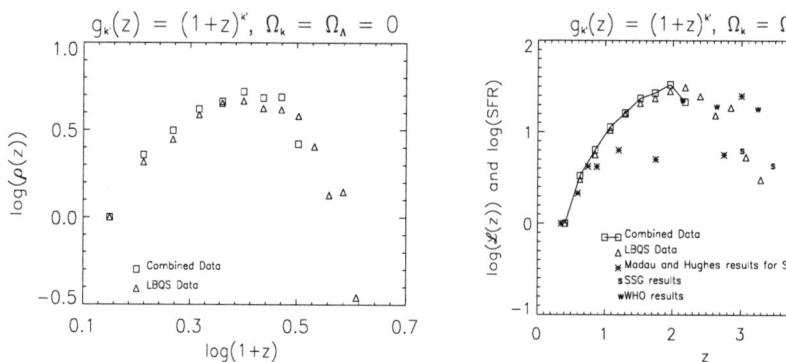

Figure 3. Left Panel: The density evolution $\rho(z)$ vs redshift for the Einstein - de Sitter cosmological model. Clearly, neither the pure luminosity evolution model with $\rho(z) = $ const., nor a simple evolutionary form $\rho \propto (1+z)^\mu$ will be adequate to describe the density evolution. The squares show the results for the combined data in the region $0.3 < z < 2.2$ and the triangles show the results for the LBQS data in the region $0.3 < z < 3.3$. Right Panel: The luminosity density $\mathcal{L}(z)$ vs redshift for the Einstein - de Sitter model. The squares and triangles give our results for the combined data and the LBQS data, respectively. The high redshift results of Schmidt et al. (1995) and Warren et al. (1994) are given by the letters "s" and "w", respectively. The vertical normalizations are arbitrary. All of the above results indicate that $\mathcal{L}$ peaks somewhere in the region $z \approx 2$. The star formation rate (SFR) as a function of redshift, as found by Madau (1997) and Hughes et al. (1998), is given by the asterisks. The results of Hughes et al. (the point at $z = 3$ which in reality extends from $z = 2$ to 4) give a much higher SFR at high redshift than the results of Madau and is in better agreement with the quasar results.

References

Boyle, B. J., Fong, R., Shanks, T., & Peterson, B. A. 1990, M.N.R.A.S., 243, 1.
Boyle, B. J. 1992, in *Texas/ESO-CERN Symp. on Rel. Astro., Cosmology and Particle Physics*, eds. Barrow, J. D., Mestel, L. & Thomas, P., Ann. N. Y. Acad. of Sci., 647, 14.
Burbidge, G. R., & O'Dell, S.L. 1973, Ap.J., 183, 759.
Caditz, D., & Petrosian, V. 1990, Ap.J., 357, 326.
Caditz, D., Petrosian, V., & Wandel, A. 1991, Ap. J. (Letters), 372, L63.
Cavaliere, A., & Padovani, P. 1988, Ap.J. (Letters), 333, L33.
Cavaliere, A., & Vittorini, V. 1998, to appear in *The Young Universe*, Eds. S. D'Odorico, A. Fontana & E. Giallongo ASP Conf. Series 1998, v2; astro-ph/9802320.
Cristiani, S., La Franca, F., & Andreani, P., et al. 1995, A. & A. Suppl., 112, 347.
Efron, B., & Petrosian, V. 1992, Ap.J., 399, 345.
Efron, B., & Petrosian, V. 1998, JASA, in press; astro-ph/9808334.
Goldschmidt, P., Miller, L., La France, F., & Cristiani, S. 1992, M.N.R.A.S., 256, 65.
Hatziminaoglou, E., Van Waerbeke, L. & Mathez, G. 1998,
Hewett, P. C., Foltz, C. B., & Chaffee, F. H. 1995, A.J., 109, 1498.
Hughes, D. H. et al. 1998, Nature, 394, 241
La Franca, F., & Cristiani, S. 1996, invited talk in *Wide Field Spectroscopy* (20-24 May 1996, Athens), Eds. M. Kontizas et al.; astro-ph/9610017.
Lynden-Bell, D. 1971, M.N.R.A.S., 155, 95.
Lynds, R. C., & Petrosian, V. 1972, Ap.J., 175, 591.
Lynds, R. C., & Wills, D. 1972, Ap.J., 175, 531.
Madau, P. 1997, to appear in *The Hubble Deep Field*, Eds. M. Livio, S. M. Fall, & P. Madau, STScI Symposium Series; astro-ph/9709147.
Maloney, A. & Petrosian, V. 1999, Ap.J., in press, vol 518; astro-ph/9807166.
Marshall, H. L. 1985, Ap.J., 299, 109.
Miyaji, T. Husinger, G. & Schmidt, M. 1998, astro-ph/9809398.
Petrosian, V. 1974, Ap.J., 188, 443.
Petrosian, V. 1992, in *Statistical Challenges in Modern Astronomy*, Eds. E. D. Feigelson & G. J. Babu, (New York: Springer-Verlag), p. 173.
Schmidt, M. 1968, Ap.J., 151, 393.

Schmidt M., Schneider, D. P. & Gunn, J. E. 1995, A.J., 110, 68.
Shaver,P.A., Hook, I. M., Jackson, C. A., Wall, J. V., & Kellermann, K. I. 1998, to appear in *Highly Redshifted Radio Lines*, eds. C. Carilli, S. Radford, K. Menten, G. Langston, (PASP: San Francisco); astro-ph/9801211.
Warren, S. J., Hewett, P. C., & Osmer, P. S. 1994, Ap.J., 421, 412.

Blazars: Clues to jet physics

Rita M. Sambruna

Pennsylvania State University, 525 Davey Lab, University Park, PA 16802

Abstract.

Being dominated by non-thermal emission from aligned relativistic jets, blazars allow us to elucidate the physics of extragalactic jets, and, ultimately, how energy is extracted from the central black hole. Crucial information about jet structure is provided by the spectral energy distributions from radio to γ-rays, their trends with luminosity, and correlated multifrequency variability. Since blazar jets have broad implications for all radio-loud (and possibly radio-quiet) AGNs, we also need to understand their circumnuclear structure, especially the details of the physical and dynamical conditions of the highly ionized gas on sub-pc scales, which could be directly related to jet formation and radiative power. Eventually, the bulk of information provided by blazars will help us clarify the origin of the radio-loud/radio-quiet AGN dichotomy, one of the most outstanding open issues of extragalactic astrophysics.

1. The importance of blazars for all AGNs

Blazars (including BL Lacertae objects and quasar-like blazars) are among the most violent manifestations of activity in galaxies. Their defining properties include large luminosities emitted on short timescales, high and variable polarization degrees in optical and IR, smooth continuum emission from radio to γ-ray energies. These properties are best explained as non-thermal emission from a relativistic jet oriented close to the line of sight (Blandford & Rees 1978). Compelling evidence for relativistic beaming of the emitted radiation is provided by the strong and rapidly variable γ-ray emission observed in many blazars in recent years (Hartman 1998).

Today, the relevance of blazars for all AGNs is increasingly recognized as it becomes evident that relativistic jets are a feature of accretion in all radio-loud AGNs (Urry & Padovani 1995), and possibly also in their radio-quiet counterparts (e.g., Falcke et al. 1996), while superluminal ejecta are also present on Galactic scales (Mirabel & Rodriguez 1998). What originally made blazars appear exotic and of little appeal to the astronomical community, now turns out to be their major strength. We need to understand how jets are formed and work if we want to understand how energy is extracted from the central black hole in AGNs (Krolik, this volume). Blazars provide us with a direct probe of this fundamental issue.

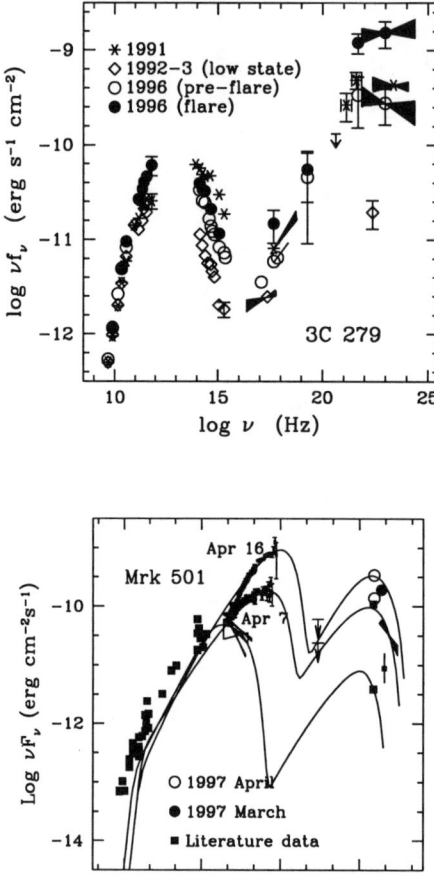

Figure 1. Spectral energy distributions and variability of the red blazar 3C279 *(top)* and blue blazar Mrk 501 *(bottom)*. The continua are characterized by two broad humps peaking at different wavelengths in the two cases, respectively. Variability is more pronounced above the peaks at both low and high energies (data from Wehrle et al. 1998 and Pian et al. 1998).

2. Spectral energy distributions: The blazar family

First clues about jet physics are provided by blazars' spectral energy distributions (SEDs). These typically exhibit two broad humps (Fig. 1): the first one peaks at IR/optical in "red" blazars and at UV/X-rays in their "blue" counterparts (Giommi et al. 1994), and is due to synchrotron emission (Sambruna et al. 1996). The second component extends from X-rays up to GeV/TeV energies, and its origin is less well understood. One promising explanation is inverse Compton (IC) scattering of ambient photons, either internal (synchrotron-self Compton, SSC; Maraschi et al. 1992) or external to the jet (External Compton, EC; Dermer & Schlickeiser 1993; Sikora et al. 1994; Ghisellini & Madau 1996).

Red and blue jets are just the extrema of a continuous population. Deep multicolor surveys are finding blazars with intermediate spectral shapes (Laurent-Meuhleisen et al. 1998; Perlman et al. 1998). Spectral trends with bolometric luminosity are observed in existing samples: from red to blue blazars the luminosity decreases and the synchrotron and IC peak frequencies decrease proportionately. Red sources also appear to have a larger γ-ray dominance and more luminous broad emission lines (Sambruna et al. 1996; Sambruna 1997; Fossati et al. 1998).

In the context of a simple homogeneous model, where a single electron population is responsible for producing both components through synchrotron and IC processes, the spectral trends can be interpreted as a change of a few jet physical parameters: the higher synchrotron peak frequencies of blue blazars call for higher magnetic fields or/and electron energies than red sources (Sambruna et al. 1996). It is also possible that red blazars have larger luminosities and lower electron energies as a result of a more efficient cooling through EC scattering (Ghisellini et al. 1998). In this case, the paradigm would be that the type of electron cooling depends on the density of the external photons, with the SED continuity hinting to a large spread of physical properties of the jet ambient medium, commensurate with the observed line properties (Scarpa & Falomo 1997).

However, the luminosity trends are most likely affected by strong biases, especially at γ-rays, where the limited sensitivity of current GeV detectors combined with large variability could exaggerate the γ-ray dominance in the most distant red blazars (and bias toward EC-dominated sources, where the γ-rays are more beamed). In blue blazars Klein-Nishina cutoffs kick in at TeV energies and the ratio of the synchrotron to IC peak is only approximately constant. Finally, recent deep surveys show the existence of an unexpected population of blue blazars with strong emission lines (Perlman et al. 1998; see also Sambruna 1997). While the new spectral trends will be tested more extensively in the future, a clear message emerges from the current data: *blazars exhibit a rich variety of spectral behaviors, which point to a diversity of jet physical conditions* rather than being due to beaming/orientation effects only (Sambruna et al. 1996; Georganopoulos & Marscher 1998; Kubo et al. 1998).

3. Correlated multiwavelength variability: Looking inside the jet

One crucial still unanswered question is the origin of the GeV and TeV emission. An alternative to the SSC and EC scenarios (§ 2) is provided by the proton-induced cascades models (PIC; Protheroe & Biermann 1997; Mannheim & Biermann 1992), where the γ-rays are direct synchrotron emission from ultra-relativistic electrons while the X-rays are emitted by less energetic electrons originating from pair cascades initiated by protons. The viability of hadronic models, at least in red blazars, may be supported by recent observations of circular radio polarization (Wardle et al. 1998). Clearly, distinguishing between the IC and PIC scenarios is of central importance to understand the jet composition and how energy is transported away from the central black hole.

Correlated multifrequency variability provides a way to this end, since the different models make distinct predictions for the relative flare amplitudes and for the time lags. In a one-zone homogeneous approximation, the synchrotron plus IC models predict: 1) strongly correlated variability of the fluxes at the low- and high-energy peaks with no time lags, since the same electron population is responsible for emitting both spectral components; 2) simple and yet precise relationships for the relative amplitudes of the synchrotron and IC flares depending (for a fixed beaming factor δ) on the change of electrons and/or seed photons (Ghisellini & Maraschi 1996); 3) accurate shapes of the synchrotron and IC flares depending on a few source typical timescales (Chiaberge & Ghisellini 1998) and parameters (Romanova & Lovelace 1997; Böttcher & Dermer 1998); 4) lags between soft and hard X-rays, reflecting the electron radiative time ($t_r \propto E^{-1/2} B^{-3/2} \delta^{-1/2}$), from which the local magnetic field B can be derived (modulo δ; Takahashi et al. 1996); 5) defined time-dependent spectra, according to how fast electrons are accelerated compared to t_r (Kirk et al. 1998).

In contrast, PIC models allow more freedom: lags of either sign between long and short wavelengths are possible, as well as more arbitrary relations in the flare relative amplitudes, because the relative energy deposited in electrons and protons is a free parameter, and so the details of the ensuing pair production are less constrained.

The target selection for multifrequency campaigns is strongly constrained by the γ-ray band, where only a handful red blazars are bright enough to be monitored at GeV energies. At TeV even fewer sources are detected, while TeV photons from distant blazars may be difficult to observe because of their interaction with the IR background (Salomon & Stecker 1994). Despite the intrinsic difficulty of such campaigns, now severely undermined by the death of the EGRET CGRO detector, a few blazars were monitored well enough to allow comparison of models to the data.

Results for red blazars. The multi-epoch SEDs of 3C279, one of the most luminous GeV sources in the sky (Fig. 1), shows indeed correlated variability at IR/optical and GeV, supporting IC models. The amplitude of the GeV variation goes quadratically with the optical flux in early campaigns, indicating a variation of the electron density as the cause of the flare in the SSC model (Maraschi et al. 1994) or a change of beaming factor in the EC model. During the 1996 campaign, however, the γ-rays varied more than quadratically than the optical/IR flux, and exactly (within 1 day) simultaneously to the X-rays (Wehrle et al. 1998), with the rapid decay time of the GeV flare favoring EC scenarios (Ghisellini & Madau 1996).

A new red candidate for multifrequency campaigns is BL Lac itself, which underwent a strong GeV and optical outburst in 1997 July (Bloom et al. 1997). The γ-ray light curve shows a strong flare possibly anticipating the optical flare by ~ 0.5 days; however, the poor γ-ray sampling prevents any firm conclusion. Interestingly, a broad Hα emission line was detected in BL Lac (Vermeulen et al. 1995), which increased in luminosity at the time of the optical/GeV outburst, suggesting that EC models are responsible for producing the flux above a few MeV. This is indeed supported by a detailed modeling of the SED (Sambruna et al. 1999).

Results for blue blazars. Mrk 501, one of only four extragalactic TeV sources (all blazars), flared dramatically at TeV and X-rays in 1997 April (Aharonian et al. 1997; Catanese et al. 1997); TeV and X-ray light curves track each other well, with no lags larger than 1 day, confirming that the TeV photons are produced by the same electrons responsible for the X-rays. A similar behavior was observed also during our recently concluded campaign in 1998 June. As shown in Fig. 1, during the 1997 TeV flare the synchrotron peak shifted forward by 2 orders of magnitude in correspondence to a dramatic flattening of the X-ray continuum, implying large acceleration events (Pian et al. 1998). Mrk 501 is an eloquent example of the dramatic powers that can be reached in blazars.

An ideal candidate for studying the energy-dependence of the synchrotron flares is PKS 2155-304, one of the brightest and most rapidly variable BL Lacs at optical through X-ray wavelengths, recently detected in TeV (Chadwick et al. 1998). Previous campaigns (Edelson et al. 1995; Urry et al. 1997) detected correlated variability at optical through X-ray wavelengths, and established that the shorter wavelengths generally vary first, with lags of the order of a few hours to a few days. The measured 1 hour delay between soft and hard X-rays in 1994 May yields $B \sim 0.1$ Gauss (Urry et al. 1997), similar to Mrk 421 (Takahashi et al. 1996).

The multifrequency experiment for PKS 2155–304 was repeated in 1996 May (Fig. 2), with better sampling in X-rays but worse at longer wavelengths. Complex variability was observed, with several short flares superposed to a longer trend. The flares become more symmetric at decreasing wavelengths, roughly consistent with a homogeneous synchrotron scenario where the radiative time at high energies is fast, and the light travel time limits both the rise and decay times (Chiaberge & Ghisellini 1998). No lags larger than ~ 1.5 hours were observed. The X-ray hardness ratios (Fig. 2) follow a general clock-wise loop with intensity, indicating that acceleration is fast and the spectrum is controlled by radiative cooling, with superposed events of anti-clockwise cycles, when cooling and acceleration are in equilibrium (Kirk et al. 1998).

What did we learn? Current multifrequency data support models which predict correlated variability at the low- and high-energy peaks in both blue and red blazars. While it proved difficult to distinguish among models based on the relative flare amplitudes, due to the large number of free parameters involved, detection of time lags and flare shape promises to be more efficient to this end, but is limited at present by sampling constraints, especially at γ-rays.

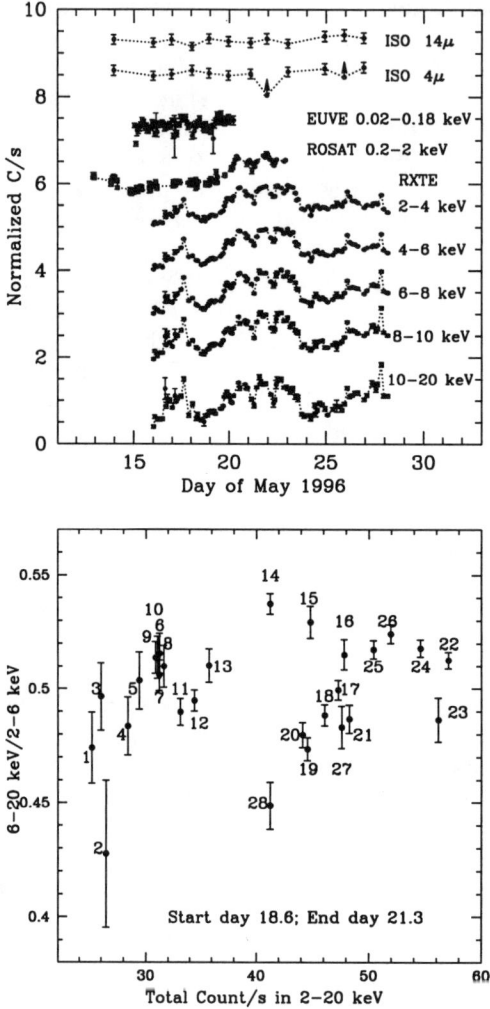

Figure 2. Multifrequency variability of PKS 2155–304 in 1996 May. *Top:* IR to X-ray light curves, normalized to their mean fluxes and arbitrarily offset (data from papers in prep. by Bertone, Brinkmann, Marshall, Sambruna). No lags are detected larger than the RXTE binning time (90 min); the flares have larger amplitudes and more symmetric shapes at increasing energies. *Bottom:* An example of X-ray hardness ratios vs. intensity, numbered chronologically. The spectral changes follow a general clock-wise cycle, with anti-clock-wise smaller loops. This is a diagnostic of the acceleration processes in the emission region.

Spectral variability at X-rays is a powerful diagnostic of the acceleration events in the jets, and can be studied well with RXTE and SAX.

4. Blazar environment: Where jets form

The study of blazar environment has recently received much attention, as the importance of the external medium for jet formation and deceleration (De Young 1993; Bicknell 1995) and for jet radiative power (Dermer & Chiang 1998) is increasingly recognized. IR and optical studies from the ground (see Heidt 1998 for a review) and with HST (Urry et al. 1999 and references therein) show that BL Lacs generally lie in poor clusters (Abell richness 0–1), and that their host galaxies are smooth, luminous ellipticals, with no differences between red and blue sources. The host galaxies of BL Lacs are very similar to those of their putative parent populations, Fanaroff-Riley I galaxies, of matching extended radio powers (Urry et al. 1999b, in prep.), supporting unified models for radio-loud AGNs. Substantial progress in the study of blazar environs will be achieved in the near future with AXAF[1] observations in X-rays, which will allow us to measure directly the medium properties (temperature, density, abundances) and probe in depth the gravitational potential.

High-energy absorption in BL Lacs. A powerful probe of the inner (pc/sub-pc scales) jet environs is provided by high-resolution observations of a few bright BL Lacs at X-ray and EUV wavelengths. Broad absorption features were detected in PKS 2155–304 and Mrk 421, the only two blazars observed at grating resolutions in soft X-rays (Canizares & Kruper 1984) and EUV (Königl et al. 1995; Kartje et al. 1997), suggesting absorption in highly ionized gas outflowing with mildly relativistic velocities ($v \lesssim 0.1c$). Ionized absorption was also detected in BBXRT and ASCA observations of a few more BL Lacs (Sambruna et al. 1997; Sambruna & Mushotzky 1998), confirming earlier claims (Madejski et al. 1992).

The origin of the absorbing medium is still a speculation. One possibility is the disk-driven hydromagnetic wind model (Königl & Kartje 1994), based on the mechanism originally proposed by Blandford & Payne (1982) to explain the origin of relativistic jets (Lovelace, this volume): clouds of gas are lifted from the accretion disk surface by the magnetic field lines until they intercept the beamed jet radiation, giving rise to absorption when crossing the line of sight. Alternatively, absorption could originate in matter swept up by the jet outer edge (Krolik et al. 1985), perhaps the same decelerating medium required by unification schemes (Bicknell 1995). At the current low spectral resolution and sensitivity there is no way to discriminate among the possible models; however, this will be trivial with the gratings onboard the future AXAF and X-ray Multi-Mirror missions, which will allow us to measure precisely the absorber properties and trace a detailed portrait of the inner jet environs.

---

[1]The Advanced X-ray Astrophysical Facility will carry, among its various detectors, a set of CCD imagers with spectroscopic capabilities, with a much narrower PSF (FWHM $\lesssim 1$ arcsec) and lower intrinsic background than ROSAT.

5. Future work

While recent years have seen gigantic progress in our understanding of the blazar phenomenon, much more work remains to be done.

We need to test more extensively the luminosity/distance spectral trends, in particular at γ-rays, with larger complete samples, fully addressing the various selection effects. More TeV observations, where we currently have only 4 detected sources, are badly needed, as well as more precise determinations of the TeV spectra in order to quantify better the position of the IC peak in blue blazars.

We need to intensify our efforts in multiwavelength monitoring of blue and (more) red blazars, in particular to measure more precisely the time lags down to the smallest accessible scales, and the flare shapes both at low and high energies. With the death of EGRET on CGRO, the short-term future of multifrequency campaigns relies only on TeV blazars.

Blazar environment studies at IR, optical, and X-ray wavelengths with current and future high-resolution satellites (HST, NGST, AXAF, XMM) will address the AGN/host galaxy connection, in particular the role of the ambient medium for jet formation and collimation, and for fueling the central black hole.

With the advent of new technical resources of improved sensitivity and resolution in a large wavelength range, the future holds great promise for blazars. Eventually, the bulk of information provided by blazars will help us shed light on the origin of jets and, ultimately, of the radio-loud/radio-quiet AGN dichotomy, one of the most outstanding open problems of extragalactic astrophysics.

Acknowledgments. This work was supported by NASA grants NAS–38252, NAG–3313, NAG5–7121, and by an IAU travel grant. I thank Meg Urry for interesting discussions, and all my colleagues at the conference for a memorable meeting.

References

Aharonian, F. et al. 1997, A&A, 327, L5
Bicknell, G. V., 1995, ApJS, 101, 29
Blandford, R.D. & Payne, D.G. 1982, MNRAS, 199, 883
Blandford, R.D. & Rees, M.J. 1978, in Pittsburgh Conference on BL Lac Objects, A.M.Wolfe, Univ. Pittsburgh Press, 1978, p.328
Bloom, S. D. et al. 1997, ApJ, 490, L145
Böttcher, M. & Dermer, C.D. 1998, ApJ, 501, L51
Canizares, C.R. & Kruper, J. 1984, ApJ, 278, L99
Catanese, M. et al. 1997, ApJ, 487, L143
Chadwick, P.M. et al. 1998, ApJ, in press, astro-ph/9810209
Chiaberge, M. & Ghisellini, G. 1998, MNRAS, subm., astro-ph/9810263
De Young, D. S. 1993, ApJ, 405, L13
Dermer, C.D. & Chiang, J. 1998, astro-ph/9810222
Dermer, C.D. & Schlickeiser, R. 1993, ApJ, 416, 458

Edelson, R. et al. 1995, ApJ, 438, 120
Falcke, H. et al. 1996, ApJ, 473, L13
Fossati, G. et al. 1998, MNRAS, 299, 433
Georganopoulos, M. & Marscher, A. 1998, ApJ, 506, 621
Ghisellini, G. et al. 1998, MNRAS, in press, astro-ph/9807317
Ghisellini, G. & Madau, P. 1996, MNRAS, 280, 67
Ghisellini, G. & Maraschi, M. 1996, in Blazar Continuum Variability, H.R.Miller, J.R.Webb, & J.C.Noble, ASP Conf. Series, Vol. 110, 436
Giommi, P., Ansari, S.G., & Micol, A. 1995, A&ASupp., 109, 267
Hartman, R.C. 1998, in BL Lac Phenomenon, Eds. A.Sillanpaa & L.O.Takalo, PASP Conf. Series, in press
Heidt, J. 1998, in BL Lac Phenomenon, Eds. A.Sillanpaa & L.O.Takalo, PASP Conf. Series, in press
Kartje, J.F. et al. 1997, ApJ, 474, 630
Kirk, J.G., Rieger, F.M., & Mastichiadis, A. 1998, A&A, 333, 452
Königl, A. et al. 1995, ApJ, 446, 598
Königl, A. & Kartje, J.F. 1994, ApJ, 434, 446
Krolik, J.H. et al. 1985, ApJ, 295, 104
Kubo, H. et al. 1998, ApJ, 504, 693
Laurent-Muehleisen, S.A. et al. 1998, ApJS, 118, 127
Madejski, G. et al. 1991, ApJ, 370, 198
Mannheim, K. & Biermann, P.L. 1992, A&A, 253, L21
Maraschi, L. et al. 1994, ApJ, 435, L91
Maraschi, L., Ghisellini, G., & Celotti, A. 1992, ApJ, 397, L5
Mirabel, I.F. & Rodriguez, L.F. 1998, Nature, 392, 673
Perlman, E.S. et al. 1998, AJ, 115, 1253
Pian, E. et al. 1998, ApJ, 492, L17
Protheroe, R.J. & Bierman, P.L. 1997, APh, 6, 293
Romanova, M.M. & Lovelace, R.V.E. 1997, ApJ, 475, 97
Salomon, H.M. & Stecker, F.W. 1994, ApJ, 430, L21
Sambruna, R.M. et al. 1999, ApJ, in press, astro-ph/9810319
Sambruna, R.M. & Mushotzky, R.F. 1998, ApJ, 502, 630
Sambruna, R.M. 1997, ApJ, 474, 639
Sambruna, R.M. et al. 1997, ApJ, 483, 774
Sambruna, R.M., Maraschi, L., & Urry, C.M. 1996, ApJ, 463, 444
Scarpa, R. & Falomo, R. 1997, A&A, 325, 109
Sikora, M., Begelman, M.C., & Rees, M.J. 1994, ApJ, 421, 153
Takahashi, T. et al. 1996, ApJ, 470, L89
Urry, C.M. et al. 1999, ApJ, in press, astro-ph/9809030
Urry, C.M. et al. 1997, ApJ, 486, 799
Urry, C.M. & Padovani, P. 1995, PASP, 107, 803

Vermeulen, R.C. et al. 1995, ApJ, 452, L5
Wardle, J.F.C. et al. 1998, Nature, 395, 457
Wehrle, A. E. et al. 1998, ApJ, 497, 178

Halton Arp and Ed Khachikian

Markarian and Seyfert Galaxies

E.Ye. Khachikian

Byurakan Astrophysical Observatory, Byurakan, 378433, Armenia

1. Historical Perspective

Since this symposium is devoted to the 90th anniversary of Prof. Ambartsumian's birthday I would like to touch briefly upon some historical aspects of V. Ambartsumian's ideas on activity of nuclei of galaxies.

Modern extragalactic astronomy achieved such high success that we have reached almost the stage of our Universe when it was only a couple of billion years old. It is very strange that some 75 years ago (an instant compared with the supposed age of the Universe) we had still no certain idea about the existence of remote galaxies. But already 10-15 years after discovering galaxies, it was a common idea that they were thoroughly formed steady systems with a rich past and no prospect of radical changes in the future. Except for classification of form of galaxies and investigation of their general photometry and colourimetry, nobody supposed that the most important part of galaxies is their central part. The role of nuclei in the evolution of galaxies was manifestly underestimated.

V. Ambartsumian was the first to pay particular attention to the significance of galactic nuclei. As was said figuratively by A. Sandage, "No astronomers would deny today, that mystery indeed surrounds the nuclei of galaxies, and Victor Ambartsumian was the first who understood what a rich reward is contained in this treasury".

In the second half of the 50's a new phase of extragalactic exploration set in, when the new concept of the basic role of the nuclei of galaxies in their life and evolution had been advanced by Ambartsumian (1956a,1956b, 1956c, 1958a, 1958b, 1961a, 1961b, 1962, 1968).

Of course the origin of this concept was not unfounded. It was preceded by a number of wonderful discoveries resulting in the revision of our notions on the world of galaxies.

In the first place was the identification of the powerful radio source Cygnus A by Baade and Minkowski with a galaxy containing two nuclei (Baade & Minkowski, 1954).

A similar picture was observed also in the radiosource Perseus A (NGC 1275). The role of two papers was also significant: Haro(1956) discovered 44 galaxies, unusually blue in colour; and, specially, in 1943 Seyfert's paper, now regarded as classical, should be singled out for mention.

The galaxies which he investigated are distinguished by the high luminosity of their nuclei, and, more importantly, by the width of the Balmer emission lines. The great width of the emission lines indicates that the turbulent motions of gas clouds in the nuclei of those galaxies, subsequently termed "Seyfert", at times attain a velocity of over 3000 km/s. Now it seems quite strange and surprising that this very important paper of Seyfert was not duly taken into account in the succeeding twenty years or so. It was only after Ambartsumian's idea concerning the activity of nuclei of galaxies had been made public that the astrophysicists returned to that paper, and a regular study of the Seyfert galaxies was started. On the basis of analysing these facts, Ambartsumian came to the idea of activity of nuclei of galaxies, which manifests itself mainly in the following forms :

1. Outflow of ordinary gas matter (in form of jets or clouds) from the nuclear region at velocity of up to hundreds of km/s.

2. Continuous emission of flux of relativistic particles or other agents producing high energy particles,as a result of which a radio halo may form around the nucleus.

3. Eruptive ejections of gas matter (M 82 type).

4. Eruptive ejections of concentrations of relativistic plasma (NGC 4486, 5128, etc.).

5. Ejection of compact blue condensations with an absolute magnitude of the order of luminosity of dwarf galaxies (NGC 3561, IC 1182). Here the division of the nucleus into two or more comparable components is also presumed, initiating the formation of multiple galaxies.

The presence of one or several of these phenomena allows us to call a galaxy active.At present a number of types of objects are considered as active: radio galaxies, QSOs, Seyfert galaxies, Lacertides, UV-excess galaxies, Blazars, Liners.

Let us come back to the phenomenon of radiogalaxies. Baade and Minkowski explained this phenomenon as a result of accidental collision of two galaxies. V.A. Ambartsumian was the first and only astronomer who in his works (1956b, 1956c, 1958a, 1961b, 1962), convincingly showed that in the case of radio-galaxies we have not a collision, but just activity of nuclei of galaxies, which brings ejection of matter from the nucleus, and in some cases as a result of this activity a radio-galaxy is originated.

I am lucky that I had the chance in 1961 in Berkeley to be present at the XI General Assembly of IAU, where V.Ambartsumian gave a public talk on the idea of activity of galaxies. The interest in his talk was so high that the conference hall and corridors were full of people (even outside, where microphones had been mounted). During this Assembly he was elected as President of the IAU.

It was really a revolutionary and extraordinary idea. It is said that in the Solvey conference in 1958 W.Baade even accused him of idealism and noted that for a scientist from the Soviet Union to speak about ejection and activity of nuclei of galaxies looked very strange. But just a couple of years later American astronomers Sandage and Lynds (1963) discovered the explosion in the center of the galaxy M82. Almost the same time in 1963 QSOs were discoverd. Actually, as Ambartsumian noted,they were just the naked active nuclei, which radiate an unusually high quantity of energy, highest among known cosmic objects. Now we know that many QSO are really surrounded by a stellar population. Thus, Ambartsumian's idea has been confirmed at first by observations of American astronomers.

A new stage in the extragalactic field had been started: the era of active galaxies. The majority of large observatories over the world started to find new active galaxies (AG).By the leading of V.Ambartsumian the Byurakan observatory also engaged itself in activities aimed at discovering galaxies with active nuclei. Ambartsumian and Shahbazian (1954) were the first to show the existence of blue ejections and condensations associated with contiguous active elliptical galaxies. Subsequently Stockton (1968) showed that those objects are in fact associated with galaxies and display emission spectra similar to the associations.

Then on the initiative of V.A.Ambartsumian, B.E.Markarian started in Byurakan in the mid-sixties observations of the sky with a view to detecting galaxies with anomalous spectra, using the 40" Schmidt telescope with an objective prism of the same diameter.

2. The First Byurakan Survey

The first Byurakan survey (FBS) is the most famous work done with this telescope. More than 2000 photographic plates covering about 17000 square degrees of sky were obtained. Each plate contains low dispersion spectra (2500A/mm near H_β) of more than 15 000 objects. As a result, 1500 galaxies with strong UV-excess were discovered. In 1978 Markarian with his co-workers began the Second Byurakan Survey (SBS). The limiting magnitude of objects increased from 17^m (for FBS) up to about $19^m.5$ for SBS. The SBS covers about 1000 square degrees.

I was lucky to be the first to observe almost all galaxies from the first Markarian list of UV-galaxies with the largest optical telescopes of the USA. I would like to emphasize once more that the detailed spectral investigations of these objects indicated (Khachikian 1968, Weedman & Khachikian 1968, 1969) that over 85% of them have emission lines, their intensity being directly dependent on the value of UV-excess. One can conclude that the presence of a strong ultraviolet continuum is closely associated with the formation of the emission spectrum and the more intense the continuous spectrum in the visible ultraviolet is, the more intense are the emission lines (see Fig 11.6 in Osterbrock, 1989).

It became also evident that the spectra of those objects differ, nevertheless, essentially from each other as to the excitation degree of the emission lines and their widths. Moreover, they turned out to differ sharply in morphological characteristics as well: one can come across the blue galaxies of Haro, the compact galaxies of Zwicky, the N type galaxies, spiral and irregular galaxies among the Markarian objects. Quite important is the discovery of the Seyfert galaxies and quasars among those objects. As far back as 1968 I also demonstrated that on the basis of slitspectra, Markarian galaxies can be classified in five groups(Khachikian 1968):

1. Narrow line spectra both in emission and absorption.

2. Narrow, strong emission lines only.

3. Strong and diffuse emission lines; [OIII] lines much stronger than the hydrogen lines (Seyfert type 2).

4. Very broad hydrogen lines, narrow forbidden lines (Seyfert type 1).

5. No strong emission lines (BL Lac type).

These results were presented at the first international conference on Seyfert galaxies and related objects in 1968 (Tuscon, Arizona, USA), where I called these objects "Markarian galaxies". Further spectral investigations of Markarian galaxies from both Byurakan Surveys have been carried out intensively in Byurakan and in many other observatories.

These searches show that among Markarian objects there are representives of all forms of activity predicted by V.Ambartsumian: QSOs, Seyfert galaxies, BL Lac objects, galaxies with jets, blue compact galaxies, double nuclei galaxies and so on. But the most important is, as was shown by Weedman and Khachikian (1971a), that 10% of Markarian galaxies turned out to be Seyfert galaxies.

The number of Seyfert type galaxies was extremely increased thanks to study of Markarian objects. In the original paper of Seyfert there are only 6 galaxies of Seyfert type. But now more than 1000 of these type of galaxies are known!

On the base of detailed spectral investigations of a number of Seyfert type galaxies Weedman and Khachikian (1971a) have shown that Seyfert galaxies clearly are divided into two types:

1. Galaxies with very broad hydrogen lines, and narrow forbidden lines (Seyfert type 1).

2. Galaxies with very broad both hydrogen and forbidden lines (Seyfert type 2).

There is also quite precise difference in relative intensities of some emission lines. In table 1 the relative intensities and equivalent widths of lines of typical Sy galaxies from the Markarian list are presented.

So, the second difference of Sy1 and Sy2 is the intensity ratio of emission lines $[OIII]$ to H_β, which is less than unity for Sy1 and more than 10 for Sy2.

Therefore it is very strange sometimes to read in articles about narrow line Sy galaxies. There is no Sy galaxy with narrow lines! I would like to dwell here upon three subjects. a) morphology of Markarian and Seyfert galaxies; b) double nuclei AG; c) variability in the spectrum of AG.

As in this Symposium are presented many articles on Markarian and Seyfert galaxies in the range of radio, IR, UV, FUV, X- and Gamma ray, I shall concentrate on the optical properties of these objects.

a). The detailed morphological description of Markarian and Seyfert galaxies has been done by Khachikian (1987). It was shown that the activity is not correlated with the morphology of the whole galaxy: most important is the structure of the central part of galaxies. The majority of AG according to morphology of the central part can be divided into the following groups: i) starlike galaxies; ii) galaxies with starlike nucleus (in general spirals); iii) double nuclei galaxies; iv) multinuclei galaxies; v) galaxies with bulges. Some AG show jets starting from the nucleus. Representatives of these groups are shown in Fig1-Fig6 in Khachikian (1987).

Table 1. From Weedman & Khachikian (1969)

ion	λ	Markarian numbers							
		Sy2				Sy1			
		1	3	6	34	9	10	42	52
[S II]	6717+ 6731	3.30	6.11	0.62	2.85	-	-	-	1.29
[NII] H_α [NII]	6583+ 6562+ 6548	11.8	17.9	9.33	9.85	4.61	5.20	5.66	9.41
[OIII]	5007+ 4959	11.4	15.5	2.53	15.0	0.58	0.83	0.32	0.91
H_β	4861	1.0	1.0	1.0	1.0	1.0	1.0	1.0	1.0
He II	4686	-	0.25	-	-	-	0.19	-	-
OIII H_γ	4363+ 4340	0.68	0.89	-	0.77	0.54	0.57	0.48	0.48
[NeIII]	3869	1.08	1.34	0.16	1.57	-	0.17	-	-
[OII]	3726+ 3729	1.87	2.98	0.37	3.54	-	0.24	-	1.95
$H_\beta/10^{40}erg/s$		2.1	5.8	18.6	14.2	72.5	36.2	2.8	1.1
		equivalent widths							
[SII]	6717+ 6731	50	92	16	44	-	-	-	33
[NII] H_α [NII]	6583+ 6562+ 6548	177	425	236	128	177	240	112	190
[OIII]	5007+ 4959	217	840	135	276	35	56	7	20
H_β	4861	17	36	43	21	59	61	21	19
He II	4686	-	8	-	-	-	10	-	-
[OIII] H_γ	4363+ 4340	11	41	-	15	23	27	8	8
[NeIII]	3869	24	88	7	37	-	8	-	-
[OII]	3726+ 3729	41	-	15	78	-	11	-	31

3. Galaxies with Multiple Nuclei

As it was mentioned above, the radio galaxy Cygnus A has two nuclei. It is interesting to note that the majority of AG turn out to be double nuclei. It is necessary to stress that in addition to double nuclei there are galaxies with three and more nuclei(or nuclear type formations). It is known also that each of the nuclei of double nucleus galaxies can themselves consist of two components (Mark.273, Knapen et al.1998). Therefore the opinion concerning the nature of double nucleus AG is relevant to the multinuclei AG as well. The existence of galaxies with two condensations in the centre having Seyfert type spectra (Mark.266, 463, 673, 789) is the most definite evidence of the possibility of galaxies with double nuclei in general. However, double nuclei galaxies are not an unusual phenomenon among AG. They are particularly common among UV galaxies. In Table 2 are some physical parameters: the apparent and absolute photographic magnitudes of the components of the nucleus, distance between components in arcseconds and kiloparsecs (H=75 km/s.Mpc) and differences of their radial velocities for six AG with double nuclei are presented.

Table 2. From Khachikian (1987)

Mark.No	m(pg)	M(pg)	d″	d(kpc)	v(km/s)
266 a	17.5	-17.8	12	6.5	127
b	17.8	-17.5			
273 a	17.5	-18.4	4.3	3.2	-
b	18.2	-17.7			
463 a	17.0	-19.5	4.5	4.3	50
b	17.2	-19.3			
673 a	16.2	-19.6	5.3	3.7	166
b	16.2	-19.6			
739 a	16.2	-19.1	6.6	3.8	85
b	17.0	-18.3			
789 a	16.0	-19.5	4.1	2.5	2
b	18.0	-17.5			

From Table 3, where the sizes of nuclei, distance between them and difference of radial velocities of Sargent-Searle double nuclei objects are presented, one can see that for them we have the same picture.

The number of double and multi-nucleus AG is increasing all the time. Zwicky compact galaxies with emission spectrum, many radio galaxies, so-called isolated giant HII regions (Sargent & Searle objects) or Superassociations (SA) are double nucleus objects. Among the galaxies from FBS more than 100 double nucleus galaxies were discovered. It is ridiculous that in the 60's a few people spoke about the merging of galaxies, but now most of them explain any complicated nuclei of galaxies without hesitation as a merger. As is known now a lot of double and multiple nuclei galaxies, it seems that the Universe instead of red shifting is collapsing. Therefore I am sure that many of this type of nuclei are not a result of merging. No doubt, many of them (if not the majority) are real double nucleus galaxies. That is, they are probably not a result of merging or

Table 3. From Shaver & Chen (1985)

Object	d''	d(kpc)	v(km/s)
0128-531 A	2.2	7.8	55
B	0.9		
1030+073 A	1.1	4.3	25
B	0.8		
1247-043 A	0.2	1.3	130
B	0.3		
1247-051 A	0.05	0.7	20
B	0.3		

interaction of two independent galaxies. Note that the following observational data are difficult to reconcile with the hypothesis of gravitational merging:

1. the discovery of Seyfert type double nuclei galaxies because of their rarity among galaxies (Khachikian, Petrosian, Sahakian 1978, 1980);

2. the discovery of twin-objects with quite identical spectra and morphology similar to two isolated SA being considered as one galaxy with double nuclei each of which is SA (Arp, Heidmann & Khachikian 1974, Khachikian 1979);

3. the discovery of numerous double nuclei galaxies.

Aditionally, the excellent articles of Mazzarella and Boroson (1993) and Nordgren et al. (1995) support very much this idea.

It seems to me that one of the important current problems of AG is the nature of their double and multiple nuclei: is it the result of merging of galaxies (two or more) or evolution of the central part of an individual galaxy. From the time of Kant and Laplace up to the present, the majority of theorists, as well as observers, believe that the Universe develops in a direction from concentrations of diffuse matter to the denser states. Perhaps V.Ambartsumian was the first who declared the opposite point of view. As far back as the end of the 40's, he stated the revolutionary idea that evolution in the Universe goes from the dense condition of matter to the rarefied one. It seems to me that many observational data speak in favour of this point of view. The existence of double and multi nuclei galaxies is a good As it was mentioned above one of the forms of activity is the ejection from the nuclei of active galaxies of isolated clouds with different contents. Further, both spectroscopic and morphological investigations show that the nuclei of some AG are variable and undergo irregular changes in brightness and in spectra. In the end of the 60's and the beginning of the 70's it became clear that in central parts of same AG take place physical processes leading to ejection of huge amounts of matter from the nucleus of AG. The small sizes of AG nuclei (AGN), about 10 arcsec, don't permit detecting such a new gas formation by means of direct observation in the near surroundings of them even in radio wavelengths. Therefore the only possibility for the investigation of these physical events is detailed spectrophotometrical observations with comparatively high dispersion. The most effective are investigations in the optical.

The appearance of additional new emission components of Hydrogen lines in the spectrum of AGN first was discovered in 1969 by Khachikian and Weedman (1970,1971b). During one year (between February 1968 and January 1969) in the spectrum of Markarian 6, which was a Sy2 galaxy, new broad emission components of Hydrogen lines H_α, H_β and H_γ were detected. Their blue-shift velocity corresponds to 3000 km/sec. In January 1970 the intensity of the H_β component was equal to 50% of the basic H_β line. These observations have been confirmed by many authors (Adams 1972, Notni et al. 1973, Adams & Weedman 1975, Chuvaev 1991, Khachikian 1973, Khachikian et al.1982). It is known now many active objects with double Hydrogen line structure in the spectrum: 3C 390.3, NGC 1097, NGC 1566 (Lynds 1968, Pastoriza & Gerola 1970, Alloin et al.1986). There are some models which have been suggested to explain this phenomenon (Khachikian & Weedman 1971, Khachikian 1973, Pronik 1987, Zheng et al. 1990, Eraclous & Halperin 1992, Chen et al 1989). All these models do not give a complete explanation of the phenomenon. In Ambartsumian et al. (1998) a new fairly simple model is suggested, which gives quantitative accordance with observational data. In Fig.1 the schematic picture of the model is shown.

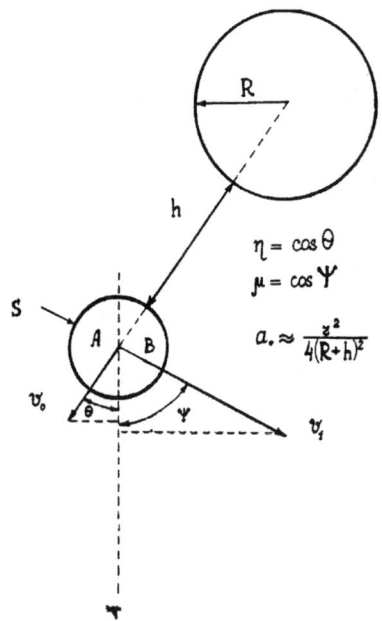

Figure 1.

From the nucleus of AG a compact formation S is ejected with the velocity Vo. At some distance from the surface of the AG it exploded. As a result of this explosion a globular gas cloud S, mainly consisting of Hydrogen atoms, is formed. Similar to the situation in planetary nebula, the L_c-quanta of AG lead to ionization of Hydrogen atoms in S. As a result of recombination and following cascade transitions the subordinate Hydrogen lines are formed. On the whole the additional components of subordinate lines arise. The shift of additional lines relative to that of the nucleus of AG is explained by the speed of ejection of S from AGN. As for the widths of additional lines they depend on the velocity of expansion of S. The simple estimation presented in this work shows that comparable small size and mass of S cloud and an acceptable value of speed of its expansion can explain the phenomenon.

The terms "active galaxies", "active nucleus", introduced by Ambartsumian, are now generally accepted in science, although some scientists (mostly younger one) have no idea about that. Therefore, his followers sometime have to remind us about the tremendous impact of V.Ambartsumian in science, in particular in extragalactic astronomy. Ambartsumian's idea on the activity of galaxies exerted tremendous influence on the further development of extragalactic astronomy and stimulated numerous studies in many observatories over the world. In the well-known Volume of the U.S. Academy of Sciences "The Heritage of Copernicus", commemorating the quinquecentennial of the birth of Copernicus, this Ambartsumian idea is considered as a Copernican type of revolutionary idea, which has changed our notion about the nature of galaxies.

References

Abramian, M. G., Fridman A. M., & Khachikian E. Ye., 1994, In "The Cold Universe". Moriond Astrophysics Meeting, Les Arcs, France, p.357

Adams, T. F., 1972, ApJ, 172, L 101

Adams, T. F., & Weedman, D. W., 1975, ApJ, 199, 19

Adams, T. F., 1972, ApJ, 172, L 101

Alloin, D., Potal, D., Fosbury, R. A., Freeman, K., Phillips, M. M. 1986, ApJ, 207, L 147

Ambartsumian, V. A., 1956a, Proceedings of V Conference on Problems of Cosmogony, Moscow.

Ambartsumian, V. A., 1956b, Izv. Acad. Sci. Arm. SSR, serie Phys. Math. Sci., 9, 23

Ambartsumian, V. A., 1956c, Dokladi of AS of Armenia, 23, 161

Ambartsumian, V. A., 1958a, Izv. Acad. Sci. Arm. SSR, serie Phys. Math. Sci., 11, 9

Ambartsumian, V. A., 1958b, "La structure et l'evolution de l'univers", Solvey Conference, p.241, Bruxelles, ed.by R.Stoops

Ambartsumian, V. A., 1961a, " The problems of extragalactic resurch",General Assamble of IAU, Berkeley

Ambartsumian, V. A., 1961b, AJ, 66, 536

Ambartsumian, V. A., 1962, The problems of Cosmogony, 8, 3

Ambartsumian, V. A., 1968, The Problems of Evolution of the Universe,Yerevan, p. 85

Ambartsumian, V. A., & Shahbazian, R. K., 1957, C. R. Acad. Sci. Arm., 25, 185

Ambartsumian, V. A., Khachikian E. Ye., & Yengibarian, N. B., Astrofizika, 1998, 41, 321

Arp, H. C., Heidmann, J., & Khachikian, E. Ye., 1974, Astrofizika, 10, 7
Baade, W., Minkovski, R., 1954, AJ, 119, 206
Chen, K., Halperin, J., & Filippenko, A. V., 1989, ApJ, 339, 742
Chuvaev, K. K., 1991, Izv. Crimean. Astr. Observ., 83, 194
Eraclous, M., & Halperin, J., 1992, in "Testing the AGN Paragdigm". Ed. by S. Holt
Haro, G., 1956, Bol. obs. Tonantzintla, 14, 8
Khachikian, E. Ye., 1968, AJ, 73, 891
Khachikian, E. Ye., 1973, Astrofizika, 9, 139
Khachikian, E. Ye., 1979, "Stars and Star Systems", Uppsala, p.107
Khachikian, E. Ye., 1987, IAU Symp. No121. Ed. by Khachikian et al., Byurakan, p.65
Khachikian, E. Ye., Petrosian, A. R., & Sahakian, K. A., 1978, Astrofizika,
Khachikian, E. Ye., Petrosian, A. R., & Sahakian, K. A., 1980, Astrofizika, 16, 621
Khachikian, E. Ye., & Weedman, D. W., 1970, Astron. Zircul. No591, 2. 14, 69
Khachikian, E. Ye., Popov, V. N., Yegiazarian, A. A., 1982, Astrofizika, 18, 541
Khachikian, E. Ye., & Weedman, D. W., 1971, Astrofizika, 7, 389
Khachikian, E. Ye., & Weedman, D. W., 1971, ApJ, 164, L109
Knapen, J. H., Laine, S., Yates, J. A., Robinson, A., Richards, A., M., Doyon R., & Nadeau D., 1998, ApJ(in press)
Lynds, C. R., 1968, AJ, 73, 888
Lynds, C. R., & Sandage, A. R., 1963, AJ, 137, 1005
Mazzarella, J. M., & Boroson, T. A. 1993, ApJS,85, 27
Nordgren, T. E., Helou, G., Chengalur, J. N., Terzian, Y., & Khachikian, E. Ye. 1995, ApJS, 99, 461
Notni, P., Khachikian, Ye. E., Butslov, M. M., & Gevorkian, G. T., 1973, Astrofizika, 9, 39
Osterbrock, D., 1989, in "Astrophysics of Gaseous Nebulae and Active Galactice Nuclei"
Pronik, I. I., 1987, in IAU Symp. No 121. Ed. by Khachikian et al., Buyrakan, p. 169
Sargent, W. L. W., 1970, AJ, 159, 765
Seyfert, C., 1943, AJ, 97, 28
Shaver, P. A., Chen, J.-S., 1985, A&A, 148, 443
Stockton, A., 1968, AJ, 73, 887
Weedman, D. W., & Khachikian, E. Ye., 1968, Astrofizika, 4, 587
Weedman, D. W., & Khachikian, E. Ye., 1969, Astrofizika, 5, 113
Zheng, W., Binette, L., Sulentic, J., 1990, ApJ, 365, 115

Active Galactic Nuclei and Related Phenomena
IAU Symposium, Vol. 194, 1999
Yervant Terzian, Daniel Weedman, Edward Khachikian, eds.

Feeding the Central Engine in Giant Radio Galaxies

I. F. Mirabel[1,2] & O. Laurent[1]

[1] Service d'Astrophysique. Centre d'Etudes de Saclay. 91191 Gif/Yvette, France.
[2] Instituto de Astronomía y Física del Espacio. c.c. 67, suc 28. (1428) Buenos Aires, Argentina.

Abstract.
Giant radio galaxies are thought to be massive ellipticals powered by accretion of interstellar matter onto a supermassive black hole. Interactions with gas rich galaxies may provide the interstellar matter to feed the active galactic nucleus (AGN). To power radio lobes that extend up to distances of hundreds of kiloparsecs, gas has to be funneled from kiloparsec size scales down to the AGN at rates of ~ 1 $M_\odot$ yr^{-1} during $\geq 10^8$ years. Therefore, large and massive quasi-stable structures of gas and dust should exist in the deep interior of the giant elliptical hosts of double lobe radio galaxies. Recent mid-infrared observations with ISO revealed for the first time a bisymmetric spiral structure with the dimensions of a small galaxy at the centre of Centaurus A (Mirabel et al. 1999). The spiral was formed out of the tidal debris of accreted gas-rich object(s) and has a dust morphology that is remarkably similar to that found in barred spiral galaxies. The observations of the closest AGN to Earth suggest that the dusty hosts of giant radio galaxies like CenA, are "symbiotic" galaxies composed of a barred spiral inside an elliptical, where the bar serves to funnel gas toward the AGN.

1. A barred spiral at the centre of Centaurus A

The mid-infrared and submillimeter observations of the dust emission in CenA revealed a bisymmetric structure of 5′ (~ 5 kpc for a distance of 3.5 Mpc) in total length (Mirabel et al. 1999). Figure 1 shows that in contrast to the optical dark lanes which show a wide and somewhat chaotic distribution, the structure of the mid-infrared emission is remarkably thin, smooth and bisymmetric. The emitting dust is less extended and displaced from the most prominent optical dark lanes. This displacement is not due to major differences between the spatial distributions of the cold and very warm dust components. In a three dimensional tilted and warped disk, projection effects play an important role, and the optical appearance of the dark lanes in a luminous ellipsoidal system may be affected by relatively small amounts of cold dust in the outer parts of the bending disk that are located in the foreground side of the luminous ellipsoidal distribution of stars.

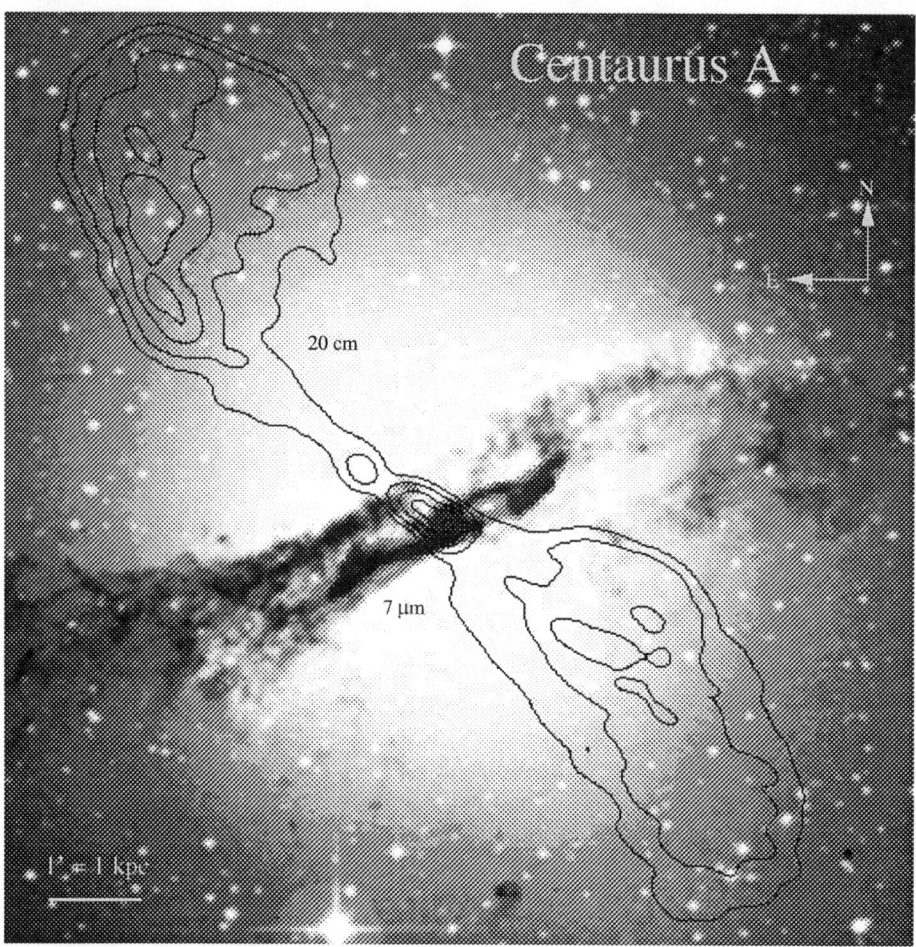

Figure 1. The ISO 7 μm emission (dark structure) and VLA 20 cm continuum in contours (Condon et al. 1996), overlaid on an optical image from the Palomar Digital Sky Survey. The emission from dust with a bisymmetric morphology at the centre is about 10 times smaller than the overall size of the shell structure in the elliptical (Malin et al. 1983) and lies on a plane that is almost parallel to the minor axis of its giant host. Whereas the gas associated to the spiral rotates with a maximum radial velocity of 250 km s^{-1}, the ellipsoidal stellar component rotates slowly approximately perpendicular to the dust lane (Wilkinson et al. 1986). The synchrotron radio jets shown in this figure correspond to the inner structure of a double lobe radio source that extends up to 5° ($\sim$ 300 kpc) on the sky. The jets are believed to be powered by a massive black hole located at the common dynamic center of the elliptical and spiral structures.

The structures seen in CenA with higher angular resolution by means of the mid-infrared observations with ISO are inside the general distribution of the molecular gas (Eckart et al. 1990; Rydbeck et al. 1993). Figure 2 shows that the interpretation of the bisymmetric structure at the centre of CenA as a barred spiral is fully consistent with the morphology of the dust lanes observed in galaxies classified as barred spirals, and the kinematics observed in CO data. Furthermore, the interpretation of a barred structure is consistent with theoretical models (Athanassoula, 1992) that predict shocks at the leading edge of bars, producing an arc-like appearance of the warm dust. The barred spiral is a dynamic instability, i.e. a density wave in the warped disk of gas and dust.

ISOCAM could not resolve features smaller than $\sim 5''$ but K($2.2\,\mu$m) band polarization (Packham et al. 1996) due to the passage of starlight through clouds that contain electrons and/or aligned small grains can be used to trace with higher angular resolution the preferential distribution of dust in the innermost central region. The position angle of the polarization in Figure 2 suggests the presence of a secondary bar, or "nuclear" bar of gas of few hundred parsecs in size. This presumed secondary bar inside the primary bar could be the dynamical instability that brings gas towards the supermassive black hole, as proposed in the theoretical model by Shlosman et al. (1989). In fact, molecular gas absorption has been detected in front of the compact nuclear source at millimeter wavelengths, and it has been proposed that the absorption at redshifted velocities represents gas falling into the centre (Israel et al. 1991).

It is believed that the same accretion event(s) that began tearing apart a gas-rich object(s) also created the faint stellar and gaseous shells observed around CenA. It is known that the accretion of a small disk galaxy by a massive elliptical would lead to the complete tidal disruption of the former as it spirals inward in the potential of the elliptical galaxy. In this process the gas decouples from the stars and sinks more readily to the centre, forming a new disk out of the gaseous component alone. In CenA the overall angular momentum of the newly formed disk is not aligned with the major axis of the elliptical. Therefore, the gaseous disk is subject to torques forcing it to warp. Although gas is still settling towards the central regions, the morphological and dynamical symmetry of the spiral indicates that it is a stable structure and not a transient feature. Rotating at 250 km s^{-1}, it must have undergone several full rotations depending on how long after the initial encounter it took the gas to settle into the central disk we now see.

Using near infrared photometry as well as the kinematics of the gas, a disk-to-total mass ratio within the turnover radius of the rotation curve of the order of 10^{-2} is obtained. N-body simulations would rule out that a low-mass stellar bar has formed spontaneously in the disk, and survived at steady-state. However these simulations consider stellar disks, while here we are dealing with a gaseous one. Using the much lower velocity dispersion (typically 5-10 km s^{-1} in the gas, compared with 50 km s^{-1} for a stellar disk and 145 km s^{-1} for the spheroidal component in CenA (Wilkinson et al. 1986, Eckart et al. 1990) a much lower Toomre's Q parameter (of the order of 1) is derived for the gaseous disk in CenA. Therefore, a gaseous disk is much more self-gravitating than a stellar one of similar mass, and the bar in CenA can be in a quasi-steady state and might have formed spontaneously, or be driven.

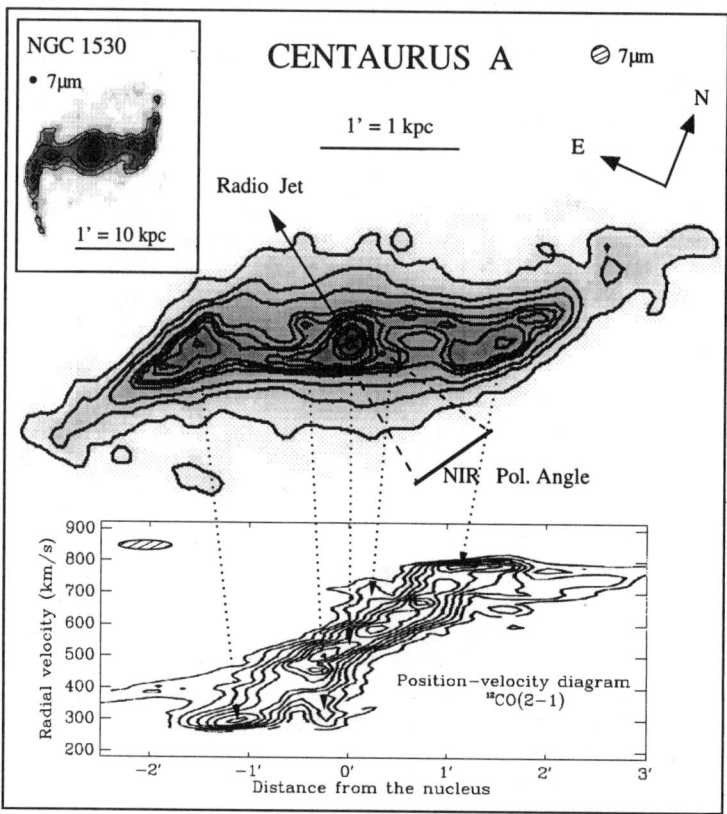

Figure 2. ISO 7 μm image and ^{12}CO(2-1) position-velocity map (Quillen et al. 1992) of the central region of CenA. Note the similarity in morphology with the 7 μm image of the prototype barred spiral NGC 1530. In NGC 1530 the plane that contains the bar is tilted by ∼ 55° to the line of sight, whereas in CenA it is tilted by ∼ 18°. The overall structure exhibited by the 7 μm emission from CenA is that of a barred spiral with a primary bar extending ∼ 1′ in radius from the nucleus, connected in its outer ends to trailing spiral arms. The strongest 7 μm emission from the primary bar is along its leading edge in what takes the form of two slightly curved arcs, where shocks and density enhancements take place. The inner ends of these two arcs are connected to what may be a secondary nuclear bar whose position angle is defined by the NIR K(2.2 μm) band polarization (Packham et al. 1996). On the plane of the sky the radio jets appear perpendicular to the innermost polarization angle. The kinematics of the gas in the lower panel is consistent with a barred spiral, where the bar rotates as a rigid body within 70″, whereas at radii larger than ± 70″ the gas exhibits the differential rotation (flat rotation curve) typical of galactic disks.

2. The symbiotic galaxy Centaurus A: a template for giant radio galaxies

Heckman et al. (1986) have shown that radio galaxies with giant radio lobes have peculiar optical morphologies. In this respect, CenA is not an exception and could serve as a well-positioned template to examine in detail the clues to the origin and evolution of activity in early-type radio galaxies with the same radio morphology, namely, with giant double radio lobes. Prominent dust bands are frequently observed in the hosts of this type of radio galaxy. Fornax A, the second nearest giant radio galaxy after CenA, exhibits clear signs for the merger of gas-rich galaxies with a dusty early type galaxy. Cygnus A, the prototype radio galaxy with double morphology, is crossed by prominent optically dark bands that contain $\sim 10^8$ $M_\odot$ of dust (Robson et al. 1998). It has been shown that in dusty radio galaxies with double radio structure, the dust is usually found perpendicular to the radio axis (Kotani & Ekers 1979; van Dokkum & Franx 1995) which suggests a connection between the mechanism leading to these double radio morphologies and the rotation axis of the dusty disk.

The specific mechanism in rapidly rotating disks of gas and dust that brings fuel to the central engine in radio loud AGNs has been difficult to probe observationally for several reasons. First, galaxies like CenA are at greater distances (for instance, Fornax A is 5-10 times and Cygnus A is $\sim$ 70 times more distant than CenA), and at those distances it is difficult to see the detailed morphology of dust and gas on scales $\leq$ 100 pc. Second, at optical and near-infrared wavelengths the light from the old stellar population with a giant ellipsoidal distribution overwhelms any emission from dust and newly formed stars in the deep interior. It is in the mid-infrared that the emission from very warm dust can be better traced, and to this end we had to wait for the unprecedented capabilities of ISOCAM. Third, the observation of the cold gas distribution by means of millimeter observations of weak molecular line emission on top of the strong continuum of powerful radio galaxies is a difficult task.

The need for the presence of bars in AGNs has been questioned because the ocurrence of bars in Seyfert galaxies is not higher than in normal galaxies (McLeod & Riecke, 1995; Ho et al. 1997; Mulchaey & Regan, 1997). Furthermore, using HST images in the optical and near-infrared, no signatures of dust *absorption* with barred structures have been found in Seyfert 2's (Regan & Mulchaey, 1999). However, it is difficult to trace the presence of dust inside bulge and ellipsoidal systems using only optical and near-infrared observations because at those wavelengths the stellar emission dominates. Mirabel et al. (1999) have shown that the true structure of the dust in the deep interior of CenA can only be revealed by the emission at wavelengths longer than $3\,\mu$m. In CenA the dust absorption derived from colors between the near-infrared and the optical (Quillen, 1993) only trace the foreground side of the structure shown in Figures 1 and 2.

It is believed that the AGN activity in Seyfert galaxies is much more sporadic than in giant radio galaxies and that Seyferts may require accretion masses with sizes that are orders of magnitude smaller than those needed to power the giant lobes of radio galaxies. The observation of *emission* from the accreting fuel in the central regions of Seyfert galaxies may have to await the high angular resolution and sensitivity of the NGST and MMA/LSA.

The barred spiral at the centre of CenA has dimensions comparable to that of the small Local Group galaxy Messier 33. It lies on a plane that is almost parallel to the minor axis of the giant elliptical. Whereas the spiral rotates with maximum radial velocities of ~ 250 km s^{-1}, the ellipsoidal stellar component seems to rotate slowly (maximum line-of-sight velocity is ~ 40 km s^{-1}) approximately perpendicular to the dust lane. The genesis, morphology, and dynamics of the spiral formed at the centre of CenA are determined by the gravitational potential of the elliptical, much as a usual spiral with its dark matter halo. On the other hand, the AGN that powers the radio jets is fed by gas funneled to the center via the bar structure of the spiral. The spatial co-existence and intimate association between these two distinct and dissimilar systems suggest a "symbiotic" association.

3. Conclusions

1) The observation of dust emission from the centre of CenA opens the more general question on whether the hosts of giant radio galaxies are symbiotic galaxies composed of spirals at the centre of giant ellipticals.
2) The true structure of dust in the deep interior of ellipsoidal stellar systems is better traced by *emission* from dust at wavelengths longer than 3 microns, rather than by absorption using colors from near-infrared and optical photometry.
3) The accreting matter in Seyferts is likely to have masses and sizes that are several orders of magnitude smaller than in giant radio galaxies. To trace the accreting gas and dust in Seyferts, the angular resolution and sensitivity of the NGST and MMA/LSA are needed.

References

Athanassoula E., 1992, MNRAS, 259, 345
Condon J.J., Helou G., Sanders D.B., Soifer B.T., 1996, A&AS, 103, 81
Eckart A., Cameron M., Genzel R. et al., 1990a, ApJ, 365, 522
Heckman, T.M. et al., 1986, ApJ, 311, 526
Ho L., Filippenko A. & Sargent W., 1997, ApJ, 487, 591
Israel F.P., van Dishoeck E.F., Baas F. et al., 1991, A&A, 245, L13
Kotanyi C.G., Ekers R.D., 1979, A&A, 73, L1
Malin D.F., Quinn P.J., Graham J.A., 1983, ApJ, 272, L5
McLeod K.K. & Rieke G.H. 1995, ApJ, 441, 96
Mirabel I.F., Laurent O., Sanders D.B. et al. 1999, A&A, 341, 667
Mulchaey J.S. & Regan M.W., 1997, ApJ, 482, L135
Packham C., Hough J.H., Young S. et al., 1996, MNRAS, 278, 406
Quillen A.C., de Zeeuw P.T., Phinney E.S., Phillips T.G., 1992, ApJ, 391, 121
Quillen A.C., Graham J.R., Frogel J.A., 1993, ApJ, 412, 550
Regan M.W. & Mulchaey J.S. 1999, AJ, in press
Robson E.I., Leeuw L.L., Stevens J.A., Holland W.S., 1998, MNRAS, 301, 935

Rydbeck G., Wiklind T., Cameron M. et al., 1993, A&A, 270, L13
Shlosman I., Frank J., Begelman M.C., 1989, Nature, 338, 45
van Dokkum P.G., Franx M., 1995, AJ110, 2027
Wilkinson A., Sharples R.M., Fosbury R.A., Wallace P.T., 1986, MNRAS, 218, 297

P. Véron

On the QSO Content of the First Byurakan Survey and the Completeness of the Palomar Green Survey

P. Véron[1], A. M. Mickaelian[2], A. C. Gonçalves[1] and M.-P. Véron-Cetty[1]

[1] *Centre National de la Recherche Scientifique, Observatoire de Haute Provence, 04870 St. Michel l'Observatoire, France*

[2] *Byurakan Astrophysical Observatory, Byurakan 378433, Republic of Armenia*

Abstract. The second part of the First Byurakan Survey is aimed at detecting all bright ($B < 16.5$) UV-excess starlike objects in a large area of the sky. By comparison with major X-ray and radio surveys we tentatively identified as QSOs 11 FBS objects. We made spectroscopic observations of nine of them. We found six new QSOs bringing the total number of known QSOs in this survey to 42.

By comparison with the Palomar Green (PG) QSO survey, we found that the completeness of this last survey is of the order of 70% rather than 30–50% as suggested by several authors.

1. Introduction

The surface density of bright QSOs ($B < 17.0$) is still very poorly known. The PG survey (Schmidt & Green 1983) covering an area of $10\,714$ deg^2 led to the discovery of 69 QSOs brighter than $M_B = -24$ ($H_o = 50$ km s^{-1} Mpc^{-1}) and $B = 16.16$, corresponding to 0.0064 deg^{-2}. However, several authors have concluded that it could be incomplete by a factor 2 to 3.

The First Byurakan Survey (FBS), also known as the Markarian survey, was carried out in 1965–80 by Markarian et al. (1989). It covers $17\,000$ deg^2 and is complete to about $B = 16.5^m$. It has been used by Markarian and his collaborators to search for UV excess galaxies; it can also be used for finding UV-excess starlike objects. Such a program — the second part of the FBS — has been undertaken in 1987 (Abrahamian et al. 1990; Mickaelian 1994). Its main purpose is to take advantage of the large sky area covered to get a reliable estimate of the surface density of bright QSOs.

At the present time, $4\,109$ deg^2 have been searched and a catalogue of $1\,103$ blue stellar objects (715 at $|b| > 30°$) has been built. 434 spectroscopic identifications (398 stars and 36 QSOs) are already known. Photoelectric UBV photometry for 113 FBS objects has been published. For all except one, $U - B < -0.50$.

2. Comparison with X-ray and radio surveys

2.1. The ROSAT All-sky Survey Bright Source Catalogue (RASS-BSC)

We have cross-correlated the list of 1 103 UV-excess objects with the RASS-BSC (Voges et al. 1996). We have found seven coincidences with FBS objects of unknown nature.

2.2. The WGA ROSAT catalogue of point sources (WGACAT)

The WGACAT has been generated using the *ROSAT* PSPC pointed data publicly available as of September 1994. It contains more than 45 600 individual sources (White et al. 1994). We have found two coincidences with FBS objects of unknown nature.

2.3. The NRAO VLA sky survey (NVSS)

The NVSS covers the sky north $\delta = -40°$ at 1.4 GHz (Condon et al. 1998). It contains almost 2×10^6 discrete sources stronger than S = 2.5 mJy. We have found two coincidences with FBS objects of unknown nature.

3. Observations

Spectroscopic observations of nine of the eleven objects coincident either with a X-ray or a radio source were carried out with the Observatoire de Haute Provence 1.93 m telescope. Six (listed in Table 1; the spectra are shown in Fig. 1) turned out to be QSOs, while three are stars.

4. Discussion

4.1. Completeness of the FBS

The Catalogue of mean UBV data on stars (Mermilliod & Mermilliod 1994) contains 102 stars in the FBS area, with $11.0 < V < 16.5$ (bright stars are saturated on the FBS plates and are therefore missed) and $U - B < -0.50$; 53 are included in the FBS catalogue suggesting that, in this magnitude interval, the completeness of the FBS is 52% (53/102). The completeness of the survey increases from $\sim$ 20% at $U - B \sim -0.6$ to $\sim$ 80% for $U - B < -1.0$. The QSO $U - B$ colour changes with z; but most of the changes, at least for $z < 2.2$, are due to the presence of an emission line in one of the two filters. The $U-B$ colour of the continuum is in the range $-0.9 < U - B < -0.7$. Slitless spectroscopic surveys are sensitive to the colour of the continuum, unaffected by the emission lines. There are 58 known stars with $14.0 < V < 16.5$ and $U - B < -0.70$ in the FBS area; 39 (67%) have been found by the FBS; we shall adopt this value for the completeness of the FBS.

The PG and FBS samples have about 2 250 deg^2 in common. Out of the 1 103 FBS objects, 618 are within the PG fields, 276 being in the PG sample. Thirty-six PG objects have not been found in the FBS. So 88% (276/312) of the PG objects have been discovered. The 36 undiscovered objects have been

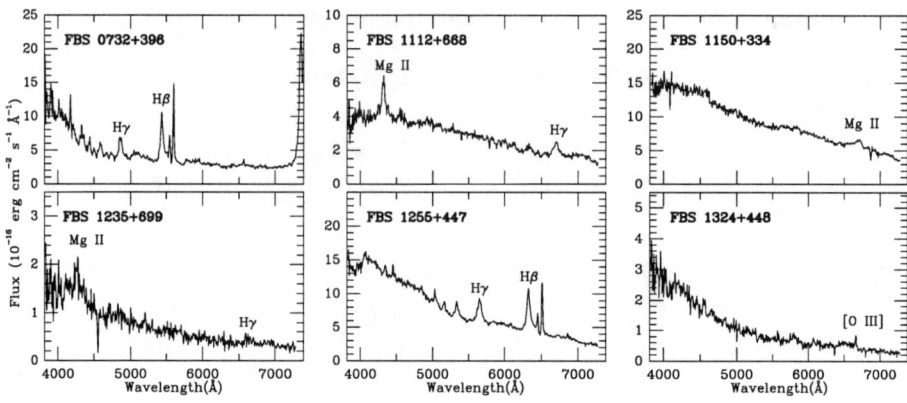

Figure 1. Spectra of the six newly discovered QSOs.

examined on the FBS plates; 24 have a weak UV excess (the PG survey finds a significant fraction of stars with $U - B \sim -0.4$, while the FBS is relatively insensitive for $U - B > -0.7$). From this, we conclude that the FBS is $\sim 95\%$ complete for $U - B < -0.5$. This is significantly larger than the 67% success rate obtained from the UBV stars; it is probably due to the fact that, in principle, the PG survey contains only objects brighter than $B = 16.2$. There are 25 PG QSOs in the FBS area, 23 of them have been found, confirming that the FBS is very efficient in discovering bright QSOs.

Table 1. Newly discovered QSOs

FBS name	Mag.	O		(1)	(2)		b	z
0732+396	16.0	14.70	X	10	20	N	25.1	0.118
1112+668	17.0	16.53	X	10	4	Y	47.9	0.544
1150+334	16.2	16.30	R	0.8	1.2	Y	76.0	1.40
1235+699	17.9	17.96	x	5	4	Y	47.4	0.522 ?
1255+447	16.5	16.48	X	10	13	Y	72.6	0.300
1324+448	17.0	18.09	X	8	9	Y	71.1	0.331 ?

X: in the RASS-BSC; x: in the WGACAT; R: in the NVSS; (1): error of the *ROSAT* or VLA position (in arcsec); (2): distance between the FBS and *ROSAT* or VLA positions (in arcsec); Y: in the PG area; N: not in the PG area; O: APS *O* magnitudes.

4.2. The AGN content of the FBS

Thirty-four FBS objects are listed as QSOs in the Véron-Cetty & Véron (1998) catalogue. Two more have been shown to be QSOs by Hagen et al. (1998) and six have been identified in the present paper. There are therefore 42 known QSOs in the FBS, 41 being at high galactic latitude ($|b| > 30°$). At high galactic latitude, all FBS objects associated with a RASS-BSC source have been identified. Among them, there are 25 QSOs. As about 60% of all PG QSOs are RASS sources, and assuming that this is true for the FBS QSOs, we should have a total of about 43 QSOs in the FBS catalogue. This suggests that the number of QSOs still to be found in the FBS catalogue is very small as 42 are already known.

All 114 AGNs from the PG survey have been observed at 5 GHz with the VLA (Kellerman et al. 1989); 35 (30%) have been detected with a flux density larger than 3 mJy. The same fraction (12/40) of the known FBS QSOs have been detected in the NVSS survey, confirming that the number of QSOs in the as yet spectroscopically unobserved FBS objects is small and probably cannot exceed 10, as the fraction of radio-detected QSOs would then drop below 25% and be significantly lower than the corresponding fraction for the PG survey.

4.3. Completeness of the PG survey

In principle the PG survey selected all objects with $U - B < -0.46$ and brighter than $B \sim 16.2$; however, the $U - B$ colour was measured with a relatively large error (0.24^m rms) which induced an incompleteness estimated at around 12%. Moreover, in the interval $0.6 < z < 0.8$, the strong Mg II $\lambda 2800$ emission line is in the B filter which gives a much redder $U - B$ colour than for neighbouring redshifts, resulting in an incompleteness of 28% in this range (Schmidt & Green 1983).

The catalogue of mean UBV data on stars (Mermilliod & Mermilliod 1994) contains 283 stars in the magnitude range $12.0 < B < 16.5$ and with $U - B < -0.40$ in the full 10 714 deg^2 area of the PG survey; 190 are included in the PG catalogue. Twenty-four stars photoelectrically observed because they were in the PG catalogue have been ignored; the overall completeness of the PG survey is therefore 64% (166/259). 67% (162/241) of the stars brighter than $B = 16.2$ were found in the PG survey, while only 22% (4/18) of the weaker stars were. The completeness of the PG survey rises from about 55% at $U - B = -0.60$ to 80% at $U - B < -1.0$. For PG QSOs, the completeness should not be less than $\sim 70\%$.

An independent estimate of the completeness of the PG survey comes from the comparison with the RASS-BSC. Of the 25 PG QSOs in the FBS area, 13 are brighter than $B = 16.16$ and have been detected by *ROSAT*. Twelve of them have been found in the FBS. All FBS objects associated with a *ROSAT* source, located in the PG area, have been spectroscopically identified; among them, there are eleven non-PG QSOs. No more than two are suspected to have been possibly bright enough to have been detected by the PG survey. This suggests a completeness of about 87% (13/15) with however a large uncertainty due to the smallness of the sample.

Goldschmidt et al. (1992), Köhler et al.(1997) and La Franca & Cristiani (1997) have claimed that the PG survey is incomplete by a factor of 2 to 3, but their samples are quite small (5, 8 and 7 objects, respectively); in addition, we do not know if there is an offset between their magnitude scales and that of the PG survey (the mean differences between the PG and photoelectric B magnitudes for 190 stars is equal to -0.02^m). Therefore, their results should be considered as tentative.

5. Building a "complete" QSO survey based on APS O magnitudes

Because of their variability, it is an impossible task to compare directly two QSO surveys of the same region of the sky made at different epochs. However we now have, for a large fraction of the sky, the possibility to extract from the

APS database, for any object, the O magnitude as measured on the Palomar Sky Survey plates with an accuracy of about 0.2^m (Pennington et al. 1993). By doing this for all known QSOs found in the course of a number of different surveys of the same area of the sky, we may hope to reach as near as possible from an ideal survey complete to a well defined limiting magnitude.

We have extracted the APS O magnitudes, when available, for all objects in the QSO catalogue (Véron-Cetty & Véron 1998) brighter than $B = 17$, with $M_B < -24.0$ and $z < 2.15$, located in the $2\,400$ deg^2 of the FBS at $|b| > 30°$. Whenever this O magnitude exists, we give it the preference. There are 18 such QSOs with $O < 16.0$ (from a sample of 105 UV-excess stars, we have found that $\langle O - B \rangle = -0.17$ with a rms error of 0.25^m), and seven with $B < 16.2$. Thus our "complete" sample contains between 18 and 25 QSOs brighter than $B = 16.2$, or 0.0075 to 0.010 deg^{-2}, i.e. 1.2 to 1.6 times larger than the PG surface density. It should be possible, when the APS database will be completed, to get the O magnitudes of the seven objects for which they are not yet available.

Acknowledgments. A. M. Mickaelian is grateful to the CNRS for making possible his visit to OHP. A. C. Gonçalves acknowledges the *Fundação para a Ciência e a Tecnologia*, Portugal (PRAXIS XXI/BD/5117/95 PhD. grant).

References

Abrahamian, H. V., Lipovetsky, V. A. & Stepanian, J. A. 1990, Astrophysics, 32, 14

Condon, J. J., Cotton, W. D., Greisen, E. W. et al. 1998, AJ, 115, 1693

Goldschmidt, P., Miller, L., La Franca, F. & Cristiani, S. 1992, MNRAS, 256, 65P

Hagen, H.-J., Engels, D. & Reimers, D. 1998, A&AS, (in press)

Kellerman, K. I., Sramek, R., Schmidt, M., Shaffer, D. B. & Green, R. 1989, AJ, 98, 1195

Köhler, T., Groote, D., Reimers, D. & Wisotzki, L. 1997, A&A, 325, 502

La Franca, F. & Cristiani, S. 1997, AJ, 113, 1517

Markarian, B. E., Lipovetsky, V. A., Stepanian, J. A., Erastova, L. K. & Shapovalova A. I. 1989, Commun. Special Astrophys. Obs., 62, 5

Mermilliod, J.-C. & Mermilliod, M. 1994, Catalogue of mean UBV data on stars, Springer-Verlag

Mickaelian, A. M. 1994, *Discovery and Investigation of Blue Stellar Objects of the First Byurakan Survey*, Ph.D. thesis, Byurakan

Pennington, R. L., Humphreys, R. M., Odewahn, S. C., Zumach, W. & Thurmes, P. M. 1993, PASP, 105, 521

Schmidt, M. & Green, R.F. 1983, ApJ, 269, 352

Véron-Cetty, M.-P. & Véron, P. 1998, ESO Scientific Report N°18

Voges, W., Aschenbach, B., Boller, T. et al. 1996, IAU circ. 6420

White, N.E., Giommi, P. & Angelini, L. 1994, IAU Circ. 6100

Active Galactic Nuclei and Related Phenomena
IAU Symposium, Vol. 194, 1999
Yervant Terzian, Daniel Weedman, Edward Khachikian, eds.

2.5–11.6 μ Spectrophotometry and Imaging of AGNs

J. Clavel, B. Schulz, B. Altieri, P. Barr, P. Claes, A. Heras, K. Leech, L.Metcalfe, A. Salama

ISO Science Operations Centre, Astrophysics Division, ESA Space Science Dept.,P.O. Box 50727, 28080 Madrid, Spain

Abstract. We present low resolution spectrophotometric and imaging ISO observations of a sample of 58 AGN's over the 2.5–11.6 μ range. The data strongly support unification schemes and set new constraints on models of the molecular torus.

1. Introduction

According to the "unified model" of Active Galactic Nuclei (AGN), Seyfert 1 and Seyfert 2 galaxies (hereafter Sf1 and Sf2) are the same objects viewed at a different angle: Sf1's are observed close to face-on such that we have a direct view to the Broad emission Line Region (BLR) and the accretion disk responsible for the strong UV-Optical-X-ray continuum, whereas Sf2's are seen at an inclination such that our view is blocked by an optically thick dusty torus which surrounds the disk and the BLR (e.g. Antonucci 1993). This model makes specific predictions. In particular, the UV photons from the disk which are absorbed by the grains in the torus should be re-emitted as thermal radiation in the IR. Several arguments constrain the torus inner radius to be of the order of $\sim$ 1 pc in which case the dust temperature should peak to about 700–1000 K and give rise to an emission "bump" between $\simeq$ 2 and 15 μ (Pier and Krolik 1992). The model also predicts that the silicate 9.7 μ feature should appear preferentially in absorption in Sf2's and in emission in Sf1's. In order to test these predictions and better constrain the model, we initiated a programme of mid-IR observations of a large sample of AGN's. Throughout we use $H_0 = 75 \, \mathrm{km \, s^{-1} Mpc^{-1}}$; $q_0 = 0$.

2. Observations

A sample of 57 AGN and one non active "normal" SB galaxy was observed with the PHT and CAM instruments on board the Infrared Space Observatory (ISO; Kessler et al. 1996). The sample is drawn from the CfA hard X-ray flux limited complete sample but lacks the most well known objects (e.g. NGC 4151) which were embargoed by ISO guaranteed time owners. On the other hand, the sample was enriched in bright Sf2's. We caution that our sample is therefore not "complete" in a statistical sense. It is about equally divided into Sf1's (28 sources, including 2 QSO's) and Sf2's (29), where we define Sf1's as all objects of type $\leq$ 1.5 and Sf2's those whose type is $>$ 1.5. The mean and $r.m.s.$ redshift are 0.047 ± 0.083 and 0.016 ± 0.013 for Sf1's and Sf2's. For each object, the

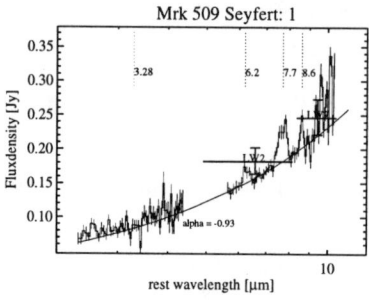 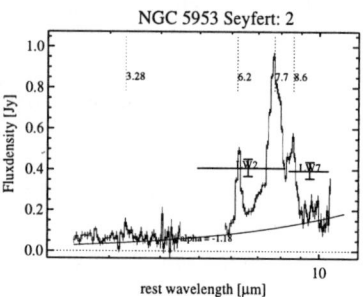

Figure 1. Two representative spectra of a Sf1 (left) and a Sf2 (right) with error bars. The two data-points marked LW2 and LW7 indicate the flux from the CAM images with its error and the filter wavelength range. The best-fit power-law continuum is also shown

data-set consists of CAM images obtained through filters at 6.75 and 9.63 μ, together with 2.5–11.6 μ spectra obtained immediately before with the PHT-S low resolution (3000 km s^{-1}) spectrograph. The images consists of arrays of 32 × 32 pixels (i.e. 96 × 96") with an effective resolution (FWHM) of 3.8" and 4.5", for the 6.75 and 9.63 μ filters, respectively. The exposure times were 200 s, sufficiently long to ensure stabilisation of the detectors. For the spectra, on-source measurements were alternated with sky measurements at a frequency of 1/256 Hz, with a chopper throw of 300". The on-source exposure times varied between 512 s and 2048 s, depending on the source brightness. The spectrograph aperture is 24" × 24".

3. Calibration and data reduction

The CAM images were reduced and calibrated using standard procedures of the CAM Interactive Analysis (CIA) package. Nuclear fluxes were obtained by integrating all the emission in a circle of 5–6 pixels radius (15–18"). Their accuracy, mainly limited by flat-fielding residuals, is ±10 %. Radial profiles were also computed and compared to that of point like calibration stars. For all sources but 4, the AGN are unresolved at the $\simeq$ 4" resolution of ISOCAM.

Because PHT-S was operating close to its sensitivity limit, special reduction and calibration procedures had to be applied. After a change of illumination, the responsivity of the Si:Ga photoconductors immediately jumps to an intermediate level. This initial jump is followed by a characteristic slow transient to the final level. At the faint flux limit, this time constant is extremely long, and in practice only the initial step is observed in chopped-mode. The spectral response function for this particular mode and flux-level was derived directly from observations of a faint standard star HD 132142 whose flux ranges from 0.15 to 2.54 Jy. The calibration star observation was performed with the same chopper frequency and readout-timing as the AGN observations. The S/N of the PHT-S spectra was considerably enhanced by two additional measures: i) the 32-s integration ramps were divided into sub-ramps of 2 sec and no de-glitching (removal of cosmic ray hits) was performed at ramp-level; ii) after slope-fitting

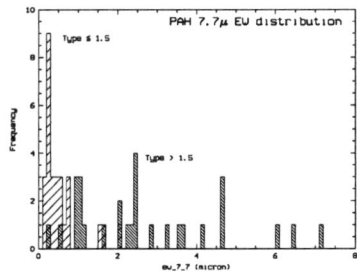 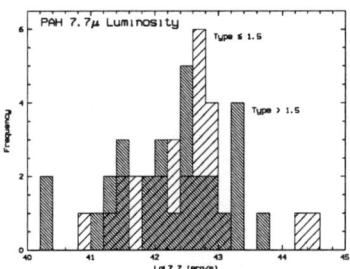

Figure 2. Distribution of PAH 7.7 μ EW (left) & luminosities (right) for Sf1 (spaced hatching at +45°) & Sf2 (fine hatching, -45°).

and de-glitching at slope-level, the maximum of the distribution of the slopes was determined by fitting a gaussian to the histogram. The resulting PHT-S fluxes should be accurate to within ±10 %. For all sources but 4, the agreement between the CAM and the PHT-S flux is excellent which confirms the reliability of our calibration. The 4 slight mismatches occur for the 4 spatially resolved sources.

4. The difference between Sf1 and Sf2

As illustrated in fig 1, the mid-IR spectrum of a typical Sf1 (Mrk 509, left) is markedly different from that of a Sf2 (NGC 5953, right): Sf1's have a strong continuum well approximated by a power-law ($F_\nu \propto \nu^\alpha$) of average index and r.m.s. dispersion $\langle \alpha \rangle = -0.84 \pm 0.24$ and weak emission features. By contrast, most Sf2's display only a weak continuum together with very strong emission features. These features have well defined peaks at at 6.2, 7.7 and 8.6 μ, usually ascribed to Polycyclic Aromatic Hydrocarbon (PAH) bands. In many galaxies of adequate S/N and redshift, the blue side of the strong 11.3 μ PAH feature is also detected as a sharp rise in flux toward the long wavelength end of the PHT-S array (see e.g. fig 3.b). The mean spectral index of the Sf2's, $\langle \alpha \rangle = -0.75 \pm 0.56$ is consistent with that of Sf1's. The EW distribution of the strongest PAH band at 7.7 μ is shown in fig 2.a. It clearly illustrates that PAH's are systematically weaker in Sf1's than in Sf2's where EW's extend up to $\sim 7\,\mu$. A two-tail KS test confirms that Sf1's and Sf2's have statistically different EW distributions at the 4×10^{-8} confidence level. The mean (±r.m.s) 7.7 μ PAH equivalent width of Sf1's is $\langle \mathrm{EW_{PAH}} \rangle = 0.53 \pm 0.47\,\mu$, 5 times smaller than that of Sf2's, $\langle \mathrm{EW_{PAH}} \rangle = 2.72 \pm 1.83\,\mu$. As can be seen from fig 2.b, the distribution of the 7.7 μ PAH *luminosity* is however the same in Sf1's and Sf2's, at the 52 % confidence level (KS test). The mean (±r.m.s) PAH luminosity of Sf1's is $\langle \log \mathrm{L_{PAH}} \rangle = 42.44 \pm 0.80$ erg s^{-1}, not statistically different (at the 26 % confidence level) from that of Sf2's, $\langle \log \mathrm{L_{PAH}} \rangle = 42.22 \pm 0.87$ erg s^{-1}.

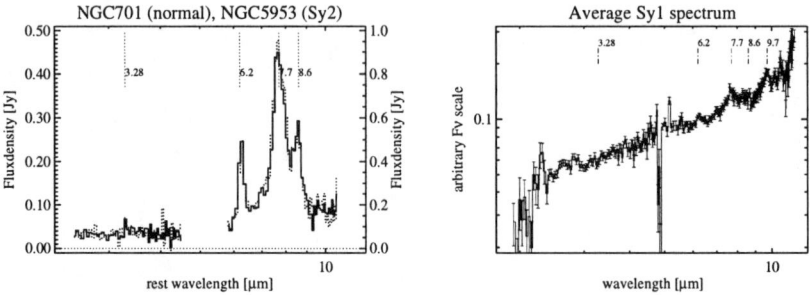

Figure 3. Left: the spectra of the non-active galaxy NGC 701 (heavy line) and of the Sf2 galaxy NGC 5953 (dotted-line). Right: The average Sf1 spectrum of the 28 type ≤ 1.5 galaxies in the sample

5. Implication for unification schemes

The continuum flux at a fiducial wavelength of 7 μ was read-out from the best-fit power-law. The resulting mean ($\pm r.m.s.$) continuum logarithmic luminiosities for Sf1's and Sf2's are $\langle \log \nu L_{\nu,7} \rangle = 43.73 \pm 0.86$ and 42.80 ± 0.79 erg s^{-1}, respectively, implying that the mid-IR continuum of Sf2's is $\sim$ 9 times less luminous, on the average, than that of Sf1's. Together with the difference in PAH EWs, these results imply that *the prime distinguishing feature of Sf2's in the mid-IR is that their continuum is depressed relative to that of Sf1's*. The above result is broadly consistent with unification schemes in that the mid-IR continuum, which is directly visible in face-on objects (i.e. Sf1's), is largely extinguished in edge-on objects (i.e. Sf2's). It further implies that the torus is opaque to its own mid-IR radiation. Assuming that face-on objects suffer zero extinction, one can use the average PAH EW ratio $\langle R \rangle$ in Sf1's and Sf2's to infer the mean Sf2 extinction. The average ratio for the strongest PAH band at 7.7 μ is $\langle R \rangle = 5.2 \pm 3.5$, where the error quoted reflects the r.m.s. dispersion of Sf2 EW's. This implies that the Sf2 continuum suffers on the average from 1.8 ± 0.7 magnitudes of extinction at 7.7 μ, i.e. $A_v = 89 \pm 37$ magnitudes (Rieke and Lebofsky 1985). For a normal gas to dust ratio, this corresponds to an average X-ray absorbing column, $N_H = 2.0 \pm 0.8 \times 10^{23}$ cm^{-2} (Gorenstein 1975). The latter is in good agreement with the mean Sf2 absorbing column as measured directly from X-ray data by Mulchaey et al. (1992), $N_H = 1.6^{+8.6}_{-1.3} \times 10^{23}$ cm^{-2} or Smith and Done (1996), $N_H = 1.0 \pm 1.3 \times 10^{23}$ cm^{-2}. The large dispersion in A_v and N_H presumably reflects the spread in Sf2 torus viewing angles, i.e. from grazing incidence to completely edge-on.

It also suggests that PAH emission is isotropic and arises from outside the torus, either in the Narrow Line Region or in the ISM of the bulge. To check the origin of the PAH features, we have observed the nucleus of the "normal" (i.e. non active) SB galaxy NGC 701. Its PHT-S spectrum is plotted in fig 3.a together with that of the Sf2 nucleus NGC 5953. The mid-IR spectrum of NGC 701 – fairly typical of a "normal" galaxy (Helou et al 1998) – is virtually indistinguishable from that of NGC 5953. As a matter of fact, the 7.7 μ PAH EW in NGC 701, $5.88 \pm 0.05 \mu$ and luminosity, $\log L_{PAH} = 42.048 \pm 0.009$ erg s^{-1}, are well within the range spanned by Sf2's. We therefore conclude that the *PAH emission is unrelated to the active nucleus and arise from the ISM in the bulge.* As can be seen from fig 3.a, the continuum of NGC 5953 is also indistinguishable from that of NGC 701. This further suggests that in Sf2's with 7.7 μ PAH EW's in excess of $\simeq 5\,\mu$, the torus emission is totally suppressed, at least up to 11.6 μ, and the faint residual mid-IR continuum arises entirely from outside the active nucleus, i.e. from stars shortward of $\sim 3\,\mu$ and from very small ISM dust particles at longer wavelengths. In such objects, the PAH EW only provides a lower limit to A_v and N_H. From the 29 Sf2's, only 3 have EW$_{PAH} \geq 5\,\mu$. This suggests that about 10 % of Sf2's suffer from extinction in excess of 125 visual magnitudes, sufficient to block-out the mid-IR continuum. These extreme Sf2's are presumably those where the torus symmetry axis lies in the plane of the sky.

Ten Sf2 galaxies have 7.7 μ PAH EW $\leq 1.6\,\mu$, in the range occupied by Sf1's (fig 2). Among these, four have been observed in spectropolarimetry and *all 4 display broad-lines in polarized light*. These are Mrk 3, NGC 7674, IRAS 05189-2524 and NGC 4388 (Heisler et al. 1996). Conversely, none of the three Sf2's with 7.7 μ PAH EW $> 1.6\,\mu$ for which spectropolarimetric or IR spectroscopic data exist (Mrk 266, NGC 5728, NGC 1097) have "hidden" broad lines. This confirms the suggestion by Heisler et al that those Sf2 which have "hidden" BLR (seen in spectropolarimetry or in direct IR spectroscopy) are those for which our line of sight (LOS) grazes the torus upper surface such that we can see the reflecting mirror but not the BLR. The mid-IR continuum most likely originates from thermal emission by hot dust grains located on the inner wall of the torus. Hence, the fact that the "hidden" BLR Sf2's are the same sources which display a Sf1-like mid-IR continuum further constrains the mirror and the torus inner wall to be in neighbouring regions. It is conceivable that the mirror is the wall itself or a wind of hot electrons boiled-off the torus. Mrk 3 has an X-ray absorbing column of $10^{24}\,\text{cm}^{-2}$ (Turner et al 1997) corresponding to an extinction of 125 magnitudes, sufficient to block the mid-IR radiation. The fact that it is nevertheless visible up to 11.6 μ indicates that the X-ray source is embedded further down the throat of the torus than the mirror, and the wall is emitting the mid-IR continuum.

6. The Silicate 9.7 μ features and further constraints on the torus

Fig. 3.b shows the weighted mean Sf1 spectrum obtained by averaging the rest wavelength spectra of all 28 type ≤ 1.5 AGN's. The Silicate 9.7 μ feature appears *in emission* with an equivalent width $\langle\text{EW}_{9.7}\rangle = 0.254 \pm 0.009\,\mu$. This immediately *rules out models with very large torus optical depths*. In the model of Pier and Krolik (1992) for instance, the strength of the Silicate feature is calculated as a function of inclination i and of the vertical and radial Thomson optical depth τ_z and τ_r, respectively. Reading from their figure 8, models with $\tau_z \geq 1$ and/or $\tau_r \geq 1$ are ruled-out as they would predict the Silicate feature in absorption. For an average Sf1 inclination $\cos i = 0.8$, the best fit to $\langle\text{EW}_{9.7}\rangle = 0.254 \pm 0.008\,\mu$ suggests $\tau_r \simeq 1$ and $0.1 \leq \tau_z \leq 1$. A unit Thomson optical depth corresponds to a column density $N_H \simeq 10^{24}\,\text{cm}^{-2}$. While these figures are somewhat model dependent, it is reassuring that they agree with our independent estimate of N_H based on the PAH EW ratio. The PAH bands are so strong in Sf2's that placing the continuum at 9.7 μ becomes a subjective decision. The mean Sf2 spectrum shows a weak maximum at 9.7 μ and a shallow minimum near 10 μ. In the absence of longer wavelength data, one can only set a provisional upper limit of 0.32 μ to the Silicate EW in Sf2's, whether in absorption or in emission.

References

Antonucci, R. 1993, ARA&A, 31, 473
Gorenstein, P. 1975, ApJ, 198, 95
Helou, G. et al. 1998, preprint
Heisler, C.A. et al. 1996, Nature, 385, 20
Kessler, M.F., Steinz, J.A., Anderegg, M.E., et al. 1996, A&A, 315, L27
Mulchaey, J.S., Mushotzky, R.F. & Weaver, K.A. 1992, ApJ, 390, L69
Oliva, E. et al 1998, A&A329, 210
Pier, E.A. & Krolik, J.H. 1992, ApJ, 401, 99
Rieke, G.H. & Lebofski, M.J. 1985, ApJ, 288, 618
Roche, P.F. et al. 1991, MNRAS, 248, 606.
Smith, D.A. and Done, C. 1996, MNRAS, 280, 355
Turner, T.J. et al 1997, ApJ488, 164

Active Galactic Nuclei and Related Phenomena
IAU Symposium, Vol. 194, 1999
Yervant Terzian, Daniel Weedman, Edward Khachikian, eds.

The Second Byurakan Survey: Faint Markarian Galaxies, AGN and QSOs

J. A. Stepanian

*Special Astrophysical Observatory RAS Nizhnij Arkhys,
Karachai-Cherkessia, 357147 Russia. e-mail: jstep@sao.ru*

Abstract. On the basis of the Second Byurakan Survey (SBS) we have produced the largest and most homogeneous new complete samples of faint Markarian Galaxies and bright QSOs. The volume of reliable investigation of UVX galaxies reaches to $\sim 200 - 250 Mpc$,- ten times deeper than the FBS.

New reliable data for the surface density of bright QSOs in mag. range $16.05 < B < 17.45$ and redshift $0.3 < z < 2.2$ are presented. The complete sample of bright$(B < 17^m5)$ SBS QSOs contains 93 QSOs in these ranges. The new data allow us to assume that bright QSOs fast evolution might be perhaps the result of a selection effect.

1. Introduction

The tremendous success of the Markarian survey has initiated a number of other extragalactic thin–prism surveys, initiatint a new direction in extragalactic astronomy–systematic searches for peculiar objects using low-dispersion spectroscopy. In 1974 we undertook the new survey – the Second Byurakan Survey(SBS). The primary goal of the SBS is to spread the Markarian survey as deep as possible to obtain a large, well–defined sample of AGNs and QSOs that are selected in a reasonably uniform fashion.

2. BYURAKAN SURVEYS (FBS and SBS).

The First Byurakan Survey (FBS). Begining in the mid–1960s and continuing through 1980, the first large-scale objective-prism survey for galaxies with blue and UVX in their continuum radiation was conducted by Markarian.

The 1500 Markarian galaxies contained in the FBS have provided the principal base from which the major types of AGNs have been discovered, classified, and studied in detail by numerous workers. FBS resulted in a complete sample of AGNs down to limiting magnitude 15^m2. Markarian galaxies comprise 10% of field galaxies and about 10% of Markarian galaxies turned out to be Sy galaxies, so 1% of field galaxies were found as Sy galaxies.

The Second Byurakan Survey (SBS). The SBS which is the continuation of the Markarian survey, aimed to reach fainter limiting magnitudes. First list of SBS objects was published in 1983 (Markarian & Stepanian 1983).Totally there are published seven lists (Stepanian 1994, and reference therein).

SBS started in 1974, and plate searching was completed in 1991. A total area of 1000 deg^2 confined by the contiguous strip defined by $7^h40^m < \alpha < 17^h15^m$, $+49° < \delta < +61°$ has been observed. The SBS is being conducted with the same 40-52 inch Schmidt telescope, but with various objective prisms in combination with more modern(in 1974)IIIaJ and IIIaF emulsions sensitized in heated nitrogen(Stepanian 1984). The use of both emulsions extends the wavelength range of sensitivity, increase the uniformity of discoveries, and permits the acquisition of spectra for AGNs down to $m_{pg} \sim 19^m5$. A selection of 1700 galaxies (~ 950 UVX and ~ 750 ELG without significant UV excess) and 1800 stellar objects with an excess ultraviolet emission is the main result of the SBS survey.

So far, the nature of 464 new QSOs and Sy galaxies, 830 galaxies with narrow emission lines and 810 galactic stars, the vast majority of which are hot WD, have been confirmed. Hundreds of new close-binaries, and pairs of faint UVX galaxies were found. The present status of SBS survey is briefly illustrated in table 1.

Table 1. The Second Byurakan Survey. Present (1998) status.

Spectroscopy	~ 2200	objects		photometry	~ 400
Stellar objects:	~ 1200	spectra;QSOs	- 355	stars	~ 810
BALQSO	- 15	Radio sources	- 110	WD	~ 300
DampQSO	- 5	X-ray sources	- 100	subdwarf	~ 200
Ly forest	- 15	Gamma-ray	- 2	HBB+NHB	~ 100
BLLac	- 5	BLAZAR	- 3	F/G	~ 100
Abs-Line QSOs	- 35	IRAS	- ?	Continual	- 25
Grav. lenses	- 3			C2+dMe	- 45
Other	- 278			CV	- 25
				Composite	- 15
Galaxies	~ 1000	slit spectra		Classified	~ 500
Seyfert gal.	- 109	SBN	- 170	IRAS sources	~ 520
Sy1	- 52	BCDG	- 130	Radio sources	~ 200
Sy1.5	- 8	HII	- 50	X-ray	~ 50
Sy2	- 49	ELG	- 480	Close-binaries	~ 150
LINERS	- 41	Abs.gal.	- 20		

The redshift distribution of SBS and Markarian galaxies, QSOs and AGN for total and complete ($B < 17^m5$) samples, as well as the plots of number-counts for FBS, SBS and field galaxies, and $LogN(<B) - B$ are shown in fig.1.

3. The results and discussion:

SBS galaxies. The compilation of the surface densities of SBS, some other samples and field galaxies are given in table 2.

The completeness level of SBS galaxies, as seen from fig.1, is about $\sim 90\%$ for $m \leq 16^m.5$ and $\sim 70\%$ for $m \leq 17^m.0$ galaxies.

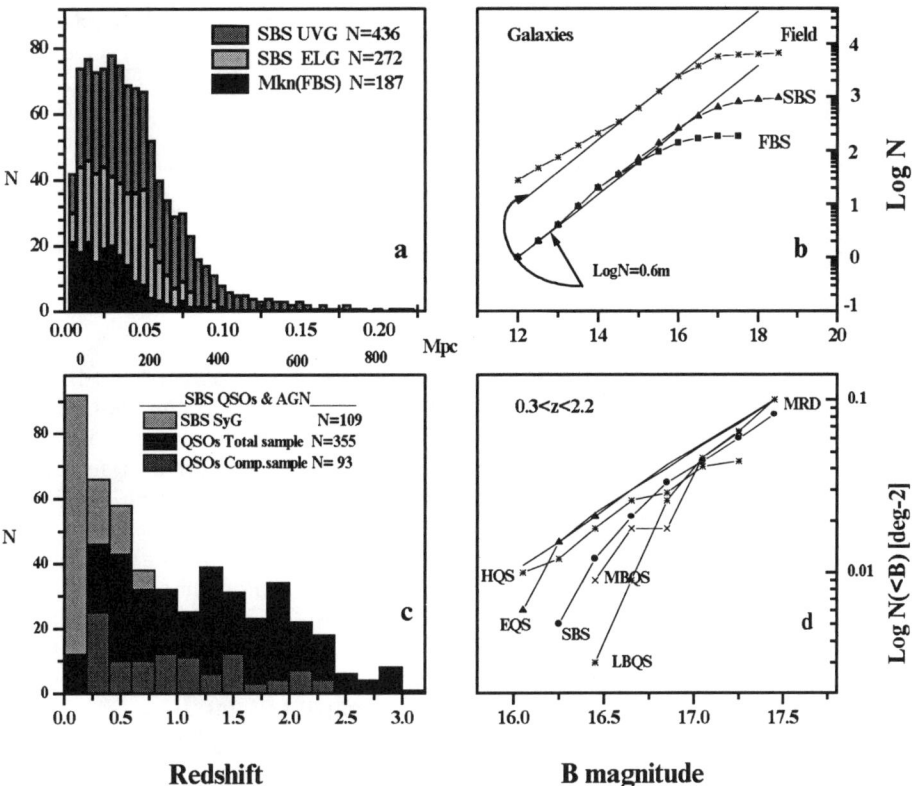

Figure 1. The redshift distribution of SBS and FBS galaxies(a); SBS QSOs and AGN, total and complete samples(c); the plots of number-counts for SBS and field galaxies(b); and $LogN(<B) - B$.(d).

Now it is well known, that the relative number of Sy and UVX galaxies among field galaxies consist of about 1% and 10% respectively. However, practically all these data were obtained by the use of the samples of nearby and accordingly bright objects in a space volume of about 100-120 Mpc. What is the relative number of Sy and UVX galaxies in a distance scale greater than 120 Mpc, is their nature, power, morphology changed?; what is the connection between the distant Sy galaxies and nearby QSOs?. Some of the mentioned questions may be answered on the base of SBS data.

1. The relative number of SBS AGN and the proportion of UVX galaxies among field galaxies in the volume of 200-250 Mpc consist of about 10-12%, which is the same as in FBS ($\sim$ 100 Mpc), but the number of QSOs and Liners is much higher.
2. About $\sim$ 30% of the total sample, and $\sim$ 50% of complete ($m < 16.5$) sample of SBS galaxies are IRAS sources. The proportion of IRAS galaxies among the field galaxies can be estimated no more than $\sim$ 30%. Radio sources among SBS galaxies consist of about $\sim$ 12%.

Table 2. The surface density of SBS, other samples and field galaxies.

Name of sur.	Area	Number	15.0	15.5	16.0	16.5	17.0	Compl.
FBS	17000	1500	0.09					15.2
Tololo	1225	201	0.16					15.0
Wasilewski	825	96	0.12					15.0
UM	667	349	0.12			0.52		16.5
Case	184	161	0.12	0.23	0.30	0.47	0.72	16.5
SBS	1000	1700	0.12	0.25	0.49	0.79	1.10	16.5
FIELD	1000	10000	0.6	1.3	2.5	5.0	10.0	

3. Perhaps no voids exist in the distant scale of about 200 Mpc in SBS area ($\sim 1000\ deg^2$). The data of hundreds of bright ($\leq 15^m5$) and thousands of faint ($m \leq 17^m0$) galaxies doesnt exist in any database. This conclusion is made on the basis of "direct" selection on SBS objective prism plates. Field galaxies in SBS area consist of $\sim 5000 (m < 16.5)$ and $\sim 10000 (m < 17.0)$ galaxies. The redshifts are aviable for about $\sim 2500 (m < 16.5)$ of them, the main part of which is SBS data. This data makes the size of known voids much smaller ($< 20 Mpc$), which allow us to predict, that after obtaining redshifts for the remainder ~ 2500 ($m < 16.5$) galaxies, the voids may be filled.

4. The complete sample of faint Markarian galaxies from SBS gives a possibility to construct the first complete sample of faint $B \leq 17^m5$ AGN, to increase the volume of reliable investigation to 200-250 Mpc (ten times deeper than in FBS), and investigate the Large-scale structure of the Universe in a distant scale of about 400-500 Mpc.

Bright QSOs and the problem of completeness. The zero point of LogN-B is principal for any quasar evolution model. In the last years a number of papers have been published suggesting that the standard two power-law model of a pure luminosity evolution may not be an adequate representation of the data. The very steep number-magnitude relation which is important evidence for cosmological evolution of the QSO population comes mainly from BQS(Schmidt & Green 1983), and AAT survey(Boyle at al.1990) data. There is a significant difference between the published data for the bright end.It is worth noting that many workers transform the original Bj magnitudes to the standard Johnson B band by assuming B-V=0.3 for the QSOs and applying a mean transformation using the Blair & Gilmore color equation Bj=B-0.28(B-V) for stars.

To investigate this question, we undertake direct photometric CCD measurement for bright($b < 17^m5, 0.3 < z < 2.2$) QSOs in complete samples of surveys, the data of which as a rule are used for the bright end of $Log N(< B) - B$, and evolution models of QSOs. There are only a few of them: -BQS(Schmidt & Green 1983), MBQS(Mitchell et al.1984), LBQS(Hewett et al. 1995), and the new survey EQS(Goldshmidt et al. 1992). The BQS is well investigated, so we study the SBS and other three.

There are 8 objects in EQS, 26 in MBQS, 80 in LBQS and 93 in SBS. For SBS objects part of the photometic data are published (Chavushian et al. 1995), the next part including the LBQS, EQS and MBQS data are in preparation. The number of objects, the original and corrected data for integral surface density and the most reliable data(MRD) for bright ($B < 17.45$) QSOs in the redshift range $0.3 < z < 2.2$ are given in table 3. The CCD data of the recent survey HQS(Kohler et.al 1997) are also presented.

Table 3. The corrected integral surface density of bright QSOs.
$0.3 < z < 2.2$ $16.05 < B < 17.45$.

< B	EQS		MBQS		LBQS		HQS	SBS	MRD
	n orig.	corr.	n orig.	corr.	n orig.	corr.	n CCD	n CCD	
16.05	4 0.012	2 0.006	1 0.009				6 0.01	3 0.003	0.011
16.25	6 0.018	5 0.015	1 0.009				7 0.012	5 0.005	0.015
16.45	8 0.024	7 0.021	2 0.018	1 0.009	2 0.004	1 0.003	11 0.018	12 0.012	0.021
16.65			3 0.027	2 0.018	6 0.013	4 0.009	16 0.026	21 0.021	0.030
16.85			4 0.037	2 0.018	14 0.031	12 0.026	18 0.029	33 0.033	0.040
17.05			6 0.055	5 0.046	25 0.055	21 0.046	25 0.041	44 0.044	0.055
17.25			8 0.073	7 0.065	39 0.086	31 0.068	27 0.044	60 0.060	0.073
17.45					73 0.161	49 0.11		82 0.082	0.1

1. The "converted" B magnitudes used in EQS, MBQS and LBQS as a rule are brighter than standard Johnsons B magnitudes.
2. The gradient of LogN-B decrease from 0.98 to $\sim 0.67 \pm 0.04$. There is a flattening of the bright part of the QSO LF at low redshift. The models of the optical luminosity function for luminous QSOs need a smaller amount of cosmological evolution. The Baldwin effect clearly becomes to zero.

Acknowledgments. This work was supported by the research grants No. 97-02-17168 and 1.2.2.2 from the Russian Foundation for Basic Research and from State Program "Astronomy" respectively.

References

Boyle,B.J.,Fong,R.,Shanks,T., Peterson,B.A.1990, MNRAS, 243, 1

Chavushian,V.H., Stepanian,J.A., Balayan,S.K., Vlasyuk,V.V.
    1995, Astron.Letters. 21, 894

Goldshmidt,P, Miller L, La Franca,F,Cristiani, S.1992, MNRAS,256,65

Kohler,T., Grote,D., Reimers,D. & Wisotski.L. 1997, A&A.325,502

Hewett,P.C., Folz,C.B., Chaffee,F.H. 1995, AJ, 109, 1498

Mitchell K.J., Warnock A., Usher P.D. 1984, ApJ, 287, L3

Markarian,B.E., Stepanian,J.A. 1983, Astrofizika, 19, 639

Stepanian, J.A., 1994, Doctoral dissertation, Nizhnij Arkhys

Stepanian,J.A., Astron. Journal, XX, 1985, 62, 1211

Schmidt, M.R, Green, R.F. 1983, ApJ, 269, 352

Active Galactic Nuclei and Related Phenomena
IAU Symposium, Vol. 194, 1999
Yervant Terzian, Daniel Weedman, Edward Khachikian, eds.

Investigation of a New Sample of $IRAS$ Galaxies

A. M. Mickaelian, S. A. Hakopian, S. K. Balayan

Byurakan Astrophysical Observatory, Byurakan 378433, Republic of Armenia. E-mail: aregmick@bao.sci.am, susanaha@bao.sci.am, bal@moon.yerphi.am

Abstract.

A new sample of some 900 faint $IRAS$ galaxies will be constructed after optical identifications of all $IRAS$ point sources from PSC and FSC in an area with 1 500 deg^2 at high galactic latitudes. The identifications are being made on the basis of FBS low-dispersion spectra, DSS images and infrared colours. Some 320 $IRAS$ sources have been identified and 180 galaxies have been found. Spectral observations of these objects revealed new AGN and luminous infrared galaxies.

1. Introduction

More than 250 000 infrared point sources have been detected by the Infrared Astronomical Satellite ($IRAS$) at 12μ, 25μ, 60μ and 100μ bands. They are cataloged in the $IRAS$ Point Source Catalog (PSC) and the $IRAS$ Faint Source Catalog (FSC). Systematic cross-identifications have been made for these sources with the main available catalogs, including catalogs of bright stars, galaxies, QSOs, variable stars, planetary nebulae, late-type stars, radio sources, UVX objects, emission stars, as well as radio and other IR catalogs. The associations are made by coincidence (or proximity) of the coordinates (depending on the coordinate accuracy of the given catalog). These "blind" identifications result in cases, where several objects (stars, galaxies, etc.) are identified with a single IR source. Moreover, half of the $IRAS$ sources remain without any identification yet and their physical nature is unknown.

2. Construction of samples and optical identification of IR sources

In order to make possible further study of the $IRAS$ sources, it is necessary to identify their optical counterparts. Various samples of $IRAS$ galaxies have been constructed. As it appeared, the $IRAS$ galaxies make up a rather interesting population, including AGNs, luminous and ultraluminous IR galaxies, which are important for the study of star-formation phenomena.

Beginning with the first published data of the $IRAS$ surveys, works on construction of various samples of $IRAS$ galaxies began. Lonsdale & Helou (1985) published the catalog of known QSOs and galaxies which were associated with $IRAS$ sources. On the basis of the cross-identifications the $IRAS$ Bright

Galaxy Sample (*BGS*) of 330 galaxies was made up by Soifer et al (1987, 1989 and other papers). On the basis of the *IRASPSC*, Spinoglio & Malkan (1989) and on the basis of the *IRASFSC*, Rush et al (1993) have constructed a 12μ galaxy sample. Rowan-Robinson et al (1991) constructed an *IRAS* 60μ galaxy catalog, covering 82% of the sky. Strauss et al (1990) and Fisher et al (1995, and other papers of the series) on the basis of the *IRAS* data constructed a complete all-sky sample of objects brighter than 1.2 Jy at 60μ. The sample consists of 5014 objects, including 2649 galaxies. More than 20 different samples of *IRAS* galaxies are known already.

As the IR data are very useful for investigation of objects, there are many works on optical identifications of *IRAS* sources in different interested regions (e.g. star clusters), galaxy samples, etc. Some works are devoted to identifications of all the *IRAS* sources in limited regions.

There are numerous works on optical identifications of *IRAS* sources in separate regions. Johnson & Klemola (1987) investigated 56 unidentified *IRAS* point sources in the Draco cloud region. They found probable counterparts inside the error-ellipse fields, including both galaxies and stars. Sutherland et al 1991, using the APM Galaxy Survey, have generated a collection of finding charts for 4614 sources with non-stellar colors in the *IRASFSC* south of $\delta=-17.5°$ and most of them were identified with galaxies. A program of radio/IR identifications was carried out by Condon et al (1995). By the coincidence of the radio and IR coordinates for all the *FSC* sources with $F_{60} > F_{12}$ they constructed a sample of 354 probable extragalactic sources. Using the optical data these authors looked for the optical counterparts of these radio-loud FIR galaxies and QSOs.

The *OPTID* database (optical identifications for *IRAS* sources) also gives numerous possible optical counterparts for each *FSC* source (Conrow et al), but the real identifications are not made. This database is better for use in the Southern Hemisphere, as in the Northern Hemisphere, only objects with optical magnitudes $<15^m$ are given and the main part of galaxies are not included.

3. The FBS database and the program of mass-identification of IRAS sources

The largest objective prism survey is the First Byurakan Survey (*FBS*), carried out with the Byurakan 1-m Schmidt telescope and 1.5° objective prism by Markarian et al (1989). It consists of 1133 4°×4° fields and covers an area of more than 17 000 deg² at high galactic latitudes ($|b|>15°$). All the Northern Hemisphere and part of the Southern Hemisphere ($\delta>-15°$) is covered at these galactic latitudes. The dispersion is 1800Å/mm near H_γ and the spectral range is 3400-6900Å, so that one can notice some absorption and emission lines (such as Balmer lines, molecular bands, He, N_1+N_2, broad emission lines of QSOs etc.), follow the spectral distribution and compare the blue and red parts of the spectrum. These possibilities give a chance of recognition of the types of objects and their preliminary rough classification. Since 1987 surveys of blue stellar objects and red stars are being carried out on the basis of this observational material (Abrahamian et al 1990, Mickaelian 1994, and references therein).

For cosmological studies it is rather desirable to have a complete sample of optical counterparts of $IRAS$ sources at high galactic latitudes. Hence, an optical identification program of all IRAS sources (mass identifications) on a large area is required. A program of mass optical identifications of IRAS sources on the basis of low-dispersion spectra is conducted for the first time since 1994 (Mickaelian 1997). The area with $+61° < \delta < +90°$ (at galactic latitudes $|b| > 15°$) was chosen with a total surface of 1 500 deg^2, where the FBS plates have deeper limiting magnitudes than the average. Positions of all the $IRAS$ sources are examined on the $POSS$ O and E charts, FBS plates and DSS images. IR colors provide additional information on the possible nature of the optical counterpart, but the latter is selected by means of recognition of the definite object, which can be an IR source. The ultimate selection of the optical counterpart is made on the basis of many parameters, including the IR colors, optical images, optical magnitudes and colors, and of course the low-dispersion spectra. These objects are not numerous among the whole stellar population, and in 90% of cases the selection is made confidently.

4. Building a new sample of IRAS galaxies

During the past 3 years, the following results have been obtained. All the unidentified $IRAS$ point sources for the selected region have been extracted from the $IRASPSC$ and FSC. They are checked for associations by the NED and $SIMBAD$ databases. It appears that there are several known optical objects (mainly stars) near some of the selected sources. So it is an additional task to clarify why these objects have not been associated with the IRAS sources even in the updated versions of the catalogs. In all 1517 $IRASPSC$ and FSC sources are included in the identification program.

At present optical identifications have been made for 317 IR point sources out of 341 in an area of 382 deg^2 (Mickaelian 1997, Mickaelian & Gigoyan 1998, and references therein). Optical coordinates, deviations from the $IRAS$ positions are determined, V magnitudes, $B - V$ colors are estimated, and rough classification is made for all the objects. There are 133 late-type stars, 7 planetary nebulae, 8 QSO candidates, 140 galaxies, and 29 multiple galaxies and small groups among the identified optical counterparts. There is no optical counterpart near the positions of 27 sources even in the DSS, and taking into account their IR colors are typical for galaxies, they must be very faint galaxies in optical wavelengths. QSO candidates are selected by their low-dispersion spectra, but the QSO-like spectral distribution is approximate.

The identified galaxies can be considered as the most interesting counterparts of the IR sources. For them angular sizes and position angles are determined, morphological classification, and comments on structural and environmental anomalies, as well as on low-dispersion spectra (UV-excess, emission lines etc.) are made in addition to the above mentioned main parameters. On the basis of their low-dispersion spectra and compact structure, 21 are suspected to be Sy galaxies. 30 more have also compact and bright bulges with very faint peripheries, 2 galaxies have bright HII regions in their spiral arms, 3 have double and multiple nuclei, and $IRAS\ 16102+6345$ has a jet-like feature. Multiple galaxies have various appearances and structures. In many cases a galaxy can

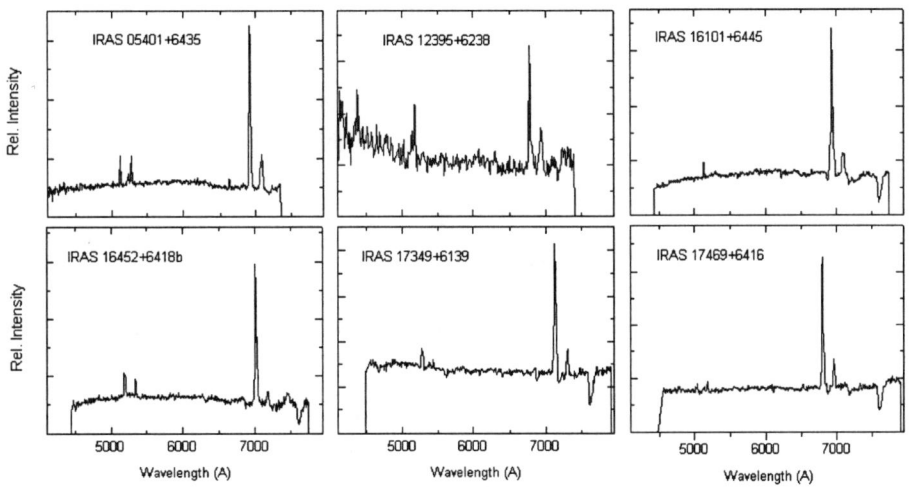

Figure 1. Slit spectra of 6 IRAS galaxies

be considered as a satellite of another, brighter one. Interaction features (tails, bridges) are present as well. Further identifications will complete the sample of *IRAS* galaxies in the region.

5. Observations and classification

The purposes of spectral observations are the investigation of spatial distribution of *IRAS* galaxies and the IR luminosity function. Besides, we will distinguish the most interesting objects (AGN, high luminosity IR galaxies, and composite spectra objects) and study them in more detail.

Observations have been made during 1997 and 1998 at the 6-m telescope of the Russian Special Astrophysical Observatory (Mickaelian et al 1998, Balayan et al 1999). The long-slit spectrograph UAGS at the prime focus with a 580×530 CCD is used (Afanasiev et al 1995). The dispersion is 5.8 Å/px and the spectral range 4000-8000ÅSpectra of 42 galaxies (including multiple ones), responsible for 33 *PSC* sources, have been obtained and reduced. Redshifts for all observed galaxies are measured, and their activity types are estimated, based mainly on well-known emission-line relations (Baldwin et al 1981, Veilleux & Osterbrock 1987). The redshifts are in the range $0.02<z<0.09$, calculated infrared luminosity is in the range $10^{10}<L_{ir}<10^{12}$. Within the scheme constructed of HII, LINER, Composite AGN and Sy2 type classes, most objects are of HII nature, 3 are LINERs and 2 are of Composite AGN nature, namely *IRAS 16118+6231* and *IRAS 12395+6238* (Veron et al 1997). Some examples of spectra of the observed objects are given in Fig. 1.

The observations and data reduction are being continued.

6. Conclusions

The optical identification program will produce a new $IRAS$ galaxy sample in a region of $1\,500$ deg^2. It will include all PSC and FSC sources having extragalactic optical counterparts. After the fulfillment of the program of identifications, the following studies are planned:

1) study of the sample contents. In particular the availability of optical and IR (at 4 bands) data for many galaxies will help to clarify the classification principles as well (relation between IR luminosity and activity types),

2) detailed optical study (morphology: study of interactions, fine structure of central regions; and spectroscopy: determination of redshifts and activity classes) of the newly identified galaxies and groups of galaxies,

3) radio millimeter observations,

4) overall statistical study of the area, including previously identified $IRAS$ sources.

5) comparison with the existing samples of $IRAS$ galaxies and their statistical conclusions on space density, evolution, star-formation rate and status of the IR galaxies.

We hope to obtain the real distribution pattern of extragalactic IR sources in the local Universe. The large amount of interesting objects (active galaxies, interacting systems, QSOs) among the $IRAS$ sources allow important investigations in this field, as their study brings understanding of evolutionary phenomena and processes taking place in galaxies.

Acknowledgments. A.M.Mickaelian is grateful to P.Veron for useful discussions and invitation to Observatoire de Haute-Provence, which helped to fulfill an important part of this work.

References

Abrahamian H. V., Lipovetski V. A., Stepanian J. A., 1990, Astrofizika 32, 29

Afanasiev V. L., Burenkov A. N., Vlasyuk V. V., Drabek S. U., SAO technical report 1995, V. 234

Baldwin J., Philips M., Terlevich R., 1981, PASP 93, 5

Condon J. J., Anderson E., Broderick J. J., 1995, AJ, 109, 2318

Conrow T., Lonsdale C., Evans T., Fullmer L., Moshir M., Yentis D., Wolstencroft R., MacGillivray H., Egret D., Chester T., The Faint Source Survey Optical Identification Database (OPTID), at IPAC

Fisher K. B., Huchra J. P., Strauss M. A., Davis M., Yahil A., and Schlegel D., 1995, ApJS 100, 69

Balayan S. K., Hakopian S. A., Mickaelian A. M., Burenkov A. N., 1999, Pisma v AZh, in press

Johnson H. M., Klemola A. R., 1987, ApJ S 63, 701

Lonsdale C. J., Helou G., 1985, Catalogued Quasars and Galaxies from the IRAS Survey

Markarian B. E., Lipovetski V. A., Stepanian J. A., Erastova L. K., Shapovalova A. I., 1989, Commun. Special Astrophys. Obs., 62, 5

Mickaelian A. M., Discovery and Investigation of Blue Stellar Objects of the First Byurakan Survey, Ph. D. thesis, Byurakan, 1994

Mickaelian A. M., 1997, Astrofizika 40, 1

Mickaelian A. M., Gigoyan K. S., 1998b, Astrofizika 41, in press

Mickaelian A. M., Hakopian S. A., Balayan S. K., Burenkov A. N., 1998, Pisma A.Zh., 24, 1

Osterbrock D. E., Tran H. D., Veilleux S., 1992, ApJ 389, 196

Rush B., Malkan M. A., Spinoglio L., 1993, ApJS 89, 1

Rowan-Robinson M., Saunders W., Lawrence A., Leech K., 1991, MNRAS 253, 485

Soifer B. T., Sanders D. B., Madore B. F., Neugebauer G., Danielson G. E., Elias J. H., Lonsdale C. J., Rice W. L., 1987, ApJ 320, 238

Soifer B. T., Boehmer L., Neugebauer G., Sanders D. B., 1989, AJ 98, 766

Spinoglio L., Malkan M. A., 1989, ApJ 342, 83

Strauss M. A., Davis M., Yahil A., Huchra J. P., 1990, ApJ 361, 49

Sutherland W. J., McMahon R. G., Maddox S. J., Loveday J., Saunders W., 1991, MNRAS, 248, 483

Veilleux S., Osterbrock D. E., 1987, ApJS 63, 295

Veron P., Goncalves A. C., Veron-Cetty M.-P., 1997, A&Ap 1997, 319, 52

Active Galactic Nuclei and Related Phenomena
IAU Symposium, Vol. 194, 1999
Yervant Terzian, Daniel Weedman, Edward Khachikian, eds.

Galaxies in Selected Fields of the Second Byurakan Survey

S. A. Hakopian, S. K. Balayan

Byurakan Astrophysical Observatory, Byurakan 378433, Armenia

Abstract.
The current state of investigation of galaxies in seven fields of the Second Byurakan Sky Survey (SBS) is presented. These fields have been selected by the results of completeness estimation of the samples of galaxies in 65 fields.

Observations of the SBS faint candidate galaxies are carried out to complete spectroscopy of galaxies in the selected fields. Currently in one SBS field, with coordinates of center $\alpha=15^h30^m$ and $\delta=+59°$, the spectra of all galaxies have been obtained and reduced. Besides the redshift and spectral classification, these data allow estimates of the quality of object selection in the Second Byurakan Survey at faint magnitudes.

1. Introduction

Due to an improved observational technique (Stepanian 1994a), the Second Byurakan Sky Survey (Markarian et al 1983, Markarian & Stepanian 1983) extends a small part (1 000 deg^2) of the First Byurakan Survey (Markarian 1967) to deeper magnitudes, up to 19^m-19.5^m. In particular, differently from the First Byurakan Survey, during the Second Survey three objective prisms, namely 1.5, 3, and 4 degrees have been used. In total, photographic plates for 65 fields, each 16 deg^2, have been obtained with the 1.5° prism. Only ten of these fields have been observed with the two other prisms also (Stepanian 1994b).

The relative uniformity of samples of objects selected by eye inspection of photographic plates of the survey fluctuates from field to field. It depends mainly on the number of plates obtained in the field and the limiting magnitudes. For the subsequent more detailed researches of samples of galaxies, we selected seven fields, for which:
  - photographic plates have been obtained with the three objective prisms;
  - the limiting value 19^m-19.5^m has been reached;
  - large amount of follow-up slit spectroscopy has been carried out.

2. The samples of galaxies in selected SBS fields

Among 3 000 objects selected and catalogued during the SBS (Stepanian 1994a) more than 1 200 are galaxies with UV-excess, emission features and other peculiarities.

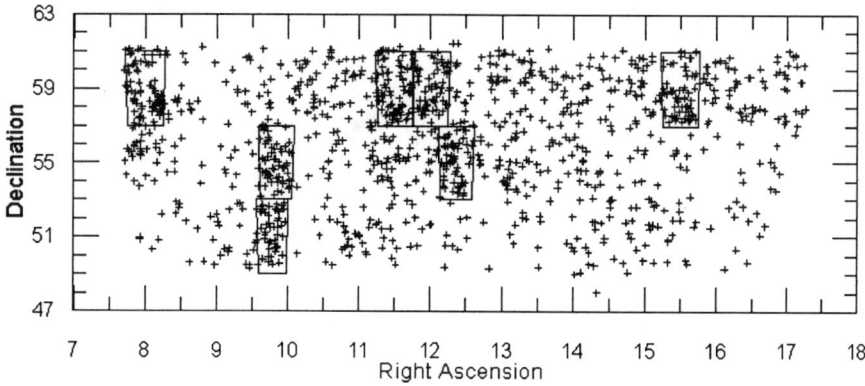

Figure 1. Distribution of SBS galaxies in the sky and location of seven selected fields.

Fig.1 illustrates the distribution of all SBS galaxies in the sky and location of the seven fields. More precise coordinates of the field centers are given in the second column of Table 1. The next two columns show accordingly the number of galaxies selected in each sample and those with known redshift.

Table 1. Main data on the selected fields

N	R.A. (1950)	Dec.	Number of Galaxies	With known redshift	m_{comp} (of V/V_{max} test)
1	08^h00^m	$+59°$	88	74	16.0
2	09^h47^m	$+51°$	54	38	18.0
3	09^h50^m	$+55°$	67	48	16.5
4	11^h30^m	$+59°$	96	81	15.5
5	12^h00^m	$+59°$	72	20	18.0
6	12^h22^m	$+55°$	74	49	17.5
7	15^h30^m	$+59°$	68	68	18.0
Total	112 deg²		519	378	16.0

For the given numbers of galaxies we applied the classical V/V_{max} test. In the last column of the Table 1 the values of completeness level are given for each sample of galaxies and their total. Fig. 4c shows the distribution of available apparent magnitudes, used during the calculations, defined in most cases with an accuracy of 0.5^m.

The areas of the seven fields together compose only a tenth of the SBS. At the same time more than forty percent of the whole sample of SBS galaxies, approximately 500, have been selected in this part. We note that only one small cluster of galaxies, Abell634, is in the indicated bounds but not reflected in our sample. All of the above about the good quality of photographic plates and

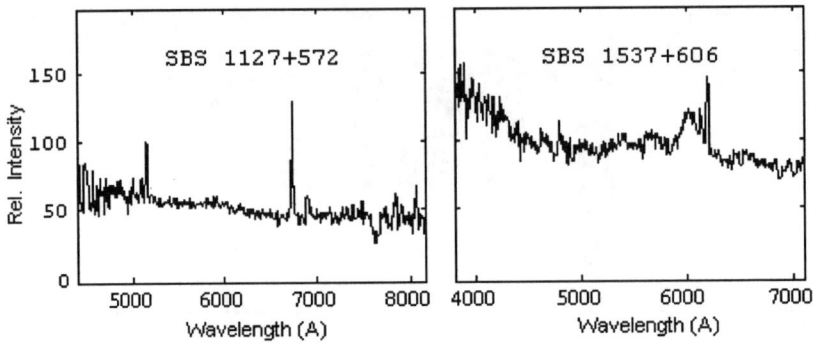

Figure 2. a) The first CCD spectra obtained at the Byurakan 2.6-m telescope. b) Spectra of the Sy1 galaxy obtained at the Russian 6-m telescope.

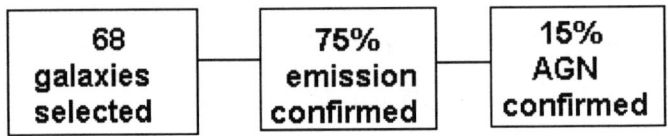

Figure 3. Schematic illustration of the sample content.

comparatively high percent of selected galaxies in the seven fields shows that their complete slit spectroscopy will help to clarify potentialities of the Survey.

3. Spectral observations in the seventh field

We carried out spectral observations from 1996 trying to obtain missing slit spectra for galaxies in the selected fields at the 6-m telescope of the Russian Special Astrophysical Observatory, and from March 1998 with the 2.6-m telescope of the Byurakan Observatory. One of the first CCD spectra obtained with the 2.6-m telescope with a grism is illustrated in Fig. 2a.

Up to now we have obtained 90 spectra of the 70 faintest galaxies in the samples with the longslit spectrograph at the 6-m telescope (Hakopian & Balayan 1997, Balayan & Hakopian 1999). By adding 35 of them to the previous spectroscopy (Stepanian et al 1993a, 1993b), the field numbered as seventh became the first field of the SBS with complete slit spectroscopy of its galaxies. The analysis of data obtained for this field sample is especially interesting, because it is one of the three with the highest level of completeness (see Table 1).

The follow-up spectroscopy confirms the emission nature of approximately 75% of this subsample of galaxies, with 15% of AGN galaxies among them. Fig. 3 schematically illustrates these results.

One of the objects, *SBS 1537+606* (Fig. 2b), turned out to be one of the most distant Sy1 type galaxies in the Survey, with redshift $z = 0.2378$. In all there are seven AGNs in the seventh field - four of Sy1 and three of Sy2 type.

The most probable values of the Survey objects' redshifts, to be confirmed by follow-up observations, can be estimated, knowing the limits of spectral ranges obtained with combinations of prisms and photographic emulsions used (Stepanian 1994b), and the main criteria of object selection. In particular, all objects with redshifts up to 0.05, that is 200 Mpc at a Hubble constant of 75 $\mathrm{km\,s^{-1}\,Mpc^{-1}}$, could be selected by eye inspection, if even one of the emission-lines of hydrogen H_α, H_β and/or oxygen 5007Å (N_1), is noticeable on the low-dispersion spectra.

New slit spectra of the faintest galaxies have revealed a number of absorption-line galaxies. According to Markarian et al (1986), all these objects were selected with the erroneous assumption to have one of the above mentioned emission lines. Figure 4a shows the redshift distribution of all galaxies in the seventh field, including those with absorption lines (shaded).

4. Current results

Figure 4b shows the distribution of all known redshifts of galaxies in selected fields, including the seventh. Their apparent magnitudes are given by the unshaded part of the histogram, presented by Figure 4c. Figure 4d shows the distribution of absolute magnitudes, calculated with value of the Hubble constant 75 $\mathrm{km\,s^{-1}\,Mpc^{-1}}$. A comparison of distributions given can be done also with those for the whole sample of *SBS* galaxies, using published data (e.g. Carrasco et al, 1997).

We found counterparts in different spectral ranges for about 120 galaxies in selected fields, using mainly the *CATS* database (Verkhodanov et al 1997). Here we shall dwell only on identifications of galaxies of AGN nature, namely 26 galaxies of Sy type.

There are 14 Sy1 and 12 Sy2 type galaxies in seven fields. All of the Sy2 objects radiate in radio and some of them in IR ranges. Eight Sy1 objects have counterparts in the *X-ray* range and some of them in other spectral ranges, too. It is interesting that the remaining 6 Sy1 galaxies, which have no counterparts in any other spectral ranges, are the most distant in this subsample. It seems that there is some threshold at redshifts about 0.12, after which our Sy1 galaxies are not discovered in other parts of the spectrum. Taking into w account that such a distance effect is apparent only for a small sample, and could be affected by the lack of spatial resolution of *X-ray* telescopes, we have to learn more about their morphology and space orientation for more comprehensive conclusions.

5. Summary

We have completed slit spectroscopy of galaxies in one of the seven fields of the Second Byurakan Survey, selected for further investigations. The faint tail has been observed for the first time, enabling the evaluation of quality of object selection at limiting magnitudes of the Survey. The relatively high level of completeness confirms the comparison of data obtained for three subsamples of

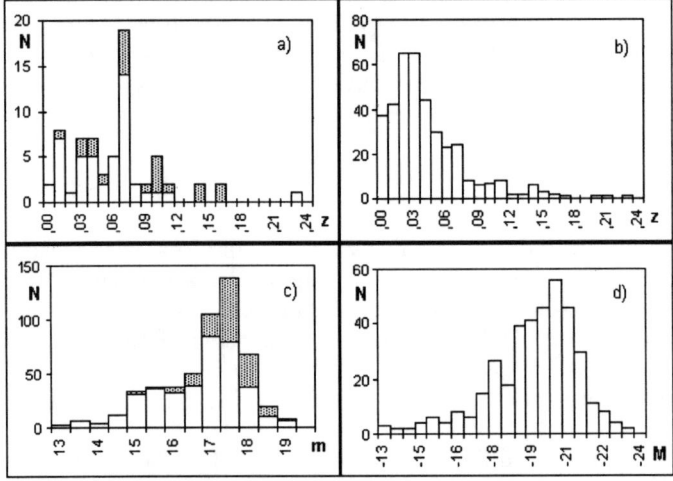

Figure 4. a) Redshift distribution for 68 galaxies in the seventh field; absorption-line galaxies are shaded, b) Histogram of known redshifts of 378 galaxies in selected fields. c) The distribution of apparent magnitudes of all 519 galaxies selected in seven fields; those with known z are unshaded. d) The distribution of absolute magnitudes of 378 galaxies in selected fields.

SBS galaxies - in the seventh field, in the seven fields, and in the whole SBS area. The value of the redshift distribution peak of the seventh field sample, shifted as compared with two others, and the unusually high number of Sy type galaxies ($0.4 per sq.deg.$) can be considered as an upper limit of the potentialities of the SBS.

Acknowledgments. The authors are very thankful for useful advice to Dr. J. A. Stepanian, and for discussions with Dr. S. N. Dodonov and Prof. E. Ye. Khachikian.

References

Balayan S. K., & Hakopian S. A. 1999, Astrofizika, in press

Carrasco, L., Serrano, A., Tovmassian, H. M., Stepanian, J. A., Chavushian, V. H., & Erastova, L. K. 1997, AJ, 113, 1527

Hakopian S. A., & Balayan S. K. 1997, Astrofizika, 40, 169

Markarian, B. E. 1967, Astrofizika, 3, 55

Markarian, B. E., Lipovetsky, V. A., & Stepanian, J. A. 1983, Astrofizika, 18,29

Markarian, B. E., & Stepanian, J. A. 1983, Astrofizika, 19, 639

Markarian B. E., Stepanian, J. A, & Erastova L. K. 1986, Astrofizika, 25, 345

Stepanian, J. A., Lipovetsky, V. A., Erastova, L. K., Shapovalova, A. I., & Hakopian, S. A. 1993a, Bull. Spec. Astrophys. Obs., (Izv. SAO), 35, 24

Stepanian, J. A., Lipovetsky, V. A., Erastova, L. K., & Hakopian, S. A., Izotov, Yu. I., & Guseva, N. G. 1993b, Bull. Spec. Astrophys. Obs., (Izv. SAO), 35, 38

Stepanian, J. A. 1994a, Doctoral dissertation, Nizhnij Arkhys

Stepanian, J. A. 1994b, H. T.MacGillivray et al.(eds), Astronomy from Wide-Field Imaging, p.731

Verkhodanov, O. V., Trushkin S. A., Andernach H., & Chernenkov V. N. 1997, The CATS database to operate with astrophysical catalogs.In Proc. of the "Astronomical Data Analysis Software and Systems VI", ed. G.Hunt & H. E.Payne, ASP Conference Series. V. 125, P. 32

Active Galactic Nuclei and Related Phenomena
IAU Symposium, Vol. 194, 1999
Yervant Terzian, Daniel Weedman, Edward Khachikian, eds.

Identification of Faint Markarian Galaxies with IRAS Sources

V. T. Ayvazyan

*Special Astrophysical Observatory RAS Nizhnij Arkhys, Karachai-Cherkessia, 357147 Russia. Electronic mail: ayvo@sao.ru
Armenian State Pedagogical Institute, Yerevan, Khanjian 5, 375010, Armenia.*

Abstract.
Identifications of IRAS sources with faint Markarian galaxies on the base of the Second Byurakan Survey were made. It is shown that about 30% of SBS galaxies are also IRAS sources. The list of newly identified objects is presented.

1. Identification of faint Markarian galaxies with IRAS sources

The Second Byurakan Survey (SBS) was aimed to reach fainter limiting magnitudes (Stepanian et al. 1990 and reference therein) in comparison with the FBS (Markarian 1967).

Between 1974 and 1991 a total area of 1000 square degrees of the sky was observed down to the limiting magnitude $19\overset{m}{.}5$. This area is confined by the contiguous strip defined by $7^h40^m < \alpha < 17^h15^m$, $+49° < \delta < +61°$. A selection of nearly 1700 galaxies and about 1800 stellar objects with an excess ultraviolet emission is the main result of the SBS survey (Stepanian 1994).

Infrared data for nearly 500 SBS galaxies were obtained from the NED (*NASA/IPAC Extragalactic database*), IPSS (*IRAS Point Source Survey*) and IFSS (*IRAS Faint Source Survey*). About 50 SBS galaxies are newly identified with IRAS sources. The difference of coordinates less than 1' has been used for preliminary identification. The presence of optical objects has been analysed in the 1' circle with the centres of IRAS coordinates. Then the SBS galaxies as well as the other optical objects up to the limit of the Digitized Palomar Sky Survey which were located inside of this circle were analyzed.

The difference of coordinates for the above mentioned 500 SBS IRAS galaxies also may be used for preliminary identification. The plot of these differences is shown in Fig.1a, where nearly 80% of objects are inside of the box of 20''. The similar plot for newly identified objects is shown in Fig.1b.

The use of 20'' criteria for newly identified objects show that 25 objects may be identified as SBS IRAS sources. All these galaxies are the only nearest objects around IRAS sources. The remaining 25 SBS galaxies require additional identification.

So, the total amount of SBS IRAS galaxies contains 525 objects that compose about 30% of all SBS galaxies, 5% of these sources are newly identified. The list of these 25 objects is presented in Table 1.

Acknowledgments. The author thanks J. A. Stepanian for a formulation of the task and repeated discussion of this paper and N. Serafimovich for her help in drawing up the article. This work was supported by the research grants No. 97-02-17168 and 1.2.2.2 from the Russian Foundation for Basic Research and from State Program "Astronomy," respectively.

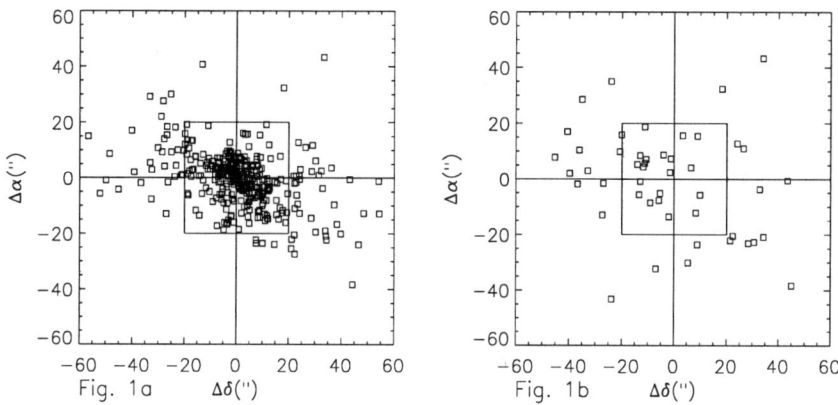

Figure 1. The plot of the difference of coordinates between SBS galaxies and IRAS sources.

Table 1. The list of 25 newly identified objects.

SBS name	IRAS name	SBS name	IRAS name
0906+502	IRASF09067+5015	1339+559	IRAS 13397+5555
0933+524	IRASF09333+5227	1410+504	IRASF14107+5028
0943+563A	IRASF09437+5620	1418+540	IRASF14187+5406
1001+584	IRASF10016+5824	1423+600	IRASF14235+6000
1020+610	IRAS 10204+6100	1512+583	IRASF15122+5823
1050+505	IRASF10502+5032	1519+508A	IRASF15195+5050
1050+573	IRASF10507+5723	1528+577B	IRASF15288+5747
1115+540A	IRASF11154+5401	1535+547	IRASF15353+5443
1115+540B	IRAS 11154+5401	1551+593B	IRASF15512+5923
1115+588	IRASF11159+5853	1600+565	IRASF16005+5632
1123+550	IRAS 11236+5503	1609+490	IRASF16094+4902
1125+581	IRAS 11258+5806	1626+596	IRASF16263+5941
1139+572	IRASF11392+5718		

References

Markarian, B.E. 1967, Astrofizika, 3, 55

Stepanian, J.A., Lipovetsky, V.A., & Erastova, L.K. 1990, Astrofizika, 32, 441
Stepanian, J.A., 1994, Doctoral dissertation, Nizhnij Arkhys

The Study of Extended Radio Galaxies

M.A.Hovhannisyan

Byurakan Astrophysical Observatory, Byurakan, 378433, Armenia

The jets of radio galaxies have an essential role in the formation of their extended components. However, in general, the physical characteristics of the jets differ from the characteristics of the extended components (the spectra, polarisation, density and so on).

Detailed observations show that the total jets are in the same shapes as the jets in the core of the galaxies. The study of radio spectra show that the spectra of the core are not all flat. Half of them have steep spectra ($\sim 45\%$), but the majority of spectra of the galaxies with radio jets are flat ($\sim 90\%$). This proves that an origin of the jets is connected with the younger cores of galaxies, having flat spectra.

However, the jet formation processes are proceeding only in specific physical conditions, for instance, the energies of magnetic field and relativistic electrons must be of the same order. We can draw the final statement that the radio jets are connected with the earlier stage of evolution of galaxies. But in some cases, due to a slow dissipation, the jets can live a relatively long time. Therefore, there are only a small number of galaxies which have radio jets with steep spectra.

References

Artyukh, V. S., & Hovhannisyan, M. A. 1988, Pisma AZh, 14, 706

Artyukh, V. S., & Hovhannisyan, M. A. 1993, AZh, 70, 443

Artyukh, V. S., Hovhannisyan, M. A., & Tyul'bashev, S. A. 1994, Pisma AZh, 20, 178

Artyukh, V. S., Hovhannisyan, M. A., & Tyul'bashev, S. A. 1994, Pisma AZh, 20, 258

Hovhannisyan, M. A. 1995, Astrofizika, 38, 692

The Variability of Optically Selected QSO Sample

Keliang Huang and Hongnan Zhou

Physics Department, Nanjing Normal University, Nanjing 210097, PRC

1. Introduction

Variability may be a common property of AGNs. It provides a powerful tool to study the physics of AGNs. For QSOs, variability is a suitable criterion for selecting candidates. Many variability studies on quasar samples have been made (e.g. Netzer and Sheffer 1983, User et al. 1983, Pica and Smith 1983, Wampler and Ponz 1985, Neugebauer et al. 1989, Huang et al 1990, Cristiani et al. 1990, Giallongo et al. 1991, Cimatti et al. 1994, Treverse et al. 1994, Hook et al. 1994, Cristiani et al. 1996, Axetxage et al. 1997). In some studies, a large percentage of quasars is found to have detectable variation. However, other studies find no variability evidence. Many studies also discuss the correlations of variability with luminosity or redshift. Some interesting results have been made. However, up to now, there are no positive conclusions. Further study on variability of quasar samples is obviously needed. In this paper, we report on a variability study of 32 members at the Medium Bright Quasar Sample (Mitchell et al. 1984). The study is actually a continuation of the previous study (Huang et al. 1990).

2. Observational Data

The previous study (Huang et al. 1990) is conducted with the help of an uniform series of blue-sensitive filtered plates taken with the Palomar Schmidt telescope spanning the years 1978-1981. The variability is characterized by a canonical standard deviation for each time series and an extremum statistic Q.

Since 1993, we made CCD photometry on the 32 members of the MBQS, using the 1 m telescope of Yunnan Observatory and the 1.56 m telescope of Shanghai Observatory. B mag. of the 32 quasars is obtained by use of standard procedure.

3. The Results and Discussion

We obtained the B mag. of 32 quasars in each year over 1993-1996. Q-statistic is still used to characterize the variability. Combining the previous results over 1978-1981, the Q-values and maximum amplitude of variation (ΔB_{max}) can be obtained.

We find that 12 quasars, i.e. about 40 % of the total quasars, vary over an 18 yr. baseline if $Q \geq 5$ is taken as variability criterion.

Using the data of 12 variable quasars, we can analyse the relationships between variability and luminosity and between variability and redshift. Taking $q_0 = 0$, $H_0 = 50$ km sec^1 Mpc^{-1}, we find that variability is strongly correlated with absolute magnitude. The regression results are:

$$(\Delta M)_{max} = 0.026 M_{max} + 1.29, (\gamma = 0.72) \quad (1)$$

Here $(\Delta M)_{max}$ and M_{max} are the maximum variation amplitude and the maximum absolute magnitude for the 12 quasars. γ is the regression coefficient. The regression is significant at $\alpha = 0.01$. Also we can use Q instead of $(\Delta M)_{max}$ and make regression. The result is:

$$Q = 0.44 M_{max} + 17.80, (\gamma = 0.85) \quad (2)$$

The regression is also significant at $\alpha = 0.01$.

Figure 1 shows $((\Delta M)_{max}, M_{max})$ and (Q, M_{max}) for the variable quasars.

The results above show that there is a strong anticorrelation between variability and luminosity, in the sense that more luminous quasars show less variability. It confirms previous results.

Figure 2 shows $((\Delta M)_{max}, Z)$ and (Q, Z) for all variable quasars. Similarly, we make regression analysis. We find:

$$(\Delta M)_{max} = -0.092 Z + 0.69, (\gamma = 0.67) \quad (3)$$

$$Q = -1.71 Z + 7.71, (\gamma = 0.85) \quad (4)$$

Both regressions are significant at $\alpha = 0.05$.

The results above seem to show there is an anticorrelation between variability and redshift, in the sense that more distant quasars show less variability. However, the apparent magnitudes of MBQS have a maximum difference less than 2 mag. and their redshifts spread over 0.06-2. Therefore, more distant quasars are generally more luminous for MBQS. The relationship between variability and redshift obtained above reflects partly the relationship between variability and luminosity. Eliminating the effect of magnitude-redshift, we find no significant correlation between variability and redshift. More detailed analysis is in progress.

Acknowledgments. This work is partly supported by National Climbing Program and NSF of China.

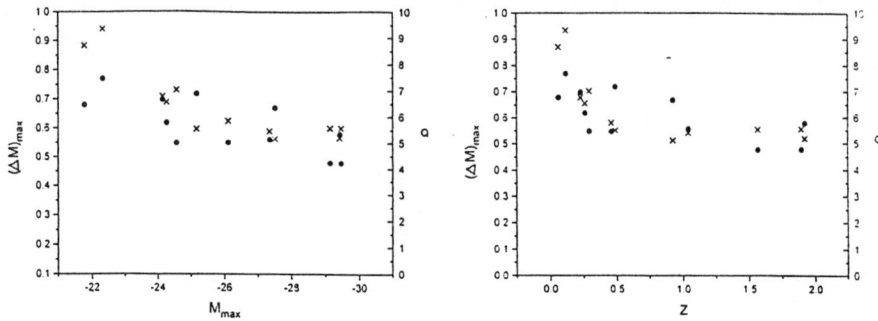

Fig.1 $(\Delta M)_{max}(\bullet)$ .Vs. M_{max} and Q (×) .Vs. M_{max} Fig.2 $(\Delta M)_{max}(\bullet)$ .Vs. Z and Q (×) .Vs. Z

References

Axetxage et al., 1997, MNRAS, 286, 271
Cimatti, A., Zomorani, G. and Marano, B., 1993, MNRAS, 263, 236
Cristiani, S., Vio, R., and Andreasni, P., 1998, AJ, 100, 56
Cristiani, S., Trentini, S., et al. 1996, A&A, 306, 395
Giallongo, E., Trevese, D., Vagnetti, F., 1991, ApJ, 337, 345
Hook, I.M., et al., 1994, MNRAS, 268, 305
Huang, K.L., Mitchell, K.J., and User, P.D. 1990, ApJ, 362, 33
Mitchell, K.J. eta l., 1984, ApJ(letters), 273, L59
Neugebauer, G., et al. 1989, AJ, 97, 957
Netser, H. and Sheffer, T., 1983, MNRAS, 203, 935
Pica, A.J. and Smith, A.G., 1993, ApJ, 272, 11
Trevese, D., Kro, R.G., et al., 1994, ApJ, 433, 494
Usher, P.D. et al., 1983, ApJ, 269, 73
Wamper, E.J and Ponz, D., 1985, ApJ, 298, 448

Active Galactic Nuclei and Related Phenomena
IAU Symposium, Vol. 194, 1999
Yervant Terzian, Daniel Weedman, Edward Khachikian, eds.

Photometric Monitoring of Selected Quasars: The Highly Luminous Quasar HS 1946+7658

B.M. Mihov, R.S. Bachev, A.A. Strigachev, L.S. Slavcheva, G.T. Petrov

Institute of Astronomy, 72 Tsar. chausse Blvd., 1784 Sofia, Bulgaria
e-mail: bmihov@astro.bas.bg

Abstract. We present and discuss Johnson – Cousins V and R photometry of the highly luminous quasar HS 1946+7658 ($z = 3.051$). We found no evidence for strong variability during one year of monitoring.

1. Introduction

The quasar HS 1946+7658 is the most luminous quasar discovered up to now – it has an absolute magnitude of -31.1 ($H_0 = 50\,\mathrm{km\,s^{-1}\,Mpc^{-1}}$, $q_0 = 0$, Galactic reddening taken into account) at a rest wavelength of 1450 Å (Hagen et al. 1992). Hagen et al. (1992) found that the quasar had in 1986 approximately the same brightness as in 1991. Kuhn et al. (1995) obtained V magnitude of the quasar to be 16.17 ± 0.02 that is close to the approximate V magnitude derived by Hagen et al. (1992). These data points are too sparse but it seems that the quasar is not strongly variable. We need a lot of data points in order to say more about the variability of HS 1946+7658. To solve this problem, we carried out an HS 1946+7658 monitoring program since July 1997.

2. Observations and Data Reduction

Observations were performed using a 1024×1024 Photometrics CCD camera attached at the RC focus of the 2-m telescope at the NAO – Rozhen. Standard broad-band Johnson – Cousins V and R filters were used. Several frames per night in each filter were usually taken; the seeing was better than $2\rlap{.}''5$ during all nights. The frames were bias subtracted, flat-fielded and cosmic rays cleaned using procedures under MIDAS.

We made relative photometry – the quasar flux was measured relative to six field stars, in order to be independent of the photometrical conditions. No absolute calibration of the magnitudes of the comparision stars was made. The instrumental magnitudes of the quasar and comparision stars were obtained through aperture photometry with an aperture radius equal to the estimated FWHM of the frame. The magnitude differences $\delta_{klmn} = mag^{(quasar)}_{kmn} - mag^{(star)}_{klmn}$ (here k counts the frames obtained during a given night, l counts the comparison stars, m is V or R filter and n means epoch of the observation) obtained for a given set of values for l, m and n were weighted averaged over k. The variations of these averaged diferences relative to its mean value (over n), namely the quantities $\Delta\delta_{lmn} = \delta_{lmn} - \bar{\delta}_{lm}$, were plotted together for all l. We found that

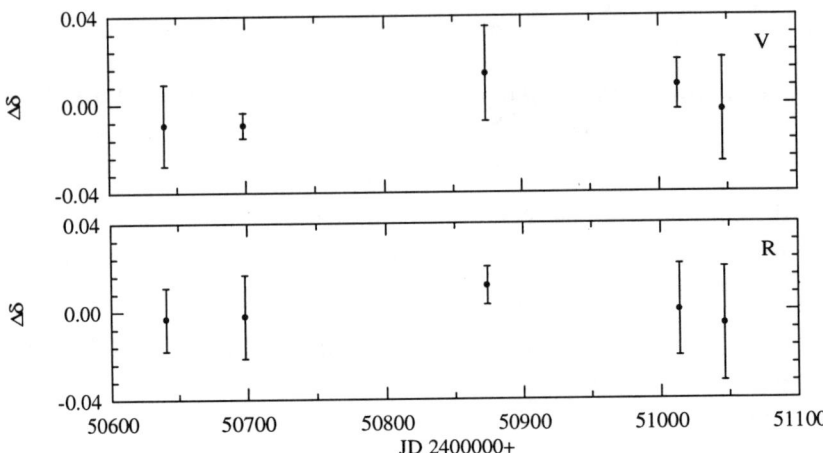

Figure 1. The quantity $\Delta\delta_{mn}$ for V and R bands.

$\Delta\delta_{lmn}$ for two of the comparison stars showed large deviations and we discarded these stars. Finally, for each epoch n, $\Delta\delta_{lmn}$ were weighted averaged over l to produce the final result $\Delta\delta_{mn}$ that is plotted on Fig. 1 for V and R bands. The error bars plotted are calculated from $\sigma^2 = \sigma^2_{\mathrm{phot}} + \sigma^2_{\mathrm{star}}$ where σ^2_{phot} is the error obtained from aperture photometry, and σ^2_{star} is max $|(\Delta\delta_{lmn} - \Delta\delta_{mn})|$.

3. Discussion

Our photometric measurements in both bands are consistent with an almost constant flux: there is slight evidence for a weak brightening at the beginning of 1998 followed by fading (Fig. 1). Both light curves are well fitted by a third order polynomial. Due to the lack of absolute calibration we cannot compare our data with the above cited photometry of HS 1946+7658 in order to follow the long-term photometrical behaviour of the quasar. In the future we plan to make an absolute calibration of the magnitudes of the comparision stars and to include data points from the 0.6-m telescope at the AO – Belogradchik, as well as to obtain deep images of the field around HS 1946+7658 (see Tripp et al. 1996).

References

Hagen, H.-J., Cordis, L., Engels, D., Groote, D., Haug, U., Heber, U., Köhler, Th., Wisotzki, L., Reimers, D. 1992, A&A, 253, L5

Kuhn, O., Bechtold, J., Cutry, R., Elvis, M., Rieke, M. 1995, ApJ, 438, 643

Tripp, T.M., Lu, L., Savage, B.D. 1996, ApJS, 102, 239

Active Galactic Nuclei and Related Phenomena
IAU Symposium, Vol. 194, 1999
Yervant Terzian, Daniel Weedman, Edward Khachikian, eds.

Monitoring of 1–22 GHz Instantaneous Spectra of 550 Compact Extragalactic Objects in 1997–1998

Y.Y. Kovalev[1], N.A. Nizhelsky[2], Yu.A. Kovalev[1], V.N. Sidorenkov[3], M.G. Mingaliev[2], A.V. Bogdantsov[2]

[1] *Astro Space Center of the Lebedev Physical Institute, Profsoyuznaya 84/32, Moscow, 117810 Russia*

[2] *Special Astrophysical Observatory, N. Arkhyz, 357147 Russia*

[3] *Sternberg Astronomical Institute, Universitetskij pr. 13, Moscow, 119899 Russia*

Abstract. First results of six frequency observations at five epochs in March, 1997, — April, 1998, and a statistical analysis are presented.

1. Results

Results presented of 1997–1998 instantaneous spectral observations at 31, 13, 7.6, 3.9, 2.7, & 1.4 cm of a full sample from the Preston et al. (1985) survey with correlated flux at 13 cm more than 0.1 Jy at declination $-30° < \delta < +42°$ are part of the long-term monitoring program (details in Kovalev, 1998).

Results are presented on Figure 1 for quasars, BL Lacs, and other objects (radio galaxies, unidentified, etc.). We have constructed the "average" spectra for each subsample using averaged values of spectral indices obtained. All spectra were transferred to the rest frame of the sources before calculations, excluding the sources with unknown redshift value from the analysis. It has been done for the list of "EGRET objects" from Mukherjee et al. (1997) as well. Each of the spectra can be presented as a sum of a spectrum of an extended optically thin component, dominating at 1–4 GHz for all spectra excluding BL Lac's, and a spectrum of a compact component, dominating at frequencies above 4 GHz (similar to the earlier samples, see Kovalev, 1998). The well defined HF turnover for BL Lacs, quasars and EGRET objects is in the range of 12–25 GHz. One can see that the BL Lacs average spectrum is the most flat. Only the BL Lacs spectrum is increasing with frequency above 1 GHz. This might indicate that for the BL Lacertae objects the relative contribution of the optically thin extended LF component to the total emission is the smallest one. On the contrary, the spectrum of galaxies is dominated by the extended component emission. It can be explained by the orientation and relativistic beaming effects of a compact jet in AGNs. Other explanation can account for the different parent populations.

A contribution of the compact component to the total averaged spectrum is the biggest one for the EGRET sample (excluding complex BL Lacs spectrum). We have also analyzed the speed of variability (like Valtaoja & Teräsranta, 1996) and have found that the EGRET sources have the greatest at $\lambda > 4$ cm, next are the BL Lacs. This supports the models which involve a connection between HF radio emission (from a compact jet) and γ emission (by Compton or synchrotron mechanisms, see e.g. for review von Montigny et al., 1995).

Dagkesamansky (1970) and Gopal–Krishna & Steppe (1982) pointed out the relation between spectral indices and the flux density for meter-wavelength samples of extragalactic sources. We have found such a correlation as well, but only for the 13 cm sample of objects — for the subsample of galaxies (see Table 1). QSOs and BL Lacs do not display such correlation, having a relatively large contribution of the compact component to the spectrum of the total emission (see also Figure 1). This result extends the α–$\lg(F_\nu)$ dependence to the high frequency sample and confirms the previous results obtained from the low frequency ones.

Table 1. A value of a coefficient k (and its error in brackets) for a linear fit like $\alpha = k \lg(F_{2.3}) + b$, $F_\nu \sim \nu^\alpha$.

Type of objects	Number of objects	$\alpha_{2.3-3.9}$	$\alpha_{3.9-7.7}$	$\alpha_{7.7-11.1}$	$\alpha_{11.1-21.7}$
Quasars	317	−0.13 (0.07)	−0.15 (0.07)	−0.16 (0.07)	−0.05 (0.06)
BL Lacs	56	−0.03 (0.12)	0.03 (0.11)	−0.04 (0.11)	−0.28 (0.10)
G	154	−0.57 (0.11)	−0.52 (0.10)	−0.52 (0.11)	−0.47 (0.08)

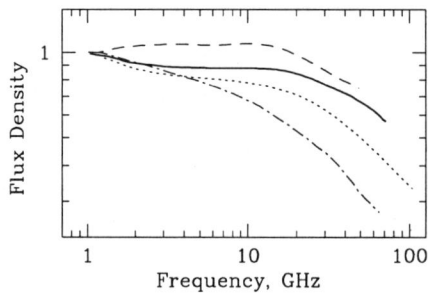

Figure 1. "Averaged" normalized spectra for quasars (a dotted line), for BL Lacertae objects (a dashed line), for other objects (a dot–dashed line), and EGRET objects (a solid line), in the rest frame.

References

Dagkesamansky, R. D. 1970, Nature, 226, 432
Gopal–Krishna, & Steppe, H. 1982, A&A, 113, 150
Kovalev, Yu. A. 1998, Bull. SAO, 44, 50
Preston, R. A., et al. 1985, AJ, 90, 1599
Valtaoja, E., & Teräsranta, H. 1996, A&AS, 120, 491
von Montigny, C., et al. 1995, ApJ, 440, 525
Mukherjee, R., et al. 1997, ApJ, 490, 116

Active Galactic Nuclei and Related Phenomena
IAU Symposium, Vol. 194, 1999
Yervant Terzian, Daniel Weedman, Edward Khachikian, eds.

The Continuum Spectra of the Core and Hotspots of Cygnus-A in the Millimetre and Submillimetre

L.L. Leeuw and E.I. Robson

Centre for Astrophysics, University of Central Lancashire, Preston, UK

Joint Astronomy Centre, 660 N. A'ohoku Place, Hilo, Hawaii, USA

Abstract.
Submillimetre imaging (350 μm to 850 μm) and millimetre photometry (1.35 mm and 2 mm) observations, obtained with SCUBA (Holland et al. 1998), are used: (1) to investigate electron aging for synchrotron emission and (2) to determine the dust content in Cygnus-A.

Cygnus-A (3C405) is a cD galaxy with V=15 at z=0.0567 and shows spectacularly symmetrical radio lobes and jets. It has a FRII-type radio structure: powerful radio lobes with $L > 10^{35}$ W at 178 MHz which are edge-brightened with prominent 'hotspots'. For $H_0 = 75$ km^{-1}s^{-1}Mpc^{-1}, the distance is 227 Mpc giving a plate-scale of 1.1 kpc/arcsec. To most AGN pundits Cygnus-A is the archetypal 'quasar' on the plane of the sky with the central AGN buried in a dusty molecular torus.

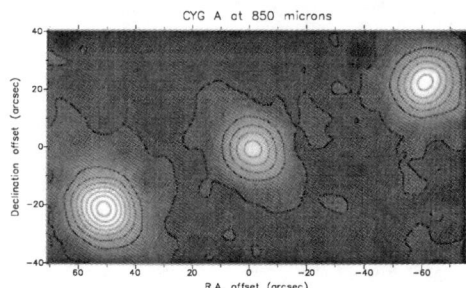

Figure 1. The hotspots and central core at 850 μm. The rms is 40 mJy/beam.

The hotspots show a well defined power-law spectrum between 1 GHz and 700 GHz. The spectral indexes ($S_\nu \propto \nu^\alpha$) between 140 GHz and 677 GHz are $\alpha = -1.04 \pm 0.01$ and $\alpha = -0.99 \pm 0.01$ for hotspots A (northern) and D (southern), respectively. The lack of spectral steepening means no electron aging to the highest frequency of ~ 677 GHz. If we assume an equipartition magnetic field energy of 30 nT, this gives a lifetime for the radiating electrons at 450 μm of

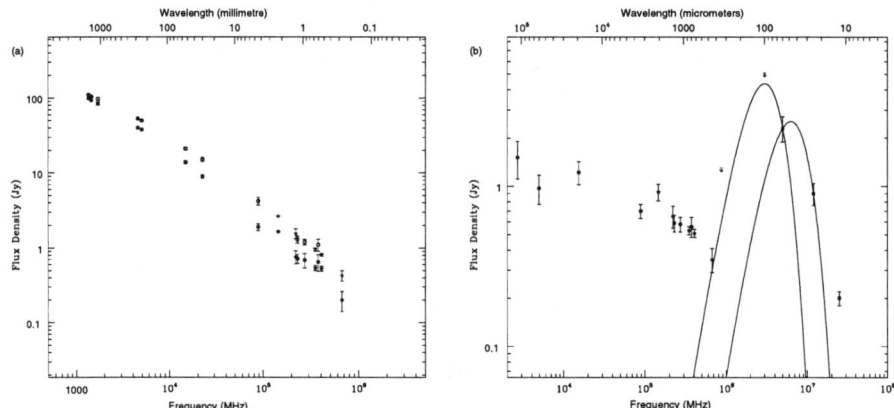

Figure 2 The spectral energy distribution of (a) the hotspots A (open circles) and D (solid circles) and (b) the central core. The SCUBA data points are indicated by asterisks and stars. All the upper limits are 3σ. The two curves in (b) represent emission from greybodies with temperatures of 37K and 85K with an emissivity index, β, of 1.3 (Robson et al. 1998).

$< 10^4$ y. However, it is known from 3 mm interferometry observations that the hotspots are 2 - 3 kpc across. Therefore the electron diffusion speed is either $\sim$ c (uncomfortable), or more likely non-localized relativistic particle acceleration is taking place.

The Core synchrotron spectral index is much flatter with a value of -0.6 ± 0.1. Re-analysis of the *IRAS* HIRES data products confirm previous measurements at 25 and 60 μm but show that the revised 100 μm upper limit is not helpful in determining dust parameters. With our new submillimetre data, we constrain the non-thermal contribution to *IRAS* and *ISO* fluxes. In particular with our photometry value at 450 μm, and the two *IRAS* measurements, we constrain the dust temperature between 37 K and 85 K and corresponding dust masses of $1.0 \times 10^8 M_\odot$ and $1.4 \times 10^6 M_\odot$, respectively. Our results are consistent with new, better-constrained (because of a good number of data points at the critical part of the spectrum) dust temperature of 52 K and dust mass of $5 \times 10^6 M_\odot$ obtained by Haas et al. 1998 using new *ISO* data from 60 μm to 180 μm. Further details and analysis of the SCUBA data are presented in Robson et al. 1998.

References

Hass M. et al. 1998, ApJ 503:L109-113
Holland W.S. et al., 1998, MNRAS, in press
Robson E.I., Leeuw L.L., Stevens J.A, Holland W.S., 1998, MNRAS, in press

Hidden Quasars in Ultraluminous Infrared Galaxies

H. D. Tran, M. S. Brotherton, S. A. Stanford, & W. van Breugel

Institute of Geophysics & Planetary Physics, Lawrence Livermore National Laboratory, Livermore, CA 94550

Abstract. Many ultraluminous infrared galaxies (ULIRGs) are powered by quasars hidden in the center, but many are also powered by starbursts. A simply diagnostic diagram is proposed that can identify obscured quasars in ULIRGs by their high-ionization emission lines ([O III]$\lambda 5007$/H$\beta \gtrsim 5$), and "warm" IR color ($f_{25}/f_{60} \gtrsim 0.25$).

Ultraluminous infrared galaxies (ULIRGs, $L_{IR} > 10^{12} L_\odot$) are an important constituent of our local universe, with luminosities and space densities similar to those of QSOs (Soifer et al. 1987). This led to the suggestion that ULIRGs could contain infant quasars enshrouded in a large amount of dust (Sanders et al. 1988). On the other hand, they may also represent energetic, compact starbursts (Condon et al. 1991). Understanding the dominant energy input mechanism in these ULIRGs – whether it is obscured quasars or intense bursts of star formation – has been the main issue concerning their nature. In order to better understand the energy sources of ULIRGs and the relationship between AGN and starburst activity, we started a spectropolarimetric survey to search for hidden broad emission lines in a sample of ULIRGs that were identified in the cross correlations between the *IRAS Faint Source Catalog* (*FSC*) and those of the FIRST (Becker et al. 1995) and Texas (Douglas et al. 1996) radio surveys.

Using the 10-m Keck II telescope, we obtained spectropolarimetric observations of one ULIRG selected from a sample identified in the FIRST-*FSC* correlation (FF sources, Stanford et al. 1998), and two ULIRGs from the Texas-*FSC* correlation (TF sources, Dey & van Breugel 1994). The infrared properties of these galaxies are listed in Table 1. Our results show that only the

Table 1. Infrared Properties

Object	z	m	$\log(L_{IR}/L_\odot)$	f_{12}	f_{25}	f_{60}	f_{100}	f_{25}/f_{60}
FF J1614+3234	0.710	19.1	13.2	< 0.065	< 0.055	0.174	< 0.54	< 0.31
TF J1020+6436	0.153	19.0	12.1	< 0.095	< 0.062	0.86	1.24	< 0.072
TF J1736+1122	0.162	18.0	12.3	< 0.081	0.196	0.484	< 3.31	0.404

high-ionization Seyfert 2 galaxy TF J1736+1122 is highly polarized, displaying a broad-line spectrum visible in polarized light (Fig. 1a). The other two objects, TF J1020+6436 and FF J1614+3234, exhibit spectra dominated by a population of young (A-type) stars similar to those of "E + A" galaxies. They are unpolarized, showing no sign of hidden broad-line regions. The presence of young starburst components in all three galaxies indicates that the ULIRG phe-

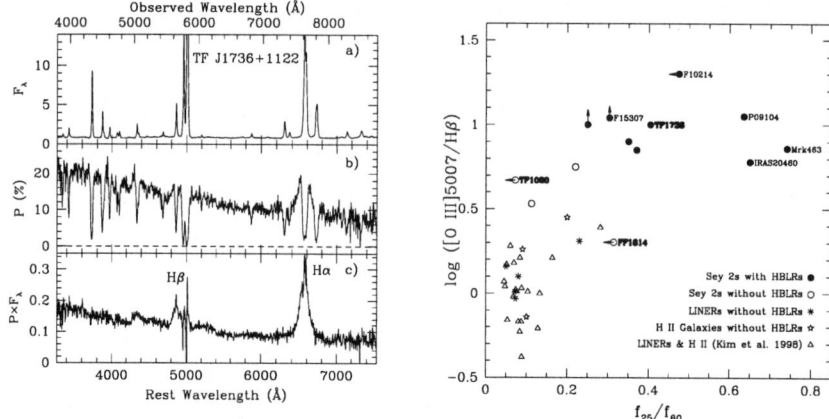

Figure 1. *Left panel:* Spectropolarimetry of TF J1736+1122. (a) Total flux spectrum, (b) observed degree of polarization, and (c) polarized flux spectrum. *Right panel:* [O III] $\lambda 5007/\mathrm{H}\beta$ versus IR color f_{25}/f_{60} for narrow-line ULIRGs in which HBLRs have been searched for. Also plotted in open triangles are all ULIRGs classified as H II and LINERs in Kim et al. (1998). Note the clear tendency for ULIRGs with HBLRs to have warmer IR color and higher excitation spectrum. The opposite holds for LINERs and H II galaxies.

nomenon encompasses both AGN and starburst activity, but the most energetic ULIRGs do not necessarily harbor "buried quasars".

It is of interest to see if it could be determined, from optical spectroscopic and *IRAS* photometric data alone, whether an ULIRG harbors a genuine quasar or is powered by a starburst. Figure 1b shows a plot of the line ratio [O III] $\lambda 5007/\mathrm{H}\beta$(narrow), which can serve as an indicator of the ionization level, versus the infrared color f_{25}/f_{60}. As can be seen, there is a clear tendency for higher-ionization and warmer Seyfert 2 ULIRGs to show hidden broad-line region (HBLR) indicative of a "buried quasar." Furthermore, Seyfert ULIRGs without HBLRs lie in a region of the diagram similar to that occupied by H II and LINER galaxies, none of which has been found to have HBLRs.

Acknowledgments. Research at IGPP/LLNL is performed under the auspices of the US Department of Energy under contract W-7405-ENG-48.

References

Becker, R. H., White, R. L., & Helfand, D. J. 1995, ApJ, 450, 559
Condon, J. J. et al. 1991, ApJ, 378, 65
Dey, A., & van Breugel, W. 1994, in Mass-Transfer Induced Activity in Galaxies, ed. I. Shlosman, p. 263
Douglas, J. N. et al. 1996, AJ, 111, 1945
Kim, D.-C., Veilleux, S., & Sanders, D. B. 1998, ApJ, in press, astro-ph/9806149

Sanders, D. B. et al. 1988, ApJ, 325, 74
Soifer et al. 1987, ApJ, 320, 238
Stanford, S. A., Stern, D., van Breugel, W., De Breuck, C. 1998, ApJ, in prep.

The Variability of QSOs in the Optical Band

D. Trèvese and A. Bunone

Istituto Astronomico Università di Roma "La Sapienza", via G.M. Lancisi 29, I-00161, Roma, Italy

R.G. Kron

Fermi National Accelerator Laboratory, MS 127, Box 500, Batavia, IL 60510

Constraints on the emission mechanism of AGNs can be provided by the variability of their spectral energy distribution (SED). Recently Di Clemente et al. (1996) have shown that the positive correlation of QSO variability with redshift can be due to a hardening of the spectrum in bright phases, coupled with the increase of the rest-frame frequency of the (fixed) observing band, for increasing redshift. Direct evidences of slope changes in the SEDs of a limited number of individual AGNs have been provided by Cutri et al. (1985), Edelson et al. (1990), and Kinney et al. (1991). In the following we present some preliminary results of a direct measure of variations of the SED slope in the complete, magnitude limited, sample of QSOs of the Selected Area 57 (Koo, Kron & Cudworth 1986; Trèvese et al. 1989 (**T89**)). The data are derived from two sets of prime focus plates of the SA 57, in the U, B_J, F and N bands, obtained with the 4-m telescope at the Kitt Peak National Observatory at two epochs separated by one year. Photometric methods and signal-to-noise optimization are described in T89 and Trèvese et al. (1994). The quasar sample, of $<z>\simeq 1.4$, consists of 33 objects from T89, plus the brightest 3 members of the the Bershady, Trèvese and Kron (1998) sample of extended objects selected on the basis of variability. In Figure 1a $\Delta\alpha$ is reported versus the changes $\Delta log f_v$ in the B_J band and shows positive correlation, indicating a hardening of the spectrum in the bright phase. A special care is needed to avoid spurious $\Delta\alpha - \Delta log f_v$ correlations (see Massaro & Trèvese 1996). The correlation coefficient is r=0.46 with a probability P(>r)=0.995, after the exclusion of one deviant point, whose inclusion would produce a higher correlation. The slope of the linear regression of $\Delta\alpha \Delta log v$ is b=1.9. Assuming that the spectra are, dominated by the big blue bump (BBB) in the sampled spectral region, around $\lambda \approx 2000$ Å, we can use the simple approximation of a single black-body spectrum. To check the hypothesis that both the slope and brightness changes are caused by a temperature variations only, we derive the black-body temperatures from the SED slope as deduced from a linear fit of the $log f_v$ v.s. $log v$ relation, based on the U, B_J and F data. The average slope (excluding the two highest and lowest z objects) is $\langle\alpha\rangle$=-0.4±0.6 and the average temperature is $T \approx 25000$ K. For a black-body of fixed emitting surface and varying temperature T, the changes of the local SED slope and the relevant luminosity variations are related by $(d\alpha/dT)/(dlog B_v/dT) \equiv f(x)$, where $x \equiv hv/kT$. The slope b of the linear regression of $\Delta\alpha$ v.s. $log f_v$ of Figure 1a is compared with the function $f(x)$

in Figure 1b, where $\alpha(x)$ for a black-body is also plotted with the range of x

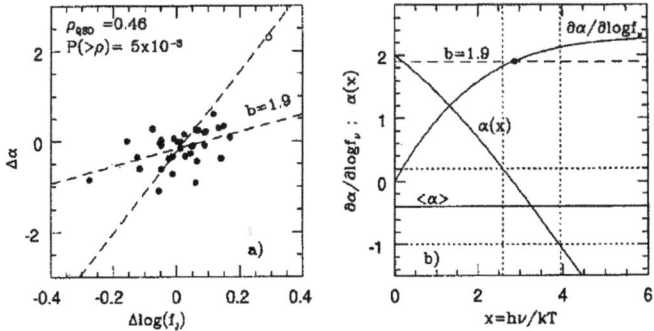

Figure 1. a) Slope changes $\Delta\alpha$ versus $\Delta log f_v$ in the B_J band. The straight lines represent the linear regressions, after the exclusion of one deviant point (open circle). b) The functions $f(x) \equiv d\alpha/dlog B_v$, and $\alpha(x)$ for a black-body, versus $x \equiv h\nu/kT$. The values of the slope b (see Figure 1a), the average $\langle\alpha\rangle$ and the dispersion of the slopes are represented by the horizontal lines. The relevant range of x values is indicated by the vertical dashed lines.

values of the sample determined by the dispersion of α. We can conclude that $d\alpha/dlog B_v$ and α are consistent for $x \approx 3$. For an average sampling wavelength $< \lambda > \approx$ 2000 Å of the sample, this corresponds to an average temperature $T \approx 2.5 \times 10^4$ of a black-body of fluctuating temperature.

The above results show that: **i)** slope changes of the SED and intensity variation in the optical-UV are at least not inconsistent with temperature variations of an emitting black-body; **ii)** aside from any specific interpretation, the relation between the variability of slope and intensity found in our complete QSO sample, provides a new constraint on the physical mechanisms responsible of variability.

References

Bershady, M.A., Trèvese, D., & Kron, R.C.: 1998, ApJ, 496, 103.
Cutri, R.M., Wisniewski, W.Z., Rieke, G.H., & Lebofski, H.J.: 1985, ApJ, 296, 423.
Di Clemente, A., Natali, G., Giallongo, B., Trèvese, D., Vagnetti, F.: 1996, ApJ, 463, 466.
Edelson, R.A., Krolik, J.H., Pike, G.F.: 1990, ApJ, 359, 86.
Giallongo, B., Trèvese, D., Vagnetti, F.: 1991, ApJ, 377, 345.
Kinney, A.L., Bohlin, R.C., Blades, J.C., & York D.G.: 1991, ApJS, 75, 645.
Koo, D.C., Kron, R.G., Cudworth, K.M.: 1986, PASP, 98, 285.
Massaro, E., & Trèvese, D.: 1996, A&A, 312, 810.

Trèvese, D., Pittella, G., Kron;R.G., Koo, D.C., Bershady, M.A.: 1989, AJ, 98, 108. (**T89**)

Trèvese, D., Kron, R.G., Majewski, S.R., Bershady, M.A., Koo, D.C.: 1994, ApJ, 433, 494.

III. AGN THEORY AND MODELS

Malcolm Longair

Willem Baan and Michael Dopita

Active Galactic Nuclei and Related Phenomena
IAU Symposium, Vol. 194, 1999
Yervant Terzian, Daniel Weedman, Edward Khachikian, eds.

Energy Sources and Physical Processes in Active Galaxies

Daniel W. Weedman

Department of Astronomy and Astrophysics, The Pennsylvania State University, State College, PA, 16802 USA

Abstract.

The galaxy Markarian 231 is reviewed as a prototype that includes all physical processes associated with active galaxies. It contains both a starburst and an AGN which are highly obscured by dust. Its luminosity at various wavelengths is used to summarize how such a galaxy would appear at high redshift, as observed with missions such as HST, AXAF, and SIRTF. The conclusion is that Markarian 231 provides an excellent example for comparing low redshift and high redshift phenomena.

1. Introduction

It is exciting and rewarding to look back personally over 30 years of research on active galaxies that was stimulated by discoveries at the Byurakan Observatory. The Markarian survey was the initial extragalactic survey that attempted to sort objects spectroscopically. It was undertaken with commendable scientific motivation, which was to locate in other galaxies examples of ejections from galactic nuclei that Ambartsumian felt were the key to understanding galaxy formation. The survey proved to be a treasure trove for locating galaxies with intense, non-stellar sources of luminosity in their nuclei, galaxies similar to the very small number of Seyfert galaxies known at the time. It was these Markarian galaxies which provided vital connecting links to the characteristics of quasars. In the decades since, the Markarian galaxies have provided prototypes for many objects classified as "active galactic nuclei" (AGN).

Markarian's survey not only found many examples of active nuclei, but also discovered vigorous regions of star formation that could be identified with the "superassociations" championed by Ambartsumian. Now labelled "starburst" galaxies, these locations of intense star formation are a vital form of galaxy activity, especially important when attempting to understand how galaxies form primordially. Because we have learned that the luminosities of starbursts can exceed the luminosities arising from central nuclei, and that starbursts are often intimately associated with other forms of activity in and near the centers of galaxies, it is appropriate to include both starbursts and AGN when summarizing the physical processes and energy sources in "active galaxies".

The luminosity function of the Markarian starburst galaxies gave us an early census of star formation within the nearby universe (Huchra 1977). Since then, a great deal of observational effort has developed an overall picture of star formation as it can be seen through the universe back to redshifts of approximately four (Madau et al. 1996). Similarly, the Markarian survey yielded the Seyfert galaxy luminosity function in the nearby universe for comparing with quasars elsewhere (Weedman 1986). The survey initiated the low dispersion spectral survey technique that led eventually to assembling the quasar data which define quasar evolution. Stimulated by the Markarian surveys, similar objective prism surveys with Schmidt Telescopes began at Cerro Tololo and the Anglo-Australian Observatory (Smith 1983). In addition to finding galaxies, these surveys showed the surprising ease of finding quasars based on the emission lines seen in these spectra. Similar surveys were then extended to fainter magnitudes using wide-field grism spectroscopy with larger telescopes, and the combined results definitively show the evolution of the quasar luminosity function (Schmidt, Schneider and Gunn 1995).

The Markarian galaxies came at a crucial time because they provided an initial list for subsequent study in the ultraviolet and infrared wavelengths that soon became accessible from space. As a consequence, this exceptionally broad wavelength coverage led to observational definitions of highly diverse phenomena. Now, these can be unified, and it is possible to review all of the fundamental processes that have been discovered within active galaxies by describing a single Markarian galaxy which shows examples of all such processes. This galaxy is Markarian 231, and my review of physical processes will use Markarian 231 as a case study to demonstrate the exceptional range of observational techniques and conclusions gained in three decades of study. As a guide to interpreting future observations, I will also summarize how this prototype galaxy would appear to the Hubble Space Telescope (HST), the Advanced X-Ray Astrophysics Facility (AXAF), and the Space Infrared Telescope Facility (SIRTF), if located at a redshift of 3.

From the perspective of an observer of Markarian galaxies, the most dismaying development was the eventual realization that the ultraviolet luminosity which we observed is only a small part of what is intrinsically produced. Dust is taking away most of the ultraviolet. It is probably fortunate that the infrared observers stumbled on starburst galaxies after Markarian. Had the extent of dust obscuration been known, an observing proposal to look for galaxies based on their ultraviolet continuum would have been skeptically received! I will review the evidence for these absorbing effects to emphasize that future progress will depend crucially on our abilities to understand and penetrate this obscuration.

2. Markarian 231 and Physical Processes in Active Galaxies

This single Markarian galaxy can be used as a prototype to illustrate all of the crucial physical processes and energy sources in and near galactic nuclei. Originally, this galaxy seemed very strange to Ed Khachikian and me when defining the Seyfert 1 galaxies. It was the only one known with conspicuous absorption lines (spectrum in Khachikian and Weedman 1974). Now, we know that it has virtually all of the varied characteristics associated with differently classified active galaxies; the fact that a straightforward and consistent interpretation of all these characteristics can be summarized is an indication of the unity among the diverse processes of active galaxies. Markarian 231 contains young and old starbursts, contains an AGN with broad line region that is a strong Fe II emit-

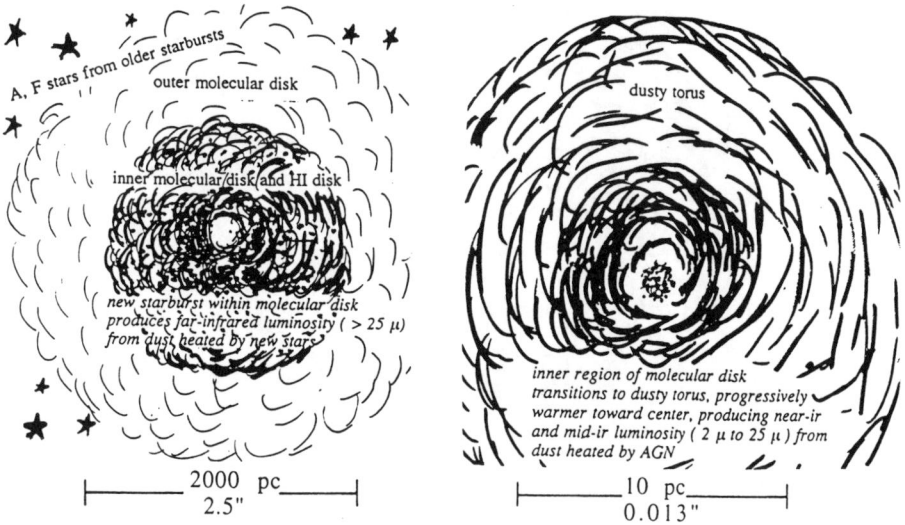

Figure 1. Markarian 231 on Kpc and pc scales.

ter, has extensive molecular clouds with super-starbursts, and is obscured by extensive dust absorption. As a consequence of the latter, it is a strong source from near infrared through far infrared and is a prototype for ultra-luminous infrared galaxies (ULIRG - Sanders et al. 1987). Like several such objects, it produces an OH megamaser (Baan, Haschick and Henkel 1992). Its AGN shows the X-ray, ultraviolet, and optical continuum characteristically attributed to a hot accretion disk. Complex, blue-shifted absorption in the optical and ultraviolet produces the characteristic spectrum of a BAL quasar, and it shows a variable radio jet from the nucleus. Markarian 231 has something for everyone! Because of the extensive observational material available at all wavelengths from X-ray through radio, it is possible to deduce simplified cartoons which consistently interpret the varied observational phenomena and which illustrate them to scale.

Markarian 231 is a large, irregular galaxy that is presumably disturbed because of a galaxy interaction (Hamilton and Keel 1987). This interaction or a previous one stimulated earlier starbursts, as revealed by the early star spectrum characterizing much of the outer regions (Boksenberg et al. 1977). Although its outer regions extend to tens of Kpc, it is in the central region in which the bulk of the luminosity is generated. My first view has a scale size of 3 Kpc, sketched in Figure 1. The key to this part of Markarian 231 is the large mass of molecular and HI gas that has accumulated within the central 3 Kpc. Downes and Solomon (1998) and Carilli, Wrobel and Ulvestad (1998) thoroughly describe this gas disk and emphasize that this is the fundamental source of most of the galaxy's luminosity. Downes and Solomon conclude that 4×10^9 $M_\odot$ of new stars are present in the current starburst within the molecular disk; Carilli et al. estimate these stars to be forming at the rate of 60 $M_\odot yr^{-1}$. The farinfrared luminosity of this starburst dominates the bolometric luminosity, being

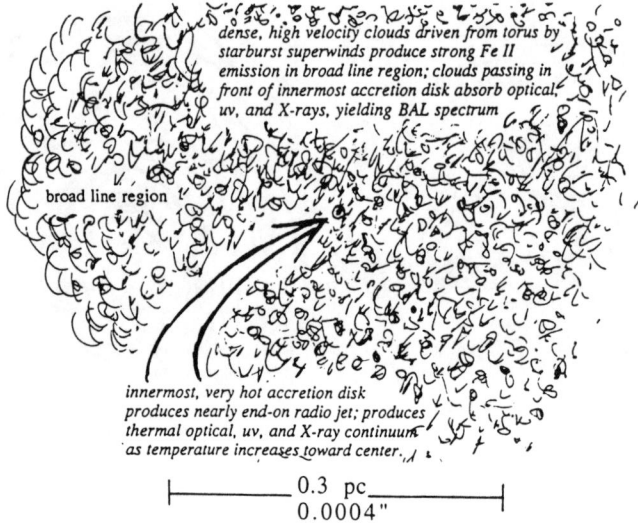

Figure 2. Markarian 231 on sub-pc scales.

responsible for about two-thirds of the 10^{46} erg s^{-1} which this ultraluminous galaxy generates.

The inner region of the molecular disk transitions to a "dusty torus" which is close enough to the AGN to be heated by it (also in Figure 1). Hot dust in the torus produces near infrared and mid infrared luminosity (Krabbe et al. 1997, Miles et al. 1996). Such a torus and its orientation are responsible for explaining the differences between Sy 1 and Sy 2 galaxies in various unified schemes.

Within the nearly face-on torus is the broad line region and the innermost accretion disk, as in Figure 2. The torus is a source of dense clouds, driven off by starburst superwinds. These clouds driven into the broad line region are unusually dense compared to those in many Sy 1, so they yield strong Fe II emission (Lawrence et al. 1997) while causing conspicuous absorption effects when they pass in front of the X-ray, uv, and optical continuum source. Because the clouds are boiling out above the center of the torus and in front of the accretion disk, the cloud absorptions show blueshifts relative to the galaxy, explaining the BAL spectral characteristics (Lipari, Colina and Macchetto 1994).

Markarian 231 has been detected in X-rays, in both ROSAT and ASCA bandpasses (Lawrence et al. 1997, Turner 1998), and in the ultraviolet, using HST (Smith et al. 1995). In these wavebands, also, this galaxy shows indicators of both starburst and AGN components. The soft X- ray ROSAT HRI detection (0.5 - 2 keV) is dominated by the extended starburst, but the higher energy ASCA flux (2 - 10 keV) seems to arise from the nucleus. The ultraviolet continuum from the nucleus is polarized but at shorter wavelengths is dominated by unpolarized light from the starburst.

An unresolved but variable radio source is observed in the center, with an extension implying a radio jet (Carilli et al. 1998, Smith, Lonsdale and Lonsdale 1998). The radio jet is that characteristic which is considered to be most intimately associated with a massive black hole. The total radio flux of 240 mJy is divided evenly between the AGN component and a more extended, presumably starburst component.

The unique spectral characteristics of Markarian 231, initially so puzzling, are thereby explained because the torus is face on while extending close enough to the AGN to produce obscuring effects. The AGN is heavily obscured but not so obscured as to hide its fundamental character. Light from the inner accretion disk and broad line region is not completely obscured, so it is classifiable as a Seyfert 1 galaxy. But the extensive amount of dense material close to the nucleus, and the energizing effects in the torus caused by the starburst, produce unusually thick and high velocity absorbing clouds which appear in front of the innermost accretion disk. Had there been less obscuring material in the inner torus, Markarian 231 would have been a more normally appearing Sy 1. Had the torus been more edge on, it would have been seen as a Sy 2. Had the torus been so thick as to block all signs of the AGN, Markarian 231 would have been classified as purely starburst. That the AGN dominates some characteristics and the starburst dominates others illustrates why it is often difficult to determine the ultimate source of luminosity for highly obscured objects.

3. Markarian 231 at High Redshift

Being a ULIRG makes Markarian 231 a feasible prototype for the optically-faint galaxies detected by the SCUBA submillimeter camera (Hughes et al. 1998) and by the Infrared Space Observatory (ISO) far-infrared surveys (Kawara et al. 1998). Some such detections probably are of dusty galaxies at z > 2. It is crucial to understand the dust content and energy sources of the high-redshift universe, so it is important to summarize how Markarian 231 would appear at such redshifts. (Markarian 231 has z = 0.0422. Results which follow are all based on a cosmology with $\Omega = 0.2$ and $\Lambda = 0$; if $\Omega = 1$, all fluxes given for redshift 3 would be brighter by a factor of 2.3.)

An object like Markarian 231 would be marginally detectable by AXAF at a redshift approaching 3. The 2-10 keV luminosity is approximately 10^{43} erg s^{-1} (Turner 1998). At redshift of 3, this would be redshifted into the 0.5-2.5 keV band; within this band, AXAF should have sensitivity in long pointings of about 3×10^{-16} erg cm^{-2} s^{-1}. Markarian 231 would have X-ray flux in this band of between 1.2×10^{-16} and 4×10^{-16} for 3 > z > 2. Because it is the rest-frame 2-10 keV X-rays that are observed at high redshift, an X-ray detection of an obscured galaxy at such redshifts would be a strong indicator of the presence of an AGN.

The ultraviolet flux as observed at rest frame wavelength of 2030 Å is approximately 1.3×10^{-15} erg cm^{-2} s^{-1} $Å^{-1}$ (Smith et al. 1995), all arising from an unresolved region. This rest wavelength corresponds to the wavelength at which a galaxy with redshift 3 would be observed in the F814 filter of the Hubble Deep Field (HDF). The flux of Markarian 231 at such a redshift would be 1.7×10^{-20} erg cm^{-2} s^{-1} $Å^{-1}$. As an unresolved object, this flux would fall within an area of about 9 pixels in the HDF. The surface brightness within a 9 pixel area would be 1.2×10^{-18} erg cm^{-2} s^{-1} A^{-1} $arcsec^{-2}$. This is similar to the maximum surface brightnesses of the known high redshift starburst objects in the HDF (Lowenthal et al. 1997), as tabulated by Weedman et al. (1998). The difference, however, is that those starbursts are much larger in size and are easily resolved in the HDF. Markarian 231 would be a "blank-field" source based on ground-based optical images, because the flux at 8140 Å would correspond to a magnitude fainter than 27. This is consistent with the lack of optical detections of some SCUBA and ISO sources, if those sources are attributed to objects like Markarian 231 at z > 2.

With SIRTF, spectrometry will be available for $5\mu < \lambda < 40\mu$ and can determine redshifts based on broad PAH and silicate features. The ISO mission has shown us these strong spectral features in active galaxies (Moorwood et al. 1996). These features could be followed to high redshifts in order to identify very obscured galaxies whose redshifts cannot be determined optically. Within the NASA/IPAC Extragalactic Database, Markarian 231 has 8.4 μ flux of 1.1 Jy. Using this to determine a rest-frame luminosity, at z = 2 it would have observed 25 μ flux of 0.47 mJy; this flux should be within the range of SIRTF's spectrometer. The same rest-frame luminosity by z = 3 would have observed 33 μ flux of 0.21 mJy, which will be marginal.

An important use of radio observations will be to determine accurate positions, in search of faint optical identifications, for sources detected in the far infrared or submillimeter. Taking the total flux of Markarian 231 as 240 mJy at 20 cm with a flat spectral index (Smith et al. 1998), the 20 cm flux would reduce to 0.046 mJy at z = 3; this would be readily within current VLA detection capabilities.

4. Effects of Dust Obscuration

Most of what we know about starbursts and quasars at high redshift is based upon their appearance in the rest frame ultraviolet. This is troublesome because of the evidence for substantial dust obscuration. Many efforts have been made to apply dust extinction corrections to the observed ultraviolet spectra of both nearby and distant starburst galaxies by considering the shapes of these spectra (e.g. Meurer et al. 1997, Pettini et al. 1998). These corrections indicate that ultraviolet observations underestimate by factors of five or more the total luminosity of starbursts. A beautiful picture of Centaurus A from HST (http://oposite.stsci.edu/pubinfo/pr/1998/14/) indicates why applying such dust corrections is problematical; the starlight you correct is only the starlight you can see. Extinction corrections are applied to partially obscured but still visible stars. This is only a lower limit on the correction because of the many stars that remain completely hidden in the dust clouds.

If sufficient infrared observations are available to deduce the bolometric luminosity of the starburst, including dust re-radiation, the total luminosity can be compared to what emerges in the ultraviolet and can yield an empirical fraction for this emergent ultraviolet luminosity. In large measure because of the many Markarian galaxies detected both with IUE and IRAS, this comparison can be made for a sample of galaxies with detections in both wavebands. Using the bolometric fluxes from Schmitt et al. (1997), the IUE ultraviolet fluxes from Kinney et al. (1993), and the relation between ultraviolet and bolometric fluxes for star formation models such as those in Meurer et al. (1997), the intrinsic and observed ultraviolet fluxes can be deduced and compared for a sample of 24 local starburst galaxies, as described by Weedman et al. (1998). This result shows that, as a median value, only 10% of the intrinsic ultraviolet luminosity escapes from these starburst galaxies. This sample is subject to selection effects, the most dominant of which is the requirement of having a detected ultraviolet flux. Galaxies so obscured that IUE cannot see them are not included, so this sample underestimates the amount of dust obscuration for the entire local population of starburst galaxies.

Because of the large obscuration that can be demonstrated in local starbursts and is suspected in high redshift starbursts, it is obvious that determinations via rest-frame ultraviolet observations of the star formation rate (and, by similar arguments, of the AGN rate) are highly uncertain. This means that our census of physical processes and energy sources in the high redshift universe is woefully incomplete. It is crucial to determine how obscured are those objects already discovered at high redshifts. It is also necessary to know if other dusty objects at high redshifts are similar to objects which can be dimly seen via their rest-frame ultraviolet, or if the dusty objects are primarily a separate population. Gaining such a census requires infrared observations.

Acknowledgments. We have learned an extraordinary amount about active galactic nuclei and about star formation since those early days forty years ago when Victor Ambartsumian and Beniamen Markarian decided to hunt for unusual galaxies. I thank them and their Armenian colleagues for that contribution to astronomy and wish they could know the progress which has come about because of the pioneering ideas and observations originating at Byurakan Observatory.

References

Baan, W.A, Haschick, A., & Henkel, C. 1992, AJ, 103, 728

Boksenberg, A., Carswell, R., Allen, D., Fosbury, R., Penston, M. & Sargent, W. 1977, MNRAS, 178, 451

Carilli, C.L., Wrobel, J.M. & Ulvestad, J.S. 1998, AJ, 115, 928

Downes, D. & Solomon, P.M. 1998, ApJ, in press

Hamilton, D. & Keel, W. 1987, ApJ, 321, 211

Huchra, J.P. 1977, ApJ, 217, 928

Hughes, D.H. et al., 1998, Nature, 394, 241

Kawara, K. et al., 1998, A&A, in press

Khachikian, E. Ye. & Weedman, D.W. 1974, Ap.J., 192, 581

Kinney, A.L., Bohlin, R.C., Calzetti, D., Panagia, N. & Wyse, R.F.G. 1993, ApJS, 86, 5.

Krabbe, A., Colina, L. Thatte, N., & Kroker, H. 1997, ApJ, 476, 98

Lawrence, A., Elvis, M., Wilkes, B.J., McHardy, I., & Brandt, N. 1997, MNRAS, 285, 879

Lipari, S., Colina, L. & Macchetto, F. 1994, ApJ, 427, 174

Lowenthal, J.D. et al. 1997, ApJ, 481, 673

Madau, P., Ferguson, H.C., Dickinson, M.E., Giavalisco, M., Steidel, C.C. & Fruchter, A. 1996, MNRAS, 283, 1388

Meurer, G.R., Heckman, T.M., Leitherer, C., Kinney, A., Robert, C., & Garnett, D.R. 1995, AJ, 110, 2665.

Meurer, G., Heckman, T., Leitherer, C., Lowenthal, J., & Lehnert, M. 1997, AJ, 114, 54.

Miles, J.W., Houck, J.R., Hayward, T.L., & Ashby, M.L.N. 1996, ApJ, 465, 191

Moorwood, A.F.M., Lutz, D., Oliva, E., Marconi, A., Netzer, H., Genzel, R., Sturm, E., & de Graauw, Th.

Pettini, M., Kellogg, M., Steidel, C.C., Dickinson, M., Adelberger, K.L., & Giavalisco, M. 1998, ApJ, in press

Sanders, D.B., Young, J.S., Scoville, N.Z., Soifer, B.T. & Danielson, G.E. 1987, ApJ, 312, L5.

Schmidt, M., Schneider, D.P., & Gunn, J.P. 1995, AJ, 110, 68

Schmitt, H.R., Kinney, A.L., Calzetti, D., & Storchi-Bergmann, T. 1997, AJ, 114, 592

Smith, H.E., Lonsdale, C.J., & Lonsdale, C.J. 1998, AJ, 492, 137

Smith, M.G. 1983, in Proceedings of the 24th Liege Colloquium, Quasars and Gravitational Lenses, Liege: Institut d'Astrophysique, 4

Smith, P.S., Schmidt, G.D., Allen, R.G. & Angel, J.R.P. 1995, ApJ, 444, 146

Turner, T.J. 1998, ApJ, in press

Weedman 1986, Quasar Astronomy, Cambridge: Cambridge University Press, 107

Weedman, D.W., Wolovitz, J.B., Bershady, M.A., & Schneider, D.P. 1998, AJ, 116, 1643

Active Galactic Nuclei and Related Phenomena
IAU Symposium, Vol. 194, 1999
Yervant Terzian, Daniel Weedman, Edward Khachikian, eds.

What Physics Drives the Unified Model?

Michael A. Dopita

Mt. Stromlo & Siding Spring Observatory, Institute of Advanced Studies, The Australian National University, Australia

Abstract. The Unified Model holds that the aspect-dependent effects primarily determine the apperance of the active galactic nucleus that we observe. However, additional parameters will be needed to fully unify the different tribes of AGN. Three parameters; aspect, accretion rate into the nuclear regions, and the evolutionary status of the central black hole probably hold the key to "grand" unification schemes.

1. What is the Unified Model?

The Black Hole (BH) paradigm of Active Galactic Nuclei (AGN) is now universally accepted, and explains the structure of AGN in terms of an accretion disk which may or may not include a relativistic jet emerging in the poleward direction. The "unified model" was developed to clarify the relationship between the various sub-classes of AGN (Rowan-Robinson 1977; Lawrence and Elvis 1982; Antonucci and Miller 1985; Lawrence 1987) and has been summarised in two excellent reviews (Antonucci 1993 and Urry and Padovani 1995) which are essential reading. The basic idea behind the unified model is that orientation effects exercise a profound influence on the appearance of the AGN, and determine the sub-class into which it will be categorised. The "strong" unification hypothesis; that all AGN classes are fundamentally the same phenomenon seen at different orientations, has not proven successful. However, in its more restricted form; that some classes of object transform into other classes as a function of viewing angle, it seems secure. The cause of the orientation effect is believed to be a geometrically thick and optically-thick dusty molecular torus located about the equatorial plane of the AGN which may obscure it from direct view.

A second orientation effect is found in the radio-loud categories, for which the unification concept was orginally developed (Orr & Browne 1982; Barthel 1991). AGN which are observed close to the polar direction display apparently super-luminal motions, rapid variability, intense power-law continua, flat radio spectra and high linear polarisation all due to relativistic beaming effects associated with the jet. The BL Lac objects, and the optically-violently variable (OVV) or flat-spectrum radio quasars (FSRQs) all seem to fit into this category (Blandford and Rees, 1978; Blandford and Konigl 1979). Combining these two effects, the following transformations apply. Seyfert 1 ⇔ Seyfert 2; QSOs ⇔ Ultraluminous IRAS Narrow-line Galaxies; BL Lac / Blazar⇔ Double lobe radio Elliptical Galaxy.While these are satisfying as far as they go, there is still no way to fully unify the various tribes of AGN. In particular, there is no explanation of the radio-loud : radio-quiet dichotomy. Clearly, a parameter other than orientation must play a role. It has been variously argued that radio loudness is related to host galaxy type (Smith et al. 1986), black hole spin (Blandford 1990; Wilson and Colbert 1995), or to differences in the rate of nuclear feeding (Rees et al. 1982; Baum, Zirbel and O'Dea 1995).

2. How Can the Dusty Torus Exist?

The dusty torus is usually depicted as a fat doughnut surrounding, or forming the outer edge of, the inner thin accretion disk. Such a configuration is, of course, not dynamically consistent with a steady state, since if it is composed of a continuous gaseous medium, it will collapse to a thin disk on a timescale of order half the orbital timescale. Nor, as pointed out by Krolik & Begelman (1988), can it be supported by internal gas pressure, since, in order for it to be geometrically thick, it would have to have a temperature of order 10^6 K. Any dust present would then be very rapidly destroyed. Physically self-consistent models for a thick torus depend upon the existence of an accretion flow towards the central engine.

According to Krolik & Begelman (1988) the dusty gas is confined into a large number of self-gravitating clouds in dynamic balance between mergers and tidal shearing. Cloud-cloud collisions drive a net inward flow. The mean column density of the clouds in this model is very large ($\sim 10^{24}$ cm^2). However this seems incompatible with obervations of rotational support of the dense regions of the accretion flow (Jackson et al. 1993), and the relatively low line-of-sight column densities ($\sim 10^{23}$ cm^2) inferred for the thick torus (Dopita et al. 1998). The X-ray absorption region must be located closer to the central engine.

The alternative picture, that the thick torus simply represents a large-scale accretion flow of a continuous medium toward the nucleus, is more compatible with observation. In this model, the inner boundary of the torus is determined by the dust sublimation point. Near the mid-plane the rotationally-supported gas will fall almost perpendicular to the accretion disk, and is brought to rest by passage of the gas through a symmetric pair of accretion shocks. Thus, such a structure produces a dense, geometrically thin accretion disk embedded in the mid-plane, and extending within, the inner edge of the thick torus.

3. What is the Role of Accretion?

Merger events act on the gas through torques and dissipation to dump a large fraction ($\sim 10^{10} M_\odot$) into the nuclear regions (*e.g.* Solomon, Downes and Radford 1992). These processes have been amply confirmed by theoretical calculations (Barnes and Hernquist 1991, 1992, 1996). At the point where orbital support of the gas becomes important (within 300 pc of the centre) a strong accretion shock is formed, which is likely to trigger a major nuclear starburst within a dynamical timescale of $\sim 10^8$ years.

Matter dumped into the central kiloparsec of a galaxy has still to lose a great deal of angular momentum before it can be fed down into the AGN. However, star formation itself provides an efficient means of re-distributing the angular momentum. See, for example, Collin (this volume), or Bekki (1995). If these processes work down to small enough scales within the accretion disk, then we may expect a symbiosis between the amount of gas fed down to the AGN, and the amount of gas converted to stars in the accretion flow. This may provide a physical basis for the Kormendy relationship between the stellar bulge mass, and the BH mass, which seems to apply to both E and S galaxies.

As far as the unified model is concerned, the appearance of the AGN should depend critically upon the rate of accretion into the nuclear regions. For sub-Eddington accretion we would obtain the classic Shakura and Sunyaev (1973) thin disk, and there should be no optically-thick accretion torus. Such disks allow the free escape of the relativistic jet from the nuclear region, but are ineffective in producing either a broad-line or a extended narrow line emission region. The radio jets in such systems are characterised by efficient relativistic boosting, and can therefore give rise to Blazar phenomena. The sub-Eddington (and possibly, advective) nature of the accretion will ensure that such sources are of relatively low luminosity, and of low radio power. Such objects might be identifiable with the FR I radio sources as postulated in the unified models of radio sources (Urry and Padovani 1995).

At very high (super-Eddington) accretion rates into the nuclear regions, the accumulation of dusty matter in the nuclear regions tends to obscure the central engine. This may be identified with the dusty "thick" torus of the unified models. Indeed, Dopita et al. (1998) have demonstrated that if the thick dusty accretion torus is an optically thick accretion flow, it must be highly super-Eddington. If the flow of gas into the broad-line region is also super-Eddington, much of the matter entering these regions must be lost in the form of a thermal or magnetic wind (*e.g.* Chiang and Murray, 1996; Lovelace, this volume). At high enough mass-loss rates, this wind will become optically-thick to electron scattering, providing a hot electron-scattering photosphere to reprocess the hard radiation (EUV, X- and γ-rays) from the central BH and the innermost portions of the accretion disk. This would serve to obscure the central engine from direct view except in the polar directions where the relativistic jets may escape. Such a toroidal reprocessing photosphere would also provide enough absorption to explain the weakness of Seyfert 2 galaxies in X-rays (Mushotzky, Done & Pounds 1993), and its inner surface could also be used for K-α scattering. In this model, an AGN seen in an intermediate angle (but outside the "thick" dusty torus) would appear as either a QSO or as a Sy 1 galaxy, depending on the BH mass.

Models with a thermal radiation-driven wind allow the possibility of direct interaction between the thermal and relativistic winds. This interaction is favoured when the accretion into the BLR is highly super-Eddington with respect to the nuclear BH. When such interaction occurs below the reprocessing photosphere, it will lead to mass entrainment into the jet, and a slowing of the jet to highly sub-relativistic speeds. This is likely to be the cause of radio-loud: radio-quiet dichotomy, as discussed below. Consider a simple (toy) parametric form for the mass accretion rate into the nuclear core region;

$$\dot{M}(t) = \dot{M}_{av} \exp[-t/\tau_{acc}] \times (1 - \exp[-t/\tau_{acc}]) \qquad (1)$$

with only one characteristic timescale, the accretion timescale:

$$\tau_{acc} = \lambda \tau_{ff} = \lambda R^{3/2}/(GM)^{1/2} \qquad (2)$$

For $R \approx 10$ kpc and $M \approx 10^{10} M_\odot$, $\tau_{acc} \approx 1.5 \times 10^8 \lambda$ years. The accretion rate is super-Eddington provided:

$$\dot{M}_{BH} = M_{BH}/\tau_{BH} \tag{3}$$

where τ_{BH} is the growth timescale for the BH accreting at the Eddington limit. This can be calculated from the luminosity:

$$L_{BH} = \phi \dot{M}_{BH} c^2 = \theta L_{edd} = \theta 4\pi c G M_{BH}/\kappa \tag{4}$$

where ϕ is the fraction of the rest mass energy radiated by matter falling into the BH. For a Schwartzschild BH $\phi = [1 - (2^{3/2}/3)]$. Typically $\phi \sim 0.1$. The factor θ is the fraction of the Eddington Luminosity produced by this accretion, and is assumed ~ 1. Thus, the growth timescale of the BH is:

$$\tau_{BH} = M_{BH}/\dot{M}_{BH} = \frac{\kappa c}{4\pi G}\left(\frac{\phi}{\theta}\right) \approx 4.4 \times 10^8 \left(\frac{\phi}{\theta}\right) \; years \tag{5}$$

inserting numerical values, $\tau_{BH} \approx 5.10^7$ years. Thus, the growth timescale of the BH is of the same order than, or somewhat smaller than τ_{acc}, so that appreciable growth of the BH (from $10^{5-7} M_\odot$ to $10^{8-9} M_\odot$) can occur during a merging/dynamical collapse event. During this event, molecular clouds are presumably being converted rapidly to stars in a massive nuclear starburst, so that the growth of the BH is determined by the competition between inflow, accretion, and star formation.

4. An Accretion Disk Photosphere?

During super-Eddington accretion and possibly even later, the radiation pressure from the central BH would be sufficient to drive a thermal or magnetic wind. If this wind is accelerated by radiation pressure of the central source, the detailed theory has already been developed in a series of papers by Murray and his collaborators (Murray and Chiang, 1995; Murray et al. 1995; Chiang and Murray, 1996). For sufficient outflow rates the radiatively driven wind will be both dense, and optically thick to the escape of X- and γ- ray photons from the central source so that the continuum observed is produced by a reprocessing photosphere dominated by electron scattering opacity. In this case, the momentum flux in the wind is given by:

$$\dot{M}_w v_w = \left(\frac{\eta}{c}\right)\left(\frac{\Omega}{4\pi}\right) L_{BH} \tag{6}$$

where Ω is the solid angle covered by the optically-thick radiatively-driven wind, subtended at the BH, and η is the effective number of scatterings per photon. In stars, this factor is greater than unity; typically $\eta \sim 3 - 4$; however, in Wolf-Rayet stars it may rise even higher. In such radiatively-driven winds, the outflow velocity is similar to the escape velocity at the base of the outflow. In the AGN case the wind originates at the inner edge of the region of super-Eddington flow, and has a velocity of order 3×10^4 km s^{-1}. The photospheric radius occurs when the electron scattering optical depth is of order unity:

$$r_{phot} = \kappa \left(\frac{\Omega}{4\pi}\right)^{-1} \left(\frac{\dot{M}_w}{4\pi v_w}\right) \sim 1.0 M_7 \text{ light days} \qquad (7)$$

where κ is the electron scattering opacity. This is consistent with variability studies and reverberation mapping analysis of nearby Sy 1 galaxies (Maoz, 1994). The effective temperature of the electron-scattering photosphere is given by Stefan's Law, with the radiation density diluted by a factor $\Psi \sim 10$ with respect to a Black-Body emitter;

$$T = \left(\frac{4\pi c^2 \Psi}{\sigma \kappa^2 \eta^2}\right)^{1/4} v_w L_{BH}^{-1/4} \sim 37000K \left[\frac{\Psi}{10}\right]^{1/4} \left[\frac{\eta}{3}\right]^{-1/2} v_4 L_{45}^{-1/4} \qquad (8)$$

Thus, when the wind velocity is of order 20,000 km s^{-1} the effective photospheric temperature is high enough ($\sim 10^5$ K) to provide a "big blue bump" in the continuum spectrum which, thanks to the high outflow velocity and the electron scattering, should provide only weak and broad photospheric lines. Such electron scattering dominated extended atmospheres will show only a weak Lyman Limit discontinuity. The spectral distribution of such an electron scattering photosphere is not characterised by a simple temperature. In general, such atmospheres give a power-law below the peak in emergent flux corresponding to the temperature given by eqn. (8), and roll off sharply above this peak. The power-law slope depends on the effective curvature of the atmosphere, *i.e.* on the ratio $\dot{M}_w/v_w$ (Abbott & Conti, 1987). AGN atmospheres should have little curvature, giving a rather flat power-law at lower energies.

This photosphere is extended on both sides of the accretion disk, and illuminates and photoionises the surface layers or the infalling matter, giving rise to the broad-line region. The effective source temperature decreases for higher BH luminosities, providing lower ionisation conditions in the more luminous AGN. This is presumably the explanation of the Baldwin Effect in QSOs (Baldwin 1977, Baldwin *et al.* 1978, Kinney, Rivolo and Koratkar 1990). The wind equations imply that:

$$L_{BH} = \frac{4\pi \epsilon c G M_{BH}}{\kappa \eta} = \left(\frac{\epsilon}{\eta}\right) L_{Edd} \qquad (9)$$

Consistency therefore requires $\epsilon \sim \eta$ if the BH luminosity is to be of the same order as the Eddington Limit. The mechanical energy flux in the wind is simply related to the bolometric luminosity of the central object:

$$L_{Mech} = \left(\frac{\eta v_w}{2c}\right)\left(\frac{\Omega}{4\pi}\right) L_{BH} = 0.05 \left[\frac{\eta}{3}\right]\left[\frac{\Omega}{4\pi}\right] v_4 L_{BH} \qquad (10)$$

Finally, we can relate the mass flux in the wind to the mass flux onto the BH:

$$\left(\frac{\dot{M}_w}{\dot{M}_{BH}}\right) = \left(\frac{\eta \phi c}{v_w}\right)\left(\frac{\Omega}{4\pi}\right) = 9 \left[\frac{\phi}{0.1}\right]\left[\frac{\eta}{3}\right]\left[\frac{\Omega}{4\pi}\right] v_4^{-1} \qquad (11)$$

Thus, the mass flux into the radiatively driven wind will dominate over the mass flux into the BH until the accretion disk becomes thin ($\Omega/4\pi \sim 0.1$).

5. The Jet Model

Let us adopt the jet–black hole symbiosis model of Falke and Biermann (1995) and Falke, Malkan and Biermann (1995):

$$L_{jet} = \gamma_{jet} \dot{M}_{jet} c^2 \approx 0.3 L_{BH}; \quad 3 \leq \gamma_{jet} \leq 10 \tag{12}$$

It therefore follows immediately that:

$$\left(\frac{\dot{M}_{jet}}{\dot{M}_{BH}}\right) = \left(\frac{\phi}{3\gamma_{jet}}\right) \tag{13}$$

Thus, the mass flux in the jet is only about 1% of the mass flux into the BH, and more importantly, is only about 0.1% of the mass flux in the radiatively-driven wind, in the super-Eddington regime. This implies that any mixing between the jet and the thermal wind will be critical in determining whether the jet can escape or is mixed with thermal matter within the radio photosphere of the wind (determined by its radio free-free opacity). In the first case the object will be radio-loud – this is the case usually considered. In the second case, the very luminous synchrotron emission, and the synchrotron Compton losses all occur below the photosphere, and what emerges is a fast thermal wind (seen in some objects as a Broad Absorption Line QSO) which carries the energy flux to excite the narrow-line emission. This has only a small synchrotron component and produces the faint radio-emitting bubbles seen in radio-quiet objects such as Seyferts (Bicknell et al. 1998). Bicknell, Dopita and O'Dea (1997) have shown that the jet energy flux is related to the synchrotron power at frequency ν through an efficiency factor:

$$\kappa_\nu = \left(\frac{5-\delta}{8-\delta}\right) f(t, \gamma_0, B, \nu, \alpha) = c E_0^{2\alpha} F_e B^{(\alpha+1)} \nu^{-\alpha} t \tag{14}$$

where δ is the index of the power law in the density distribution of the galactic medium (~ 2), t is the age of the source, B the magnetic field in the radio lobes, E_0 is the lower energy limit of the relativistic electrons, α is the spectral index, and F_e is the fraction of the internal energy of the plasma contained in relativistic electrons. In the super-Eddington phase, mixing with the thermal wind gives $F_e \sim 10^{-3}$. On the other hand, highly sub-Eddington accretion allows for the free escape of the relativistic plasma so $F_e \sim 1$. However, radio-quiet objects have lobe luminosities which are typically weaker by a factor of a thousand than their radio-loud counterparts. Thus, for a given jet energy flux, this difference in the entrainment factor is in itself sufficient to explain the difference between the radio-quiet, and the radio-loud cases, respectively, provided that the entrainment has occurred below a radio photosphere.

The transition between the radio-quiet and radio-loud cases presumably comes about at the time that the accretion rate into the BLR decreases to the point that the thermal wind starts to clear in the polar regions and entrainment into the radio jet is no longer enough to slow this to transonic speeds. This allows for transition objects displaying *both* BLRs and relativistic jets (radio loud QSOs).

6. What is the Role of Evolution and Aspect?

In the case of a major merger event between two massive gas-rich systems, the initial phase should be a galaxy-wide starburst producing a luminous infrared galaxy with HII-like characteristics. In the later phases, gas flow toward the nucleus feeds the BH and outflow could then excite the Seyfert 2 - like emission seen in some of the very distant ultraluminous IRAS galaxies (Rowan-Robinson, et al. 1991, 1993; van Ojik 1994). Such mergers will initially be radio-quiet, since the supply of gas into the nuclear regions is sufficient to choke off high-luminosity radio jets, but as nuclear accretion rates fall, radio jets might later escape to produce a high z radio source with its strong narrow-line emission region (Meisenheimer, Hippelein & Neeser 1994; McCarthy, Spinrad and van Breugel 1995). Such mergers of gas-rich systems would have been much more common in the early universe and especially at the epoch of formation of the Abell clusters. Since this epoch was so early, $z \sim 2-4$, the merging gas-rich systems may not even have had time to form regular spiral galaxies. However, let us call them 'spirals' for convenience. It is therefore reasonable to propose the following evolutionary scenario for 'primary' merger events giving rise to the QSOs and the majority of elliptical galaxies.:

$$\begin{aligned} \text{Sp} + \text{Sp} &\Rightarrow \text{Starburst} \\ &\Rightarrow \text{Postmerger} + \text{RQQSO} \\ &\Rightarrow \text{E} + \text{massive BH} + \text{radio jets (?)} \end{aligned}$$

In the event of a later merger event:

$$\begin{aligned} \text{E} + \text{Irr} &\Rightarrow \text{Dust-lane E} + \text{Nuclear starburst} \\ &\Rightarrow \text{Dust-lane E} + \text{GPS radio source} \\ &\Rightarrow \text{Dust-lane E} + \text{FR II radio source} \\ &\Rightarrow \text{Dust-lane E} + \text{FR I radio source} \end{aligned}$$

Alternatively, the case of M87 suggests that low-power radio sources may be fed by a subsequent accretion episode at sub-Eddington rates, such as feeding through a cooling flow:

$$\text{E} + \text{Cooling flow} \Rightarrow \text{E} + \text{LINER} + \text{FRI} \quad or \quad \text{E} + \text{Blazar}$$

Thus, evolution determines the nature of the AGN, but aspect determines the type of AGN we see from outside.

7. Conclusions

The arguments given here (for an extended account, *see* Dopita 1997), establish the feasibility of the hypothesis that galactic mergers determine, as much as does orientation, the nature and evolutionary stage of the AGN that we observe. Such mergers also produce a massive star burst, linking them with luminous and ultra-luminous infrared galaxies. The size of the BHs produced depend on the total gas supply. For much of the early growth phases, the supply of gas into the region of influence of the BH is likely to be super-Eddington. This is capable of producing an optically-thick radiatively-driven wind, which may choke off the relativistic jets in this phase, producing a radio-quiet AGN.

For sub-Eddington accretion rates, such as found in either elliptical mergers or in cooling flows, the radio jets can freely escape from the thin accretion disk, so radio-loud AGN and relativistically-boosted BL Lac /Blazer phenomena will be preferentially produced in this case.

Acknowledgments. I wish to thank Ski Antonucci, Geoff Bicknell, Stefi Baum, Anuradha Koratkar, Ralph Sutherland, Pete McGregor, Charlene Heisler, Agris Kalnajs, Mark Whittle and Wil van Breugel for many stimulating conversations and physical insights in the preparation of this work. I also wish to thank the Australian Dept. of Industry, Science and Tourism (DIST) for support under an IS&T major grant.

References

Abbott, D. C., & Conti, P. S. 1987, Ann. Rev. A & A, 25, 113.
Antonucci, R. 1993, Ann. Rev. A& A, 31, 473
Antonucci, R. & Miller, J.S. 1985, ApJ, 297, 621.
Baldwin, J. A. 1977, ApJ, 214, 679.
Baldwin, J. A., Wampler, E. J., & Gaskell, C. M. 1978, Nature, 273, 431.
Barthel, P. D. 1991, Advances in Space Research, 11, 231.
Barnes, J. E., & Hernquist, L. 1991, ApJ, 370, 65.
Barnes, J. E., & Hernquist, L. 1992, Ann. Rev. A&A, 30, 705.
Barnes, J. E., & Hernquist, L. 1996, ApJ, 471, 115.
Baum, S. A., Zirbel, E. I. & O'Dea, C. P. 1995, ApJ, 451, 88.
Bekki, K. 1995, MNRAS, 276, 9.
Bicknell, G.V, Dopita, M.A. & O'Dea, C. 1997, ApJ 490,202.
Bicknell, G.V et al. 1998, ApJ, 495, 680.
Blandford 1990, in *Active Galactic Nuclei*, eds. T. J.-L. & M. Mayor Saas-Fee Advanced Course 20, (Berlin: Springer), p161.
Blandford, R.D. & Rees, M. J. 1978, in *Pittsburgh Conference on BL Lac Objects*, ed. A. N. Wolfe (Pittsburgh: U.Pittsburg Press), p328.
Blandford, R. D. & Konigl, A. 1979, ApJ, 232, 34.
Chiang, J. & Murray, N. 1996, ApJ, 466, 704.
Dopita, M. A. 1997, PASA, 14, 230.
Dopita, M. A., et al. 1998, ApJ, 498, 570.
Falcke, H., & Biermann, P.L. 1995, A&A., 293, 665..
Falcke, H., Malkan, M.A. & Biermann, P.L. 1995, A&A., 298, 375.
Kinney, A. L., Rivolo, A. R., & Koratkar, A. P. 1990, ApJ, 357, 338.
Krolik, J.H. & Begelman, M.C. 1988 ApJ, 329,702.

Lawrence, A. 1987, PASP, 99, 309.
Lawrence, A. & Elvis, M. 1982, ApJ, 256, 410.
McCarthy, P. J., Spinrad, H., & van Breugel, W. 1995, ApJ, 447, 77.
Maoz, D. 1994, in *Reverberation Mapping of the Broad-Line Region in AGN*, eds. P. M. Gondhalekar, K. Horne, & B. M. Peterson, ASP Conf. Ser, 69, 95.
Meisenheimer, K., Hippelein, H. & Neeser, M. 1994, *The First Stromlo Symposium : The Physics of Active Galaxies*, eds. G. V. Bicknell, M. A. Dopita & P. J. Quinn, ASP Conf. Ser., 54, 397
Murray, N., & Chiang, J. 1995, ApJ, 459, L105.
Murray, N. et al. 1995, ApJ, 451, 498.
Mushotzky, R. F., Done, C., & Pounds, K. A. 1993, Ann. Rev. A&A, 31, 717.
Orr, M. J. L., & Browne, I. W. A. 1982, MNRAS, 200, 1067.
Rees, M. et al. 1982, Nature, 295, 17.
Rowan-Robinson, M. 1977, ApJ, 213, 638.
Rowan-Robinson, M. et al. 1991, Nature, 351, 719.
Rowan-Robinson, M. et al. 1993, MNRAS, 261, 513.
Shakura, N. I., & Sunyaev, R. A. 1973, A&A, 24, 337.
Smith, E.P. et al. 1986, ApJ, 306, 64.
Urry C.M, & Padovani, P. 1995, PASP, 107, 803.
van Ojik, R., et al. 1994, A&A, 289, 54.
Wilson, A. S., & Colbert, E. J. M. 1995, ApJ, 438, 62.

Magnetohydrodynamic Origin of Jets from Accretion Disks

R.V.E. Lovelace,

Department of Astronomy, Cornell University, Ithaca, NY 14853; rvl1@cornell.edu

G.V. Ustyugova

Keldysh Institute of Applied Mathematics, Russian Academy of Sciences, Moscow, Russia, 125047, ustyugg@spp.Keldysh.ru

A.V. Koldoba

Institute of Mathematical Modelling, Russian Academy of Sciences, Moscow, Russia, 125047

Abstract. A review is made of recent magnetohydrodynamic (MHD) theory and simulations of origin of jets from accretion disks. Many compact astrophysical objects emit powerful, highly-collimated, oppositely directed jets. Included are the extra galactic radio jets of active galaxies and quasars, and old compact stars in binaries, and emission line jets in young stellar objects. It is widely thought that these different jets arise from rotating, conducting accretion disks threaded by an ordered magnetic field. The twisting of the **B** field by the rotation of the disk drives the jets by magnetically extracting matter, angular momentum, and energy from the accretion disk. Two main regimes have been discussed theoretically, hydromagnetic winds which have a significant mass flux, and Poynting flux jets where the mass flux is negligible. Over the past several years, exciting new developments on models of jets have come from progress in MHD simulations which now allow the study of the origin - the acceleration and collimation - of jets from accretion disks. Simulation studies in the hydromagnetic wind regime indicate that the outflows are accelerated close to their region of origin whereas the collimation occurs at much larger distances.

1. Introduction

Powerful, highly-collimated, oppositely directed jets are observed in active galaxies and quasars (see for example Bridle & Eilek 1984), old compact stars in binaries (Mirabel & Rodriguez 1994), and emission line jets in young stellar objects (Mundt 1985; Bührke, Mundt, & Ray 1988). A broad spectrum of ideas and models have been put forward to explain astrophysical jets (see reviews by Begelman, Blandford, & Rees 1984 and Bisnovatyi-Kogan 1993). The matter is thought to go to the jet from the inner region of an accretion disk surrounding

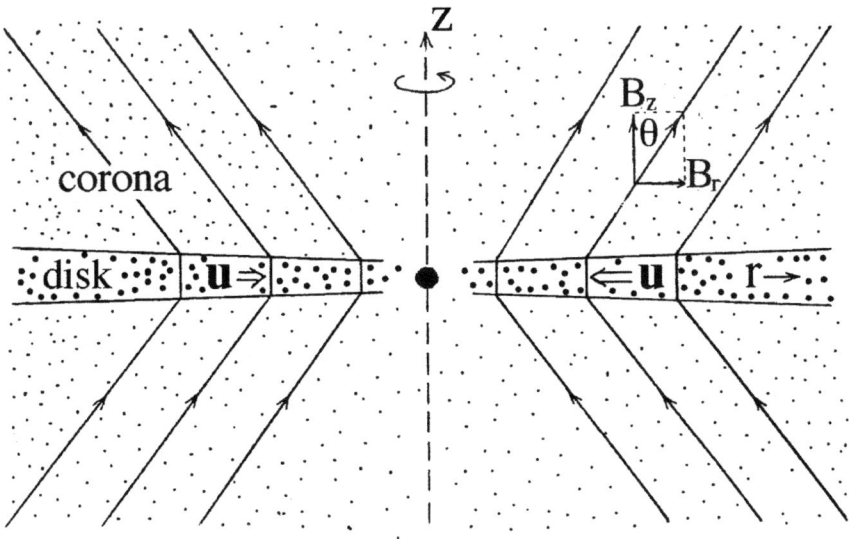

Figure 1. Sketch of an accretion disk threaded by a magnetic field for conditions which may lead to hydromagnetic jet formation.

the compact object - a star or black hole. The disk matter must then be accelerated to a velocity higher that the escape velocity from the central object. Further, the jet matter should have sufficient momentum to propagate through surrounding inter-stellar or intra-galactic matter out to large distances.

An ordered magnetic field is widely thought to have an essential role in jet formation from a rotating accretion disk. Two main regimes have been considered in theoretical models, the *Poynting flux regime* where the energy outflow from the disk is carried mainly by the electromagnetic field with the energy carried by the matter small, and the *hydromagnetic regime* where the energy is carried by both the electromagnetic field and the kinetic flux of matter. Poynting flux models for the origin of jets were proposed by Lovelace (1976) and Blandford (1976) and studied further by Lovelace et al. (1987) and Colgate & Li (1998). In these models the rotation of a Keplerian accretion disk twists a poloidal field threading the disk, and this results in outflows out of the disk which carry angular momentum (in the twist of the field) and energy (in the Poynting flux) away from the disk, thereby facilitating the accretion of matter.

Important questions to be answered by the jet models include the following: **1.** What is the main driving force pushing matter into the jet? **2.** What determines the mass flow rate in the jet $\dot{M}_j$, and what fraction is this of the accretion rate? **3.** What physics determines the asymptotic speed or Lorentz factor of the bulk flow? **4.** What determines the collimation of the jet and at what distance from the central object does the jet become collimated? **5.** What is the acceleration mechanism of leptons to Lorentz factors $\gamma \sim 10^2 - 10^3$ in the radio jets? Observations of Blazars indicate that $\gamma \sim 10^3 - 10^5$ in some objects.

The focus of recent work has been the hydromagnetic regime of jet formation for the geometry sketched in Figure 1. A strong case for hydromagnetic jets as an explanation of jets in protostellar systems emerges because the temperature of the inner regions of these systems is insufficient to permit driving by thermal or radiation pressure (Königl & Ruden 1993). Part of the investigations have been analytical or semi-analytical and outgrowths of the self-similar solution of Blandford & Payne (1982) (Pudritz & Norman 1986; Königl 1989; Pelletier & Pudritz 1992; Contopoulos & Lovelace 1994). The outflows in this model are often referred to as "centrifugally driven" owing to the driving force close to the disk: If the poloidal magnetic field lines diverge from the disk surface (making an angle with the z-axis of more than 30°), then the sum of the gravitational and centrifugal forces is in the $+z$ direction for an MHD fluid particle which initially tends to maintain a constant angular rotation rate. The self-similar models are unsatisfactory in the respect that they must be cutoff at small cylindrical radii, $r \leq r_{min}$. This is the most important region of the jet flow. Observations of optical stellar jets (Mundt 1985) reveal jet velocities $\sim 200 - 400$ km/s, which are comparable to the Keplerian disk velocity close to the star's surface. This suggests that the jets originate from the inner region of the disk close to the star (Shu et al. 1988; Pringle 1989).

The limitations of analytical models has motivated efforts to study jet formation using MHD simulations. Simulation studies of hydromagnetic jet formation have addressed two main regions: The jet formation region where the matter enters with sub slow-magnetosonic speed and exits with super fast-magnetosonic speed. The second region includes the disk and the problem of the Velikhov (1959) - Chandrasekhar (1981) - Balbus-Hawley (1998) instability and the resulting 3D MHD turbulence. A number of studies have addressed the coupled problem of the disk and near jet regions (Uchida & Shibata 1985; Shibata & Uchida 1986; Stone & Norman 1994; Bell and Lucek 1995;). MHD simulations of the near jet region have been carried out by several groups (Bell 1994; Ustyugova et al. 1995, 1998; Koldoba et al. 1995; Romanova et al. 1997, 1998; Meier et al. 1997; Ouyed & Pudritz 1997).

Here, we first review recent results on MHD simulations of hydromagnetic jet formation and later discuss the Poynting flux regime. Section 2 discusses general considerations of hydromagnetic outflows. Sections 3 discusses MHD simulations which give non-stationary and stationary hydromagnetic outflows. Sections 4 describes the Poynting flux regime which remains to be fully explored by simulations. Section 5 gives the conclusions.

2. Basics of MHD Outflows from Disks

The main forces which drive a hydromagnetic outflow from a disk threaded by a magnetic field are the centrifugal force and the magnetic pressure gradient force. If disk has a hot corona, the pressure gradient may also be important. We neglect the radiative force but in this regard see Phinney (1987). Accreting matter of the disk carries magnetic field inward thus generating a B_r component of the magnetic field as sketched in Figure 1. On the other hand, rotation of the disk acts to generate a toroidal component of the field B_ϕ (< 0 if $B_z > 0$).

For a sufficiently inclined magnetic field (θ in Figure 1 sufficiently large), outflows can result from the centrifugal force (Blandford & Payne 1982) and/or the magnetic pressure gradient force ($-\nabla_z B_\phi^2/(8\pi)$) (Lovelace et al. 1989, 1991; Koupelis & Van Horn 1989). This depends on the ratio of energy densities at the base of the outflow at the inner radius of the disk denoted r_i. Thus, the main parameters are $\varepsilon_{th} = (c_s/v_K)_i^2$ and $\varepsilon_B = (v_A/v_K)_i^2$, where v_K is Keplerian velocity, c_s the sound speed, v_A the Alfvén speed, and the i subscript indicates evaluation on the surface of the disk at its inner radius, $r = r_i$. For $\varepsilon_B \sim 1$ the outflow is magnetically driven, whereas for $\varepsilon_{th} \sim 1$, the flow is thermally driven.

Processes in the disk are of course coupled to the outflows (Lovelace et al. 1994, 1997; Falcke, Malkan, & Biermann 1995). However, it is difficult to simulate both regions simultaneously because the time scales of the accretion and outflow are in general very different. The accretion is much slower. On the other hand the processes in the disk may involve the small scale MHD instability of Chandrasekhar, Velikhov, Balbus, and Hawley, and therefore require high spatial resolution. Stone & Norman (1994) attempted to simultaneously simulate the internal MHD dynamics of a disk and MHD dynamics of outflows. This proved impractical because essentially all of the spatial resolution was needed for treating the unstable dynamics of the disk. Also, there was the problem that the initial configuration was far from equilibrium. Simulation of the internal MHD disk dynamics has led several groups to the problem of simulating 3D MHD turbulence in a sheared flow of a patch of a disk (for example, Hawley et al. 1995; Brandenburg et al. 1995). This is a much larger project than that of understanding MHD outflows. At the same time it is widely thought, and observations of cataclysmic variables support the view, that the disk turbulence - including MHD turbulence - can be modeled approximately using the Shakura (1973), Shakura & Sunyaev (1973) "alpha" viscosity model (Eardley & Lightman 1975; Coroniti 1981). In contrast with the internal disk dynamics, there is theoretical and simulation evidence that the outflows can be treated using axisymmetric (2D) MHD (Blandford & Payne 1982; Lovelace et al. 1991; Ustyugova et al. 1995). Here, we consider outflows from a disk represented as a boundary condition. This approach has subsequently been adopted by other groups (Meier et al. 1997; Ouyed & Pudritz 1997). This treatment of the disk is justified for outflows from a disk where the accretion speed is small compared with the Keplerian speed (Ustyugova et al. 1995).

3. Numerical Simulations of MHD Outflows

In order to test the analytical models of stationary outflows, MHD simulations of flows from a disk treated as a boundary condition have been carried out by a number of groups.

3.1. Non-Stationary Outflows

The initial magnetic field configuration was chosen so that the magnetic field was significantly inclined to the disk ($\theta > 30°$) over most of the disk surface (Ustyugova et al. 1995; Koldoba et al. 1995). The simulations involve solving the complete system of ideal non-relativistic MHD equations using a Godunov-type code assuming axisymmetry but taking into account all three components

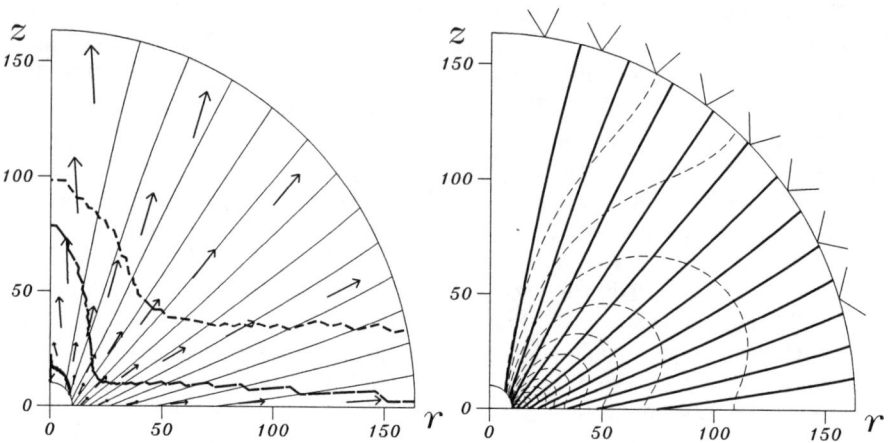

Figure 2. Simulations results for stationary MHD outflow obtained using a spherical coordinate system (Ustyugova et al. 1998). The solid lines represent the poloidal magnetic field, and the arrows the velocity vectors. The dashed lines in the left-hand plot represent the slow magnetosonic surface (the lowest dashed line), the Alfvén surface (the middle line), and the fast magnetosonic surface (the top line). In the right-hand panel, the dashed lines are surfaces of constant toroidal current density, while the lines on the outer boundary are the projections of the fast magnetosonic Mach cones.

of velocity and magnetic field. Matter of the corona was initially in thermal equilibrium with the gravitating center. At $t = 0$, the disk is set into rotation with Keplerian velocity and at the same time matter is pushed from the disk with a small poloidal velocity equal to a fraction of the slow magnetosonic velocity ($v_p = \alpha v_{sm}$, with $\alpha = 0.1 - 0.9$). A relatively high temperature and small magnetic field was considered. We found that at the maximum of the outflow, matter is accelerated to speeds in excess of the escape speed and in excess of the fast magnetosonic speed within the simulation region ($\sim 100 r_i$). The acceleration is due to both thermal and magnetic pressure gradients. The outflow collimates within the simulation region due to strong amplification, 'wrapping up' of the toroidal magnetic field and the associated pinching force.

However, the outflows are *not stationary*. The matter flux grows to a peak and then decreases to relatively small values. The strong collimation of the outflow reduces the divergence of the field away from the z-axis ($\theta < 30°$) and this "turns off" the outflow of matter and leads to flow velocities less than the escape speed. Thus, this simulation is an example of a temporary outburst of matter to a jet. Unfortunately, this type of flow has a significant dependence on the initial conditions.

3.2. Stationary Outflows

More recently, stationary magnetohydrodynamic outflows from a rotating accretion disk have been obtained by time-dependent axisymmetric simulations by

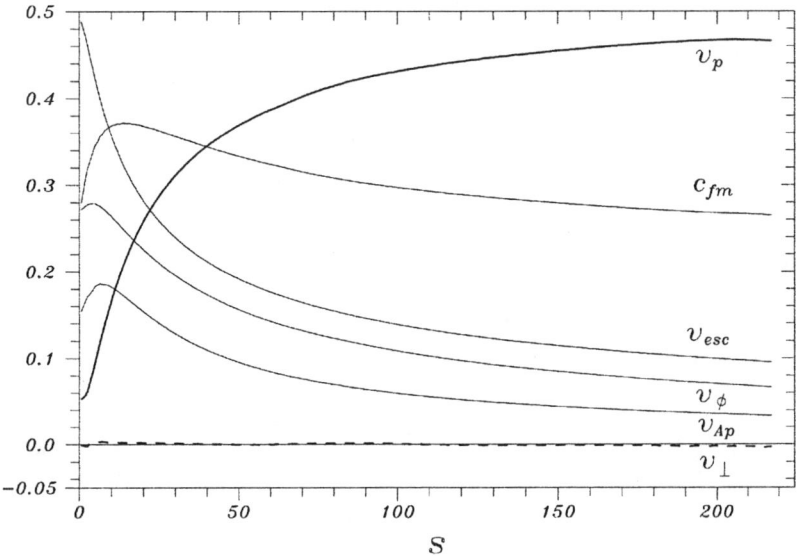

Figure 3. Dependences of different velocities on distance s measured in units of r_i from the disk along the second magnetic field line away from the axis in Figure 2 (Ustyugova et al. 1998). This field line "starts" from the disk at $r \approx 6r_i$ where it has an angle $\theta \approx 28°$ relative to the z-axis. The velocities are measured in units of $\sqrt{GM/r_i}$. Here, v_p is the poloidal velocity along the field line and $v_\perp$ is the poloidal velocity perpendicular to the field line. Also, v_{Ap} is the poloidal Alfvén velocity, c_{fm} is the fast magnetosonic velocity, and v_{esc} is the local escape velocity.

Romanova et al. (1997) and systematically analyzed by Ustyugova et al. (1998). The initial magnetic field in the latter work was taken to be a split-monopole poloidal field configuration (Sakurai 1987) frozen into the disk. The disk was treated as a perfectly conducting, time-independent density boundary $[\rho(r)]$ in Keplerian rotation which is different from our earlier specification of a small velocity outflow (§3.1, Ustyugova et al. 1995). The outflow velocity from the disk is determined self-consistently from the MHD equations. The temperature of the matter outflowing from the disk is small in the region where the magnetic field is inclined away from the symmetry axis $(c_s^2 \ll v_K^2)$, but relatively high $(c_s^2 \lesssim v_K^2)$ at very small radii in the disk where the magnetic field is not inclined away from the axis. We have found a large class of stationary MHD winds. Within the simulation region, the outflow accelerates from thermal velocity $(\sim c_s)$ to a much larger asymptotic poloidal flow velocity of the order of $0.5\sqrt{GM/r_i}$, where M is the mass of the central object and r_i is the inner radius of the disk. This asymptotic velocity is much larger than the local escape speed and is larger than fast magnetosonic speed by a factor of ~ 1.75. The *acceleration distance* for the outflow, over which the flow accelerates from ~ 0 to, say, 90% of the asymptotic speed, occurs at a flow distance of about $80 r_i$. The flows are approximately *spherical outflows*, with only small collimation within the simulation region. The *collimation distance* over which the flow becomes collimated (with divergence less than, say, $10°$) is *much larger* than the size of our simulation region. Close to the disk the outflow is driven by the centrifugal force while at all larger distances the flow is driven by the magnetic force which is proportional to $-\nabla (rB_\phi)^2$, where B_ϕ is the toroidal field.

The stationary MHD flow solutions allow us (1) to compare the results with MHD theory of stationary flows, (2) to investigate the influence of different outer boundary conditions on the flows, and (3) to investigate the influence of the shape of the simulation region on the flows. The ideal MHD integrals of motion (constants on flux surfaces discussed by Lovelace et al. 1986) were calculated along magnetic field lines and were shown to be constants with accuracy 5−15%. Other characteristics of the numerical solutions were compared with the theory, including conditions at the Alfvén surface.

Different outer boundary conditions on the toroidal component of the magnetic field can significantly influence the calculated flows. The commonly used "free" boundary condition on the toroidal field leads to artificial magnetic forces on the outer boundaries, which can give spurious collimation of the flows. New outer boundary conditions which do not give artificial forces have been proposed and investigated by Ustyugova et al. (1998).

The simulated flows may also depend on the shape of the simulation region. Namely, if the simulation region is elongated in the z-direction, then Mach cones on the outer cylindrical boundaries may be partially directed inside the simulation region. Because of this, the boundary can have an artificial influence on the calculated flow. This effect is reduced if the computational region is approximately square or if it is spherical as in Figure 2. Simulations of MHD outflows with an elongated computational region can lead to *artificial* collimation of the flow.

Recent simulation studies have treated MHD outflows from disks with more general initial **B** field configurations, for example, that where the poloidal field has different polarities as a function of radius (Hayashi, Shibata, & Matsumoto 1996; Goodson, Winglee, & Böhm 1997; Romanova et al. 1998). The differential rotation of the foot-points of **B** field loops at different radii on the disk surface causes twisting of the coronal magnetic field, an increase in the coronal magnetic energy, and an opening of the loops in the region where the magnetic pressure is larger than the matter pressure ($\beta \lesssim 1$) (Romanova et al. 1998). In the region where $\beta \gtrsim 1$, the loops may be only partially opened. Current layers form in the narrow regions which separate oppositely directed magnetic field. Reconnection occurs in these layers as a result of the small numerical magnetic diffusivity. In contrast with the case of the solar corona, there can be a steady outflow of energy and matter from the disk surface. We find that the power output mainly in the form of a Poynting flux. Opening of magnetic field loops and subsequent closing can give reconnection events which may be responsible for X-ray flares in disks around both stellar mass objects and massive black holes (Hayashi et al. 1996; Goodson, Winglee, & Böhm, 1997; Romanova et al. 1998).

4. Poynting Flux Jets

In a very different regime from the hydromagnetic flows discussed in §3, a Poynting flux jet transports energy and angular momentum mainly by the electromagnetic fields with only small contributions of the matter (Lovelace et al. 1987; Colgate and Li 1998). A steady Poynting jet - sketched in Figure 4 - can be characterized in the lab frame by its asymptotic (large distance) magnetic field $B_\phi = -B_0[r_o/r_j(z)]$ at the jet's edge, $r = r_j(z)$, where r_0 is the jet's radius at $z = 0$ and B_0 is the poloidal field strength at this location. The electric

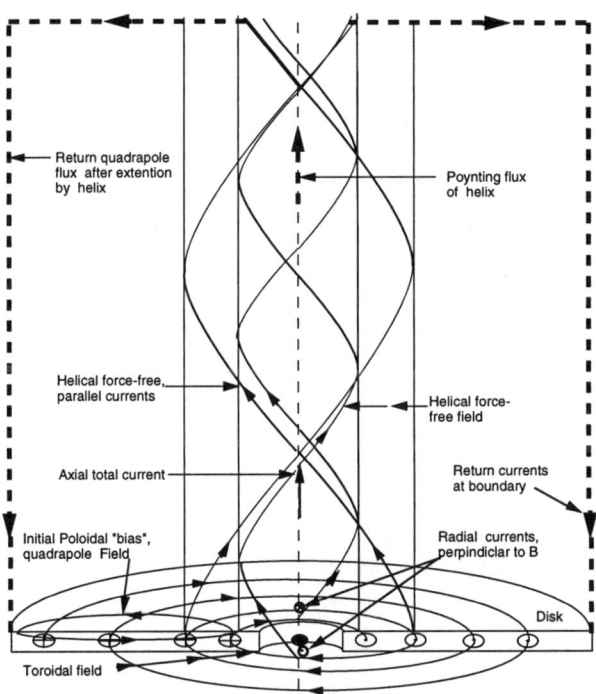

Figure 4. Field configuration of a Poynting flux jet from Colgate and Li (1998).

field in the jet is $\mathbf{E} = -\mathbf{v} \times \mathbf{B}/c$ and consequently the energy flux or luminosity of the jet is $L_j = vB_0^2 r_0^2/8 \sim 4 \times 10^{45} \text{erg/s} (v/c)(r_0/10^{14}\text{cm})^2 (B_0/10^4 \text{G})^2$. Propagating disturbances in such field dominated jets provide a simple, but self-consistent physical model for the gamma ray flares observed in Blazars (Romanova & Lovelace 1997; Levinson 1998; Romanova these proceedings). Owing to pair production close to the black hole, the main constituent of a Poynting flux jet may be electron-positron pairs.

5. Conclusions

MHD simulations carried out by a number of groups over the last several years support the idea that an ordered magnetic field of an accretion disk can give powerful outflows of matter, energy, and angular momentum. The studies so far have been in the hydromagnetic regime and find asymptotic flow speeds of the order of the maximum Keplerian velocity of the disk. In contrast, observed VLBI jets in quasars and active galaxies point to bulk Lorentz factors of order 10 - much larger than the disk Lorentz factor. This may be a result of the relativistic dynamics not included here, but more likely it reflects the fact that these jets are in the Poynting flux regime. Also, these jets may involve energy extraction from a rotating black hole (Blandford and Znajek 1977; Livio et al. 1998).

Acknowledgments. It is a pleasure to thank the organizers for a most memorable meeting. This work was supported in part by NSF grant AST-9320068 and NASA grant NAG5 6311. The Russian authors were supported in part by RFFI Grant 96-02-17113. Also, the research described here was made possible in part by Grant No. RP1-173 of the U.S. Civilian R&D Foundation for the Independent States of the Former Soviet Union.

References

Balbus, S.A., & Hawley, J.F. 1998, Rev. Mod. Phys., 70, 1
Bisnovatyi-Kogan, G.S. 1993, in *Stellar Jets and Bipolar Outflows,* L. Errico & A.A. Vittone, eds. Dordrecht: Kluwer, 369
Begelman, M.C., Blandford, R.D., & Rees, M.J. 1984, Rev. Mod. Phys., 56, 255
Bell, A.R. 1994. Phys. Plasmas, 1, 1643
Bell, A.R., & Lucek, S.G. 1995, MNRAS, 277, 1327
Blandford, R.D. 1976, MNRAS, 1976, 465
Blandford, R.D., & Payne, D.G. 1982, MNRAS, 199, 883
Blandford, R.D., & Znajek, R.L. 1977, MNRAS, 179, 433
Brandenburg, A., Nordlund, A., Stein, R.F., & Torkelson, U. 1995, ApJ, 446, 741
Bridle, A.H., & Eilek, J.A. (eds) 1984, in *Physics of Energy Transport in Extragalactic Radio Sources,* Greenbank: NRAO
Bührke, T., Mundt, R., & Ray, T.P. 1988, A&A, 99
Chandrasekhar, S. 1981, *Hydrodynamic and Hydromagnetic Stability,* (New York: Dover)

Colgate, S.A. & Li, H. 1998, in *Proc. of VII International Conference and Lindau Workshop on Plasma Astrophysics and Space Physics,* Lindau, Germany
Contopoulos, J., & Lovelace, R.V.E. 1994, ApJ, 429, 139
Coroniti, E.V. 1981, ApJ, 244, 587
Eardley, D.M. & Lightman, A.P. 1975. ApJ, 200, 187
Falcke, H., Malkan, M.A., & Biermann, P.L. 1995, A&A, 298, 375
Goodson, A.P., Winglee, R.M., & Böhm, K.H. 1997, ApJ, 489, 390
Hawley, J.F., Gammie, C.F., & Balbus, S.A. 1995, ApJ, 440, 742
Hayashi, M.R., Shibata, K., & Matsumoto, R. 1996, ApJ, 468, L37
Koldoba, A.V., Ustyugova, G.V., Romanova, M.M., Chechetkin, V.M., & Lovelace, R.V.E. 1995, Ap&SS, 232, 241
Königl, A. 1989, ApJ, 342, 208
Königl, A., & Ruden, S.P. 1993, *Protostars and Planets III,* E.H. Levy and J. Lunine, Tucson: Univ. of Arizona Press, 641
Koupelis, T., & Van Horn, H.M. 1989, ApJ, 342, 146
Levinson, A. 1998, ApJ, 507, 145
Livio, M., Ogilvie, G.I., & Pringle, J.E. 1998, ApJ, submitted
Lovelace, R.V.E. 1976, Nature, 262, 649
Lovelace, R.V.E., Mehanian, C., Mobarry, C.M., & Sulkanen, M.E. 1986, ApJ Suppl., 62, 1
Lovelace, R.V.E., Wang, J.C.L., & Sulkanen, M.E. 1987, ApJ, 315, 504
Lovelace, R.V.E., Mobarry, C.M., & Contopoulos, J. 1989, in *Accretion Disks and Magnetic Fields in Astrophysics,* ed. G. Belvedere (Dordrecht: Kluwer), 71
Lovelace, R.V.E., Berk, H.L., & Contopoulos, J. 1991, ApJ, 379, 696
Lovelace, R.V.E., Romanova, M.M.,& Contopoulos, J. 1993, ApJ, 403, 158
Lovelace, R.V.E., Romanova, M.M., & Newman, W.I. 1994, ApJ, 437, 136
Lovelace, R.V.E., Newman, W.I., & Romanova, M.M., 1997, ApJ, 424, 628
Meier, D.L., Edgington, S., Godon, P., Payne, D.G., & Lind, K.R. 1997, Nature, 388, 350
Mirabel, I.F., & Rodriguez, L.F. 1994 Nature, 371, 46
Mundt, R. 1985, in *Protostars and Planets II,* D.C. Black and M.S. Mathews, eds. Univ. of Arizona Press, Tucson, 414
Oyed, R. & Pudritz, R.E. 1997, ApJ, 482, 712
Pelletier, G., & Pudritz, R.E. 1992, ApJ, 394, 117
Phinney, E.S. 1987, in *Superluminal Radio Sources,* J.A. Zensus, & T.J. Pearson, Cambridge: Cambridge Univ. Press, 301
Pringle, J.E., 1989, MNRAS, 236, 107
Pudritz, R.E., & Norman, C.A. 1986, ApJ, 301, 571
Romanova, M.M., & Lovelace, R.V.E. 1997, ApJ, 475, 97
Romanova, M.M., Ustyugova, G.V., Koldoba, A.V., Chechetkin, V.M., & Lovelace, R.V.E. 1997, ApJ, 482, 708

Romanova, M.M., Ustyugova, G.V., Koldoba, A.V., Chechetkin, V.M., & Lovelace, R.V.E. 1998, ApJ, 500, 703

Sakurai, T. 1987, PASJ, 39, 821

Shakura, N.I. 1973, Soviet Astron., 15, 377

Shakura, N.I., & Sunyaev, R.A. 1973, A&A, 24, 337

Shibata, K., & Uchida, Y. 1986, PASJ, 38, 631

Shu, F.H., Lizano, S., Ruden, S.P., & Najita, J. 1988, ApJ, 328, L19

Stone, J.M., & Norman, M.L. 1994, ApJ, 433, 746

Uchida,Y, & Shibata, K. 1985, PASJ, 37, 515

Ustyugova, G.V., Koldoba, A.V., Romanova, M.M., Chechetkin, V.M., & Lovelace, R.V.E. 1995, ApJ, 439, L39

Ustyugova, G.V., Koldoba, A.V., Romanova, M.M., Chechetkin, V.M., & Lovelace, R.V.E. 1998, ApJ, in press (astro-ph/9812284)

Velikhov, E.P. 1959, J. Exp. Theo. Phys., 36, 1398

Active Galaxies and Cosmic Evolution

Malcolm S. Longair

Cavendish Astrophysics Group, Cavendish Laboratory, Madingley Road, Cambridge CB3 0HE.

Abstract. The evolutionary history of active galaxies is surveyed and contrasted with recent observations of the evolution of star and element formation rates with cosmic epoch. The problem of taking proper account of the effects of dust obscuration in optical and ultraviolet observations is reviewed. Recent submillimetre surveys have been used to derive the star formation rate as a function of cosmic epoch independent of dust obscuration. These observations suggest that the star formation rate peaked at redshifts $z \sim 2 - 4$, similar to the maximum in the evolution of the populations of active galaxies. The inference of these observations for theories of galaxy formation is discussed.

1. Introduction

Many of the features of the cosmological evolution of active galaxies are now quite well understood and the big challenge is to relate these features to the evolution of galaxies in general. I will therefore summarise briefly these features and then describe evidence on the evolution with cosmic epoch of star formation in galaxies which is crucial for understanding the relation between the evolutionary properties of active galaxies and galaxies in general. The key problem which bedevils many aspects of these studies is the rôle of dust, but there are now ways by which this can be circumvented. The ultimate question we need to answer is whether or not it all hangs together.

2. The Cosmological Evolution of Active Galaxies

It is simplest to begin with the evolution of the radio source population because in many ways these results are less severely affected by insidious selection effects as compared with optically-selected samples. Their great advantage is that the samples are selected according to the radio properties of these active galaxies and so are unaffected by obscuration by dust. An indication of the current state of the art is provided by the models described by Dunlop (1998). In Fig. 1, two examples of models of the evolution of the overall luminosity function of radio galaxies and radio quasars are shown which are consistent with a very wide range of observations of the number counts at different frequencies and the redshift distributions of complete samples of sources at relatively high flux densities. These examples make the key point that there was a maximum in the

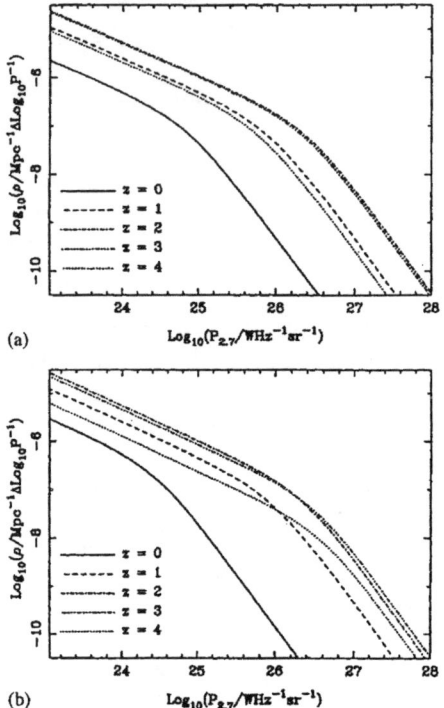

Figure 1. Two examples of the forms of evolving radio luminosity function which can account for the radio source counts and available redshift and identification data (Dunlop 1998). In model (a), the change in the radio luminosity function is described by pure luminosity evolution. In model (b), the changes in the radio luminosity function involve in addition negative density evolution at large redshifts. In both cases, the changes in the form of the radio luminosity function are described in terms of the number densities of sources per unit comoving volume (Dunlop 1998)

radio source activity of these active galaxies at $z \sim 2-3$ and a decline at greater redshifts.

A key issue is the cut-off suggested by the models in Fig. 1. A test of the reality of the cut-off is provided by the redshift distribution of complete samples of radio sources in deep source samples, the results of the 6C-B2 survey and the Leiden-Berkeley Deep Survey being shown in Fig. 2. It can be seen that the inferred redshift distributions are consistent with both models shown in Fig. 1. In particular, in models in which there is no redshift cut-off, the numbers of sources at large redshifts would far exceed the numbers observed, particularly in the Leiden-Berkeley Deep Survey. It therefore seems unambiguous that the radio quasars and radio galaxies flourished at redshifts $z \sim 2-3$ and that this activity declined at greater redshifts.

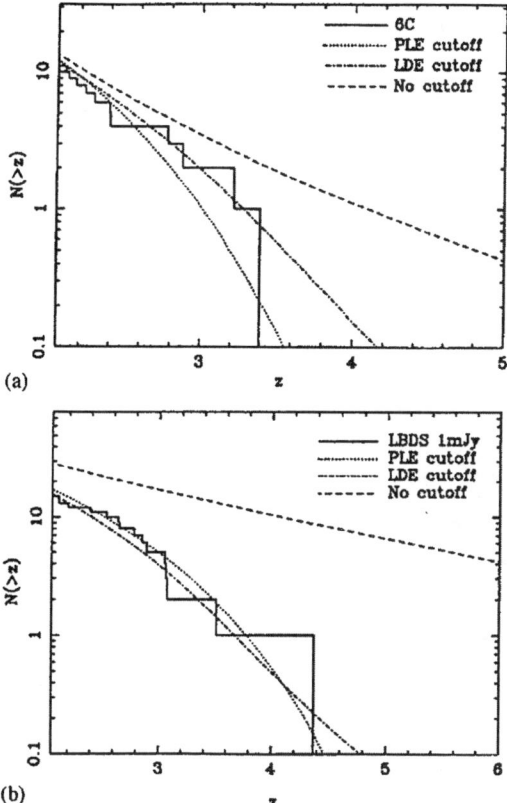

Figure 2. (a) The observed high-redshift cumulative redshift distribution for the complete 6C/B2 samples of Eales, Rawlings and their colleagues (Eales and Rawlings 1996), compared with the large redshift predictions of the two evolution models shown in Fig. 1, and a model in which the comoving luminosity function of the radio sources is constant at redshifts $z \geq 2$. (b) The same comparison for the Leiden–Berkeley Deep Survey (Dunlop et al. 1995). For more details, see Dunlop (1998).

A similar picture emerges from studies of complete samples of optically selected quasars. The problem with the optical observations is to ensure that the samples are complete – this is highly non-trivial at redshifts $z > 2$. The surveys of Boyle et al. (1991) and Shaver et al. (1996) are consistent with the same type of evolutionary behaviour found in the samples of radio sources (Fig. 3). The cut-off at large redshifts has been indicated by the deep surveys of Warren et al. (1994), Kennefick et al. (1995) and Schmidt et al. (1995), as well as by the survey of Shaver et al. (1996).

Although many detailed questions still need to be answered, the parent galaxies associated with these active systems are all massive systems, as indicated by the success of unification schemes for the radio galaxies and radio quasars in bright radio source samples and, more directly, by the similarity of the

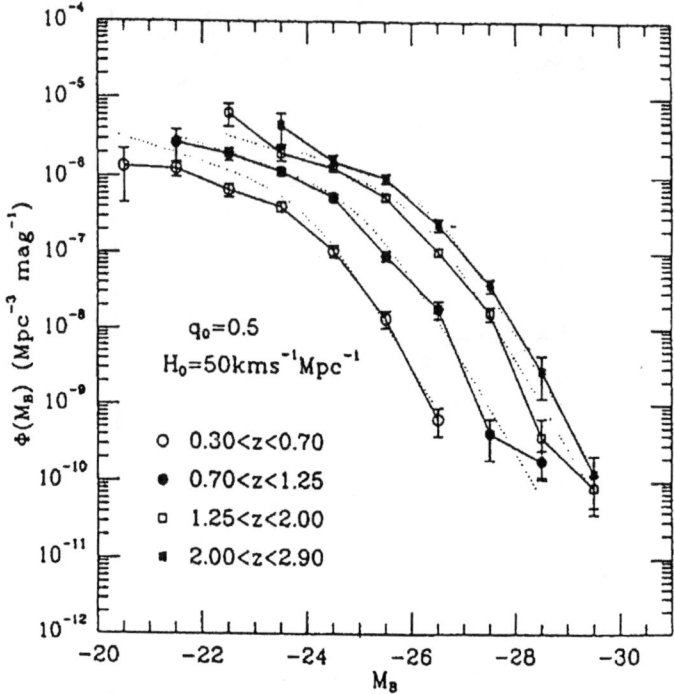

Figure 3. The evolution of the comoving optical luminosity function of radio-quiet quasars in the redshift interval $0.3 < z < 2.9$. The redshift bins have been selected to correspond to equal intervals in $\log(1+z)$. The luminosity functions have been derived from complete samples of quasars and include almost 1000 quasars. The faint dotted lines show the the expectations of a luminosity evolution model in which the luminosities of the quasars change with cosmic epoch as $L(z) \propto (1+z)^{3.5}$ in the redshift interval $0 < z < 2$ and $L(z) =$ constant for redshifts $2 < z < 2.9$ (Boyle et al. 1991).

absolute magnitudes of the parent bodies of radio galaxies, radio-loud quasars and radio-quiet quasars (Taylor et al. 1996).

It is likely that the number counts of X-ray selected active galaxies display the same form of cosmological evolution, although the redshift information is as yet not as extensive as that in the radio and optical wavebands (Hasinger et al. 1996).

3. The Evolution of Element and Star Formation Rates with Cosmic Epoch

There has been remarkable progress in defining element and star formation histories from a wide range of observations over recent years. There are however problems and it is simplest to illustrate these using the analysis of Pei and Fall

(1995), which has been widely quoted. In their programme, they attempt to relate the chemical history of interstellar matter in galaxies, as defined by studies of Lyman-limit absorption line systems in high redshift quasars, to the emission history of star formation, as defined by the characteristic blue continua of starforming galaxies. As they show elegantly, these histories are related through the equations of cosmic chemical evolution. The results of their studies repay close study and are illustrated in Fig. 4.

In all four panels of Fig. 4, the dashed lines show the observations, uncorrected for the effects of dust obscuration, whereas the solid lines show how these results change once account is taken of the effects of dust obscuration. Panel (b) shows how the density parameter in neutral hydrogen changes with redshift. The observed points are taken from the studies of Lyman-α absorption clouds by Storrie-Lombardi *et al.* (1996) which show that the density parameter in neutral hydrogen at redshifts $z \sim 3 - 4$ is of the same order as that in the stars in galaxies at the present epoch. It can be seen that the fraction of the baryonic mass in the form of neutral hydrogen in galaxies decreases to only about one twentieth of the overall mass in stars in galaxies by the present epoch. At the same time, panel (c) shows that the abundances of the heavy elements at $z \sim 2$ must have been about an order of magnitude less than the values observed at the present epoch. A good example of the type of analysis, from which this information is derived, is presented by Pettini *et al.* (1996).

The equations of cosmic chemical evolution can be used to predict the star formation rate as a function of cosmic epoch, given the observed rate of depletion of the neutral hydrogen and the build up of the heavy elements. The solid curves shown in panel (d) have been widely quoted and are in reasonable agreement with the star formation rates inferred from the Hubble Deep Field, the numbers of Lyman-α drop-outs observed by Steidel and his colleagues and the analyses of the Canada-France Redshift survey carried out by Lilly and his colleagues.

It can be seen, however, that the results are strongly dependent upon assumptions made about the influence of interstellar dust upon the observations. Dust enters in a number of different ways into the analysis. For example, account has to be taken of the fact that there may be quasars missing from the samples because of dust extinction. The abundances themselves can be strongly affected by the effects of reddening. What I find interesting about these models is that the epoch at which the maximum rate of star formation took place changes according to the assumptions made about the amount of dust extinction present. There is also the key question as to whether or not all the star forming galaxies at large redshifts have been detected. The way of circumventing many of these problems is to make observations in the submillimetre waveband and this has been achieved recently.

4. Submillimetre Observations of Distant, Dusty Galaxies

The importance of observations in the submillimetre waveband are well known. Star-forming galaxies have a characteristic flat emission spectrum in the optical-ultraviolet wavebands, but the regions in which stars form are also the dustiest regions we know of in the Universe. As pointed out by Weedman at this meeting, local star-forming galaxies emit much more of their luminosities in the far-

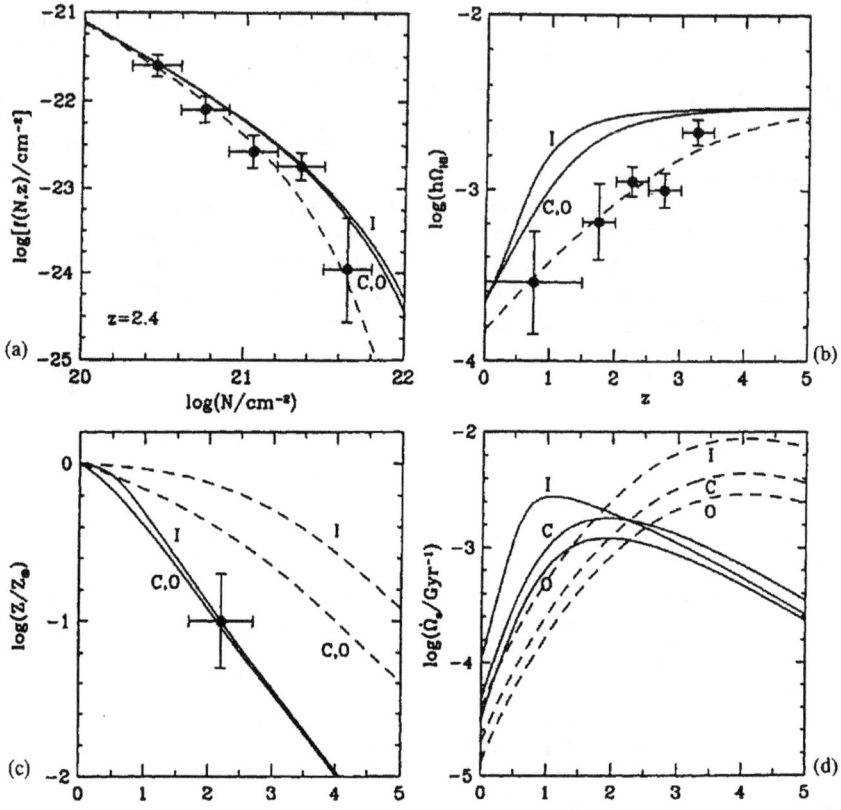

Figure 4. Illustrating the cosmic chemical evolution of models with $\Omega_g(\infty) = 4 \times 10^{-3}h^{-1}$, $\nu = 0.5$ and $\Omega_0 = 1$ (for details, see Pei and Fall 1995). (a) The distribution of HI column densities at $z = 2.4$; (b) the comoving density of neutral hydrogen as a function of redshift z; (c) the mean metallicity Z of the interstellar gas relative to its present value Z_0 as a function of redshift; (d) the predicted comoving rate of star formation as a function of redshift. The solid curves represent the true values for three models described by Pei and Fall: C – closed box model, I – inflow model, O – outflow model. The dashed curves in (a) and (b) represent the corresponding observed quantitites, while the dashed curves in (c) and (d) show the effects of neglecting dust extinction in each of the models. The data points in (a), (b) and (c) are from Lanzetta et al. (1991), Lanzetta et al. (1995) and Pettini et al. (1994), respectively.

infrared rather than in the ultraviolet region of the spectrum. This is because the dust grains, which are essential in enabling a gas cloud to collapse, absorb the optical-ultraviolet radiation, the temperature of the emission corresponding to that to which the dust grains are heated. Observations in the submillimetre waveband have the great advantage that the spectrum of dust has a very steep inverted spectrum through these wavebands and so the 'K-corrections' are large and negative (Blain and Longair 1993, 1996). The result is that a standard dusty star-forming galaxy has more or less the same flux density throughout the redshift interval $1 < z < 10$.

The great breakthrough of the last year has resulted from the commissioning of the SCUBA submillimetre common-user bolometer array on the James Clerk Maxwell Telescope in Hawaii (Holland et al. 1999). This is the first time a 'camera' has become available on a large submillimetre telescope and the results are startling. Of the various surveys already published in the literature (Smail et al. 1997, Barger et al. 1998, Hughes et al. 1998), I will concentrate upon the results from surveys of clusters of galaxies and those from the Hubble Deep Field.

The first report of a large population of distant submillimetre sources was presented by Smail et al. (1997) who observed clusters of galaxies which were known to be strong gravitational lenses. Six background sources were discovered in the fields of two clusters and their subsequent observations of five other clusters have increased the known numbers of submillimetre sources to 10 at the 4σ level and 17 at the 3σ level. The image of the Hubble Deep Field published by Hughes et al. (1998) is the deepest submillimetre image ever made and reaches the confusion limit of the JCMT. Five sources have been observed with good significance in the field.

These results have been synthesised by Blain et al. (1999) who have used all the information available on the number counts and background radiation in the submillimetre and far infrared wavebands. A number of key results come out of this analysis. First of all, the total background radiation due to sources already detected in the deep surveys is very close to the integrated background radiation in the submillimetre waveband as derived from observations by COBE. Therefore, there is unlikely to be a large population of dusty galaxies beyond those which have already been observed. Indeed, Blain et al. (1999) find that by a flux density of 0.1 mJy at 850 μm, the number counts of submillimetre sources is converging, the integral source count having slope -1.

Second, the data can be used to estimate the variation of the star formation rate with cosmic epoch. This is not a trivial exercise, because of the lack of complete redshift information for these faint submillimetre sources, but their properties are reasonably well constrained by all the information now available from the ISO satellite, as well as the properties of the IRAS population of galaxies. The results of these model computations are shown in Fig. 5. It can be seen that the inferred star formation rates exceed those derived from optical-ultraviolet observations at redshifts $z \sim 2-4$, even when the latter are corrected for the effects of obscuration. The inference is that there are substantial populations of star forming galaxies missing from the Hubble Deep Field – these must be highly obscured objects, probably at large redshifts, although probably not at redshifts greater than 5 (Smail et al. 1998).

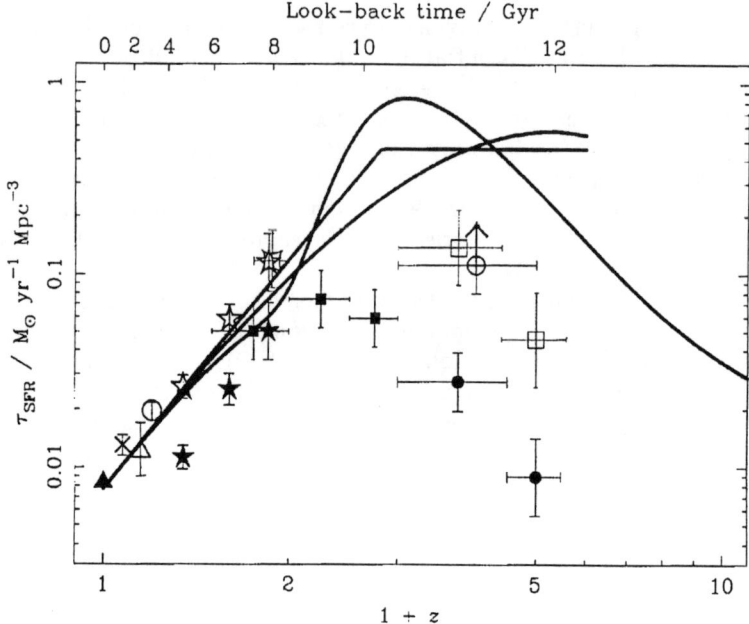

Figure 5. Comparison between the inferred star formation rate as determined by optical-ultraviolet observations and those carried out in the submillimetre waveband. The solid lines are examples of models of the star formation rate with cosmic epoch which are consistent with the number counts of faint submillimetre and far infrared sources and the limits to the background intensity due to discrete sources in the submillimetre waveband (from Blain *et al.* 1999)

These results are very important for understanding the formation of the stellar populations of galaxies. For example, they would strongly affect the rate at which the heavy elements are inferred to be built up with cosmic epoch and they have consequences for the formation of the overall population of low-mass stars. The key observations now are the determination of the redshifts of the faint submillimetre sources. It is now known that two of the background objects have redshifts of 2.5 and 2.8 (Ivison *et al.* 1998), but it is a struggle to find the identifications and then the redshifts of these distant star-forming galaxies.

5. The Active Galaxies Revisited

How do these observations impact the study of the evolution of active galaxies with cosmic epoch? There are several striking features which have come out of the observations described above which are directly related to evolution of the population of active galaxies.

1. The first is the obvious similarity between the evolutionary behaviour of the population of active galaxies and the evolution of the star formation rate with cosmic epoch. Both of these were very much greater at redshifts

$z \sim 2-4$ as compared with the present epoch. In my opinion, this is unlikely to be a coincidence.

2. The second key aspect is the direct evidence that much more of the baryonic matter at redshifts $z \sim 2-4$ was in the form of diffuse gas as compared with the present epoch. Thus, there were certainly large amounts of gas present to fuel active galactic nuclei as a by-product of the process of star formation in galaxies.

3. There is direct evidence for strong interactions between the radio jets of 3CR radio galaxies and their environments from observations of the alignment effect, the coalignment of the radio jets with the optical emission structures (Best, Longair and Röttgering 1998, Best, this volume). The origin of the optical emission is not yet fully understood, but all the viable models involve the interaction of the powerful radio jets with cool interstellar or intergalactic clouds in the vicinity of these massive galaxies.

4. We know that, in both the quasars and the radio galaxies, there must be $10^9 \, M_\odot$ black holes in their nuclei. In this regard, the remarkable diagram of Rawlings and Saunders (1991) for the 3CR radio galaxies shows that, for the most luminous radio galaxies in the sample, the rate of total energy production corresponds closely to the Eddington luminosity of $10^9 \, M_\odot$ black holes.

From the perspective of galaxy formation, the key result is that these massive black holes must have formed by the epoch when the maximum amount of star formation activity was taking place. In this regard, the analyses of Efstathiou and Rees (1988), Haehnelt and Rees (1993) and Efstathiou (1996) are of particular interest. According to the favoured hierarchical picture of the formation of galaxies and larger scale structures, the most massive objects are formed by the clustering of lower mass objects. According to their calculations, the cut-off to the quasar population at large redshifts is associated with the fact that, by a redshift of 4, there are relatively few massive galaxies and, since there is evidence that the central black hole mass is proportional to the total mass of the galaxy, there would be correspondingly fewer massive enough black holes in the nuclei of galaxies at these redshifts.

The sense of these arguments is that there is now a much closer relation between the evolution of the populations of active galaxies and studies of the formation and evolution of galaxies than in the past. It has been suspected for a long time that the origin of the strong cosmological evolution of active galaxies was related to the formation of galaxies in general, and this can now be placed on a much firmer observational and astrophysical basis.

References

Barger, A.J. et al. (1998). *Nature*, 394, 241.
Best, P.N., Longair, M.S. and Röttgering, H.J.A. (1998). MNRAS, 295, 549.
Blain, A.W. and Longair, M.S. (1993). MNRAS, 264, 509.
Blain, A.W. and Longair, M.S. (1996). MNRAS, 279, 847.

Blain, A.W., Smail, I., Ivison, R.J., Kneib, J.-P. (1999). MNRAS, (in press).
Boyle, B.J., Jones, L.R., Shanks, T., Marano, B., Zitelli, V. and Zamorani, G. (1991). In *The Space Distribution of Quasars*, (ed. D. Crampton), 191. San Francisco: ASP Conf. Series.
Dunlop, J.S. 1998. In *Observational Cosmology with the New Radio Surveys*, (eds. M.N. Bremer, N. Jackson and I. Perez-Fournon), 157. Dordrecht: Kluwer Academic Publishers.
Dunlop, J.S., Peacock, J.A. and Windhorst, R.A. (1995). In *Galaxies in the Young Universe*, (eds. H. Hippelein, K. Meisenheimer and H.-J. Röser), 84. Berlin: Springer-Verlag.
Eales, S.A. and Rawlings, S. (1996). ApJ, 460, 68.
Efstathiou, G. (1995). In *Galaxies in the Young Universe*, (eds. H. Hippelein, K. Meisenheimer and H.-J. Röser), 299. Berlin: Springer-Verlag.
Efstathiou, G. and Rees, M.J. (1988). MNRAS, 230, 5P.
Haehnelt, M.G. and Rees, M.J. (1993). MNRAS, 263, 168.
Hasinger, G., Burg, R., Giaconni, R., Hartner, G., Schmidt, M., Trümper, J. and Zamorani, G. 1993. A&A, 275, 1.
Holland, W.S. *et al.* (1999). MNRAS, (in press).
Hughes, D. *et al.* (1998). *Nature*, 394, 241.
Ivison, R.J. *et al.* (1998). MNRAS, 298, 583.
Kennefick, J.D., Djorgovski, S.G. and de Carvalo, R.R. (1995). AJ, 110, 2553.
Lanzetta, K.M., Wolfe, A.M. and Turnshek, D.A. (1995). ApJ, 440, 435.
Lanzetta, K.M., Wolfe, A.M., Turnshek, D.A., Lu, L., McMahon, R.G. and Hazard, C. (1991). ApJS, 77, 1.
Pei, Y.C. and Fall, S.M. (1995). ApJ, 454, 69.
Pettini, M., King, D.L., Smith, L.J. and Hunstead, R.W. (1996). ApJ, 486, 665.
Pettini, M., Smith, L.J., Hunstead, R.W. and King, D, (1994). ApJ, 426, 79.
Rawlings, S. and Saunders, R. (1991). *Nature*, 349, 138.
Schmidt, M., Schneider, D.P. and Gunn, J.E. (1991). In *The Space Distribution of Quasars*, (ed. D. Crampton), 109. San Francisco: ASP Conf. Series.
Shaver, P.A., Wall, J.V., Kellermann, K.I., Jackson, C.A. and Hawkins, M.R.S. (1996). *Nature*, 384, 439.
Smail, I., Ivison, R.J. and Blain, A.W. (1997). ApJ, 490, L5.
Smail, I., Ivison, R.J., Blain, A.W. and Kneib, J.-P. (1998). ApJ, 507, L21.
Storrie-Lombardi, L.J., McMahon, R.G. and Irwin, M.J. (1996). MNRAS, 283, L79.
Taylor, G.L., Dunlop, J.S., Hughes, D.H. and Robson, E.I. (1996). MNRAS, 283, 930.
Warren, S.J., Hewett, P.C. and Osmer, P.S. (1994). ApJ, 421, 412.

The Jets of Quasars 3C 345 and 1803+784

L. I. Matveenko

Space Research Institute, Profsojuznaja 84/32, Moscow 117810 Russia, e-mail:lmatveen@mx.iki.rssi.ru

A. I. Witzel

Max-Planck-Institut für Radioastronomie, Auf dem Hügel 69, D-53121 Bonn, Germany, e-mail: p0103wzl@mpifr-bonn.mpg.de

Abstract. We have studied the structures of AGN objects 3C 345 and 1803+784. The objects have one sided jets with conic helix structure, determining step and diameter of the helix. The jets are surrounded by cocoon - thermal plasma, the transparency of which determines low frequency variability and absorption of the core emission.

Quasar 3C 345 (z=0.595) and Bl Lac 1803+784 (z=0.68) are classical AGN objects. The distances of them are 2500 and 2700 Mpc; 1 mas is equal ~ 4 and ~ 5 pc accordingly. The optical and radio properties are identical and structures are typical for AGN objects: core and one side jet. The core emission dominates at short radio wavelengths. The activity in the nucleus is accompanied by strong outbursts of radio emission. These outbursts appear first, and are strongest, at mm-cm wavelengths, and propagate to low frequencies while decreasing in amplitude. The low frequency variability implies a brightness temperature of the emission region $T_b \sim 10^{16}$ K, or much greater than the limit of inverse Compton scattering - 10^{12} K (Kellermann & Pauliny-Toth 1969). The outbursts are caused by knots - clouds of relativistic plasma. The apparent motions of the knots exceed the light velocity (Cohen et al. 1976). Formation of the jet is a result of the core activity. We studied the fine structure of AGN 3C 345 and 1803+784.

The low-frequency variability of 3C 345 (Padrielli et al. 1982) has the character of "negative" outbursts with a typical time-scale of one year. The low frequency variations are anti-correlated with high and uncorrelated for nearby quasars. In 1981 - 1983 activity was accompanied by strong outbursts in the frequency band 8 - 89.6 GHz (Bregman et al. 1986). An outburst 1981.6 lasted about half a year. Its spectrum had a low frequency cut off and a spectral index $\alpha \geq 2.6$ at $f \leq 4.8$ GHz. The spectrum of the brightest compact component had $\alpha \geq 3$ at low frequencies. That suggests absorption of the synchrotron emission by an ionized medium. Optical depth of the medium was big even at 4.8 GHz, (Matveenko et al. 1996).

An analysis of fine structure of 3C 345 (Zensus 1991; Baath et al. 1992; Krichbaum et al. 1993; Unwin et al. 1992; Matveenko et al. 1996) showed that at 1983.0-1990.5 knots are injected inside $\sim 60°$ cone, central axis of which has P.A. $\sim -105°$. The knots move along conical spirals. The period and

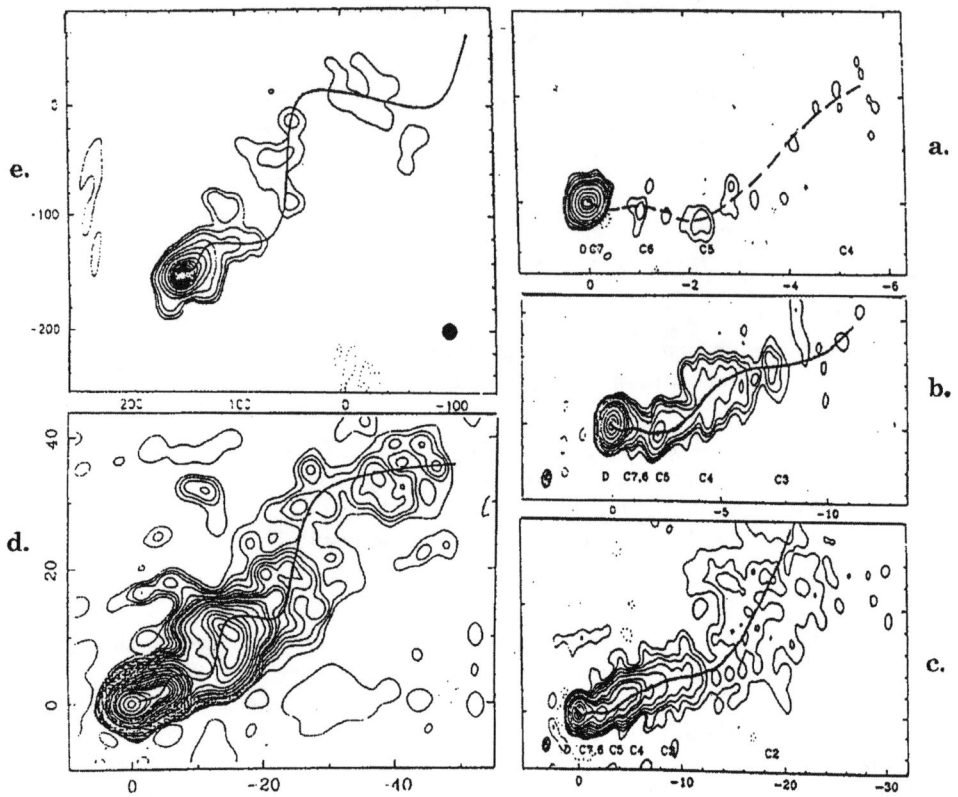

Figure 1. Images of 3C 345 at frequencies: a. 22 GHz; b. 8.4 GHz; c. 5.0 GHz (Zensus 1990); d. 1.6 GHz (Rantakuro et al. 1992); e. 0.6 GHz (Matveenko et al. 1996).

diameter of the spiral increases practically proportional to distance from the core. The direction of ejection of the knots repeated after ∼7 years. It is proposed precession period ∼7 years and the angle ∼ 30°.

The jet structure is a conical spiral too, Fig 1. Diameter of the spiral increases with distance ∼ $0.55 \cdot R^{0.9}$ and step T = $0.15 \cdot R$ until R = 200 mas, then T = $4.6 \cdot R^{0.4}$. Increasing of the diameter and the step proportional to R proposes movement of relativistic plasma flow with constant velocity. A rotating flow of relativistic plasma generates a magnetic field and self-focuses in a thin stream. Velocity of rotation is determined by injector - accretion disk. The rotation axis precesses and twists the jet around the central direction forming spiral structure, which corresponds to theoretical models (Begelman et al. 1984; Lovelace et al. 1991). The twisted velocity is equal to precession velocity. The knots are moving along a magnetic tube as a stream of relativistic plasma.

The jet axis of 3C 345 is curved, orientation of it is changed from −105° to ∼ −30° at distance 3" (Browne et al. 1982). A curvature of the axis can be due to long-term precession of the rotation axis of the accretion disk. The precession angle is ≥75°.

One of the problems for studies of fine structure is the identification of components. Positions of the compact components change with time and frequency. At the high radio frequencies a reference feature is the core - the brightest compact component. However peculiar structure of the jet has features transverse to the jet axis located at ~5, ~5, ~30 mas from the core. Locations are fairly stable and have smaller frequency dependence (Readhead et al. 1978; Biretta et al. 1986; Matveenko et al. 1992; Rantakyro et al. 1992; Unwin et al. 1992). We identified position of the core at $\lambda 49 cm$ with the compact component located at 16 mas, P.A.= 95°. Visible brightness temperature of the core is $T_b \leq 0.003 \cdot T_{peak}$ or ≥ 300 times smaller then at high frequencies that corresponds to absorption at 49 cm. At 6 cm absorption will be ≤ 4 times.

The brightest part of the 3C 345 jet has size ~5 mas at low frequencies, Fig. 1. At the period 1983.9 - 1990.8 at wavelength $\lambda = 49cm$ flux density and angular size increased ~2 times, but the brightness temperature did not change significantly and was $T_b = 0.6 \cdot 10^{12}$ K. Emission at millimetre wavelengths and UV decreased by a factor of ~2 during the same time. A stable brightness temperature of the region, limitation of $T_b \leq 10^{12}$ K, cut off steep spectrum and changed size of the emission region confirm the absorption nature of low frequency variability.

Ionized gas surrounding the core of AGN objects radiates emission lines. Electron density of the narrow line medium is $10^{4-6} cm^{-3}$ and dimension is ~ 10^{20} cm or ~8 mas for 3C 345. Absorption of the ionized medium - e^τ - is determined by optical depth:

$$\tau = 0.08 \cdot T_e^{-1.35}(1+z)^{-2.1} f^{-2.1} \int N_e^2 dl \qquad (15)$$

The optical depth at $\lambda 49 cm$ is $16 \cdot 10^{-8} ME$. A flow of relativistic plasma - jet is surrounded by thermal plasma in the form of a cocoon. The density of thermal electrons of recombination time and time of low frequency variability ~1 yr is $N_e \sim 10^5 cm^{-3}$. The pressure in the jet varies with density as $P_j \sim (R_o/R)^2$ and $N_{rme} \sim (R_{rmo}/R)^2$ (Begelman et al. 1984). The wall thickness increases with distance from the nucleus. The optical depth of the wall is $7\cdot 0 \, R^{-3}$. Optical depth of the injector region $\tau \sim 6$ at $\lambda 49 cm$ and thickness of the cocoon wall will be $l \sim 10^{-3} pc$.

Bright components are distributed in the nearest part of the 6 mas jet region. Transparency of the cocoon wall changes with distance and time. At high frequencies, the screen is transparent. However, it changes polarization. At 1980-1993 changes of emission were accompanied by variations of the polarization at $\lambda 6$ cm (Brown et al. 1994). The polarized radiation increased simultaneously with total until a definite level after which was observed anticorrelation. Polarized emission $\lambda = 6$ cm increases with distance from the core and reaches a maximum at distance $\sim 4.5 mas$. The rotation measured at λ = 18-21 cm, (beam size ~5") is RM $\sim 28 rad \, m^{-2}$ and the degree of polarization $\sim 4\%$ (Rudnick et al. 1983). The 18 and 6 cm maximum of polarized emission arises from the same region. The degree of polarization depends on bandwidth and RM. Increasing of N_{rme} increases RM and decreases polarized emission (Matveenko et al. 1996). The polarization position angle, corrected for Faraday rotation, is ~-70° and is parallel to the jet. The rotation measure

$RM \sim (R_o/R)^3$ and equals $3500 rad\ m^{-2}$ at the core region and magnetic field $B_{||} \sim 0.1 mG$.

Brightness temperature of the nearest part of the jet (R≤2 mas) is $T_b \sim 10^{12}$ K (high frequencies) and corresponds to the kinetic temperature of relativistic plasma. At distance R≤ 25 mas $T_{rmb} \sim 0.5 \cdot 10^{12}$ K (low frequencies). At R∼50 mas optical depth of the emission region is not big and $T_b \sim 10^9$ K and decreases until $\sim 10^7$ K at 1"- 4" distances. Spectral index (along spiral with the angular resolution 1 mas) of the nearest part of the spiral (2≤ R ≤ 10 mas) is equal $\alpha = -1.0 \pm 0.3$ at (λ=1.3-3.6 cm) and $\alpha = -0.5 \pm 0.3$, (λ=3.6-6 cm). Part of the jet at 10-50 mas distance has spectral index $\alpha = -0.9 \pm 0.3$ (λ=18-49 cm). Spectral index increases from a α = -0.8 at distance 1" to α = -1.5 at 4" from the core.

Low frequency variability of 1803+784 has a typical time 1.5 yr just as at high frequencies is a few days. We studied structure of object 1803+784 at 18 cm with global VLBI network 29 May 1993 with dynamic range $\leq -30 dB$. Brightness distribution - (0.3 mas) is located in two groups at distance 25 mas one from the other inside a 23° cone. The axis of the cone is curve, Fig.2. Position angle changes with distance R, $PA = -(90 + 1.5 \cdot R)°$. The injector region has jet like structure and consists of few components, peak is F = 3.54 Jy/beam. The brightness distribution with averaging beam 2 mas is shown in Fig. 2b, minimum level is 0.07% of peak. At Fig. 2c is shown strip distribution along an injector axis by a fan beam $0.3 \cdot 3.0$ mas, $PA = -10°$. We propose that the core is located left at 2.8-3 mas from the peak and has very weak brightness ($\leq 20 dB$) of the peak. The brightness increases from the core can be explained by increasing transparency of the surrounding medium. Optical depth of the medium changes approximately linearly and equals:

$$\tau = \tau_o (1 - R/R_o), \qquad (16)$$

where $\tau_o = 5.5$ the optical depth in the injector region and $R_o = 3$ mas is core location relative to the bright peak.

The electron density of the thermal plasma in the region with maximum emission at 18 cm, $R \sim 3 mas$ of the core, can be calculated the time of recombination and low frequency variability $N_e \geq 0.5 \cdot 10^5 cm^{-3}$. Thickness of the screen is $l \sim 10^{-2} pc$. Changing optical depth at $20-30\%$ explains low frequency variability.

The compact components are located from 0.1 to 25 mas of the injector. Ejections of the compact components is direction $PA \sim -(68-110)°$. The cone of ejected components is ∼40° and direction of the cone axis is - 90°. The components are moving along a conical spiral trajectory (Krichbaum et al. 1993). Angle of the cone is 34° and practically corresponds to the cone of 18 cm structure. Diameter of the spiral increases with distance of $\sim 0.6 \cdot R$ and period $T \sim R$.

Polarized emission of 1803+784 at 18-21 cm is P = 3.9%, $PA = -129°$ and rotation measure $RM = -62 rad m^{-2}$. This position angle is different from the jet orientation ∼-30° (Wrobel et al. 1987). Sign and value of RM is different from the nearest sources $RM \sim 20 rad m^{-2}$, which corresponds to RM of Galaxy. In this case 1803+784 $RM \sim -80 rad m^{-2}$. This value of rotation measure corresponds to the brightest region located at 3 mas from the injector. At 3.6

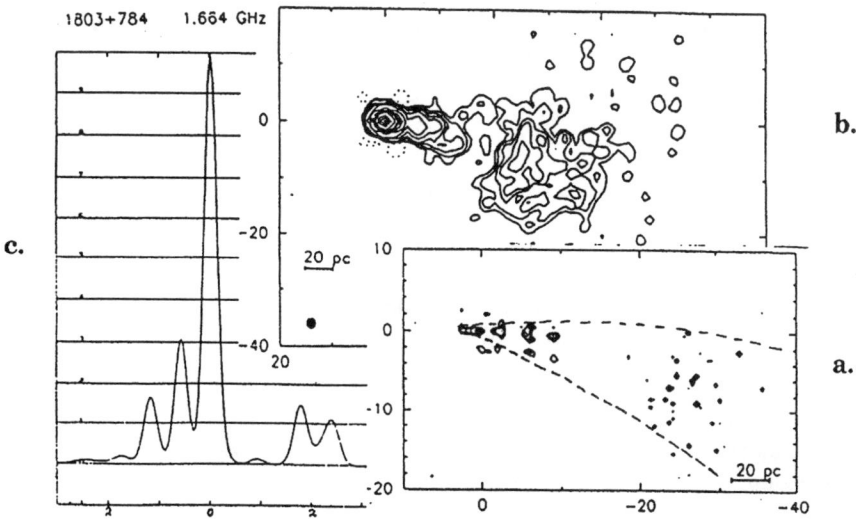

Figure 2. Images of 1803+784 at 1.6 GHz, beam size: a. 0.3 mas; b. 2 mas; c. 0.3*3.0 mas PA=-10°

cm the main polarized emission is determined by the core region, P = (15-20)% and position angle is $PA = 90°$, that corresponds to orientation of the jet.

The changing of the polarization level can be determined by changing optical depth of emission regions. A Thin source has $P_o \leq 70\%$ and orientation of E perpendicular to the magnetic field. For thick one $P_o \leq 10\%$ and E is parallel to B. At 3.6 cm P = 20%, $T_b \leq 10^{12}$ K and optical depth probably thin or practically thin, E would be perpendicular to B. At 18 cm the emission region is thick, $T_b \sim 10^{12}$ K, E would be parallel to magnetic field B. The parameters of cocoon wall at $\lambda 18cm$, R=2.8 mas: $N_e \sim 10^5$ and $l = 10^{-2}$ pc. Rotation measure is:

$$RM = 8.1 \cdot 10^5 N_e Bl, rad/m^2.$$

For RM = 80 and regular magnetic field in observer's direction (perpendicular to jet axes) is $B_{||} \sim 0.1 \mu G$.

At $\lambda 18cm$ are two regions located at distance $\sim 704 mas$. Orientation of the polarization is perpendicular (S. Aaron, private information). Perhaps changes $\sim 1.5 rad$ is determined by changing of RM(R) at the relative distance $\Delta R = 4mas$ and $\Delta RM = 46 rad/m^2$. Rotation measure in region of the maximum emission $RM \sim -80 rad/m^2$ and $RM = -34 rad/m^2$ at 4 mas.

The jets of the 3C 345 and 1803+784 have conical spiral structure, diameter and step of which are $\sim$ R. Axis of the jets have helical form. Brightness temperature of the injector $T_b \leq 7 \times 10^{12}$ K.

The jet is surrounded by a thermal plasma cocoon. Electron density of the cocoon is $N_e 10^5 cm^{-3}$, thickness of the wall $\leq 10^{-2}$.

Low frequency variability is determined by changing transparency of the cocoon.

Acknowledgments. LM is grateful to the IAU Colloquium organizers and INTAS for hospitality and financial support.

References

Baath, L. B., et al. 1992, A&A, 257, 31.
Begelman, M. C., et al. 1984. Rev. Mod. Phys., 56, 255.
Biretta, J. A., et al. 1986. apj., 308, 93.
Bregman, J. N., et al. 1986, apj., 301, 708.
Browne, I. W. A., et al. 1982, mnras., 198, 673.
Brown, L. F., et al. 1994. ApJ, 437, 108.
Cohen, M. H., et al. 1976. ApJ, 206, L1.
Kellermann, K. I., et al. 1969. ApJ, 155, L31.
Krichbaum, T. P., et al. 1993. A&A, 275, 375.
Lovelace, R. V., et al. 1991, ApJ, 379, 695.
Matveenko,L. I., et al. 1992. Pisma Astron. Zh., 18, 379.
Matveenko, L. I., et al. 1996. A&A, 312, 738.
Padrielli, L., 1982, Proc. NRAO Workshop, W.D. Cotton and S.R. Spangler(eds.),p.1.
Rantakyrö, F. T., et al. 1992. A&A, 259, 8.
Readhead, A. C. S., 1978. Nature, 276, 768.
Rudnick, L., et al. 1983. apj, 88, 518.
Unwin, S. C., et al. 1992, apj, 398, 74.
Wrobel, J. M. et al. 1987, IAU Symposium No 129, 165.
Zensus, J. A. 1991, "Extragalactic Radio Sources - From Beam to Jets", J. Roland, H. Sol, and G. Pelletier (eds.), "Cambridge University press", p.154.

Active Galactic Nuclei and Related Phenomena
IAU Symposium, Vol. 194, 1999
Yervant Terzian, Daniel Weedman, Edward Khachikian, eds.

Are Gamma-Ray Bursts Signals of Supermassive Black Hole Formation?

K. Abazajian, G. Fuller, X. Shi

Department of Physics, University of California, San Diego, La Jolla, California 92093-0350

Abstract.
   The formation of supermassive black holes through the gravitational collapse of supermassive objects (M $\gtrsim 5 \times 10^4 \, M_\odot$) has been proposed as a source of cosmological γ-ray bursts. The major advantage of this model is that such collapses are far more energetic than stellar-remnant mergers. The major drawback of this idea is the severe baryon loading problem in one-dimensional models. We can show that the observed $\log N - \log P$ (number vs. peak flux) distribution for gamma-ray bursts in the BATSE database is not inconsistent with an identification of supermassive object collapse as the origin of the gamma-ray bursts. This conclusion is valid for a range of plausible cosmological and γ-ray burst spectral parameters.

1. Introduction

We investigate aspects of a recent model for the internal engine powering γ-ray bursts (GRBs). This model produces a GRB fireball through neutrino emission and annihilation during the collapse of a supermassive object into a black hole (Fuller & Shi, 1998). This supermassive object may either be a relativistic cluster of stars or a single supermassive star. The collapse of a supermassive object provides an exceedingly large amount of energy to power the burst, much higher than the amount of energy available in stellar remnant models. In addition, the rate of collapse of these objects–when associated with galaxy-type structures–is similar to the GRB rate.

The recent observations of afterglows and galaxies associated with GRBs have secured that at least some have a cosmological origin and therefore must be extremely energetic events. The inferred redshifts of GRB 970508 and GRB 971214 suggest isotropic emission energies of $\sim 10^{52}$ and 3×10^{53} ergs (Metzger et al. 1997a, 1997b; Kulkarni et al. 1998). The energy in γ-rays alone for GRB 971214 is equivalent to 16% of the rest mass of the sun. Producing this amount of energy in γ-rays is difficult for stellar remnant collapse models, where the total amount of gravitational binding energy released in a $\sim 1 \, M_\odot$ configuration is $\sim 10^{54}$ ergs (Wijers et al. 1998). However, *if* a stellar-remnant collapse manages to concentrate energy deposition into $\sim 1\%$ of the sky, then it *may* produce a burst with the observed energies.

Supermassive black holes are abundant in the universe. They are inferred to power active galactic nuclei (AGNs) and quasars; every galaxy examined so far seems to possess a supermassive black hole in its center (van der Marel et al. 1997). These black holes could have had supermassive objects as progenitors (Begelman & Rees 1978). Two venues of supermassive black hole formation are considered: in one, a dense cluster of $1\,M_\odot$ stars is disrupted by collisions and coalesces into a central supermassive star; in the second venue, the cluster as a whole undergoes a post newtonian collapse into a supermassive object. A supermassive star may also be formed directly through the collapse of a $\sim 10^5\,M_\odot - 10^6\,M_\odot$ primordial gas cloud when cooling in these is not efficient (Peebles & Dicke 1968 and Tegmark et al. 1997).

We will discuss the formation of a GRB fireball in both collapse scenarios. We will also address how the highly variable time structure of GRBs can occur in supermassive object collapse, and how the fireball can avoid "baryon-loading", which can incapacitate the formation of a relativistic fireball. It should be noted that Prilutski & Usov (1975) previously described the emission of a GRB from magneto-energy transfer during collapse of supermassive rotators ($\sim 10^6\,M_\odot$) believed to power AGNs and quasars. The neutrino energy transfer process we discuss is not necessarily tied to AGNs or quasars, but to the formation of the black holes powering them.

2. Fireballs from Supermassive Black Hole Formation

2.1. Supermassive Star Collapse

In the first venue of supermassive black hole formation, a supermassive star undergoes a general relativistic (Feynman-Chandrasekhar) instability. A core of mass $M_5^{\rm HC} \equiv M^{\rm HC}/10^5 M_\odot$ collapses homologously and drops through the event horizon, releasing a gravitational binding energy of $\sim E_s \approx 10^{59} M_5^{\rm HC}$ erg. The mass of the homologous core can be an order of magnitude (or more) less than the mass of the initial hydrostatic supermassive star, $M_5^{\rm init} \equiv M^{\rm init}/10^5 M_\odot$.

During collapse, neutrinos are thermally emitted due to $e^\pm$ annihilation in the core. The luminosity of the neutrinos goes as the ninth power of the core temperature (Dicus 1972). This luminosity can be approximated from the product of neutrino emissivity (Schinder et al. 1987; Itoh et al. 1989) near the black hole formation point and the volume inside the Schwarzschild radius, $4 \times 10^{15}\,(T_9^{\rm Schw})^9\,(4\pi r_s^3/3)$ erg/sec, where $T_9^{\rm Schw}$ is the characteristic average core temperature in units of 10^9 K at the black hole formation epoch. In a spherical nonrotating supermassive star this is

$$T_9^{\rm Schw} \approx 12 \alpha_{\rm Schw}^{1/3} \left(\frac{11/2}{g_s}\right)^{1/3} \left(\frac{M_5^{\rm init}}{M_5^{\rm HC}}\right)^{1/6} \left(M_5^{\rm HC}\right)^{-1/2}. \qquad (17)$$

Here $\alpha_{\rm Schw}$ is the ratio of the final entropy per baryon to the value of this quantity in the initial pre-collapse hydrostatic star, and $g_s \approx g_b + 7/8 g_f \approx 11/2$ is the statistical weight of relativistic particles in the core. The characteristic free-fall timescale is labelled $t_s \approx M_5^{\rm HC}$ sec, and the characteristic radius (the Schwarzschild radius) is $r_s \approx 3 \times 10^{10} M_5^{\rm HC}$ cm. It has been shown that the ratio of the homologous core mass to the initial mass is $M_5^{\rm HC}/M_5^{\rm init} \approx \sqrt{2/5.5} \alpha_{\rm Schw}^2$ (Fuller, Woosley & Weaver 1986), so that $T_9^{\rm Schw} \approx 13 (M_5^{\rm HC})^{-1/2}$. The neutrino luminosity is

$$L_{\nu\bar{\nu}} \sim 4 \times 10^{15} \, (T_9^{\text{Schw}})^9 \, (4\pi \, r_s^3/3) \, \text{erg/sec} \approx 5 \times 10^{57} (M_5^{\text{HC}})^{-3/2} \, \text{erg/sec}. \quad (18)$$

About 70% of the neutrino emission will be in the $\nu_e\bar{\nu}_e$ channel (Woosley, Wilson & Mayle 1986).

This ample $\nu\bar{\nu}$ emission can create a fireball above the core through $\nu\bar{\nu} \to e^+e^-$. The neutrino luminosities will undergo gravitational redshift, which depresses energy deposition above the star; however, this will be compensated by increased $\nu\bar{\nu}$-annihilation from gravitational bending of null trajectories (Cardall & Fuller 1997). The neutrino emission is nearly thermal (Shi & Fuller 1998), allowing the neutrino energy deposition rate to be approximated as

$$\dot{Q}_{\nu\bar{\nu}}(r) \sim 4 \times 10^{22} \, (M_5^{\text{HC}})^{-7.5} (r_s/r)^8 \, \text{erg cm}^{-3}\text{s}^{-1}. \quad (19)$$

The total energy injected into the fireball above a radius r by this process is

$$E_{\text{f.b.}}(r) = t_s \int_r^\infty 4\pi r^2 \dot{Q}_{\nu\bar{\nu}}(r) dr \sim 2.5 \times 10^{54} \, (M_5^{\text{HC}})^{-3.5} (r_s/r)^5 \, \text{erg}. \quad (20)$$

This is an unequivocally large amount of energy. For a star where $M_5^{\text{HC}} = 0.5$, the energy of the fireball will be $\sim 10^{53}$ erg at a radius $r \sim 3r_s \sim 10^{11}$ cm. This would correspond to the energy of a GRB with isotropic emission at a redshift $z \approx 3$.

2.2. Supermassive Star Cluster Collapse

The second venue for the production of a GRB fireball during supermassive black hole formation is the collapse of a star cluster of $10^5 - 10^9 \, M_\odot$. The cluster undergoes a general relativistic instability (Shapiro & Teukolsky 1985) where collisions of $M_* \sim M_\odot$ stars could produce the neutrino emission powering a fireball. During the collapse, the stars will have relativistic speeds ($\Gamma \sim 1$) and a zero impact parameter collision of a pair will produce a typical entropy per baryon of $S \sim 10^4 \Gamma^{1/2} (g_s/5.5)^{1/4} (M_\odot/M_*)^{1/4} (V_*/V_\odot)^{1/4}$ with $T_9 \sim 1$, and where $V_*/V_\odot$ is the ratio of the stellar collision interaction volume to the solar volume. Generally, the collisions will have a non-zero impact parameter, and involve the less dense outer layers of the star, where there will be larger entropies. These entropies could be high enough ($S \sim 10^7$) to produce the pair fireball without the need for neutrino heating. The complex structure and baryon-free regions between stars can provide areas for fireballs to form with low baryon-loading. Both the collisions and neutrino emission are stochastic processes which might lead to the complex time structure of GRBs.

3. Event Rates and the $\log N - \log P$ Distribution

If all collapses occur at a single redshift, z, the observed rate is

$$4\pi r^2 a_z^3 \frac{dr}{dt_0} \frac{\rho_b F(1+z)^3}{M^{\text{init}}}, \quad (21)$$

where r is the Friedman-Robertson-Walker comoving coordinate distance of the objects, a_z is the scale factor of the universe corresponding to z (with $a_0 = 1$), t_0 is the age of the universe, $\rho_b \approx 2 \times 10^{-29} \Omega_b h^2 \,\mathrm{g\,cm^{-3}} \approx 5 \times 10^{-31} \mathrm{g\,cm^{-3}}$ (Tytler & Burles 1997) is the baryon density of the universe, h is the Hubble parameter in $100\,\mathrm{km\,s^{-1}\,Mpc^{-1}}$, and F is the fraction of all baryons in supermassive objects. With collapses occuring at $z \sim 3$ we have $r \sim 3000 h^{-1}$ Mpc, and the corresponding collapse rate is

$$0.15 F \,(M_5^{\mathrm{init}})^{-1} \sec^{-1} \sim 10^4 F \,(M_5^{\mathrm{init}})^{-1} \mathrm{day}^{-1}. \tag{22}$$

With $F \sim 0.1\%$, and with a 100% detection efficiency, we will observe one collapse per day, assuming isotropic emission. This corresponds to a density of supermassive black holes of $7\,h^2/\mathrm{Mpc}^3$, about $350\,h^{-1}$ per L_* galaxy, or $\lesssim 10\,h^{-1}$ per galaxy-scale object (i.e. including dwarf galaxies).

It is instructive, however, to estimate the rate of supermassive object collapse in terms of numbers of Lyman limit systems and damped Lyα systems. Employing a column density N_{HI} distribution per unit column density per unit absorption distance of $10^{13.9} N_{\mathrm{HI}}^{-1.74}$ (Storrie-Lombardi, Irwin & McMahon 1996), the rate of supermassive object collapse will be comparable to that of GRBs if every Lyα system with $N_{\mathrm{HI}} \gtrsim 10^{18}\,\mathrm{cm}^{-2}$ experiences one supermassive object collapse.

The relation of the number of GRBs with peak flux ($\log N - \log P$) may be able to tell us something about the distribution of GRB sources in the universe. To see if there may be a correlation between this distribution and that of supermassive object collapse, we count the number of supermassive objects through the quasar population. That is, if quasars have some characteristic lifetime, then we can say that the comoving supermassive object number density is proportional to the comoving quasar number density. There has been some recent work in the evolution of the number density of quasars (Maloney & Petrosian 1998, Shaver et al. 1998). We use the evolution of quasar number to sum the number of supermassive collapse events for a standard candle and various cosmologies. The peak flux distribution calculated from supermassive object collapse is not inconsistent with the GRB $\log N - \log P$ relation, considering the uncertainties in the quasar epoch and GRB luminosity distribution. In Figure 1, we show the case for $\Omega_m = 0.25, \Omega_\Lambda = 0, \alpha = 0.7, z_{\mathrm{th}} = 3.4$, where α is the GRB spectral index, and z_{th} is the cutoff redshift of the BATSE detector. We must be careful, however, since the luminosity function of GRBs is unknown, and recent work has shown that this can affect the observed peak flux distribution greatly (Krumholz et al. 1998).

4. GRB Time Structure and Baryon Loading

The quickly varying time structure of GRBs limits the size of the region powering the fireball to the distance light can travel during this time variation, $\sim 10^7$ cm (Piran 1998). In the first venue described above of GRB production (supermassive star collapse) the characteristic size of the emission region in this spherically symmetric model is the Schwarschild radius, $\sim 10^{10}$ cm. The supermassive star's collapse and the formation of fireball(s) will not generally be spherically symmetric, nor will convective processes be unimportant. So, we

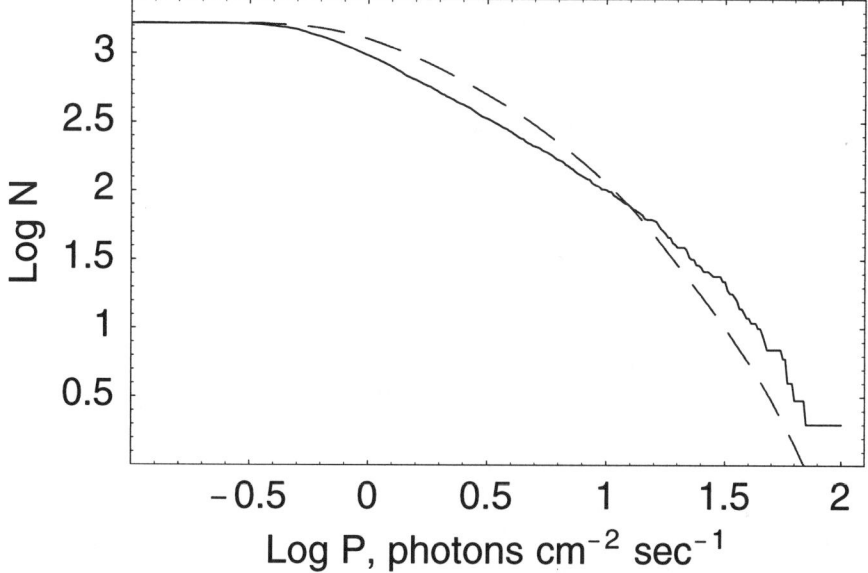

Figure 1. The BATSE 4B Catalogue number versus peak flux (solid) and the expected distribution (dashed) for supermassive object collapse associated with quasars. Here $\Omega_m = 0.25, \Omega_\Lambda = 0, \alpha = 0.7, z_{th} = 3.4$.

can say that the above energy scales will be deposited and localized by neutrino annihilation into regions outside of the core that are a fraction of the core's size. These regions will be distributed around the core, and may produce variability through superpositions and instabilities. In the second venue, 1 $M_\odot$ stars have the same physical scale, $\sim 10^{10}$ cm, but can produce variability in the means described in §2.3 or through a localization of neutrino annihilation.

Producing a small region of high photon energy density and high entropy per baryon—the GRB fireball—is a challenge for all GRB models. The supermassive star model, in the one-dimensional case, also does not deposit the needed energies in a "baryon-free" region. However, if some of the tremendous energy deposited by the collapse does find itself in an area with low baryon density, it will produce a GRB fireball of large energies. This can happen when the star is rotationally flattened, where the neutrinos deposit their energy along the "baryon-free" axis of rotation.

Acknowledgments. I would like to thank my parents, Nazar and Raisa, whose support made my attendance at this symposium possible. I would also like to thank the hospitality of Hasmig Haroutoonian and the Local Organizing Committee. This research is supported by NASA grant NAG5-3062 and NSF grant PHY95-03384 at UCSD.

References

Begelman, M.C. & Rees, M.J. 1978, MNRAS, 185, 847

Cardall, C.Y., & Fuller, G.M. 1997, ApJ, 486, L111
Dicus, D.A. 1972, Phys. Rev. D, 6, 941
Fuller, G.M., Woosley, S.E., & Weaver, T.A. 1986, ApJ, 307, 675
Fuller, G. & Shi, X. 1998, ApJ, 502, L5
Itoh, N., Adachi, T., Nakagawa, M., Kohyama, Y., & Munakawa, H. 1989, ApJ, 339, 354
Krumholz, M. et al. 1998, ApJ, 506, L81
Kulkarni, S.R., et al. 1998, Nature, 393, 35
Maloney, A. & Petrosian, V. 1998, American Astronomical Society Meeting, 192, # 16.04
Metzger, R.M., et al. 1997a, Nature, 387, 879
Metzger, R.M., Cohen, J.G., Chaffee, M.H., & Blandford, R.D. 1997b, IAU circular No. 6676
Peebles, P.J.E. & Dicke, R.H. 1968, ApJ, 154, 891
Piran, T. 1998, Physics Reports, in press, astro-ph/9810256.
Prilutski, O.F., & Usov, V.V. 1975, Ap&SS, 34, 395
Schinder, P.J., et al. 1987, ApJ, 313, 531
Shapiro, S. L. & Teukolsky, S. A. 1985, ApJ, 292, 41
Shaver, P.A., et al. 1998, to appear in: "Highly Redshifted Radio Lines", eds. C. Carilli, S. Radford, K. Menten, G. Langston, (PASP: San Francisco), astro-ph/9801211
Shi, X. & Fuller, G.M. 1998, ApJ, in press
Storrie-Lombardi, L.J., Irwin, M.J., & McMahon, R.G. 1996, MNRAS, 282, 1330
Tegmark, M., et al. 1997, ApJ, 474, 1
Tytler, D., & Burles, S. 1997, in "Origin of Matter and Evolution of Galaxies", eds. T. Kajino, Y. Yoshii & S. Kubono (World Scientific Publ. Co.: Singapore), 37
van der Marel, R.P., de Zeeuw, P.T., Rix, H., & Quinlan, G.D. 1997, Nature, 385, 610
Wijers, R.A.M.J., Bloom, J.S., Bagla, J.S., & Natarajan, P. 1998, MNRAS, 294, L13
Woosley, S. E., Wilson, J. R., & Mayle, R. 1986, ApJ, 302, 19

The Emission Line Properties of the 3CR Radio Galaxies at Redshift One: Shocks, Evolution, and the Alignment Effect

Philip Best, Huub Röttgering

Sterrewacht Leiden, Postbus 9513, 2300 RA Leiden, The Netherlands

Malcolm Longair

Cavendish Astrophysics, Madingley Road, Cambridge, CB3 0HE, UK

Abstract. The results of a deep spectroscopic campaign on powerful radio galaxies with redshifts $z \sim 1$, to investigate in detail their emission line gas properties, are presented. Both the 2-dimensional velocity structure of the [OII] 3727 emission line and the ionisation state of the gas are found to be strongly dependent upon the linear size (age) of the radio source in a manner indicative of the emission line properties of small (young) radio sources being dominated by the passage of the radio source shocks. The consequences of this evolution throughout the few $\times 10^7$ year lifetime of the radio source are discussed, particularly with relation to the alignment of the UV-optical continuum emission of these objects along their radio axis, the nature of which shows similar evolution.

1. Introduction

Powerful radio galaxies possess extremely luminous extended emission line regions, often aligned along the radio axis (e.g. McCarthy et al. 1996 and references therein). The source of ionisation of this gas has been a long standing question. Robinson et al. (1987) found that optical emission line spectra of most low redshift ($z \lesssim 0.1$) radio galaxies are well explained using photoionisation models, and a similar result was found for a composite spectrum of radio galaxies with redshifts $0.1 < z < 3$ (McCarthy 1993). Photoionisation models are supported by orientation-based unification schemes of radio galaxies and radio-loud quasars (e.g. Barthel 1989), in which the radio galaxies will host an obscured quasar nucleus. On the other hand, detailed studies of individual sources (e.g. 3C171; Clark et al. 1998) have revealed features such as enhanced nebular line emission, large velocity widths and ionisation state minima coincident with the radio hotspots, indicating that the morphology, kinematics and ionisation of the gas in some sources are dominated by shocks.

Powerful radio galaxies with $z \gtrsim 0.6$ also show enhanced optical-UV emission, which is elongated and aligned along the radio axis in a manner similar to the line emission. In recent years we have observed a sample of 28 3CR radio galaxies with $0.6 < z < 1.8$ using the HST, VLA and UKIRT, to study this aligned emission (e.g. Best et al. 1997). An important result is that the nature of the alignment effect evolves with increasing size (corresponding roughly to increasing age) of the radio source (Best et al. 1996). Small radio sources show intense strings of blue knots which track the passage of the radio jets, whilst larger sources typically have much more diffuse optical–UV emission. Clearly the passage of the radio jet has an important influence on these host galaxies.

In this contribution, the first results of a program of deep spectroscopic observations on this sample of distant radio galaxies are presented. The observations are described in Section 2. The main results are presented in Sections 3 & 4, and in Section 5 the implications of these results for both the emission line gas and the alignment effect of powerful radio galaxies are discussed.

2. Sample selection and observations

From our complete HST sample of 28 $z \sim 1$ 3CR radio galaxies (Best et al 1997), our spectroscopic studies were restricted initially to those 18 galaxies with redshifts $0.7 < z < 1.25$. Of these, 3C41, 3C65, 3C267 and 3C277.2 were not observed due to constraints of telescope time. The exclusion of these four galaxies was based upon their right ascensions, not upon source properties, and so their exclusion should not introduce any significant selection effects.

The remaining 14 galaxies were observed for between 1.5 and 2 hours each during July 1997 and February 1998, using the twin-armed ISIS spectrograph on the William Herschel Telescope. Low (~ 12Å) resolution spectra in the blue arm provided a useful observed–frame wavelength coverage of ~ 3200 to ~ 5200Å, and intermediate (~ 5Å) resolution red arm spectra sampled the rest–frame wavelength range from ~ 3500Å to ~ 4300Å. This setup covered a broad range of emission lines, allowing investigation of the ionisation state of the gas, and provided reasonably high resolution data on the strong [OII] 3727 emission line, enabling the velocity structures to be determined. Details of the dataset and the data reduction techniques are given by Best et al. (1999).

3. Emission line ratios

The line ratios of CIII] 1909 / CII] 2326 and [NeIII] 3869 / [NeV] 3426 have been determined for these galaxies, the former ratio from the ISIS blue arm data, and the latter from the red arm data (where possible) or from the literature. These line ratios are particularly useful for ionisation studies for three reasons: (i) in both cases the two lines in the ratio involve the same element, and so variations in metallicity or abundance are not important; (ii) the two lines are very close in wavelength, and so differential extinction is minimised; (iii) the theoretical predictions of photoionisation and ionisation by shocks for these line ratios are very different (see below).

Theoretical predictions for these line ratios in photoionisation models have been taken from the study of Allen et al. (1998), who generated these two line ratios (amongst others) using the MAPPINGS II code (e.g. Sutherland et al. 1993) for the simple model of a planar slab of material being illuminated by power–law spectrum of ionising radiation. For two different spectral indices of the input spectrum ($F_\nu \propto \nu^\alpha$ with $\alpha = -1$ and $\alpha = -1.4$), and two different densities of cloud ($n_e = 100$ and $1000 \, \mathrm{cm}^{-3}$), the ratios were calculated for a

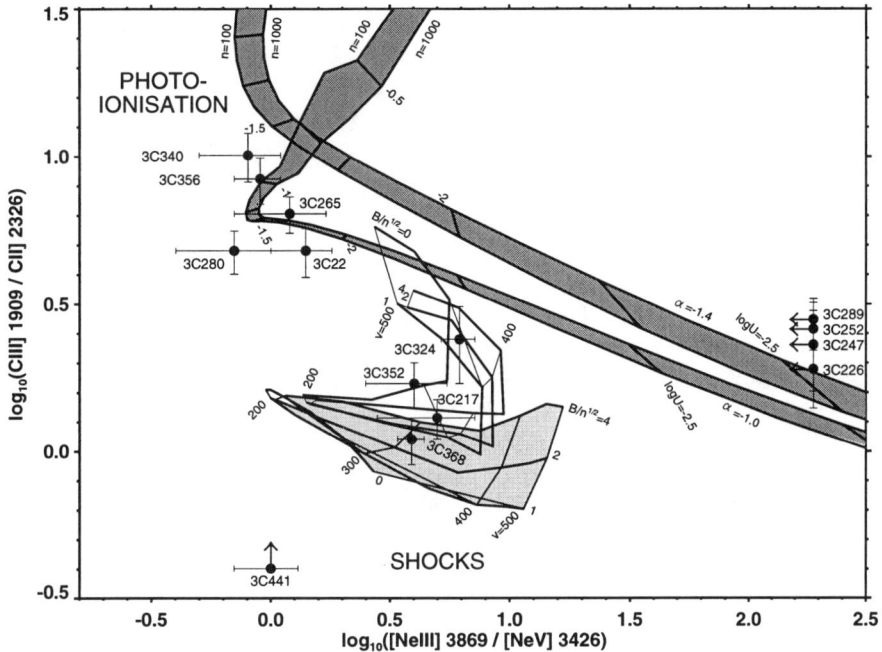

Figure 1. An emission line diagnostic plot for the 3CR radio galaxies, compared with theoretical predictions. The upper shaded regions correspond to photoionisation models, the lower shaded region to shock models, and the lower unshaded lines to shock models including a precursor region (see text for details).

wide range of ionisation parameter U, defined as the ratio of ionising photons striking the cloud to the density of the cloud $[U = (cn_{\rm H})^{-1} \int_{\nu_0}^{\infty} (F_\nu d\nu)/h\nu]$.

Line ratios for ionisation by shocks were also calculated, using the models of Dopita and Sutherland (1996). These authors tabulated theoretical line strengths for a variety of different shock velocities (150 to 500 km s^{-1}), and 'magnetic parameters' ($0 \leq B/\sqrt{n} \leq 4\,\mu{\rm G\,cm}^{-1.5}$). The line ratios were calculated both for the shocked gas, and for the combination of shocked gas with a precursor region, the latter corresponding to emission from the pre–shock gas due to ionising photons produced by the shock diffusing upstream ahead of the shock front (see Dopita and Sutherland 1996 for further discussion).

The output of these theoretical calculations are compared with the data in Figure 1. Four galaxies (3C217, 3C324, 3C352, 3C368) lie in the region of the diagram appropriate for shock ionisation, and five (3C22, 3C265, 3C280, 3C340, 3C356) lie close to the photoionisation models. The five sources plotted at the edges of the plots have no data for one of their emission lines. Interestingly, all of the four radio sources in the shock region have radio sizes smaller than 115 kpc ($\Omega = 1$, $H_0 = 50\,{\rm km\,s^{-1}\,Mpc^{-1}}$), and the five 'photoionised sources' have larger radio sizes. Smaller radio sources appear to have lower ionisation states.

4. Velocity structures of the emission line gas

The 2-dimensional [OII] 3727 velocity profiles of all 14 of the galaxies, along with their intensity distributions, are presented in Best et al. (1999). The most important aspects of those profiles are as follows.

- There is a strong inverse correlation between the FWHM of the [OII] 3727 emission and the size of the radio source (Figure 2a). The four 'shock–dominated' sources from the previous section have the highest FWHM.
- Large radio sources often have smooth 'rotation-dominated' velocity profiles, whilst those of small (lower ionisation; see above) sources are more distorted. Note that Baum et al. (1992) found that radio galaxies with redshifts $z < 0.2$ whose emission line velocities were consistent with rotation typically had a high ionisation state ([OIII] 5007 / H$\beta \gtrsim 5$); 'non-rotators' had lower ionisation states.
- The [OII] 3727 line flux, normalised by the integrated K–band flux which essentially measures the stellar mass of the galaxy, correlates inversely with the radio source size. The four 'shock–dominated' sources all have high integrated [OII] 3727 fluxes (Figure 2b).

5. Discussion

These results can be fit together in a scenario whereby the passage of the radio bow shocks through the host galaxy dominates the kinematics and ionisation of smaller (younger) radio sources, but large radio sources are more relaxed and photoionisation from the AGN dominates. As the jet passes through the emission line regions: (1) the emission line gas will be accelerated by the shock, giving rise to the larger observed FWHM, and the distorted [OII] velocity structures; (2) the shock will provide additional ionising photons, increasing the [OII] line emission; (3) the emission line clouds will be compressed by the shock, leading to a decrease in the degree of ionisation of the gas.

Together with the results of Best et al (1996,1998) that the aligned optical–UV emission is tightly associated with the radio jet, these results suggest that radio source shocks are also important for the continuum alignment effect. Shock excitation of the emission–line clouds will give rise to an enhanced contribution of nebular continuum emission in small sources (e.g. see Dickson et al 1995). Radio jet shocks may also induce the formation of massive knots of bright young stars, which will disperse and fade over the lifetime of the source. These two mechanisms would each account for both the tight alignment of the optical–UV continuum emission along the radio jet and the observed variation in the luminosity of this emission with radio size. Moreover, shocks may also be responsible for disrupting optically thick clouds along the radio jet direction and exposing previously hidden dust grains (Bremer et al 1997); this would give rise to an enhanced contribution of scattered light, distributed in a non–biconical manner.

In conclusion, these results show that the passage of the radio induced shocks through the host galaxy of powerful radio sources plays a key role in producing the emission line gas properties of these sources. Much of the continuum alignment effect may have its origin in these same shocks.

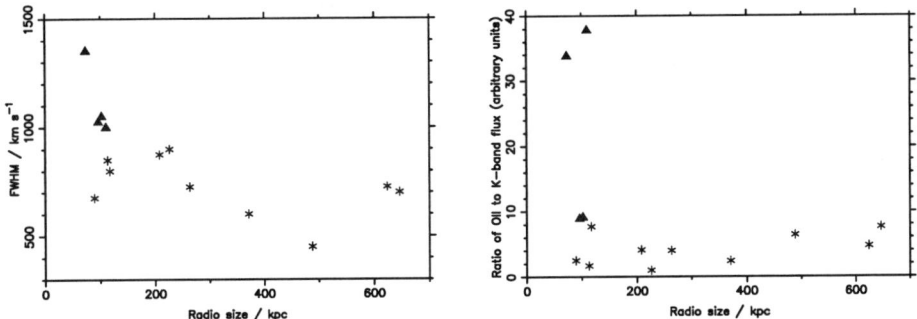

Figure 2. (a) The inverse correlation between the maximum FWHM of the [OII] 3727 emission line and the radio source size; (b) The relationship between the K-band (mass) normalised [OII] line intensity and radio size. In each case, the four sources in the shock-dominated region of Figure 1 are plotted as filled triangles, the remainder as asterisks.

Acknowledgments. The WHT is operated on the island of La Palma by the Isaac Newton Group in the Spanish Observatorio del Roches de los Muchachos of the Instituto de Astrofisica de Canarias. This work was supported in part by the Formation and Evolution of Galaxies network set up by the European Commission under contract ERB FMRX– CT96–086 of its TMR programme. We are grateful to Mark Allen for supplying the output of the MAPPINGS II code in digitised form, and to Matt Lehnert for useful discussions.

References

Allen M.G., Dopita M.A., Tsvetanov Z.I., 1998, ApJ, 493, 571
Barthel P.D., 1989, ApJ, 336, 606
Baum S.A., Heckman T.M., van Breugel W.J.M., 1992, ApJ, 389, 202
Best P.N., Longair M.S., Röttgering H.J.A. 1996, MNRAS, 280, L9
Best P.N., Longair M.S., Röttgering H.J.A. 1997, MNRAS, 292, 758
Best P.N., Carilli C.L., Garrington S.T., Longair M.S., Röttgering H.J.A. 1998, MNRAS, 299, 357
Best P.N., Röttgering H.J.A., Longair M.S., 1999, MNRAS, *submitted*
Bremer M.N., Fabian A.C., Crawford C.S., 1997, MNRAS, 284, 213
Clark N., Axon D., Tadhunter C., Robinson A., O'Brien P., 1998, ApJ, 494, 546
Dickson R., Tadhunter C., Shaw M., Clark N., Morganti R., 1995, MNRAS, 273, L29
Dopita M.A., Sutherland R.S., 1996, ApJS, 102, 161
McCarthy P.J., 1993, ARAA, 31, 639
McCarthy P.J., Baum S.A., Spinrad H., 1996, ApJS, 106, 281
Robinson A., Binette L., Fosbury R.A.E., Tadhunter C.N., MNRAS, 227, 97
Sutherland R.S., Bicknell G.V., Dopita M.A., 1993, ApJ, 414, 510

Active Galactic Nuclei and Related Phenomena
IAU Symposium, Vol. 194, 1999
Yervant Terzian, Daniel Weedman, Edward Khachikian, eds.

Accretion Disks and Star Formation

Suzy Collin, Jean-Paul Zahn

Observatoire de Paris, Section de Meudon, Place Jansen, 92195 Meudon, France

Abstract. It is generally admitted that at distances larger than $\sim 10^4$ gravitational radii accretion disks in AGN form a torus made of high velocity gaseous interacting clouds with a small filling factor. We propose here a completely different model for this region, in which unstable fragments give rise to protostars, all becoming massive stars after a rapid stage of accretion. These stars explode as supernovae, which produce strong outflows perpendicular to the disk and induce an outward transfer of angular momentum, as shown by the numerical simulations of Rozyczka, Bodenheimer and Lin (1995). So the supernovae themselves can sustain the inflow mass rate required by the AGN. Assuming that the star formation rate is proportional to the growth rate of the gravitational instabilities, one obtains a self-regulated accretion disk made of gas and stars in which the gas is maintained in a state close to gravitational instability. We show that the gaseous disk is able to support a large number of massive stars and supernovae while staying relatively homogeneous. This model could explain the high velocity metal enriched outflows implied by the presence of the broad absorption lines in quasars. It could also account for a pregalactic enrichment of the intergalactic medium, if black holes formed early in the Universe. Finally it could provide a triggering mechanism for starbursts in the central regions of galaxies.

1. Introduction

There is a large consensus that AGN are fueled via accretion disks. The observation of the "UV bump" (Shields 1978, Malkan & Sargent 1982, and many subsequent papers) argues in favor of geometrically thin and optically thick disks, possibly embedded in a hot X-ray emitting corona. Generally these disks are studied using the α prescription for viscosity introduced by Shakura and Sunayev (1973). It is well known that these "α-disks" have two serious problems at large radii: they are not able to transport rapidly enough the gas from regions located at say one parsec, and they are gravitationally unstable beyond about 0.1 parsec, precisely when the Toomre parameter Q defined as $Q = \Omega c_s/(\pi G \Sigma)$ (Toomre 1964, Goldreich and Lynden-Bell 1965) becomes of order unity (Ω is the Keplerian angular velocity, Σ is the surface density, and c_s is the sound speed).

As an illustration, Fig. 1 taken from Collin & Huré (1998) displays the Toomre radius corresponding to $Q \sim 1$, expressed in Schwarzschild radius $R_S =$

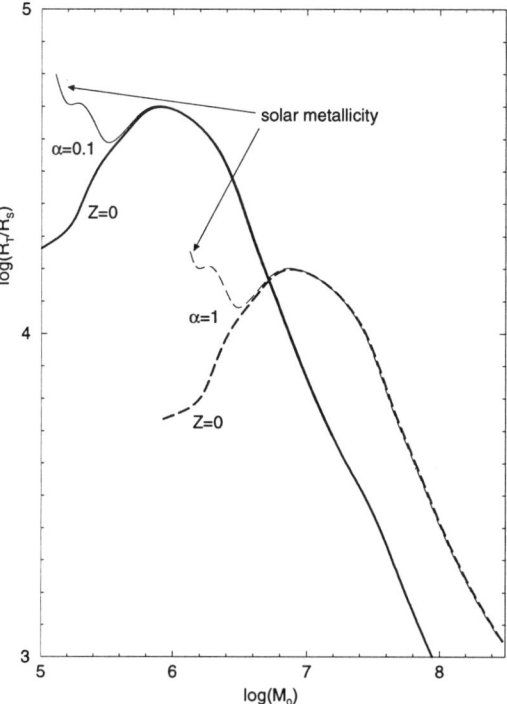

Figure 1. The Toomre radius R_T (normalized to R_S) versus the mass of the black hole M_o in units of $M_\odot$. The accretion rate is critical. $\alpha = 1$ is in dashed lines and $\alpha = 0.1$ is in solid lines. The zero metallicity case is plotted in bold lines and the solar metallicity case is in thin lines. After Collin & Huré, 1998.

$2GM/c^2 = 3\ 10^5 M/M_\odot$ cm as a function of the black hole mass in $M_\odot$ for the critical accretion rate $\dot{m} = \dot{M}/\dot{M}_{\rm crit} = 1$ (we set $\dot{M}_{\rm crit} = L_{\rm Edd}/(0.1c^2)$), for two values of the viscosity parameter $\alpha = 0.1$ and $\alpha = 1$, and for two metallicities, a solar and a zero metallicity. The Toomre radius expressed in R_S globally decreases with increasing black hole mass. It is roughly of the order of 10^3 to a few 10^4 R_S, i.e. comparable to the size of the Broad Line Region (hardly a coincidence!). These results are obtained using real opacities, but they are in good agreement with analytical expressions derived assuming constant opacity coefficient and mean mass per particle μ. For small masses ($M \leq 2 \times 10^7 \alpha^{2/5} M_\odot$) the Toomre radius scales as $\dot{m}^{-8/27} M^{-26/27}$, while for higher central masses, it scales as $\dot{m}^{4/9} M^{-2/9}$ (cf. Huré 1998).

Since the α-prescription is not valid in the gravitationally unstable region, one must find a mechanism accounting for the high rate of mass and angular momentum transport in AGN. It can be achieved by magnetic torques if large scale magnetic fields are anchored in the disk. We assume here that neither turbulent nor large scale magnetic fields are important in accretion disks. Farther from the center the supply of gas can be provided by gravitational torques or by global non axisymmetric gravitational instabilities, but this is not true in the intermediate region where the disk is locally but not globally self-gravitating.

The second problem lies in the gravitational unstability. Lin and Pringle (1987) proposed a prescription allowing for strongly unstable disks. They argued that in a self-gravitating disk, the largest size of the turbulent eddies stabilized by rotation, is Q times the disk thickness. This prescription corresponds to a large (supersonic) viscosity. On the contrary Paczyński (1978) proposed that a self-gravitating disk is maintained in a state of marginal instability (i.e. $Q = a$ few units), in which the energy dissipated by collisions between clumps prevent their collapse.

There are actually two extreme possibilities for a gravitationally unstable disk: either the unstable fragments collapse to compact bodies, and as a consequence the disk evolves rapidly towards a stellar or a non interacting system, or alternatively their collapse is stopped and the disk gives rise to small gaseous possibly interacting entities.

The suggestion of star formation in the self-gravitating region of such an accretion disk has been first made by Kolykhalov and Sunayev (1980). Begelman, Frank, and Shlosman (1989) and Shlosman and Begelman (1989) discussed in more detail the conditions for star formation at about 1-10 pc from the black hole, and its consequences on the disk. They concluded that, unless the disk can be maintained in a hot or highly turbulent state, it should transform rapidly into a flat stellar system which will be unable to build a new gaseous accretion disk and to fuel a quasar. They favor the picture of a disk made of marginally unstable randomly moving clouds with a small filling factor, where the "viscosity" of the disk is provided by cloud collisions. Here we adopt the opposite view that if an unstable fragment begins to collapse, the collapse will continue until a protostar is formed, so the disk will be made both of stars and of gas.

2. The overall star-gas disk

In Collin & Zahn (1998) we describe in detail a model of gas-star disk. We shall only recall here the salient points of this study and summarize the main results.

The model is based on the assumption that, though stars are present in the disk, it stays quasi homogeneous. It means that the supernovae shells do not overlap, a condition which is checked a posteriori for each solution. Another assumption (also checked a posteriori) is that the bulk of the gas is close to marginal instability (which implies that the density is a function only of the radius and of the central mass, as $Q = 1$ is equivalent to $\pi G \rho / \Omega^2 \sim 1$, where ρ is the midplane density of the disk). If the gas would be highly unstable (i.e. $Q \ll 1$), star formation would be too efficient and would immediately transform the gaseous disk into a stellar disk (cf. next subsection).

An important issue concerning such a marginal homogeneous disk is that when massive stars evolve, HII regions and winds create a cavity around them. One can show that the size of the cavity is limited to about one scale height of the disk, H. Therefore a new star cannot form inside a radius H of the star. One can also show that no gap is opened in the disk around the stars by induced density waves unless their mass exceeds a mass $m_{\rm gap}$ which is extremely large. So finally the gaseous disk is able to support a large number of massive stars while staying relatively homogeneous, since H is very small compared to the radius of the disk (typically $10^{-3}R$).

Finally we assume that the regions at the periphery of the disk provide a quasi stationary mass inflow during the life time of quasars or of their progenitors (for instance via global gravitational instabilities induced by merging), equal to the accretion rate on the black hole. In other words we assume that there is neither infall on the inner regions of the disk, nor a strong outflowing mass rate from the disk, except that due to the supernovae.

2.1. Star formation and evolution in the disk

There are several conditions for star formation, besides the fact that the gas should be at least marginally unstable. If the disk is marginally unstable, gravitationally bound fragments can form with a mass M_{frag} of the order of $4\rho H^3$ (Goldreich & Lynden-Bell 1965). They will then form a compact body if the formation time scale, t_{form}, and the cooling time, t_{cool}, are smaller than the characteristic mass transport time in the disk, t_{trans}. t_{form} corresponds to the maximum growth rate of the gravitation instability, which is, according to Wang & Silk (1994), equal to $\Omega^{-1} Q/\sqrt{1-Q^2}$. Unless Q is very close to unity, t_{form} is not much larger than the freefall time $t_{\text{ff}} \sim \Omega^{-1}$.

For an isothermal and spherical collapse the initial fragment gives rise to a dense core of mass m_{frag} in a time t_{collapse} corresponding to the growing rate (Chandrasekhar 1939) $\dot{M}_{\text{collapse}} \sim c_s^3/G$. The corresponding value of t_{collapse} is of the same order as t_{ff}. Actually the collapse should not be spherical, owing to the shear velocity, $\Delta V \sim H\Omega$. In a marginally unstable disk, ΔV is of the order of the sound velocity, so the collapse begins quasi spherically but it would be rapidly dominated by rotational motions unless it gets rid of a large fraction of its angular momentum. This is the case as a large proportion of the angular momentum is given to the protostellar disk, and may also be transformed into orbital motion in binaries or multiple systems. One can also show that under the tidal action of the central mass the small disk is synchronized in about a dynamical time, and this mechanism leads to the suppression of another fraction of the angular momentum. So finally the collapse must proceed as in the spherical case with a characteristic time t_{collapse}.

Once the protostar and the protostellar disk are formed, they undergo a mechanism of accretion and growing proposed by Artymowicz, Lin and Wampler (1993), for stars orbiting around the central black hole and being trapped in the disk (their mechanism is however different from ours). If the accretion would be limited to a region of radius H the accretion rate would be roughly $\dot{M}_{\text{accr}} \sim \Delta V \rho H^2 \sim \Omega m_{\text{frag}}$, and the corresponding accretion time for a star of $10M_\odot$ would be equal to $10M_\odot/(\Omega m_{\text{frag}})$. The real accretion time is shorter, as one should take into account the mass inflow coming from beyond H and swept up by the Roche lobe, which cannot be estimated simply.

A last important time is t_{migr}, the time for the stars to migrate towards the black hole owing to the mechanism of induced density waves previously mentioned and discussed by Goldreich and Tremaine (1980), Ward (1986, 1988), and Lin and Papaloizou (1986a and 1986b). One can verify that in the solutions found for the disk it is comfortably large, so the stars do not have time to migrate during their evolution. One interesting consequence is that the residual neutron stars left after the supernova explosions, if they are not ejected from the disk, can undergo other accretion phases, leading to other (presumably powerful) supernova explosions.

All these times are displayed on Fig. 2, and we will see that they satisfy the requirements for star formation.

2.2. Supernovae explosions and angular momentum transport

The expansion of a supernova in the radial direction is stopped at a radius $R_{s,max}$ when the expansion velocity is equal to the shear velocity at its edge. The shell can still expand in the azimuthal direction and will therefore be streched and sheared by the differential rotation. It will disappear after about one rotation time. The shell always reaches the disk surface, so the interior is depressurized and the expansion is driven only by momentum conservation. A large fraction of that momentum escapes from the disk, and the momentum P_{disk} supplied to the disk is equal to a fraction of the total momentum P_{tot} carried by the supernova explosion $\sim H/R_{s,max}$. Since the terminal velocity is of the order of $3P_{disk}/(4\pi \rho H R_{s,max}^2)$, one gets $R_{s,max} \sim 3P_{tot}/(4\pi \rho \Omega)$.

One can deduce the maximum rate of supernovae supported by the disk without being destroyed. As the cavities created by the blast waves are replenished roughly at the shear velocity, their lifetime is $\sim 1/\Omega$, and this rate is of the order of $\Omega(R^2/R_{s,max}^2)$. Actually this estimation is rather conservative, as the numerical simulations of Rozyczka et al (1995) show that when the cavity and the shock wave reach their maximum radial extension after a fraction of an orbital time and become strongly elongated, they are also strongly squashed, so the surface of the perturbation seems to be rapidly decreasing.

Supernovae produce a net transfer of angular momentum towards the exterior. The net angular momentum supplied by one supernova is equal to $\Delta J \sim (3/2\pi) P_{tot} R H/R_{s,max}$. If **all** the angular momentum required to sustain the accretion rate is carried by the supernovae, their rate $\mathcal{N}_{SN} \sim 2\dot{M} R^2 \Omega/\Delta J$.

2.3. Solutions

A self-regulation mechanism for the gas density has been proposed for the Galactic disk by Wang and Silk (1994), through the growth time of gravitational instabilities. According to them, the rate of gas transformed into stars is $d\Sigma_g/dt = \Sigma_g \Omega^{-1} \epsilon \sqrt{1-Q^2}/Q$, ϵ being an "efficiency" of star formation which is of the order of 0.1% for massive stars in the Galaxy. Wang and Silk have included the stellar contribution in the Toomre parameter Q, as given by a two fluid approximation. Here it should not be taken into account, as the mean distance between stars exceeds the instability length H.

Since the formation time and the lifetime of the stars are both small compared to the growth time of the black hole or to the active phase of a quasar, a stationary state of the disk can be established if there is a steady mass inflow from outward. One can determine then the star formation rate from the number of supernovae able to sustain the required accretion rate, and solve the radial disk structure in a self-consistent way.

The (vertically averaged) equations for the disk structure are the hydrostatic equilibrium equation, including the stellar contribution, and the energy equation, which should take into account 3 heating mechanisms: 1. heating by the fraction

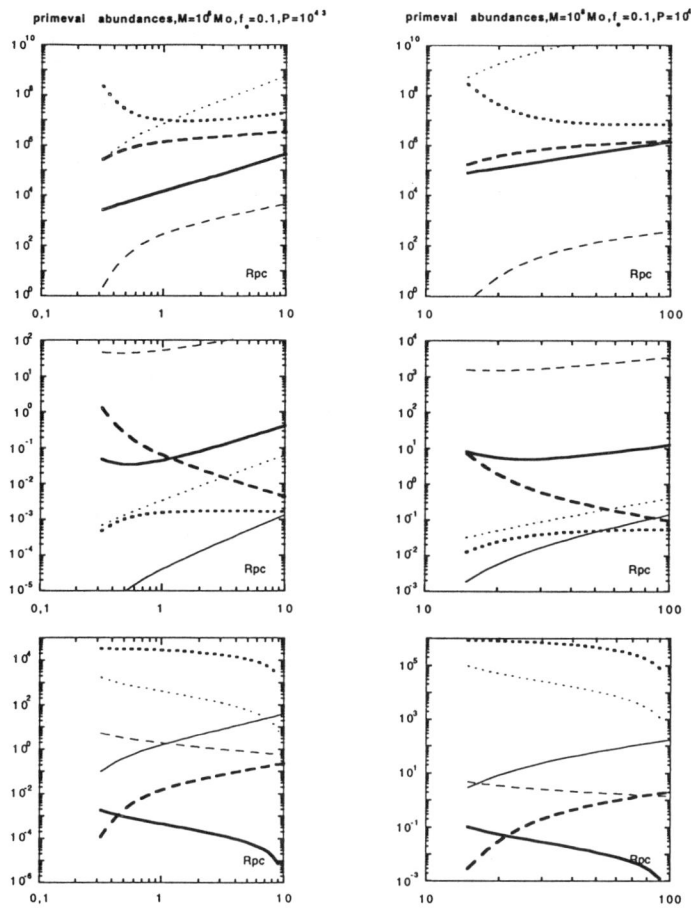

Figure 2. Solutions for the disk made of stars and gas, for primeval abundances. In the top panels are given the characteristic times in year: bold solid lines: orbital time; bold dashed lines: t_{trans}; bold dot lines: $t_{\text{accr}}(10\text{M}_\odot)$; thin dashed lines: t_{cool}; thin dot lines: $t_{\text{migr}}(10\text{M}_\odot)$. In the middle panels, are given several parameters allowing to check that the conditions required by the model are fulfilled: bold solid lines: ratio of the rate of supernovae to the maximum allowed rate; bold dashed lines: ratio of the number of stars to the maximum number allowed by the tidal effect; bold dot lines: ratio of the number of stars to the maximum number allowed by the HII regions; thin solid lines: $\epsilon\sqrt{1-Q_g}$; thin dashed line: m_{gap} in $\text{M}_\odot$; thin dot lines: $V_{\text{rad}}/V_{\text{Kep}}$. In the bottom panels, several physical parameters of the models: bold solid lines: the rate of supernovae $\mathcal{N}_{\text{SN}}$ per year, integrated from the outer edge; bold dashed lines: m_{frag} in $\text{M}_\odot$; bold dot lines: the mass in gas M_{gas} integrated from the outer edge; thin solid lines: $H/10^{16}\text{cm}$; thin dashed lines: the midplane temperature in 100K $T/100K$; thin dot lines: the number of stars N_* integrated from the outer edge.

of the stellar luminosity not absorbed in the HII regions, 2. heating by the central source, when the disk flares (as is always the case in this model), 3. heating due to the dissipation of kinetic energy through shock waves induced by supernova shells (it corresponds to the viscous heating in a standard disk). One gets then a complete algebraic system, which can be solved as a function of the central mass and of the Eddington ratio f_E (here equal to $\dot{m}$ for an efficiency factor equal to 10%). As an illustration, Fig. 2 displays the results for two solutions corresponding to primeval abundances, to an Eddington ratio equal to 0.1, and to black hole masses of 10^6 and 10^8 $M_\odot$. The luminosity of the stars is set to 10^{38} ergs s^{-1}, their average mass is 30 $M_\odot$, and we have assumed $t_{\text{lifetime}} = 3\ 10^6$ yrs and $P = 10^{43}$ g cm s^{-1}. The solutions for solar abundances are not very different.

First are displayed the characteristic times. Except for large black hole masses and accretion rates, the free fall time is about two orders of magnitude less than the mass transport time, which means that $\sqrt{1 - Q_g^2}$ can be as small as 10^{-2}. The assumption that the gas is marginally unstable is therefore self-consistent. The accretion time is at most a few 10^6 yrs, and we have seen that it is probably strongly overestimated. Finally the cooling time is very small and does not set any constraint on the collapse.

We give then several parameters: $\epsilon\sqrt{1 - Q_g}$, m_{gap}, the ratio of the radial to the Keplerian velocity $V_{\text{rad}}/V_{\text{Kep}}$, the ratio of the rate of supernovae to the maximum allowed rate, the ratio of the number of stars to the maximum number allowed by the tidal effect (if the stellar density is too large, star formation is inhibited), and the ratio of the number of stars to the maximum number allowed by the HII regions. These parameters allow to check that the conditions required by the model are fulfilled. One can also check from the product $\epsilon\sqrt{1 - Q_g}$ that a value of ϵ of $\sim 10^{-3}$ is compatible with the condition $\sqrt{1 - Q_g^2} \geq 10^{-2}$. The number of stars is smaller than the number allowed by the HII regions and by the tidal effect. The most constraining condition comes from the maximum allowed number of supernovae. For $M = 10^8$ $M_\odot$ the number of supernovae is larger than this maximum number. It is comfortably smaller only for small black hole masses or if the momentum provided by a supernova is larger than the standard value. However we have seen that the maximum rate of supernovae is underestimated, so we are confident that the cases where the rate of supernovae is slightly larger than the maximum allowed value are also viable.

Finally the figures show some interesting physical parameters: the scale height and the midplane temperature of the disk, the mass of the initial fragments, the gaseous mass of the disk, the total number of stars, and the total supernova rate. The last three quantities are integrated from the external edge. One can check that the total mass of the disk (actually of the domain where the model is valid) is much smaller than the black hole mass. The number of stars is small, corresponding to an amount of mass locked in stars of the order of that of the gas. Finally the total supernova rate, which is entirely dominated by the inner regions, varies from 10^{-3} to 1 $M_\odot$ per year. We shall come back to this point in the next section.

Although we have considered a stationary model, and shown that it can be accomodated with the picture of a gas-star disk, the most realistic scenario is that the mass inflow from the periphery of the disk is variable with time, as would be the case if it were achieved by large molecular complexes comparable to those present near the Galactic center. In this case, there will be "low states" and "high states" where the disk would be alternatively "quiescent", and "active", i.e. highly perturbed by an intense supernova activity, and star formation will occur in successive "waves" propagating from the outer to the inner regions. An inflow of matter will induce an increased gas density in an outer ring. Before the stars form, accrete and evolve to supernovae, the transfer of mass will not take place, and there will be an accumulation of gas in the ring. After a few 10^6 years, supernovae will induce mass transport towards an inner ring, while the outer ring will be cleared out of its gas until a new mass inflow. Note that during the "active" phases the mass inflow at the periphery could be super Eddington. In this case our description should be modified to take into account the non-stationarity of the process, as one would expect that the corresponding averaged momentum transport (i.e. the accretion rate) be much larger than in the stationary case, as it is not limited by a maximum allowed rate of supernovae.

3. Some implications of the model

In this model the mass accretion rate is of the order of the outflowing mass due to supernovae, although the luminous energy released in supernovae is much smaller than that of the central regions of the disk. Therefore nearby low luminosity AGN should display an increase of flux in the optical range every 10^3 years with the typical light curve of a supernova, and one should also expect to resolve spatially this supernova at a distance of 1-10pc from the central source.

This model has other consequences. Beside the fact that it solves the problem of the mass transport in the intermediate region of the disk, it could give an explanation for the high velocity metal enriched outflows implied by the presence of the Broad Absorption Lines (BALs) in quasars. This problem is discussed in details in Collin (1998), so we recall here only a few points.

There are strong observational evidences for metallicities larger than solar (or at least solar) in the central few parsecs of quasars up to $z \geq 4$ (see the recent systematic study of Hamann 1997). This enriched material is flowing out of the central regions with a high velocity, of the order of $c/30$, as observed in BAL QSOs, which constitute about 10% of the total number of radio quiet QSOs. The phenomenon is generally interpreted as an outflow existing in all quasars, but limited to an opening angle $\sim 4\pi/10$. The outflowing mass rate is quite difficult to estimate, say between 1% to 100% of the accretion rate.

Comparing the observed mass outflow with that given by supernovae, each releasing about 10 $M_\odot$ of metals out of the nucleus, one sees that the mass outflow rate due to the supernovae accounts easily for the observations. Our computations show indeed that the rate of supernovae is equal to a few 10^{-2} yr^{-1} for a quasar black hole, i.e. an outflow of metals close to the accretion rate. The observed velocities are also easily accounted for by the expanding shells, and finally the location of the phenomenon is in agreement with observations. Relatively larger enrichment in some elements like N and Fe are observed, and could be explained by the oddness of the stars formed in a particular environment. Finally the fact that the opening angle of the BAL region is equal to a small fraction of 4π is easily explained, the ejection taking place mainly in a cone aligned with the disk axis. Note that in this model the metallicity of the gas fuelling the black hole can be very small, while the observed outflow is always enriched.

A second outcome of the mechanism is to account for a pregalactic enrichment, if massive black holes are created early in the process of galaxy formation, and if galaxy formation takes place through an hierarchical scenario (Silk and Rees, 1998). Massive galaxies will retain their gas, and the supernova shells will compress the interstellar medium, trigger star formation like in the interstellar medium, and induce a starburst. This is an "inside-outside" scenario opposite to the starburst scenario of Hamann and Ferland (1992). A fraction of the enriched gas should however escape from the galaxy, not only due to the nuclear supernovae explosions, but also to the induced starburst. Small galaxies will not retain their gas, which will escape with the enriched gas produced by the supernovae. It will pollute the intergalactic medium (IGM). In particular, if the formation of the black holes precedes the formation of galaxies, it will lead to a pregalactic enrichment of the IGM. According to our computations the mass of metals ejected by the disk is of the order of the mass of the black hole itself. We can therefore estimate the **minimum** enrichment of IGM due to black holes, simply taking the integrated comoving mass density of **observed** quasars, which corresponds to about 10^{-6} of the closure density (Soltan, 1982, and further studies). If these black holes have a typical mass of 10^8 $M_\odot$, the present mechanism will provide about 10^{-6} of the closure density in metals, i.e. after mixing with the IGM, an average metallicity of a few 10^{-3} $\Omega(IGM)_{0.02} Z_\odot$, close to the metallicity observed in the $L\alpha$ forest which constitutes the main fraction of the IGM.

Finally, our Galaxy is presently not active and the black hole in the center has a small accretion rate ($< 10^{-4}$ $M_\odot$ yr^{-1}), so it has most probably accreted a large fraction of its mass (2×10^6 $M_\odot$) during an early period. The previous estimation leads to an ejection of a few $\sim 10^4$ $M_\odot$ of metals. After mixing with a hydrogen halo of 10^{11} $M_\odot$, it gives a metallicity of a few 10^{-5} solar, close to that observed in the oldest halo stars.

References

Artymowicz P., Lin D.N. and Wampler E.J., 1993, ApJ 409, 592

Begelman, M.C., Frank J., Shlosman I., 1989, in "Theory of Accretion disks", eds. F. Meyer et al., Kluwer Academic Publishers

Chandrasekhar S., 1939, "An introduction to stellar structure"

Collin S., Huré J.M., 1998, in press in A&A

Collin S., 1998, to appear in the proceedings of the Conference "From atomic nuclei to galaxies", held in Haifa, Israel, Physics Reports, Ed. O. Regev

Collin S., Zahn J.P., 1998, A&A, submitted

Goldreich P., Lynden-Bell D. 1965, MNRAS 130, 97

Goldreich P., Tremaine S., 1980, ApJ 241, 425

Hamann F., 1997, ApJS 109, 279

Hamann F., Ferland G., 1992, ApJ 391, 53
Huré J.M., 1998, in press in A&A
Kolykhalov P.I., Sunyaev R.A., 1980, Soviet Astron. Lett. 6, 357
Lin D.N.C., Papaloizou J.C.B., 1980a, ApJ 307, 395
Lin D.N.C., Papaloizou J.C.B., 1980b, ApJ 309, 846
Malkan M.A., Sargent W.L.W., 1982, ApJ 254, 22
Paczynski B., 1978, AcA, 28, 91
Rozyczka, M., Bodenheimer P., Lin D.N.C., 1995, MNRAS 276, 597
Shakura N.I., Sunyaev R.A., 1973, A&A 24, 337
Shields, G.A., 1978, Nature 272, 423
Shlosman I., Begelman M.C., 1989, ApJ 341, 685
Shlosman I., Frank J., Begelman M.C., 1989, Nature 338, 45
Silk J., Rees M.J., 1998, A&A 331, L1
Soltan, A., 1982, MNRAS 200, 115
Toomre A., 1964, ApJ 139, 1217
Wang B., Silk J., 1994, ApJ 427, 759
Ward W., 1986, Icarus 67, 164
Ward W., 1988, Icarus 73, 330

Eduard Denissyuk and Francesco Bertola

Origin of Blazar Activity

M. M. Romanova

Space Research Institute of the Russian Academy of Sciences, Moscow, Russia; and Department of Astronomy, Cornell University, Ithaca, NY 14853-6801; romanova@astrosun.tn.cornell.edu

Abstract.
Models of Blazars based on the propagation of finite discontinuities or *fronts* in the Poynting flux jet from the innermost regions of an accretion disk around a black hole are discussed. Such fronts may be responsible for short time–scale (from less than hours to days) flares in different wavebands from high frequency radioband to TeV, with delay in low radio frequencies as a result of synchrotron self–absorption. The cases of magnetic fields of one and opposite polarities across the front are investigated. We find that annihilation of magnetic field in the front leads to higher energy spectrum of leptons and possibility of strong TeV flares. Electron–positron pairs form in most cases as a result of interaction between numerous synchrotron photons and SSC photons, and constitute the majority species, compared with the ions at subparsec scales. Frequent weak outbursts may be responsible for flickering core radiation in all wavebands, while the stronger outbursts may be observed as short time–scale flares.

1. Introduction

Blazars are characterized by fast variability in different wavebands from radio to gamma and in few cases by TeV radiation (Urry & Padovani 1995; Punch 1992). Many of the objects reveal "superluminal" jets, which indicate that matter of the jet moves nearly toward us with relativistic speed (Blandford & Rees 1978). The variability of the objects may be connected with outbursts of matter and energy from the nucleus and propagation of shocks along the collimated relativistic jet (Blandford & Königl 1979; Marscher 1980). This model was further developed for investigation of different radio properties of radiogalaxies and quasars (Aller, Aller & Hughes 1985; Hughes, Aller & Aller 1985). The back extrapolated time of VLBI outbursts approximately coincides with strong optical, X–ray and gamma–ray flares (Kinman 1977; Belokon 1988; Krichbaum et al. 1995; Otterbein et al. 1998) which is in favor of this model. The idea of matter outbursts from centers of galaxies was first proposed by Ambartsumian (1958), and this has been confirmed by numerous observations.

The outbursted matter may be a normal electron–ion plasma, or it may consist mainly of electron–positron pairs. Prior to the Compton Observatory measurements, prediction of strong, collimated gamma–ray emission and electron/positron cascades in AGN relativistic jets was made by the model of Lovelace, MacAuslan & Burns (1979); Burns & Lovelace (1982). More recently, a number of theoretical models have been developed to explain the observed gamma–ray emission of AGNs. In most of the models the gamma–ray radiation is ascribed to inverse Compton (IC) scattering of relativistic electrons and possibly positrons (Lorentz factors $\gamma \sim 10^2 - 10^5$) of a jet having relativistic bulk motion (Lorentz factor $\Gamma \sim 10$) with soft photons (energies $\sim 1 - 10^2$ eV). The soft photons can arise from the synchrotron emission of the relativistic electrons in the jet as in the synchrotron–self–Compton (SSC) models (Maraschi, Ghisellini & Celotti 1992; Marscher & Bloom 1992), or from the direct or scattered thermal radiation from an accretion disk (Dermer, Schlickeiser & Mastichiadis 1992; Blandford 1993; Sikora, Begelman & Rees 1994), or from a single cloud (Ghisellini & Madau 1996). In a very different class of models, ultra high–energy protons (Lorentz factors $> 10^6$) are postulated to cause a cascade, the product particles of which produce the observed radiation (Mannheim & Biermann 1992; Protheroe & Biermann 1997).

The idea of Poynting flux outbursts of energy to the jet is based on the fact that the central regions of the disk and a black hole may accumulate strong poloidal magnetic field of the order $B \sim (10^3 - 10^4)$ G. Rotation of this configuration leads to generation of the Poynting flux, which is a "permanent machine" for matter acceleration (Blandford & Znajek 1977; Lovelace, Wang & Sulkanen 1987; Livio, Ogilvio & Pringle 1998). Recently, this idea was further developed and applied to gamma–ray Blazars by Romanova & Lovelace (1997) (hereafter RL97), Colgate & Li (1998) and by Levinson (1998). RL97 proposed that the main driving force for the observed superluminal jet components is a finite amplitude discontinuity in a Poynting flux jet. A rapid change in the Poynting jet outflow from a disk can result from implosive accretion in a disk with an ordered magnetic field (Lovelace, Romanova & Newman 1994, hereafter LRN94). Propagation of newly expelled electromagnetic field and matter from the disk with higher velocity than the old jet can lead to the formation of a pair of shock waves as in the non–relativistic hydrodynamic flows in optical jet in protostellar systems (Raga et al. 1990). Particle acceleration in the front may result from the shocks and/or from annihilation and reconnection of oppositely directed magnetic fields in the front (Romanova & Lovelace 1992, hereafter RL92; Lovelace, Newman & Romanova 1997, hereafter LNR97). Here, we consider the different aspects of the flares of Blazars interpreted in terms of discontinuities in a Poynting flux jet. We analyze different theoretical and observational aspects connected with such outbursts.

2. Magnetic Field in Jets

Accreting matter in a disk around a black hole carries with it an ordered and a chaotic magnetic field. When the matter reaches the black hole, small magnetic loops reconnect and the field annihilates. The more ordered component of the field, which can have open field lines, can be dragged into the black hole while remaining connected to more distant plasma in the corona of the disk (Figure 1). Thus, the black hole may have a significant magnetic field passing through it supported by external currents in the disk and corona (Blandford & Znajek 1977;

Macdonald & Thorne 1982). The magnetic field of the inner regions of the disk is likely to be comparable to the field in the black hole (Macdonald & Thorne 1982; Livio et al. 1998). The area of the inner region of the disk is much larger than the area of the black hole, so that the magnetic flux and the Poynting energy outflow rate of the inner regions of the disk is larger than that of the black hole (Livio et al. 1998; Lovelace et al. 1987). The magnetic field near the inner edge of the disk is deduced to be of the order of $B \sim (10^3 - 10^4)$ G (Lovelace 1976). Semi–empirical models of gamma–ray flares based on inverse Compton and/or SSC mechanisms, predict approximate values of the magnetic field in different Blazars in the regions of origin of the radiation (e.g., Sambruna, Maraschi & Urry 1996; Sambruna et al. 1997). Back extrapolated to the inner disk using $B \sim 1/r_j$ ($r_j(z)$ is the jet radius), gives a magnetic field $B = (10^2 - 10^4)$ G.

The rotating disk and black hole threaded by an ordered magnetic field generate Poynting flux outflows, or jets in which the energy density of the electromagnetic field is much larger than the matter energy density (see Lovelace, this volume).

3. Poynting Flux Jets

Poynting flux winds were first discussed by Goldreich and Julian (1968) for pulsars, and later Poynting flux jets were proposed to explain extragalactic jets by Lovelace (1976) and Blandford (1976). Solutions for Poynting flux outflows from a disk around a massive black hole were investigated by Lovelace et al. (1987).

A Poynting flux jet is self–collimated, with energy, momentum, and angular momentum transported mainly by the electromagnetic field (Lovelace et al. 1987). The collimation is due to the toroidal component of magnetic field. A steady Poynting flux jet is characterized in the lab frame by its asymptotic ($z \gg r_o$) magnetic field $B_\phi = -B_0[r_0/r_j(z)]$ and electric field $E_r = -(v_j/c)B_0[r_0/r_j(z)]$ at the jet's edge with radius $r = r_j(z)$, where r_0 is the jet's radius at $z = 0$. Also, B_0 is magnetic field at $z = 0$, and $v_j \approx c$ is jet's axial velocity. The jet radius at $z = 0$ is $r_0 \sim (1-3)r_g = (2-6)GM/c^2$, where r_g is the Schwarzschild radius, and M is the black hole mass. The energy flux (luminosity) of the $+z$ jet is

$$L_j = vB_0^2 r_0^2/8 \approx 3.0 \times 10^{43} \text{ erg s}^{-1} (v_j/c)B_3^2 M_8^2, \qquad (1)$$

where $B_3 \equiv B_0/10^3$ G, $M_8 \equiv M/10^8 M_\odot$, and $r_0 = 3r_g$.

Matter accreting in the disk will be partially expelled to the jet by the Poynting flux "machine." An important quantity is the ratio of the magnetic energy to the matter kinetic energy at the base of the jet,

$$\mu \equiv (B_0^2/8\pi)/(\rho_0 v_{j0}^2/2), \qquad (2)$$

where velocity at the base of the jet v_{j0} is assumed to be non–relativistic at this distance. The Poynting regime corresponds to $\mu \gg 1$. Note that in the Poynting flux regime, the magnetic field lines do not need to be inclined away from the z–axis in order for there to be energy outflow from the disk. This is in marked contrast with the hydromagnetic regime where the outflows require the

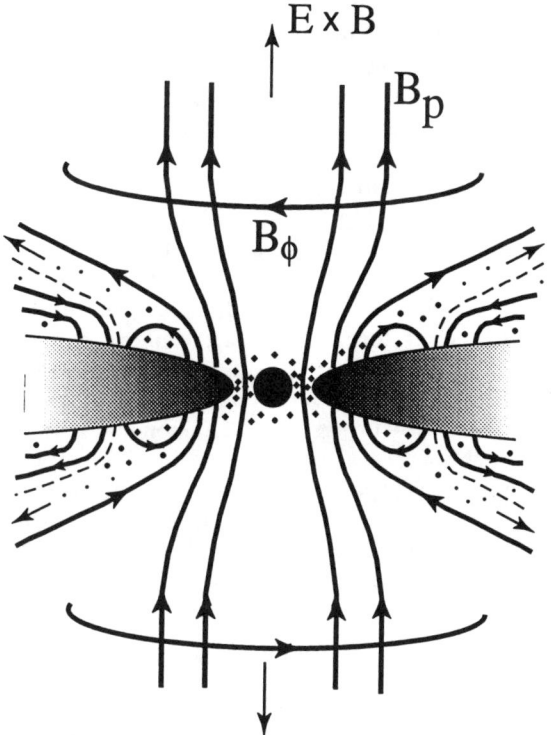

Figure 1. Sketch of magnetic field threading the inner region of an accretion disk around a black hole. A strong flux of electromagnetic energy (Poynting flux) forms as a result of twisting of the poloidal component of the magnetic field, and the fact that $\mathbf{E} = -\mathbf{v} \times \mathbf{B}/c$ in highly conducting plasma.

field to be inclined away from the z–axis by $> 30°$ (Blandford & Payne 1982; Lovelace, Berk & Contopoulos 1991; Romanova et al. 1997).

The matter flux carried by the jet is

$$\dot{M}_j = \pi r_0^2 \rho_0 v_{j0} \approx B_0^2 r_0^2/(8\mu c) \approx 5.2 \times 10^{-4} \text{ M}_\odot \text{ yr}^{-1} \ B_3^2 M_8^2/\mu. \qquad (3)$$

At say $\mu = 10$, the matter flux is $\dot{M}_j = 5.2 \times 10^{-5}$ M$_\odot$/yr, which is quite small compared with the disk accretion rate needed to fuel an AGN, $\dot{M}_{accr} \sim$ M$_\odot$/yr.

3.1. Non–stationary Accretion and Formation of Fronts

The electromagnetic energy outflow from the inner part of the accretion disk and from the black hole may be relatively steady if the disk accretion flow is steady. There may of course be small inhomogeneities connected with changes in the density and magnetic field of the accreting matter. However, if the disk is unstable, then the accretion rate increases and magnetic field also increases as a result of matter compression.

Inhomogeneity of the magnetic field threading the disk can lead to a "global magnetic instability" of the accretion disk, which is connected with angular momentum outflow to Poynting flux jets or hydromagnetic outflows (LRN94; Lubow, Papaloizou & Pringle 1994; LNR97). If the magnetic field is enhanced at some radius r in the disk, then angular momentum will be the more efficiently lost from this region of the disk's surfaces to the jets. As a result, this disk matter will accrete faster and will accumulate matter in front of it as it moves radially inward. This will in turn amplify the magnetic field and further increase the loss of angular momentum to jets. Simulations of this process have shown that a strong wave–pulse of dense matter with strong magnetic field forms in the disk and propagates inward (LRN94). In case of a relatively high turbulent or "α" viscosity, a soliton–like wave forms (LRN94), while in case of a zero viscosity disk, a shock–like wave forms (LNR97). When this wave reaches the inner part of the disk, it generates the strong outburst of energy, angular momentum and matter to the jet. The pulse brings in a stronger magnetic field to the inner regions of the jet, which increases the Poynting flux. Thus, the new jet outflow will have a larger magnetic field, and can form a *front* propagating outward along the channel of the "old", weaker jet. Also, the new jet outflow may involve a reversal of polarity of the magnetic field (Romanova et al. 1998) so that the toroidal fields of the "new" and "old" jets are opposite (LNR97).

3.2. Particle Acceleration in Propagating Front of Jet

The plasma newly launched to the Poynting flux jet propagates along the old channel of the jet forming a *front*. The front is bounded by two shock waves – the first between the front and upstream matter and the second between the front and the downstream matter. Parameters of matter in the front are those in between the "old" and "new" jets. The difference in the energy fluxes between the "old" and "new" jets necessarily goes into the acceleration of particles inside the front (RL97). It is not possible from first principles to calculate the spectrum of accelerated particles because the acceleration mechanism is not known. Relativistic MHD shock acceleration may be important (Eilek & Hughes 1990), but this is likely to accelerate mainly the ions if the plasma consists mainly of electrons and ions. From the other side, if the toroidal magnetic field reverses polarity across the front, then reconnection of magnetic field may be the dominant particle acceleration mechanism. Also, both particle acceleration mechanisms may be significantly different if the plasma consists mainly of electrons and positrons. Figure 2 shows sketch of possible configurations of the magnetic field in the front.

If accreting matter in the disk carries poloidal magnetic field of both polarities this can lead to reversals in the polarity of the toroidal magnetic field of the Poynting jet (see LNR97). This situation is sketched in Figure 2a. The case of a chaotic magnetic field (Figure 2b) may be always present in the plasma. Compression of plasma will lead to driven collisionless reconnection (Alfvén 1968; RL92).

RL97 proposed an empirical lepton spectrum motivated by observations where the energy of leptons is distributed as $f_l \sim \gamma^{-2}$ between energies γ_1, and γ_2 (in units of the rest mass energy), and is steeper $\sim \gamma^{-3}$ between energies γ_2 and γ_3. The lowest energy γ_1 was equal to that corresponding to synchrotron

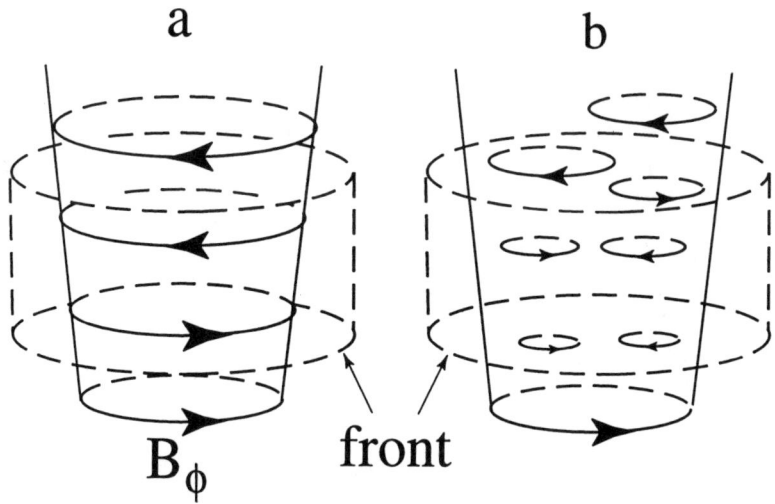

Figure 2. Sketches of possible configurations of magnetic field with opposite polarity, (a) an ordered field and (b) a chaotic field.

self–absorption. The average energy $\bar\gamma$ was calculated from the basic equations for the conservation of energy, mass and momentum. The energies γ_2, and γ_3 were derived from other considerations (see RL97). The actual spectrum f_l may be somewhat different and will be determined by details of the acceleration process(es).

The basic equations for the evolution of the front follow from the conservation of mass, momentum, and energy,

$$\frac{\partial n'_i}{\partial t'} = -\frac{\partial (n'_i v')}{\partial z'} + \left(\frac{\partial n'_i}{\partial t'}\right)_{pairs}, \qquad (4)$$

$$\frac{\partial T'_{0z}}{\partial t'} = -\frac{\partial T'_{zz}}{\partial z'} + grav + rad, \qquad (5)$$

$$\frac{\partial T'_{oo}}{\partial t'} = -\frac{\partial T'_{oz}}{\partial z'} - syn - SSC - Com, \qquad (6)$$

and an equation for the magnetic flux. Here, $T'_{\alpha\beta}$ is the stress energy tensor of the matter and the electromagnetic field, the primes denote quantities in the frame of the front, and "grav," "syn," etc denote different forces or energy fluxes discussed by RL97.

The equations were solved for different intrinsic parameters (RL97), but the example used below corresponds to a black hole mass $M = 3 \times 10^8 M_\odot$, magnetic field at the base of the jet $B_0 = 10^3$ G, initial ion/lepton ratio $f_{li} = 1$, magnetic/particle energy at the base of the jet $\mu = 15$, viewing angle $\theta = 0.2$ rad, the luminosity of background photons $L_{ph} = 10^{46}$ erg s^{-1}, their energy $\epsilon_{ph} = 10$ eV, and radius of their distribution $R_{ph} \approx 3 \times 10^{17}$ cm. The density and magnetic field ratios between the "old" and "new" matter are $n_1/n_2 \approx 0.4$ and

$B_1/B_2 \approx 0.4$. The velocities correspond to a bulk Lorentz factors $\Gamma_1 = 8$ and $\Gamma_2 = 18$ (where $\Gamma = [1 - (v_j/c)^2]^{-1/2}$). The cases of a single and reverse polarity of the magnetic field across the front were investigated and compared. After expulsion of new matter, the front accelerates up to $\Gamma \approx 12$, so that the Doppler boost factor is $\delta = 1/\Gamma[1 - (v_j/c)\cos\theta] \approx 4$. Leptons are accelerated in the front from $\gamma = 1$ to $\gamma_1 \approx 10^2$, $\gamma_2 \approx 10^3 - 5 \times 10^3$, and $\gamma_3 \approx (6 \times 10^3 - 2 \times 10^4)$. In the case of reversal polarity leptons are accelerated up to higher energies: $\gamma_2 \approx 10^4 - 10^5$, and $\gamma_3 \approx 10^5 - 10^6$. These maximum values of γ depend on the duration of expulsion of "new" matter with opposite polarity.

3.3. Energy Release Due to Magnetic Field Reconnection

Reconnection of the magnetic field may occur along the jet, in particular, in the front, where matter is compressed. The magnetic field at large distances is $B_\phi = -B_0(r_0/r_j)$. The magnetic energy–density at some distance z along the jet is $B_\phi^2/8\pi$ so that the total magnetic energy in the front is $E_m = (B_\phi^2/8\pi)(2\pi r^3)$, where we supposed that the region is a cylinder with a length equal to its diameter. If the magnetic field reverses polarity across the front, then this magnetic energy may be released entirely in the form of accelerated particles during an Alfvén time $t_A = 2r_j/v_A$, where the Alfvén velocity $v_A = c/(1+4\pi\rho c^2/B_\phi^2)^{1/2} \sim c$. Thus, the "luminosity" owing to the reconnection is

$$L_{\rm recon} \sim E_m/t_A = B_0^2 r_0^2 c/8 \sim 3.0 \times 10^{43} {\rm erg \ s}^{-1} \ B_3^2 M_8^2. \qquad (7)$$

The spectrum of leptons resulting from collisionless driven reconnection is a power law γ^{-2} for electron-ion plasma, and $\gamma^{-1.5}$ for electron-positron plasma (RL92). In our analysis of the time evolution of the front in a Poynting flux jet, we took into account the annihilation of magnetic field in the case where the field reverses polarity across the front. We found larger particle energies in the case where the field reverses polarity.

3.4. Pair Creation

The particle content of the jets is not known (Krolik, this volume). In our model, the density of synchrotron photons inside the front is typically much larger ($10^4 - 10^5$ times) than the density of the background photons (see also RL97). Interaction of the electrons with these photons produces a high density of high–energy SSC photons. Analysis of different possible mechanisms of pair creation leads to the conclusion that interaction of SSC photons with synchrotron photons is the most important process. A pair forms when $\epsilon_{syn}\epsilon_{ssc} > (m_ec^2)^2$, where ϵ_{syn} and ϵ_{ssc} are energies of synchrotron and SSC photons. For rough estimates, we can write $\epsilon_{syn} = (3/2)\gamma^2\hbar\omega_o'$, where $\omega_o' = e|B'|/(m_ec)$ is the cyclotron frequency in the front frame, $\epsilon_{ssc} = \gamma^2\epsilon_{syn}$, and $|B'|$ is the magnetic field strength in the front frame. An approximate condition for pair production is

$$\gamma \geq \gamma_{pair} \equiv (mc^2/\hbar\omega_o')^{1/3} \approx 3.5 \times 10^3(|B_3'|)^{-1/3}. \qquad (8)$$

Electron–positron recombination is negligible for the conditions considered. We observed that the pairs form at a variety of parameters of the model. In a typical case, the total number of pairs in the front N_l grows proportionally to the total

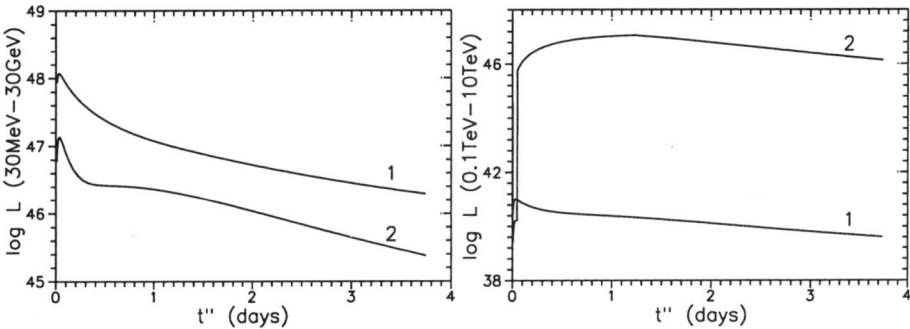

Figure 3. Luminosity of EGRET flares (left panel) and TeV flares (right panel) in erg/s for cases of (1) a magnetic field of one polarity, and (2) a field which reverses polarity across the front.

number of ions N_i accumulated by the front. Thus, their ratio $f_{li} \equiv N_l/N_i$ is almost constant during the front propagation and constitutes $f_{li} \approx 10$ for $\mu = 15$, and $f_{li} \approx 100$ for $\mu = 100$. The pair creation does not appreciably alter the energy distribution of leptons because the internal energy per particle is proportional to μ and the number of leptons is approximately proportional to μ. In RL97, pairs did not form, because at similar initial parameters, they took initial value $f_{li} = 5$ at the base of the jet. Thus, leptons had smaller initial energy per particle and the condition $\gamma > \gamma_{pair}$ was not satisfied. Summarizing, we conclude that pairs form in many cases, but only in case of very low matter energy compared to magnetic energy, $\mu \gg 1$, their number may be significant compared to the number of ions.

In case of polarity reversals of the B_ϕ field (Figure 2), the magnetic field gradually annihilates and becomes weaker in the front frame. The pairs form in the beginning of the front propagation ($z < 10^{16} M_8$ cm), when the magnetic field is strong enough to produce significant population of synchrotron and SSC photons. Later, the rate of pair creation decreases, and f_{li} decreases. The fact, that in case of polarity reversals, the internal energy of the front is larger (due to reconnection) while the number of leptons is smaller (due to weaker pair creation) leads to higher energies per lepton γ and the entire spectrum is harder.

3.5. Radiation from the Front

Accelerated leptons interact with background and synchrotron photons. Each of three processes, synchrotron, IC, and SSC, give radiation in a broad bands of energies. They cover the whole range of energies from 10^{-3} eV (radio with $\lambda \approx$ 1 mm) up to ~ 100 GeV in case of single polarity and up to $(1-10)$ TeV in case of polarity reversals. Thus, the Poynting flux outburst generates a simultaneous flare in all wavebands from high frequency radio up to high energy radiation. Here, we calculated the luminosity of flares in EGRET band (30 MeV – 30 GeV) and in TeV band for energies $(0.1-10)$ TeV (see Figure 3). The cases of one polarity (1) and two polarities (2) of the toroidal magnetic field are shown. One can see that the gamma-ray luminosity is higher in the case of a single polarity, whereas the TeV luminosity is much higher in the case of reversals of polarity of the field across the front. In case of one polarity, radiation in both energy bands is determined by SSC processes, while in case of polarity reversal, both gamma and TeV radiation are determined mainly by the IC processes (because the magnetic field becomes too weak to produce significant synchrotron and correspondingly SSC radiation). The fairly strong TeV flare in case of polarity reversals appears as a result of higher energy of leptons in the front compared with the case of a single polarity. The flare gradually decreases because the density of background photons decreases. However, the strong luminosity flare may appear later in the case of interaction with a Broad Line Region cloud (Ghisellini & Madau 1996). The flares in X-ray and optical bands are smaller, than in the EGRET band. The detailed analysis of the spectrum will be considered in the future.

In the example shown (Figure 3) it was suggested that the "new" matter was outbursted from the disk during ~ 100 periods of rotation of the inner radius of the disk, $\Delta t = 100(2\pi r_0/v_{K0})$, which corresponds to $\Delta t'' \approx 4$days in the observer's frame ($dt'' = dt(\Gamma/\delta) = dt[1 - (v_j/c)\cos\theta]$). If the source of new matter ceases earlier (see also Levinson 1998), then the flare may be much shorter, of the order of the dynamical time-scale at the inner edge of the disk, $\Delta t''_{dyn} = t_{dyn}(\Gamma/\delta) \approx 0.8(\Gamma/\delta)M_8$hr $\approx 2.4 M_8$hr. Thus, the flares shown at Figure 3, may be much shorter. The discussed model presents a definite answer regarding the amplitude of the flares, but their duration is somewhat uncertain and is determined by the processes in the inner regions of the disk. We expect that the inner regions of the disk may frequently be unstable owing to angular momentum outflow to the jet. This is an analogue of the "global magnetic instability" (LRN94; LNR97), but on smaller scales. This will lead to frequent outbursts of matter to the jet with a characteristic time, which may be as short as a few dynamical time-scales. One can expect numerous hour time-scale flares with less frequent, more powerful, day time-scale flares determined by stronger instabilities in the disk.

4. Comparison with Observations

In Blazars we are looking down the jet and this gives information not visible in other AGNs (Sambruna, these Proceedings). But *how deep can one observe the inner jet?* The interaction of gamma-ray photons with few KeV background radiation leads to electron-positron pair production (e.g., Burns & Lovelace 1982; Blandford & Levinson 1995). The corresponding γ-sphere has a radius (Takahara 1997): $r_\gamma > 1.0 \times 10^{-5}(\epsilon_{ob}/\text{GeV})L_{46}\delta_{10}^{-5}$ pc, where ϵ_{ob} is the energy of the observed photons (in GeV), and L_{46} is the total luminosity of X-ray photons. At small background luminosities, and, specifically, at small boost factors δ, the value r_γ may be very close to the radius of the black hole. On the other hand, even if $r_\gamma > r_g$, then near the black hole, the X-ray radiation is expected to be anisotropic and come from the accretion disk. Thus, interaction of gamma-ray photons with background radiation may occur only at the "walls" of the jet, which may be opaque to penetration of X-ray photons (Illarionov & Krolik 1996; Thompson 1997).

The fastest flares reported so far last a few hours as in the case of Mrk 421 (Gaidos et al. 1996), down to less than an hour in Mrk 501 (Aharonian et al.

1999). In PKS 1622-297 the duration of the flare is a few days, but the rise time of the flare is only a few hours (Mattox et al. 1997). This indicates that the size of the emitting region is very small (unless the Doppler boost factor δ is not extremely large).

The Poynting flux outbursts discussed here lead to quite small variability timescales, from less than an hour up to possibly a few days. The short flares are much more probable. The frequent small–scale flares may explain the intra-day variability in different wavebands from optics (and possibly high–frequency radio) up to GeV and TeV radiation.

What is the origin of *the long–time flares*, which may last a few weeks or even months? The long flares may be explained for example by superposition of smaller–scale flares which may occur frequently (Magdziarz, Moderski & Madejski 1997; Chiaberge & Ghisellini 1998). The overlapping of light–curves of small flares may determine the non–thermal continuum radiation in different wavebands from radio to gamma with flickering determined by the separate flares. The low–energy photons have a much longer cooling time (e.g. Atoyan & Aharonian 1997), so that the flickering of the low–energy radiation may be smoothed significantly (Chiaberge & Ghisellini 1998). However, we still cannot exclude the possibility that intraday (mm) radio variability (e.g. Wagner & Witzel 1995) of some radioquasars may be also determined by the expulsion of shocks from the region near the black hole.

The Poynting flux outburst model does not exclude the possibility of formation of longer flares at larger distances from the black hole. For example, the TeV flare in Mrk 501, may be explained by the injection of high–energy particles with $\gamma_{max} \sim 10^5 - 10^6$ with luminosity $L_j \sim 10^{41}$ erg s^{-1} to a region with the size $R \sim 10^{16}$ cm (Maschidiadis & Kirk 1997, see also Sambruna et al. 1998). These particles may originate in a shock wave or a reconnection event. This luminosity constitutes only one percent of magnetic field annihilation luminosity (eq. 7).

Correlated multiwavelength observations show near simultaneous outbursts in different wavebands from the optical to gamma–ray (e.g., Bloom et al. 1997), which support the Poynting flux model. Recent observations show clear correlation between X-ray and TeV fluxes in TeV Blazars (e.g., Pian et al. 1998), which is a strong argument in favor of SSC mechanism of TeV radiation in these sources. The TeV flares found in our model are determined by IC radiation, and hence cannot be applied to these particular sources. In these objects additional acceleration of leptons may occur in shock waves or by B field reconnection.

5. Conclusions

The main conclusions from the investigated models are:

1. The rapidly rotating inner region of an accretion disk threaded by magnetic field ($\sim 10^3 - 10^4$ G) as well as the rapidly rotating black hole can generate field dominated or Poynting flux jets.

2. Accretion disk instabilities, specifically the "global magnetic instability" of the disk, can bring matter and magnetic flux rapidly to the inner region of the disk and thereby generate strong outbursts of energy to the jet.

3. In the region between the "old" and "new" matter of the jet, a "front" forms, where particles are efficiently accelerated owing to shock waves and/or enhanced reconnection/annihilation of the magnetic field.

4. Flares in all frequency bands from IR (or mm) to GeV (and in some cases TeV) are expected to appear approximately simultaneously. In the case where the toroidal magnetic field reverses polarity across the front, strong TeV flares may occur.

5. The millimeter radio flux may have a short delay (hours) compared with other wavebands as a result of self–absorption. However, in some cases (where the magnetic field at the base of the jet is less than $\sim 10^3$ G) it may appear simultaneously with other wavebands. There is an even longer delay for lower frequency radio emission.

6. Frequent weak outbursts to the jet may overlap and determine the non–thermal continuum radiation from radio to gamma–ray band. They may be observed as intraday variability in different wavebands.

7. The strongest outbursts may appear as short time–scale flares with duration from less than an hour up to a few days, depending on accretion processes in the disk and the radiation rate of the leptons.

8. Reconnection of magnetic field leads to locally accelerated leptons and the possibility of TeV flares. Small–scale reconnection events may determine the small–scale variability at large distances from the black hole with times less than $r_j(z)/c$, where r_j is the radius of the jet.

Acknowledgments. The author thanks IAU for partial support, and the Local Organizing Committee for warm hospitality. This work was made possible in part by Grant No. RP1-173 of the U.S. Civilian R&D Foundation for the Independent States of the Former Soviet Union.

References

Aharonian, F.A., et al. 1999, A & A, 342, 69

Alfvén, H. 1968, JGR, 73, 4379

Aller, H.D., Aller, M.F., & Hughes, P.A. 1985, ApJ, 298, 296

Ambartsumian, V.A, 1958, *La structure et l'evolution de l'univers*, Solvey Conference, p. 241, Bruxelles, ed. by R. Stoops

Atoyan, A.M., & Aharonian, F.A. 1997, ApJ, 490, L149

Begelman, M.C., Blandford, R.D., & Rees, M.J. 1984, Rev. Mod. Phys., 56, 255

Belokon, E.T. 1988, Astrophysics, 27, 588

Blandford, R.D. 1976, MNRAS, 176, 465

Blandford, R.D. 1993, in Proc. Compton Symp. 1992, ed. M. Friedlander, N. Gehrel, & D.J. Macomb, New York: AIP, 553

Blandford, R.D. & Königl, A. 1979, ApJ, 232, 34

Blandford, R.D., & Levinson, A. 1995, ApJ, 441, 79

Blandford, R.D., & Payne, D.G. 1982, MNRAS, 199, 883

Blandford, R.D., & Rees, M. 1978, in *Pittsburgh Conference on BL Lac Objects*, ed. A.M. Wolfe, Univ. Pittsburgh Press, 1978, p.328
Blandford, R.D., & Znajek, R.L. 1977, MNRAS, 179, 433
Bloom, S.D., et al. 1997, ApJ, 490, L145
Burns, M.L., & Lovelace, R.V.E. 1982, ApJ, 262, 87
Chiaberge, M., & Ghisellini, G. 1998, MNRAS, astro-ph/9810263
Colgate, S.A. & Li, H. 1997, in *Relativistic Jets in AGNs*, ed. M. Ostrowski, M. Sikora, G. Madejski and M. Begelman, Poland, Kraków, p.170
Dermer, C.D., Schlickeiser, R., & Mastichiadis, A. 1992, A&A, 256, L27
Eilek, J.A., & Hughes, P.E., in *Astrophysical Jets*, ed. P.E. Hughes, Cambrudge University Press, Cambridge, p. 428
Gaidos, J.A. et al. 1996, Nature, 383, 319
Ghisellini, G., & Madau, P. 1996, MNRAS, 280, 67
Goldreich, P., & Julian, W.H. 1969, ApJ, 157, 869
Hughes. P.A., Aller, H.D., & Aller, M.F. 1985, ApJ, 298, 301
Illarionov, A.F., & Krolik, J.H. 1996, ApJ, 469, 698
Kinman, T.D. 1977, Nature, 267, 798
Krichbaum, T.P. 1995, in:*Quasars and AGN: High Resolution Radio Imaging*, ed. M. Cohen and K. Kellerman, proceedings of a Conference of the National Academy of Sciences, Irvine, CA, USA
Levinson, A. 1998, ApJ, 507, 145, 1998
Livio, M., Ogilvie, G.I., & Pringle, J.E. 1998, ApJ, in press
Lovelace, R.V.E. 1976, Nature, 262, 649
Lovelace, R.V.E., McAuslan, J., & Burns, M. 1979, in *Proceedings of La Jolla Institute Workshop on Particle Acceleration Mechanisms in Astrophysics*, ed. J. Arons, C. Max, & C. McKee (AIP, New York).
Lovelace, R.V.E., Newman, W.I., & Romanova, M.M. 1997, ApJ 484, 628 (LNR97)
Lovelace, R.V.E., Romanova, M.M., & Newman, W.I 1994, ApJ, 437, 136 (LRN94)
Lovelace, R.V.E., Wang, J.C.L., & Sulkanen, M.E. 1987, ApJ, 315, 504
Lubow, S.H., Papaloizou, J.C.B., & Pringle, J.E. 1994, MNRAS, 268, 1010
Macdonald, D.A., & Thorn, K.S. 1982, MNRAS, 198, 345
Magdziarz, P., Moderski, R., & Madejski, G.M. 1997, in: *Relativistic Jets in AGNs*, ed. M. Ostrowski, M. Sikora, G. Madejski and M. Begelman, Poland, Kraków, p.238
Mannheim, K., & Biermann, P.L. 1992, A&A, 253, L21
Maraschi, L., Ghisellini, G., & Celotti, A. 1992, ApJ, 397, L5
Marscher, A.P 1980, ApJ, 235, 386
Marscher, A.P., & Bloom, S.D. 1992, in *The Compton Observatory Science Workshop*, (ed. C.R.Shader, N. Gehrels, and B.Dennis) NASA CP-3137, 346
Mastichiadis, A., & Kirk, J.G. 1997, A & A, 320, 19

Mattox, J.R., et al. 1997, ApJ, 476, 692
Pian, E., et al. 1998, ApJ, 492, L17
Protheroe, R.J. & Biermann, P.L. 1997, APh, 6, 293
Punch, M. et al 1992, Nature 358, 477
Raga, A.C., Canto, J., Binette, L., & Calvet, N. 1990, ApJ, 364, 601
Romanova, M.M., & Lovelace, R.V.E. 1992, A&A, 262, 26 (RL92)
Romanova, M.M., & Lovelace, R.V.E. 1997, ApJ, 475, 97 (RL97)
Romanova, M.M., Ustyugova, G.V., Koldoba, A.V., Chechetkin, V.M., & Lovelace R.V.E., 1997, ApJ, 482, 708
Romanova, M.M., Ustyugova, G.V., Koldoba, A.V., Chechetkin, V.M., & Lovelace R.V.E., 1998, ApJ, 500, 703
Sambruna, R.M., Maraschi, L., & Urry, C.M. 1996, ApJ, 463, 444
Sambruna, R.M., et al. 1997, ApJ, 474, 639
Sambruna, R.M., et al. 1998, ApJ, in press, astro-ph/9810319
Sikora, M., Begelman, M., & Rees, M. 1994, ApJ, 421, 153
Takahara, F. 1997, in *Relativistic Jets in AGNs*, ed. M. Ostrowski, M. Sikora, G. Madejski and M. Begelman, Poland, Kraków, p.170
Thompson, C. 1997, in *Relativistic Jets in AGNs*, ed. M. Ostrowski, M. Sikora, G. Madejski and M. Begelman, Poland, Kraków, p.253
Wagner, S.J., & Witzel, A. 1995, Ann. Rev. Astron. Astropys., 33, 163
Urry, M.C., & Padovani, P. 1995, PASP, 107, 803

Dan Weedman and Marina Romanova

Active Galactic Nuclei and Related Phenomena
IAU Symposium, Vol. 194, 1999
Yervant Terzian, Daniel Weedman, Edward Khachikian, eds.

Velocity Fields in Spiral Galaxies[1]

A.M. Fridman[2], O.V. Khoruzhii[3]

Institute of Astronomy, Russian Academy of Sciences, 48 Pyatnitskaya St., Moscow, 109017, Russia

Abstract. Discussed are various observable features of velocity fields in spiral galaxies. Main attention is paid to giant anticyclones and motions perpendicular to the disk plane.

1. Introduction

There are two different observable manifestations of the density waves in gaseous disks of spiral galaxies. The first is well-known spiral arms — a periodical structure in the surface density. The second is a specific velocity field of a gas.

In this work we consider two features of the velocity field of a density wave. First, giant anticyclones in the disk plane located near the corotation have been discussed more then ten years beginning from the paper by Nezlin *et al.* (1986). Second, the motions along the rotation axis are only beginning to be discussed (Fridman *et al.*1998).

One can show that the anticyclones jointly with spiral arms compose a unique spiral–vortex structure. It means, in particular, that using the theory of the spiral density waves one may restore a full three-component velocity field of the gas from observed one-component line-of-sight velocity. Indeed, a method was worked out, which gave the possibility to find the velocity field in the disk plane and independently by several ways to determine the corotation radius. It offered to discover the giant anticyclones near the corotation circle in full accordance with the theory (Lyakhovich et al. 1997, Fridman et al 1997). The method include also the observational check of its basic assumption, that the spiral arms observed represent the density wave in the galaxy.

The same method was applied to investigate vertical motions of a gaseous galactic disk.

Two types of vertical motions in galactic disks are known (Fridman and Polyachenko 1984): "membrane" motions ("bending oscillations") and vertical motions in the density wave, which we shall call "vertical spiral-wave" motions.

---

[1]This work was performed under the partial support of RFBR grants N 96-02-17792 and N 96-02-19636, grant "Leading Scientific Schools" N 96-15-96648, and the grant "Fundamental Space Researches. Astronomy" for the 1998 year N 1.2.3.1, N 1.7.4.3.

[2]also Sternberg Astronomical Institute, Moscow State University, Moscow 119899, Russia

[3]also National Research Center of Russian Federation "Troitsk Institute for Innovation and Fusion Researches", Troitsk, Moscow Reg. 142092, Russia

These two types of motions have different dispersion properties (Fridman and Polyachenko 1984) and symmetries about the $z = 0$ plane. The velocities of membrane oscillations are an even function of z ($v(z) = v(-z)$), while the velocities of vertical spiral-wave motions are an odd function of z ($v(z) = -v(-z)$). Up to the current time, the main method for studying membrane oscillations has been obtaining observations of warps in galactic disks. The most convenient galaxies for such observations are those viewed nearly edge-on. However, in this case, it is not possible to say anything about the velocities in these perturbations, i.e., about their dynamical properties.

In the case of density waves, studies of vertical motions are impeded by their relative smallness (as a consequence of the smallness of the parameter kh, where k is the absolute value of the wave vector and h is the disk half-thickness). Therefore, their contribution to the observed radial velocities in galaxies is relatively important only for galaxies that are viewed nearly face-on, when the contribution of motions in the plane of the disk is suppressed (proportional to $\sin i$, where i is an inclination angle of the galaxy). Thus, studies of both types of vertical motions require analyses of the velocity fields in galaxies that are viewed face-on.

Another important circumstance is associated with the fact that, determining the vertical (z) component of the velocity of some part of a disk viewed face-on, we determine the "mean" velocity of this section over the entire "visible" thickness of the disk. If the optical depth of the disk at the wavelength considered is modest (as is true for the 21-cm line), the contributions of the near and far halves of the disk to the observed line-of-sight velocity are comparable. In this case, the observed amplitude of spiral–wave motions should be small, since the motion on different sides of the $z = 0$ plane displays mirror symmetry (if the optical depth is zero, the mean velocity is zero – the contributions of the two halves precisely compensate each other). In the case of a large disk opacity, as in the H_α line, the amplitude of spiral–wave motions will be substantially greater, since we effectively measure the velocity of the nearby outer boundary of the disk, where this velocity is maximum (the vertical velocity of the gas in the density wave grows with z). The velocities of vertical motions in membrane oscillations are virtually constant throughout the disk thickness (due to the smallness of the parameter kh); therefore, the ratio of the velocity amplitudes measured at optical and radio wavelengths should not depend significantly on the difference in the optical depths in these two observing bands. In this way, comparing the velocity fields for a galaxy viewed face-on derived using optical and radio data makes it possible to draw conclusions, not only about the velocity distribution for vertical motions, but also about the origin of these motions.

2. Giant anticyclones in spiral galaxies

To discover predicted anticyclones in spiral galaxies we worked out a method to restore the total three-component vector velocity field of the gas in a spiral galaxy from the observed (one-component) field of the line-of-sight velocities (Lyakhovich et al.1997, Fridman et al.1997 a, b). Evidently, it would be impossible without making some assumptions. The method uses two assumptions.

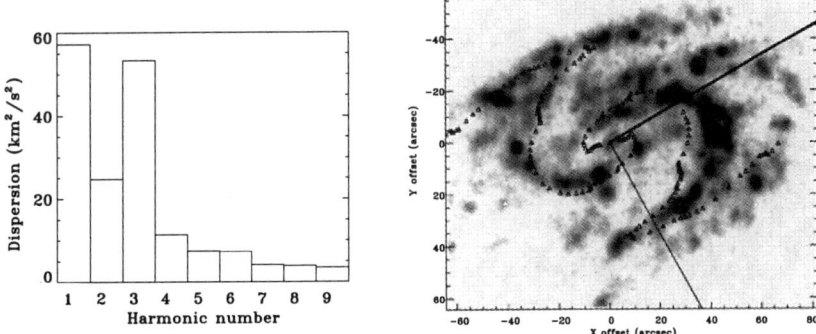

Figure 1. *Left.* Contribution of different Fourier harmonics of line-of-sight velocity field of NGC 157 in dispersion in the model of pure circular motion. The prevalence of first sine, second and third harmonics is clearly seen and demonstrates the domination of the mode with $m = 2$ in velocity perturbations in the disk. *Right.* Superposition of the modified third harmonic of the line-of-sight velocity field on the H_α image of NGC 157 (grey scale). The modified third harmonic has a form $\cos(2\varphi + \pi/2 - F_3)$, where F_3 is the phase of the original third harmonic. Triangles show the azimuth positions of the maxima of the harmonic at each radius. A good correspondence of the phase curve to the observed position of the spiral arms gives strong evidence for the non-circular velocities to be associated with the spiral structure and proves a wave nature of the spiral structure in this galaxy.

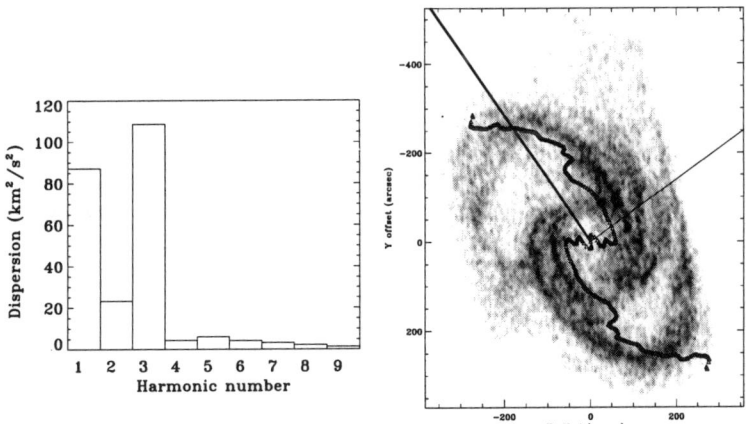

Figure 2. The same as in Fig. 2 but for the galaxy NGC 1365. The modified third harmonic is overlaid on HI image of the galaxy. The velocity field and HI image were kindly provided to us by P.O. Lindblad, description of the data see Lindblad, Lindblad, & Athanassoula 1996.

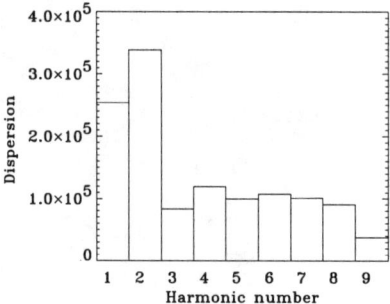 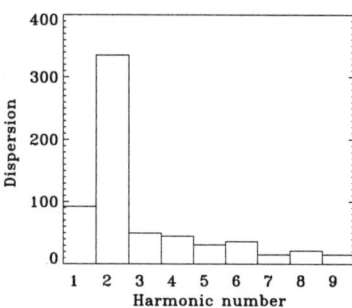

Figure 3. Contribution of different Fourier harmonics of the brightness into deviations from axisymmetry in the disks of NGC 157 (*left*) and NGC 1365 (*right*). H_α line image is used for NGC 157, and 21-cm line image is used for NGC 1365. The prevalence of second harmonic is clearly seen.

First, the observed perturbations of different disk parameters (density, velocity *etc.*) are not independent from each other, but are different manifestations of the unique self-consistent density wave. Second, in the density wave a mode with particular azimuthal number m corresponding to observed number of spiral arms dominates. The latter allows to approximate components of velocity perturbations in a simple form $V_i = C_i \cos(m\varphi - F_i)$, where C_i is an amplitude, and F_i is a phase. Remarkably, both assumptions can be checked directly from the observations (Figs. 1, 2 and 3, for details see Lyakhovich *et al.*1997, Fridman *et al.*1997 a, b). The restoration of the velocity field automatically allows determination of the corotation radius of the density wave. The method was successfully applied to several grand design galaxies: NGC 157, NGC 3893, NGC 1365 and some others. Fig. 4 shows the velocity field in the plane of the disk in the corotating frame of reference for the galaxies NGC 157 (left) and NGC 1365 (right). In both cases we can see two anticyclones between spiral arms. The location of anticyclones relative to the spiral arms depends on the behaviour of basic parameters in the corotation region and on the mechanism of the wave generation (Nezlin *et al.*1986, Baev *et al.*1987, Lyakhovich *et al.*1996). Centres of vortices lie between spiral arms in the case both of a strong centrifugal instability, when $\kappa^2 \equiv 4\Omega_0^2(1 + r\Omega'/2\Omega) < 0$, (*e.g.* Mrk 1040, Afanasiev and Fridman 1993), and of a gravitational instability, when $\kappa^2 > 0$, (e.g. NGC 157, NGC 3893, Fridman *et al.*1997b, 1999).

3. Vertical motions in spiral galaxies

To analyse vertical motions associated with density waves we chose the galaxy NGC 3631 (type SA(s)c), which has two main, rather long, spiral arms and is oriented nearly face-on ($i = 17°$). The domination of the two-armed mode is clearly seen in Fourier decomposition of the brightness of the disk both in H_α and 21-cm lines (Fig. 5). There are no objects of comparable brightness near this

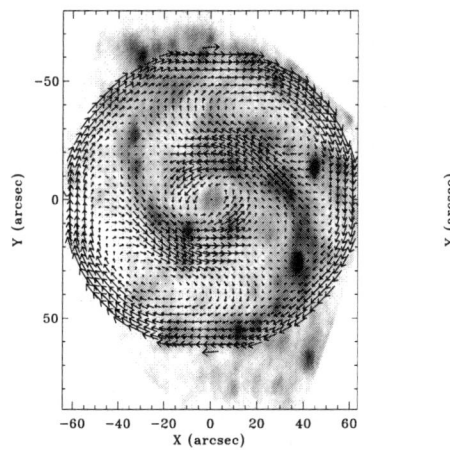 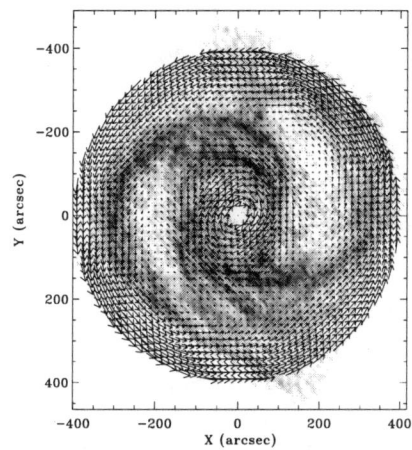

Figure 4. Restored velocity fields of NGC 157 (left) and NGC 1365 (right) in the reference frame rotating with the pattern speed overlaid on the deprojected H_α and HI images of the galaxies respectively.

galaxy, which makes it improbable that the galactic velocity field is distorted by tidal forces.

The HI observations of NGC 3631 used for the study were obtained by Knapen (1997) using the Westerbork Synthesis Radio Telescope, and have angular resolution (FWHM of the synethesized beam) $15.''2 \times 11''$.

Optical observations were obtained on the 6-m telescope of the Special Astrophysical Observatory (Fridman et al.1998). The initial velocity field derived from H_α observations has angular resolution $2''$. In order to facilitate the comparison with the radio observations, we smoothed the optical observations with a Gaussian with half-width (HWHM) of $6.''5$, which is close to the resolution of the radio observations.

In order to conduct a quantitative comparison of the velocity fields using a single method, we performed a Fourier analysis of the distribution of the line-of-sight velocities in azimuth angle in the plane of the galactic disk (measured from the line of nodes), using the "Vortex" program developed at the Institute of Astronomy (Lyakhovich et al. 1997). The size of the disk region studied ($R < 89''$) was restricted to the size of the velocity field determined by the optical data. The inclination angle of the galactic disk and the position angle of the line of nodes were taken to be the same for the optical and radio data, and to be constant over the entire region studied: $i = 17°$, PA $= 150°$ (see Knapen 1997, Table 3).

In both the optical (Fig. 6a) and radio (Fig. 6c) velocity fields, the spectra are dominated by the first three Fourier harmonics, as is natural, given the two-armed spiral structure of the galaxy (Smirnov and Sakhibov 1987, 1989, 1990; Canzian 1993; Lyakhovich et al. 1997; Fridman et al. 1997b). Motions in the galactic disk called forth by the two–armed density wave lead to the appearance of the first and third harmonics in the spectrum. The nature of the second Fourier harmonic is not entirely clear. It could be associated with asymmetry in

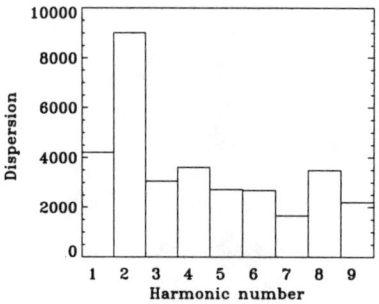 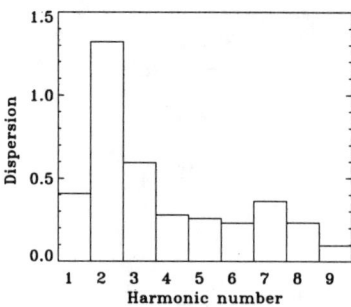

Figure 5. Contribution of different Fourier harmonics of the brightness in H_α line (*left*), and 21-cm line (*right*) into deviations from axisymmetry in the disk of NGC 3631. The prevalence of second harmonic is clearly seen.

the two-armed structure (*i.e.*, an appreciable contribution of an $m = 1$ mode) or due to vertical motions in the two–armed density wave (Lyakhovich et al. 1997, Fridman et al. 1997b). The good correlation between the lines of the maxima of the second harmonic and the shape of the spiral patterns observed for a number of galaxies (Fridman et al. 1999) argues in favor of this latter interpretation.

The results presented in Fig. 6 are consistent with the hypothesis that the second Fourier harmonic in the line-of-sight velocity field spectrum of NGC 3631 is associated with "vertical spiral-wave" gas motions (see Introduction). Indeed, in the case of the optical data, only the second harmonic has a significant amplitude, which can easily be understood if precisely this harmonic is associated with vertical motions, while the effect of projection decreases the contributions of the radial and azimuthal velocities, which are proportional to $\sin i$ (in the case of NGC 3631, by more than a factor of three: $\sin 17° = 0.3$).

At the same time, the amplitude of the second harmonic for the radio data is substantially lower than that for the optical data, which is natural according to Introduction, taking into account the lower HI optical depth. Note that the galactic disk is not completely transparent in the HI line. The inhomogeneity of the atomic hydrogen in temperature and density and the different coverages in and between the spiral arms (Brown 1997) hinders quantitative estimation of the effects of absorption.

The radial behavior of the amplitudes of the third harmonic (recalculated for the plane of the galaxy) is presented in Fig. 7, which also shows lines of constant phase corresponding to the maximum deviations. It is noteworthy that the phase behavior of the optical and radio harmonics is similar, despite the difference in their amplitudes. This suggests that the radio and optical harmonics result from a single mechanism, evidently associated with the spiral density wave.

As shown by Fridman et al. (1997), the relation between the observed third harmonic and the spiral pattern can be verified using so-called "modified" third harmonic. The latter is a two-armed spiral with phase $F_3 - \pi/2$, *i.e.*, $\propto \cos(2\varphi - F_3 + \pi/2)$, where F_3 is the phase of the original third harmonic of

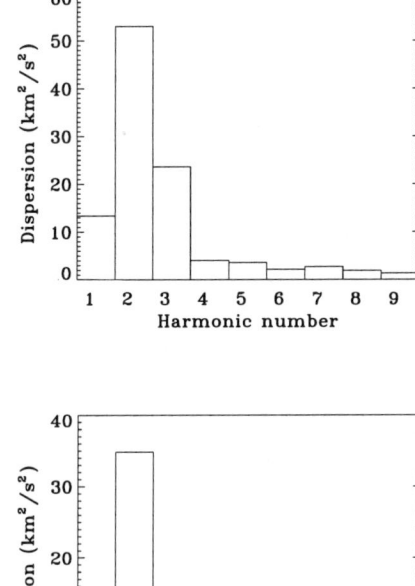

a)

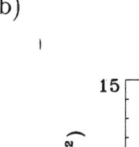

b)

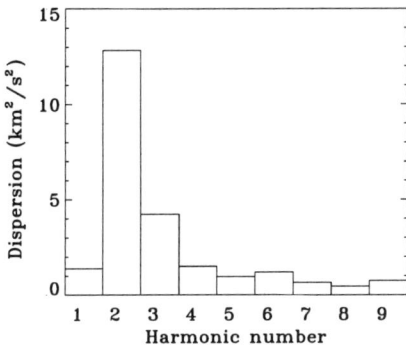

c)

Figure 6. Contribution of different Fourier harmonics of line-of-sight velocity field of NGC 3631 in dispersion in the model of pure circular motion: (a) calculated from initial H_α velocity field; (b) calculated from smoothed H_α velocity field; (c) calculated from HI velocity field. The prevalence of first sine, second and third harmonics is clearly seen and demonstrates the domination of the mode with $m = 2$ in velocity perturbations in the disk.

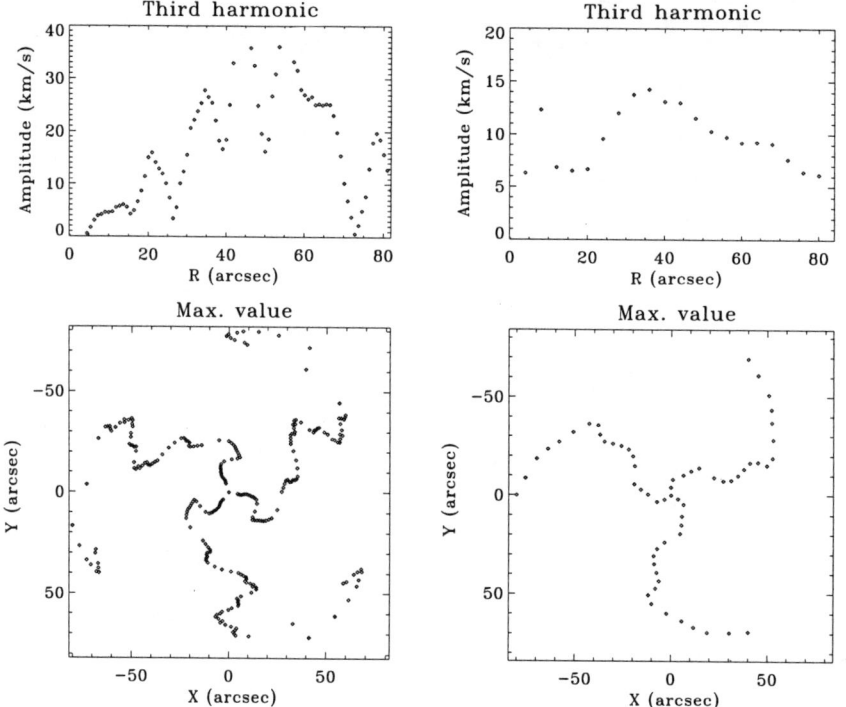

Figure 7. Radial behaviour of the amplitude (top) and the phase (bottom) of third Fourier harmonics of line-of-sight velocity. (*left*) H_α velocity field. (*right*) HI velocity field.

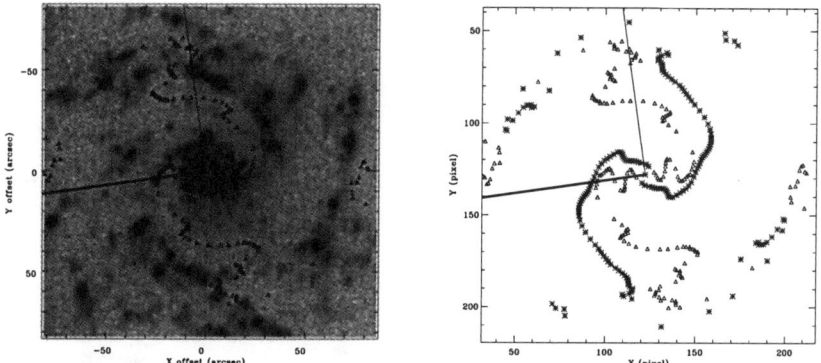

Figure 8. Superposition of the modified third harmonic of the line-of-sight velocity field of NGC 3631 on the galaxy H_α image (*left*) and on the lines of constant phase of the second harmonics of brightness (*right*). Triangles and asterisks respectively show the azimuth positions of the maxima of the harmonics at each radius.

the line-of-sight velocity field. Fig. 8 shows lines of constant phase for the modified third harmonic, corresponding to the maximum deviations, superposed on a monochromatic image of NGC 3631 in the H_α line. We can clearly see a correlation between the phase of the spiral pattern and the modified third harmonic. As shown by Fridman *et al.* (1997) this correlation proves the wave nature of the spiral structure of the galaxy. It is noteworthy that, out to distances from the center of $R \approx 40 \div 50''$, the modified harmonic is shifted from the spiral pattern by $\pi/2$, while at large distances from the center, it lies on the regions of maximum brightness. This may be due to a change in the relation between the phases of the spiral pattern and modified third harmonic near the corotation radius of the spiral structure (Fridman *et al.* 1997). Consequently, these data suggest that the corotation radius of NGC 3631 is located near $40 \div 50''$.

Acknowledgments. We are grateful to Per Olaf Lindblad who kindly provided to us the velocity field of NGC 1365.

References

Afanasiev, V.L., & Fridman, A.M. 1993, Astron. Lett., 195, 319
Baev, P.V., Makov, Yu.N., & Fridman, A.M. 1987, Sov. Astron. Lett., 13, 406
Brown, R. 1997, ApJ, 484, 637
Canzian, B. 1993, Astron. Soc. Pacif. Conf. Series, 105, 661
Fridman, A.M., Khoruzhii, O.V., Lyakhovich, V.V., Avedisova, V.S., Sil'chenko, O.K., Zasov, A.V., Rastorguev, A.S., Afanasiev, V.L., Dodonov, S.N., & Boulesteix, J. 1997a, The Twenty-first International Conference on the Unity of the Sciences, Washington, November 24-30, 1997

Fridman, A.M., Khoruzhii, O.V., Lyakhovich, V.V., Avedisova, V.S., Sil'chenko, O.K., Zasov, A.V., Rastorguev, A.S., Afanasiev, V.L., Dodonov, S.N., & Boulesteix, J. 1997b, Astroph. and Space Sci., 252, 115

Fridman, A.M., Khoruzhii, O.V., Zasov, A.V., Sil'chenko, O.K., Burlak, A.N., Moiseev, A.V., Afanasiev, V.L., Dodonov, S.N., & Knapen, J. 1998, Astron. Lett., 24, 764

Fridman, A.M., Khoruzhii, O.V., Lyakhovich, V.V., Sil'chenko, O.K., Zasov, A.V., Afanasiev, V.L., Dodonov, S.N., & Boulesteix, J. 1999, MNRAS, in press

Knapen, J.H. 1997, MNRAS, 286, 403

Lindblad, P.A.B., Lindblad, P.O., & Athanassoula E. 1996, A&A, 313, 65

Lyakhovich, V.V., Fridman, A.M., Khoruzhii, O.V., & Pavlov, A.I. 1997, Astronomy Report, 41, 447

Nezlin, M.V., Polyachenko, V.L., Snezhkin, E.N., Trubnikov, A.S., & Fridman, A.M. 1986, Sov. Astron. Lett., 12, 213

Sakhibov, F.Kh., & Smirnov, M.A. 1987, Astron. Zh., 64, 255

Sakhibov, F.Kh., & Smirnov, M.A. 1989, Astron. Zh., 66, 921

Sakhibov, F.Kh., & Smirnov, M.A. 1990, Astron. Zh., 67, 690

Active Galactic Nuclei and Related Phenomena
IAU Symposium, Vol. 194, 1999
Yervant Terzian, Daniel Weedman, Edward Khachikian, eds.

Nuclear kpc-sized Disks of Spiral Galaxies

A.V.Zasov and A.V.Moiseev

Sternberg Astronomical Institute,13 Universitetskij Prospect, Moscow 119899, Russia

Abstract. The complex structure of nuclear disks of normal spiral galaxies is illustrated with the example of five galaxies, observed at the 6m telescope. The problem of gravitational stability of nuclear disks is briefly discussed.

1. Introduction

Circumnuclear regions within a radius of a few hundred parsecs are so diverse in their properties, that they may be called the most distinctive parts of galaxies. Usually these regions cannot be considered as a simple continuation of the main disks of parent galaxies, being decoupled by their structure, angular velocity, gas content, star formation rate, or metal abundance.

In this paper we will touch upon two topics: structure of nuclear disks, and the influence of their rotation on star formation.

2. Photometrical and kinematical structures of nuclear disks

Rapidly rotating central parts of the observed galaxies as a rule do not exceed 5-15″ by size, which makes their spectral study a rather difficult task. It appears that reliable data for their kinematic properties cannot be obtained from one-dimensional spectral cuts; actually two-dimensional velocity fields are necessary to distinguish between circular and non-circular motions of gas or stars.

To illustrate the structural and dynamical properties of the nuclear disks, we will discuss briefly the observational data for the inner regions of five normal spiral galaxies: NGC 972, 1084, 4100, 6181 and 7217. All spectral observations were carried out at the 6m telescope of the Special Astrophysical Observatory in Russia (SAO RAS) in 1993 -1997. A scanning Fabry-Perot interferometer (IFP) and Multi-Pupil Field Spectrograph (MPFS) developed in this observatory were used for obtaining the velocity field of the ionized gas in the H_α emission line. Although the observations were aimed mostly to study gas motions in a global galactic scale (program "Vortex" led by A. Fridman), five objects chosen here may give bright examples of different structures and kinematic peculiarities observed in the nuclear regions of non-AGN galaxies.

The main results of observations are illustrated in Table 1. As the Table shows, all galaxies have a circumnuclear structure, containing at least two kinematic subsystems: the inner one, which may be described as a minibar or dynamically decoupled disk, and the outer subsystem – a co-planar or moderately inclined gaseous disk, which contains spiral or ring structures.

The common feature of many galaxies is the turn of kinematic major axes in their circumnuclear region. There may be two probable interpretations of this effect – a bar-like perturbation of the velocity field, and the inclination of the nuclear disk with respect to the the main disk. To distinguish between these possibilities, photometric data were involved in addition to kinematic ones.

A general approach is quite simple. In the first case (the presence of a bar) one can expect some characteristic distortion of the velocity field the of gas. Both the models of gas dynamics in a barred potential and the observations of barred galaxies show that isovelocity contours in the vicinity of a bar tend to turn along the major bar axis, so the line of largest velocity gradient (kinematic major axis) turns towards the minor photometric bar axis. Hence, photometrically and dynamically obtained position angles (PAs) of the major axes change in a different way, turning in the opposite directions. In the second case, when the plane of the circular rotating nuclear disk is not coplanar to the main disk, the photometric axis always turns parallel to the dynamical one. Observations show that both cases occur in real galaxies.

Table 1.   Kinematical structures and features of nuclear disks

NGC	Type	R(″)	R(kpc)	Structure of nuclear region	Reference
972	Sb	6	0.6	minibar and pseudoring	Zasov & Sil'chenko (1996)
		20	1.8	star-forming ring + IR spiral (?)	Zasov & Moiseev (1998)
1084	Sc	6	0.6	radial motion of the ionized gas	Moiseev et al. in prep.
		20	1.8	dynamically revealed bar	
4100	Sbc	6	0.4	blue optical ring	Moiseev et al. in prep.
		11	0.9	inclined or polar star-forming disk	
6181	Sc	5	0.8	minibar	Sil'chenko et al. (1997)
		12	2.0	co-planar disk+expanding ring or strong vertical motion	
7217	Sb	2	0.15	polar disk and ring	Zasov & Sil'chenko 1997
		10	0.8	fine spiral arms in co-planar disk	

Some comments on the chosen galaxies:

NGC 972. This is an isolated galaxy of unusual optical structure with a dusty disk and negligible bulge. A wide dust lane, bordering the bright inner part of the galaxy in the south-west side, gives some hint of the weak inclined inner disk of interstellar matter which was confirmed by measuring the orientation of the line of nodes. The galaxy is rich of molecular gas, nevertheless it possesses rather moderate star formation which is concentrated in the small nucleus and in the ring of about 2 kpc radius. The latter is not visible in the images obtained at the 1m SAO telescope in VRI bands, but clearly noticeable in the H_α line and also in the map of the distribution of the Q-parameter, which is an absorption -independent combination of V,R and I magnitudes (Zasov & Moiseev 1998). The Q-parameter map also shows a small pseudoring with radius about 0.5 ± 1 kpc, which coincides with the H_α nuclear ring, found recently by Ravindranath & Prabnu (1998).

The IFP velocity field of the inner region of about 0.6 kpc radius reveals the turn of the kinematical axis at about 30° relative to the orientation of the outer line of nodes found earlier from the observations with MPFS (Zasov &

Sil'chenko 1996). The K-band image (which was obtained in UKIRT and kindly given by Stuard Ryder to the authors) shows that the isophotes major axes turn in the opposite direction. It gives evidence of a minibar inside the nuclear ring. In between the central disk and the ring the rotation of gas is circular. The irregular structure of this region, seen in K-band, resembles a widely opened spiral. So there are two kinematic subsystems, coexisting in the inner part of this galaxy.

NGC 1084. This is a late-type galaxy with a well defined spiral structure. The velocity field analysis of the very inner bright region within 6″ (0.6 Kpc) from the centre demonstrates a fast circular rotation of gas in the plane of the main disk, in addition to less intense shifted components of H_α and [NII] line profiles, revealing non-circular (probably radial) motions. The most unusual peculiarity of gas kinematics which was found from IFP observations is the long (about 1.5 kpc size) shock front, crossing the inner part of the galaxy, where the line-of-sight velocity changes at 80-100 km/s within several arcsec perpendicular to the front line. Direct images of NGC 1084 were obtained at the SAO 1m telescope in B,V,R,I bands. Image processing showed the turn of the photometric major axis has the opposite sign with respect to the turn of the kinematic one. It agrees well with the presence of a bar-like perturbation with radius about 2 kpc. Why the presence of a any optical feature in this region is not seen in the image of this galaxy is a puzzle. It seems that the contrast of the bar potential is rather weak.

NGC 4100. This is non-barred galaxy in the Ursa Major cluster. Its rapidly rotating nucleus was found by Afanasiev et al.(1992). Analysis of the velocity field showed that in the inner 11″, or about 0.8 kpc, where a very large velocity radial gradient is present, a kinematical axis turns by 20° with respect to the outer disk. Ellipticity of the isophotes is maximal at the central 6″. The photometric axis in R,I bands turns parallel to the kinematical one, which indicates the existence of the inclined inner disk in the central 0.8-0.9 kpc. In the frame of circular gas motions the dynamically determined inclination of the main disk and the nuclear region differ by 22 ° in such a way that the nuclear disk looks more opened for the observer. There are two possible solutions for the inclination angle between the planes of two disks: 25° or 87°, depending on what side of the nuclear disk is closer to us. The blue color index and bright H_α emission gives evidence of intense SF in the nuclear disk.

NGC 6181. In addition to IFP observations, the images of this SAB(rs)c galaxy obtained at the 1m telescope were used (see Sil'chenko et al. 1997). In the very centre of this galaxy, at radius of about 0.6-0.8 kpc, both kinematic and photometric major axes turn in the opposite directions, which enables us to conclude that a small nuclear bar exists in this galaxy.

The central part of the residual velocity field, obtained by the subtraction of the observed and the expected circular velocity fields, reveals two ring-like arcs at radius 1.8 kpc, symmetrically positioned near the minor axis, where deviations from the circular rotation locally exceed 50 km/s. It is not clear what may cause such strong radial motions (or z-motions) of gas – these regions do not reveal themselves in the brightness distribution. There is also no hint of the related shock wave or the enhanced line emission. A plausible explanation is that we observe here an unusually large amplitude of oscillation of gas velocities associated with the density waves, which penetrate deep into the inner part of the disk.

NGC 7217. Contrary to the galaxies discussed above, this galaxy possesses a highly luminous spherical component. Although this galaxy has two or three optical rings, there is no bar at either visible, or near-infrared wavelengths. Observations with IFP and MPFS show that the azimuthal variation of the line-of-sight velocity gradient follows a nearly sinusoidal curve, which indicates that the gas moves along circular orbits. In the circumnuclrear region this gradient amplitude sharply increases up to 250 km/s/kpc. The significant turn of the kinematic line of nodes is also noticeable. So, in this galaxy we have an example of a sharply kinematically distinct nucleus.

Our data show a rapid decrease in the photometric PA of the major axis toward the center of the galaxy, beginning from about 4″. However, the HST measurements, taken from the NASA/ESA archive data, show that this decreasing actually occurs closer to the center – at a distance of about 1″ in such a way that the central isophotes become nearly perpendicular to the outer ones. They also increase their ellipticity toward the centre. The kinematic axis orientation is in satisfactory agreement with the photometric estimates. Therefore, the most likely explanation for the rotation of the photometric and dynamical axes is the presence of a small, strongly inclined, (probably polar) disk in the central 100-200 pc region of the galaxy. [1]

3. Nuclear disk stability and star formation

Fast rotation of nuclear disks of galaxies is a factor which tends to reduce the star formation activity due to angular momentum of collapsing gas regions, which prevents gaseous disks from being gravitationally unstable.

A flat gaseous disk of surface density $\sigma_{gas}(R)$ is gravitationally stable if the radial velocity dispersion of gas C_{gas} is high enough for the Toomre Q-parameter ($Q \sim C_{gas} \cdot \kappa(R)/\sigma_{gas}(R)$, where $\kappa(R)$ is the epicyclic frequency) to be larger than some critical value Q_c, so that $Q_c = 1$ for pure radial perturbations (Toomre criterion). In disks of spiral galaxies $Q_c = 1.5 - 2$ (Kennicutt 1989, Zasov & Bizyaev 1994). Non-WKB analysis of stability shows that the threshold for instability $Q_c \approx 1.7$ for a 'flat' rotation curve, but keeps close to 1 if the angular velocity $\Omega \approx$ const (Polyachenko et.al. 1997). It follows then, that in the case of rigid-body rotation, which usually takes place in circumnuclear regions, a higher value of the gas surface density σ_{gas} is necessary for the disk to be unstable – due to lower Q_c and higher $\kappa(R)$.

Indeed, many spiral galaxies possess dense molecular circumnuclear disks of about one kpc size, for which σ_{gas} exceeds $10^3 M_\odot/pc^2$, so a large angular velocity is necessary to stabilize the disk. However, as a rule, values of $\kappa(R)$ for them are also very high, and, as a result, the velocity dispersions C_{gas}, corresponding to $Q_c \approx 1$, remain rather low. The estimates of marginal values of C_{gas} (from the data taken from the literature) for about twenty galaxies which

---

[1]Small polar nuclear disks in normal spiral galaxies probably are not so rare: for example, their presence was claimed in NGC 2685 (Sil'chenko et al. 1998), NGC 253 (Ananthramaiah & Goss 1996) and some other galaxies.

have molecular nuclear disks shows that for most of them $C_{gas} \leq 15$km/s, which does not exceed the observed velocity dispersion of gas (Zasov 1999). This result gives evidence that in many cases nuclear molecular disks are on the threshold of gravitational stability or definitely stable. The latter is especially true for nuclear regions of galaxies poor of gas, such as NGC 7217. Nevertheless some star formation takes place even there. In NGC 7217 not only the growth of the intensity of H_α towards the centre is observed, but also, as the analysis of HST observations showed, a surprisingly well-ordered spiral-like structure exists within the inner 10" (Zasov & Sil'chenko 1997).

Even if the inner disk of NGC 7217 is marginally stable, the wavelength of growing gravitational perturbations is expected to be about several hundreds of parsecs there, whereas the observed structure presents a sort of "rippled surface" with a significantly smaller scale. It confirms that the observed pattern cannot be caused by gravitational oscillations.

Note that the small-scale spiral pattern may frequently occur in the nuclear disks, although it is usually difficult to extract it from the photometric observations with restricted angular resolution. As an example, a complex circumnuclear spiral-like structure was found in NGC 6951 (Barth et al. 1995) and NGC 488 (Sil'chenko 1999).

A possible alternative mechanism of formation of the spiral pattern is the hydrodynamical instability in gaseous disks which does not require a high surface density to develop (see the discussion by Fridman (1994)). So the presence of star formation in rapidly rotating disks may give evidence of the importance of non-gravitational mechanisms for compression of the gaseous medium there.

Acknowledgments. The authors are grateful to V. Afanasiev, J. Boulesteix, A. Burenkov, S. Dodonov, O. Sil'chenko and V. Vlasyuk for obtaining the observational data. This work was supported by grant RFBR 98-02-17102.

References

Afanasiev, V.L., Burenkov, A.N., Sil'chenko, O.K., Zasov, A.V. 1992,Soviet Ast., 36, 10
Ananthramaiah, K.R., & Goss, W.M. 1996, ApJ, 466, L13
Barth, A.J., Ho, L.S., Filippenko, A.V., Sargent, W.L.W. 1995, AJ, 110, 1009
Fridman, A.M. 1994, in "Physics of the Gaseous and Stellar Disks of the Galaxy", ed. King I.R. ASP Series, 66, 15
Kennicutt, R. 1989, ApJ, 344, 685
Polyachenko, V.L., Polyachenko, E.V., Strelnikov, A.V. 1997, AstL, 23,483
Ravindranath, S., & Prabnu, T.P. 1998,AJ, 115, 2320 1997, PAZh, 23, 551
Sil'chenko, O.K. 1998, A&A, 330, 412
Sil'chenko, O.K. 1999, PAZh, in press
Sil'chenko, O.K., Zasov, A.V.,Burenkov, A.N., Boulesteix, J. 1997,A&AS,121,1
Zasov, A.V., & Bizyaev, D.V. 1996, AstL, 22, 71
Zasov, A.V., & Moiseev, A.V. 1998, AstL, 24, 584

Zasov, A.V., & Sil'chenko, O.K. 1996, in "Barred Galaxies", Eds R. Buta et.al. ASP Series, 91, 207

Zasov, A.V., & Sil'chenko, O.K. 1997, ARep, 41, 734

Zasov, A.V., 1999, AApTrans., in press

Broad Emission Lines in AGN: Phenomenology and Models

Jack W. Sulentic

Department of Physics and Astronomy, University of Alabama, Tuscaloosa, AL, USA 35487

Paolo Marziani and Massimo Calvani

Osservatorio Astronomico, vicolo dell'Osservatorio 5, Padova, Italy

Deborah Dultzin-Hacyan

Instituto de Astronomia, UNAM, Apdo. 70-264, Mexico D. F. 04510, Mexico

Abstract.
We review the phenomenology of broad emission lines in AGN. We show that velocity displacements relative to the local rest frame of Hβ and CIVλ1549 are real. The most significant line displacement result involves a systematic blueshift for CIV, seen only in radio-quiet sources. We find some evidence that the amplitude of this displacement may correlate with source orientation. We show that disagreement between different studies of the CIV line properties is due to whether or not a narrow line component was subtracted. Finally, we consider evidence that optical and X-ray broad lines arise in an accretion disk. We show that the evidence for a disk origin is far from overwhelming for both Hβ and FeKα.

1. Introduction

The current phenomenology of emission lines in AGN suggest that three distinct emitting regions are present in at least some and possibly all sources at certain viewing angles. In other words, some AGN show inflections in their line profiles consistent with three kinematical/geometric emission zones. Figure 1 shows the profile of Hβ for the radio-quiet source PG1138+222 (see Marziani & Sulentic 1993) where we identify: a) a narrow line (NLR) component with FWHM=300 km/s, b) a broad line (BLR) component with FWHM=2700 km/s and c) a very broad (always redshifted?) line component (VBLR) with FWHM$\sim$8000 km/s. The VBLR component is not easy to study because it is so broad and frequently obscured by Fe line emission. Its reality is not in doubt (see e.g. Ferland, Korista & Peterson 1990) but its frequency of occurence is unclear. The NLR component must be subtracted from a line profile before the BLR component can be effectively studied.

We focus here on: 1) BLR properties as the most promising insight into the central source structure, 2) BLR Hβ because it is the most studied low ionization

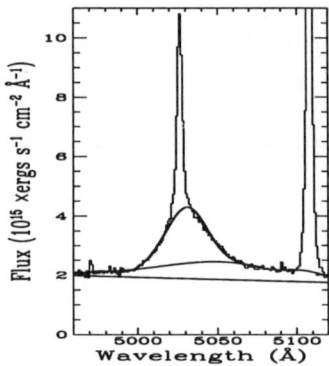

Figure 1. Hβ line profile of PG 1138+222 showing three well separated narrow, broad, and very broad components.

(LIL) line (13.6eV) that can be seen over a fairly wide redshift range and 3) BLR CIV as the least contaminated and most studied of the high ionization (64eV) lines. Our studies have been statistical in nature and, therefore, complement the reverberation studies involving a much smaller sample of sources. We ask, and attempt to answer, three BLR related questions. Do the HIL and LIL arise in the same emitting region? Are BLR internal velocity shifts real and why is there disagreement among the principal studies on this issue? Finally, do the observations support a model where the bulk of the emission arises in a region dominated by rotational motions (specifically, an accretion disk)?

2. HIL and LIL From the Same BLR?

Two years ago we reported on a direct comparison of Hβ and CIV broad line properties using FOS archival spectra for the latter and matching ground-based spectra for the former (Marziani et al. 1996). Figure 2 summarizes the velocity displacement results from that study. We measured the broad line profile displacement relative to the local source rest-frame inferred from NLR [OIII]λ5007. The most striking result we found was that CIV in radio-quiet quasars always shows a blueshift relative to the local rest frame. If we can generalize from our small sample, we suggest that the majority of AGN (~90%) should show this effect which is most simply attributed to some kind of outflow (disk wind or more collimated flow). The fact that Hβ does not show the same effect is strong evidence for a two-component BLR. We found lower order correlations that suggest the amplitude of the CIV blueshift may be related to source orientation. This result and the results of reverberation studies, which place CIV emitting clouds closer to the central engine, are easier to understand if CIV emission is concentrated in some sort of collimated outflow.

Figure 2 shows that the situation for radio-loud sources is less clear. In radio-loud sources, it is Hβ that shows large shifts rather than CIV. Our sample suggests a preponderance of redshifts but this may well reflect the small sample size because we know of radio loud sources with very large blueward displacements (e.g. 3C227; Gaskell 1983). The preliminary results suggest that BLR

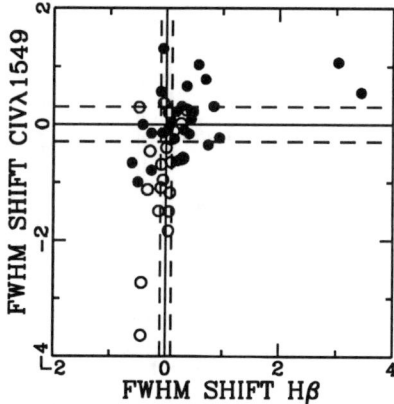

Figure 2. Distribution of Balmer line vs CIV line centroid shifts at half maximum (in units of 10^3 km s^{-1}). Open circles are radio-quiet and filled circles are radio-loud sources. Two σ error bars are shown.

structure and/or kinematics in radio-loud sources is significantly different from radio-quiet AGN.

3. Are BLR Velocity Shifts Real?

Our CIV vs Hβ comparison suggests that large velocity displacements in broad lines are concentrated in the LIL for radio-loud sources and HIL for radio-quiet ones. The LIL in radio-quiet and HIL in radio-loud objects appear to show only small displacements (typically less than 600 km/s) about the local rest frame. The displacements are equally divided between red and blue. We believe that even these line shifts are real and Figure 3 shows an example of our evidence. It shows the LIL (Hβ) profiles for three radio-quiet sources. The profiles in these sources are reasonably symmetric, making it difficult to ascribe any line shift to profile asymmetry or some other effect. At least two of these sources show clear bulk displacement of the majority of the line emission. The redshifted VBLR component affects our estimates of the amplitude of these shifts.

Studies of the CIV profile have been more conflicted, in part, because CIV rarely shows a clear inflection between BLR and NLR components. There have been several independent statistical studies of the CIV profile (Wills et al. 1993ab, Brotherton et al. 1994ab; Corbin & Francis 1994, Corbin & Boroson 1996; Marziani et al. 1996). Most important differences between these studies can be traced back to whether or not an NLR component was subtracted from the CIV profile. We subtracted such a component and find that it varies from zero to 20% of the total line flux. The NLR component may be negligible in high luminosity AGN (Wills et al. 1993a) but our sample is dominated by lower luminosity sources. Wills et al identified two-components in the profile of CIV: i) an intermediate (ILR) and ii) a VBLR line component. They find no velocity displacement for the ILR component and showed that most of its properties were identical to the NLR. We argue that this is the NLR component of CIV. It

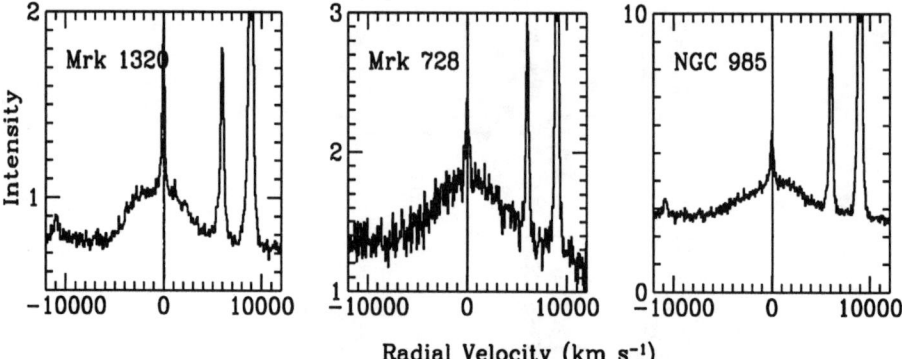

Figure 3. Examples of broad profiles with: moderate (565 km s^{-1}) blueshift (Mark 1320; left panel); small (270 km s^{-1} redshift (Mark 728; middle panel); moderate (420 km s^{-1} redshift (NGC 985) all relative to the NLR (1$\sigma \approx$60 km s^{-1}).

is slightly broader than [OIII]5007 because it is a partially resolved doublet and because it may arise in a slightly denser environment like NLR [OIII]4363. CIV is a collisional line. We interpret their (blueshifted) VBLR component as the "classical" BLR signature of CIV (note that this is not the same as the VBLR feature discussed above and in Ferland et al. 1990). While they recognize these two components, Wills et al. analyze their results in terms of the relative contributions of these two components. It is difficult to interpret data that includes the variable contribution of NLR and BLR gas.

Corbin & Boroson (1996) argue that no significant NLR CIV component exists. A comparison of sources in common with that study (Sulentic et al. 1998c) yield completely different conclusions about profile width and velocity displacement. We believe that insights into the HIL BLR can only be obtained if an NLR component is subtracted from CIV. The narrow component, if not subtracted, obscures first-order BLR properties. Corbin (1997) finds no significant (see Figure 2) CIV shifts for 2/3 of a sample of 18 sources in common with our sample (Marziani et al. 1996). This is consistent with the peak of the CIV profile being dominated by NLR emission.

4. BLR Emission From an Accretion Disk?

Many recent papers have discussed the possibility that both optical Balmer (e.g. Chen & Halpern 1989) and X-ray FeK (see e.g. Nandra et al. 1997ab) BLR emissions arise from an accretion disk. No matter how one looks at it, the double-peaked Balmer line profiles are rare. Model fits to those profiles require that we view the disk from an intermediate angle where many more such sources should be observed. Thus they are much more consistent with a biconical flow seen near pole-on where their statistical rarity is expected. Double-peaked sources cannot be the extremum of a population where an additional central component fills in the "valley" between the peaks because no population of single-peaked sources with very broad steep-sided profiles is observed. If the majority of sources do show lines that arise from a disk then the rare population of double-peaked sources are inconsistent with the disk illumination models needed to explain the majority. However one looks at it, double-peaked sources would be a miraculous minority in a disk scenario.

We have recently considered a disk illumination model (Sulentic et al. 1998a) that can simultaneously produce both broad Balmer and FeKα emission. The model profile predictions are in poor agreement with the data, and disk inclination (the only free parameter in our model) values required to fit the observed profiles are inconsistent for the two lines. Finally we showed recently (Sulentic et al. 1998b) that the FeK line is likely composed of two (Gaussian) components: i) narrow and unshifted feature with E=6.4keV and FWHM$\leq$0.25keV and ii) broad and redshifted feature with E=5.9keV and FWHM$\geq$1.6keV. Neither of these components are easily fit by the kinds of disk models that have been discussed so far.

References

Brotherton, M. et al. 1994a, ApJ, 423, 131
Brotherton, M. et al. 1994b, ApJ, 430, 495
Chen, K., & Halpern, J. 1989, ApJ, 344, 115
Corbin, M. 1997, ApJS, 113, 245
Corbin, M. & Boroson, T. 1996, ApJS, 107, 69
Corbin, M. & Francis, P. 1994, AJ, 108, 2016.
Ferland, G., Korista, K. & Peterson, B. 1990, ApJ, 363, L21
Gaskell, M. 1983, ApJ, in Proc. Liege Conf. "Quasars and Gravitational Lenses", 473.
Marziani, P. & Sulentic, J. W. 1993, ApJ,. 409, 612
Marziani, P., et al. 1996, ApJS, 104, 37
Nandra, K., et al. 1997a, ApJ, 476, 70
Nandra, K., et al. 1997b, ApJ, 477, 602
Sulentic, J., Marziani, P., & Calvani, M. 1998b, ApJ, 497, L65
Sulentic, J., et al. 1998a, ApJ, 501, 54
Wills, B. et al. 1993, ApJ, 410, 534
Wills, B. et al. 1993, ApJ, 415, 563

Active Galactic Nuclei and Related Phenomena
IAU Symposium, Vol. 194, 1999
Yervant Terzian, Daniel Weedman, Edward Khachikian, eds.

Quantized Redshifts – New Physics or Old Muddle?

W.M. Napier

Armagh Observatory, College Hill, Armagh BT61 9DG, Northern Ireland.

Abstract.
The HI redshift distribution of nearby spiral galaxies has been studied to test long-running but generally ignored claims that extragalactic redshifts are periodic or 'quantized'. The existence of the phenomenon is confirmed at an extremely high confidence level, the quantization appearing in the galactocentric frame of reference. It is proposed that the energy density of the vacuum is a local, oscillating quantity associated with large masses such as spiral galaxies. A variety of 'anomalies' should then be detectable in massive galaxies, associated with their redshifts, their ambient gravitational lensing and their dynamics.

1. Introduction

Over the past ~25 years the perceived need for 'new physics' in interpreting exotic extragalactic phenomena, long espoused by Ambartsumian, has largely receded. Although a wide variety of extragalactic redshift anomalies continues to be reported in the literature, primarily by Arp and Tifft (but also by others), these are generally ignored, or ascribed to coincidence, *post hoc* statistics or whatever. Of the 'anomalous redshift' claims, that of quantization is perhaps the least credible: thus Tifft (1976) claimed that the redshifts of galaxies in the Coma cluster are preferentially offset from each other in multiples of ~72 km/sec, Tifft & Cocke (1984) claimed that wide-profile spiral galaxies distributed over the sky have a galactocentric periodicity ~36.2 km/sec and so on.

However quantization of redshifts, while the least credible of the 'discordant' claims, is also the most testable: adequate numbers of new high-precision 21 cm data have become available since its original formulation. Thus beginning about ten years ago, the author and his colleague Dr Guthrie embarked on a programme of testing the claims (Napier et al. 1989, Guthrie & Napier 1990, 1992, 1996). To carry through this programme, we used rigorous statistical procedures and made extensive use of synthetic datasets as controls. We found that extragalactic redshifts are quantized along the claimed lines. Monte Carlo simulations yield very high formal confidence levels for the phenomenon, but here I simply present it as a straightforward observational result.

2. The Local Supercluster Galaxies

Estimates of rotation speed V_c at the Sun's distance vary widely with, for example, Feast & Whitelock (1997) deriving $V_c = 231\pm15$ km/sec from Hipparcos observations of Cepheids, Merrifield (1992) obtaining 200 ± 10 km/sec from HI kinematics, Metzger et al. finding 237 ± 12 km/sec from Cepheid kinematics, and so on. Sackett (1997), reviewing the subject, concludes that $V_c = 210\pm25$ km/sec. Adjusting for the local solar motion relative to the LSR, a middle-of-the road estimate for the solar galactocentric motion is

$$\mathbf{V}_\odot = (V_\odot, l_\odot, b_\odot) = (220\,\text{km/sec}, 90°, 0°)$$

to within ±30 km/sec or so and a few degrees. The Virgo cluster is the nearest rich cluster of galaxies, which had not been used by Tifft in the formulation of the quantized redshifts hypothesis, and so is a natural choice for testing it. Fig. 1 shows the differential redshift distribution of 48 bright spiral galaxies in the Virgo cluster, in pairs, corrected for the above solar motion. A periodicity of $\sim$71 km/sec is clearly observed (Napier & Guthrie 1997) as against $\sim$72 km/sec expected for a dense cluster. The criteria for selecting these galaxies are discussed by Guthrie & Napier (1991); in essence all bright Virgo spirals, with precisely determined HI redshifts and avoiding the core region, were employed.

In Fig. 2 the redshift differences in pairs of 97 nearby, bright spirals within the Local Supercluster are likewise plotted, again in the galactocentric frame of reference. These were selected from the catalogue by Bottinelli et al. (1990) for their precision ($\sigma \leq 3$ km/sec), rejecting those which had been employed by Tifft and colleagues in formulating the quantization hypothesis. Once more, a periodicity is clearly seen, this time one of $\sim$37.5 km/sec as against $\sim$36.2 km/sec expected. Fifty galaxies in this sample of 97 belong to groups and associations, and the periodicity was found to be strongest in the differential redshifts of the 50 galaxies in these groups. To test this, a further sample of LSC spirals was taken from data obtained with the 300-foot Greenbank telescope by Tifft and Cocke over the period 1984–1988. Of 117 'new' spirals with signal to noise ratio >10, thirty belonged to catalogued groups and associations of galaxies (Fouqué et al. 1992). These showed the same quantization, adding to the strength of the periodic signal.

The redshift distribution for the combined sample of 80 galaxies is shown in Fig. 3. Clearly, the 37.5 km/sec periodicity is indeed galactocentric, within current uncertainties. The dispersion about the best-fit periodicity (period=37.5 km/sec, phase=0°) is $\sim$8.25 km/sec, but as the data here are redshift differences taken in pairs, the intrinsic mean spread for each galaxy in the sample is $8.25/\sqrt{2} \sim 5.8$ km/sec.

A significant finding of the Guthrie/Napier study was that the strength of the 37.5 km/sec periodicity from galaxies in the real Local Supercluster is vastly in excess of that from synthetic LSCs in which only local quantization was assumed. Thus the $\sim$37.5 km/sec periodicity appears to be global rather than confined to differential redshifts within small groups and clusters; that is, the periodicity in cluster A is locked in phase with that in cluster B even although they may be at opposite ends of the Local Supercluster.

Artefacts in data selection or reduction procedures, or in radio telescopes themselves, seem unable to account for the phenomenon. A formidable problem

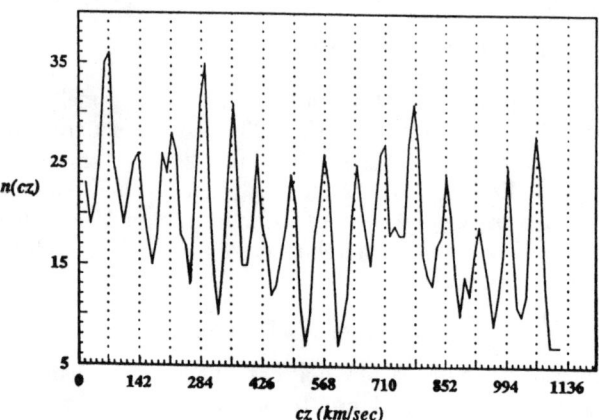

Figure 1. Differential redshifts for 48 bright Virgo spirals, in the galactocentric frame of reference, plotted in bins 11 km/sec wide.

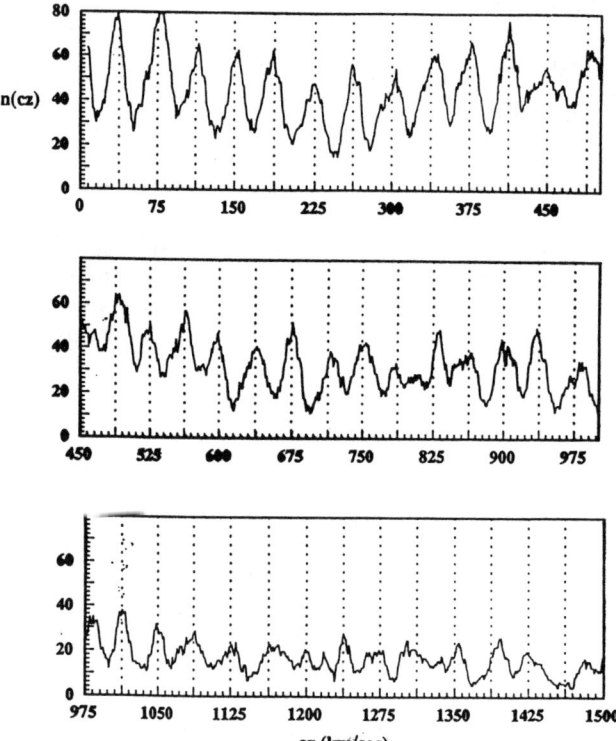

Figure 2. Differential redshifts for 97 bright spirals in the Local Supercluster, in the galactocentric frame of reference.

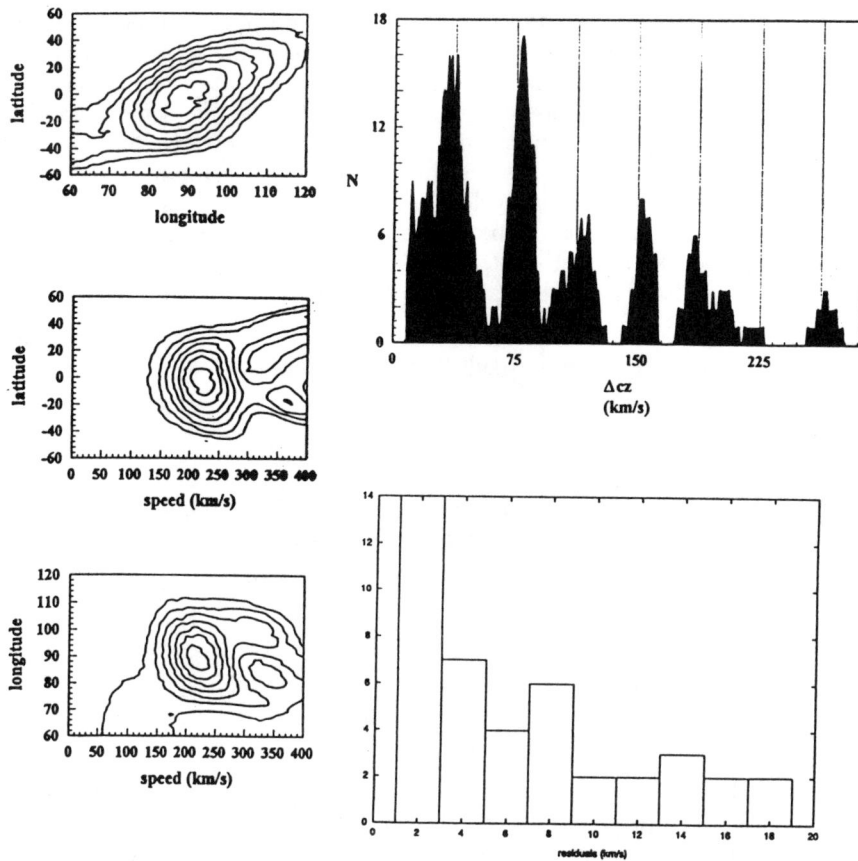

Figure 3. *Left-hand column.* Signal strength as a function of solar velocity. The signal peaks at about $V_\odot$ =220 km/sec, $l_\odot = 90°$, $b_\odot = 0°$. *Top right.* The peak signal (periodicity ∼38 km/sec). *Bottom right.* Distribution of residuals.

for any artefact theory is the galactocentric nature of the phenomenon: somehow the bug in the software, or in the telescope, must know the solar velocity around the Galaxy, and indeed have anticipated the best estimates recently obtained from Hipparcos, Cepheid and HI data. The quantization is readily detectable in individual datasets from Jodrell Bank, Greenbank 140 and 300 foot, Effelsberg, Arecibo, Westerhout and other telescopes.

3. Discussion and conclusions

The main conclusion is that extragalactic redshifts are strongly quantized. The intensely galactocentric nature of the phenomenon suggests that any underlying 'oscillating physics' should likewise be galactocentric; this presumably applies also to other massive galaxies (*cf* Schunck 1998). At a phenomenological level, one might envisage that the energy density of the vacuum, rather than being constant throughout space and time, is a local, complex variable centred on large concentrations of mass. Anthropocentric observers, adopting a constant local clock rate, would then observe extragalactic clock rates varying cyclically, with corresponding oscillations in extragalactic redshifts. Even more adventurously, a locally decaying vacuum would, to such observers, yield an apparent cosmological expansion satisfying the time dilation and Tolman surface brightness tests for expansion. This cyclic relation between local and cosmological proper times might also manifest itself through anomalous 'wiggles' in $H\alpha$ rotation curves (Schunck 1998), discretized galactic dynamics (Roscoe, this volume) and discordant redshifts between adjacent objects (Arp, this volume).

Acknowledgments. The author is indebted to Chip Arp, Victor Clube, Bruce Guthrie, Dave Roscoe and Bill Tifft for discussions.

References

Bottinelli, L., Gouguenheim, L., Fouqué & Paturel, G., 1990, A&AS, 82, 391
Feast, M. & Whitelock, P., 1997, MNRAS, 291, 683
Fouqué, P., Gourgoulhon, E., Charmaraux, P. & Paturel, G., 1992, A&AS, 93, 211
Guthrie, B.N.G. & Napier, W.M., 1990, MNRAS, 243, 431
Guthrie, B.N.G. & Napier, W.M., 1991, MNRAS, 253, 533
Guthrie, B.N.G. & Napier, W.M., 1996, A&A, 310, 353
Merrifield, M.R., 1992, AJ, 103, 1552
Metzger, M.R., Caldwell, J.A.R. & Schechter, P.L., 1998, ApJ, 115, 635
Napier, W.M., Guthrie, B.N.G. & Napier, B., 1989, in *New Ideas in Astronomy*, p. 191, eds. Bertola, F., Sulentic, J.W. & Madore, B.F., Cambridge University Press, Cambridge
Sackett, P.D., 1997, ApJ, 483, 103
Shunck, F.E., 1998, astro-ph/9802258
Tifft, W.G., 1976, ApJ, 206, 38
Tifft, W.G. & Cocke, W.J., 1984, AJ287, 492

The Nature of the UV-optical Continuum in Seyfert 2 Galaxies

Thaisa Storchi-Bergmann[1] and Henrique R. Schmitt

Instituto de Física, UFRGS, Campus do Vale, C.P. 15051, CEP 91501-970, Porto Alegre, RS, Brasil

Roberto Cid Fernandes

Departamento de Física, CFM-UFSC, Campus Universitário Trindade, CP 476, CEP 88040-900, Florianópolis, SC, Brasil

Abstract.
The nature of the UV-Optical continuum in Seyfert 2 galaxies is a matter of current debate, fundamental to which is the issue of characterizing and quantifying the stellar population contribution to the nuclear spectra. Using high S/N long-slit spectroscopy of a sample of 20 Seyfert 2 galaxies, we have applied a novel approach to investigate the nuclear stellar population in these galaxies. Our main results are: (1) the stellar populations in Seyfert 2 galaxies are varied, and in most cases **cannot** be adequately represented by an elliptical galaxy template, as done in previous works; (2) the central kpc of Seyfert 2's contain substantially larger proportions of 100 Myr stars than either elliptical galaxies or normal spirals of the same Hubble type. One important consequence of our findings is that the controversial nature of the so called "second" featureless continuum (FC2) in Seyfert 2s is most likely a result of inadequate evaluation of the stellar population.

1. Introduction

The nature of the UV-Optical continuum of Seyfert 2 galaxies has long been a matter of investigation. If, in one hand, these galaxies are known to present a bluer continuum than "normal" galaxies of the same Hubble type (this characteristic allowed them to be found in the Byurakan survey, for example), on the other hand, in the Unified Model (Antonucci & Miller 1985, Antonucci 1993), a Seyfert 1 nucleus, comprising the nuclear source and broad-line region, is hidden by an obscuring dusty molecular torus. In this scenario, the nuclear featureless continuum (FC) and broad lines are only seen via scattered light, as indeed observed in a number of polarimetric studies (Miller & Goodrich 1990; Kay 1992, Tran 1995).

---

[1] Visiting Astronomer, Cerro Tololo Inter-American Observatory. CTIO is operated by AURA, Inc. under cooperative agreement with the National Science Foundation

Nevertheless, Tran (1995) observed that, after subtraction of the stellar population contribution (usually about 70-80% at $\lambda 5500$Å) the polarization in the continuum was smaller than in the broad lines. To reconcile the two polarization values, he concluded that there were two "featureless" continua components: FC1, consisting of scattered light from the nuclear source, contributing typically 5% to the total observed continuum, and FC2, an unpolarized component contributing the remaining 15-25% of the continuum.

This study raised the new issue of the nature of the FC2 component. One possibility first proposed by Cid-Fernandes & Terlevich (1992, 1995) was the contribution from young stars in a nuclear burst of star formation. A handful of Seyfert 2's is known to have composite nuclei, classified as Starburst + Seyfert (Kinney et al. 1993), and recent studies (Heckman et al. 1997; Gonzalez-Delgado et al. 1998) have shown that the signatures of the starburst in these cases are evident in the UV-blue nuclear spectra of the galaxies. In these few Seyfert 2's, FC2 could indeed be identified with young (< 10 Myr) stars. Another possibility, proposed by Tran (1995), based on previous results by Barvainis (1993), was that FC2 is due to thermal free-free emission from a hot plasma wind. Recently, Axon, Capetti & Macchetto (1998) argued that the blue continuum can be due to free-free emission from gas shocked by the passage of a radio jet.

In this paper we discuss the results of our recent work, dedicated to investigate the nuclear stellar population and possible additional continuum in Seyfert 2 galaxies. We have used a novel approach which combines a spatially resolved study of the stellar population and spectral synthesis of the nuclear spectra.

2. The Stellar Population in Active Galaxies

In order to investigate the stellar content of Seyfert galaxies, as well as to compare it with that of other galaxies, Cid-Fernandes, Storchi-Bergmann & Schmitt (1998), have analyzed long-slit optical spectra (obtained with the CTIO 4m telescope) of 20 Seyfert 2's, 6 Seyfert 1's, 7 LINER's, 5 radio-galaxies, 1 elliptical and 3 "normal" galaxies with nuclear rings of star formation.

Our approach was to measure the continuum flux in selected windows and the equivalent widths (W's) of several stellar absorption features as a function of distance from the nuclei, at a spatial sampling of $2'' \times 2''$.

The main conclusions of this work were: (1) there is a large diversity of nuclear stellar population characteristics among Seyfert galaxies, and in most cases, differ from those of an elliptical galaxy; (2) the W's in regions of star-formation (e.g., star-forming rings), in the nuclear spectra of Seyfert 1's and for the Seyfert 2 galaxies classified as composite (SB+Sy 2) are observed to be smaller than in nearby regions, indicating a dilution by an underlying continuum, as expected; (3) for most Seyfert 2's (17 out of 20), we found no dilution in the nuclear EW's, leading to the conclusion that if there is a FC continuum present, it contributes *at most* 10% at all optical wavelengths. Conclusions (2) and (3) are illustrated in Figure 1.

The UV-optical Continuum in Seyfert 2 Galaxies

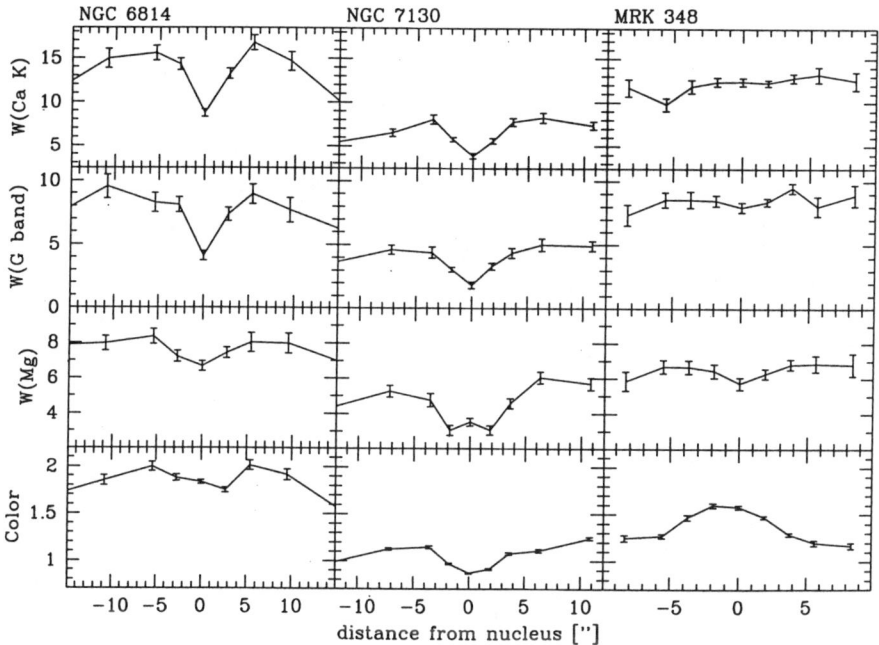

Figure 1. From top to bottom: Radial variations of the equivalent widths of the absorption features Ca II K ($\lambda3930$Å), G-band ($\lambda4301$Å), Mg I + MgH ($\lambda5176$Å) and continuum ratio $\lambda5870/\lambda4020$ for the Seyfert 1 galaxy NGC 6814 (left), the composite Seyfert + Starburst galaxy NGC 7130 (center) and the Seyfert 2 Mrk 348 (right). (Adapted from Paper I).

3. The Nature of Optical Light in Seyfert 2 Galaxies with Polarized Continuum

In a subsequent work, Storchi-Bergmann, Cid-Fernandes & Schmitt (1998) look closely at the spectra of 4 Seyfert 2's with no dilutions in the nuclear stellar absorption W's, but with previous determinations of FC contributions larger than 10%. The 4 galaxies are: *Mrk 348*, with 27% contribution at 5500Å of a nuclear FC (Tran 1995): from polarimetry, it was concluded that 5% was due to the polarized continuum (FC1), but 22% was due to FC2; *Mrk 573*, with 20% contribution at 4400Å of a nuclear FC (Kay 1994); *NGC 1358*, also with 20% contribution of FC at 4400Å (Kay 1994); and *Mrk 1210*, with 25% contribution at 5500Å (Tran 1995): from polarimetry, 6% of FC1 and 19% of FC2.

In both Kay's (1994) and Tran's (1995) work, it was assumed that the nuclear stellar population was well represented by the spectrum of a typical elliptical galaxy.

The above contributions of a nuclear FC seem to be in contradiction with the results of Cid Fernandes et al. (1998), if: (1) FC is confined to the nucleus (inner $2'' \times 2''$); (2) the stellar population does not vary within the bulge of the galaxy – it was verified that the bulges have average effective radius of $6''$ for the sample. If these two conditions are true, then the results of Cid Fernandes et al. (1998) indicate that the FC contribution to the nuclear spectra should be smaller than 10%, in contradiction with the results of previous works.

Storchi-Bergmann et al. (1998) then adopted the first hypothesis above and tested the second, extracting spectra from the bulge of the same galaxy from windows at $4''$ from the nucleus, which were used as stellar population templates for the nuclear spectra.

These extranuclear spectra were then used in combination with a small percentage ($\approx 5\%$) of FC1 (represented by a power-law spectrum) as derived by Tran (1995) to reproduce the nuclear spectrum. It was concluded that, after applying some reddening to the extranuclear spectra, this combination provided a good representation of the nuclear spectrum, with no need of FC2, as illustrated in Figure 2 for Mrk 348.

Population synthesis of these extranuclear spectra then showed that an elliptical galaxy template was only valid for Mrk 573, with the other three galaxies presenting larger proportions of younger components. It was thus concluded that the need of FC2 in these cases was a consequence of the fact that an elliptical template is not adequate to represent the stellar population of most Seyferts.

4. Spectral Synthesis of the Nuclear Region of Seyfert 2 Galaxies

In order to investigate the above result for the whole sample of Seyfert 2 galaxies, Schmitt, Storchi-Bergmann & Cid Fernandes (1998) performed spectral syntheses for the nuclear spectra of the 20 Seyfert 2's, comparing the results with those of an elliptical galaxy template. The spectral base comprised star cluster spectra of different ages (Bica, 1988), an HII continuum and a power-law component to represent the scattered component (FC1).

The results can be summarized through the contribution of the different age components to the light at $\lambda 5780$Å for the nuclear Seyfert 2 spectra compared

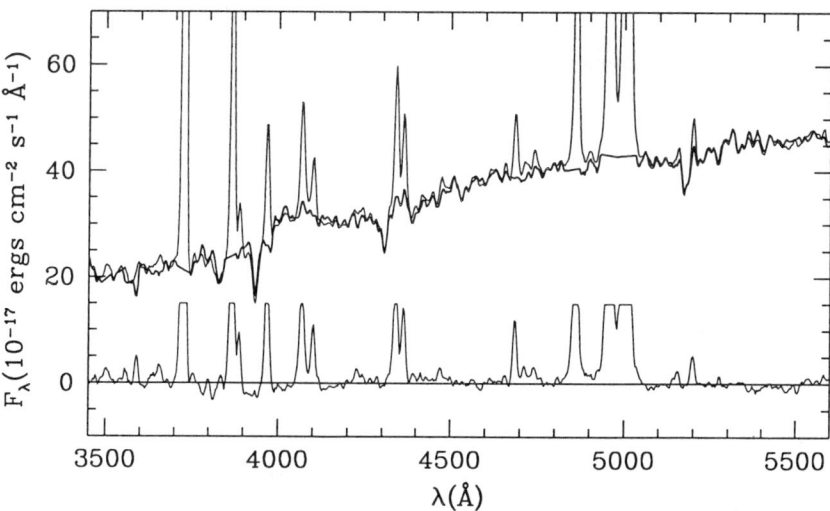

Figure 2. The Mrk 348 nuclear spectrum (thin line) is compared with the extranuclear spectrum reddened by E(B-V)=0.09, combined with 5% contribution (at $\lambda 5500$Å) of the FC1 component, represented by a $F_\lambda \propto \lambda^{-1}$ power-law (thick line). The residual between the two is plotted at the bottom, where the emission lines have been chopped for clarity.

with that for the elliptical template: 19/24 (80%) have larger contributions of stars of 100 Myrs; 7/24 (30%) have larger contributions of stars of 10 Myrs; 5/24 (20%) have larger contributions of an HII continuum.

It is also interesting to compare the above results with those for the bulge of early type spirals, the dominant Hubble types of the Seyfert 2's. From Bica (1988), it can be concluded that in only 20% of a sample of 51 early type spirals, there are contributions of stars of 100 Myrs or younger. It can be concluded that *the main difference between the stellar population of Seyfert 2's and of elliptical or early-type spiral galaxies is the large contribution from stars of 100 Myrs.*

5. Conclusions

The main conclusions of this work can be summarized as follows.

(1) Galaxies with Seyfert nuclei have a varied nuclear stellar population.

(2) In most cases this population differs from that of an elliptical galaxy. When a proper template is used, preferably from the bulge of the same galaxy, there is no need of FC2. Alternatively, if FC2 is identified with the difference between the unpolarized continuum and the elliptical template, it is due to stars with ages 0-100 Myr.

(3) The main difference between the nuclear stellar population of Seyferts and of early-type spirals and ellipticals is the larger contribution of a 100 Myr population.

(4) We could speculate that if there is a causal link between star-formation and nuclear activity, the 100 Myr star-formation timescale should be comparable to the duration of the nuclear activity cycle. Since 100 Myr is $\approx 1\%$ of the Hubble time, $\approx 1\%$ of the galaxies should be active, in agreement with recent estimates (e.g., Huchra & Burg 1992).

References

Antonucci, R. R. J. 1993, ARA&A, 31, 473

Antonucci, R. R. J. & Miller, J. S. 1985, ApJ, 297, 621

Axon, D., Capetti, A. & Macchetto, F. D. 1998, preprint

Barvainis, R. 1993, ApJ, 412, 513

Bica, E. 1988, A&A, 195, 76

Cid Fernandes, R. & Terlevich, R. 1992, in *Relationships between Active Galactic Nuclei and Starburst Galaxies*, ASP Conf. Ser., 31, Astron. Soc. Pac. (San Francisco)

Cid Fernandes, R. & Terlevich, R. 1995, MNRAS, 272, 423

Cid Fernandes, R., Schmitt, H. R. & Storchi-Bergmann, T. 1998, MNRAS, 297, 259

González Delgado, R. M. et al. 1998, ApJ, 505, 174

Heckman, T., Krolik, J., Meurer, G., Calzetti, D., Kinney, A, Koratkar, A., Leitherer, C., Robert, C. & Wilson, A. S. 1995, ApJ, 452, 549

Heckman, T. M., Gonzalez-Delgado, R., Leitherer, C., Meurer, G. R., Krolik, J., Wilson, A. S., Koratkar, A. & Kinney, A. 1997, ApJ, 482, 114 (H97)

Huchra, J. & Burg, R. 1992, ApJ, 393, 90

Kay, L. 1994, ApJ, 430, 196

Kinney, A. L., Bohlin, R. C., Calzetti, D., Panagia, N. & Wyse R. F. G. 1993, ApJS, 86, 5

Miller, J. S. & Goodrich, R. W. 1990, ApJ, 355, 456

Schmitt, H. R., Storchi-Bergmann, T. & Cid Fernandes, R. 1998, in press

Storchi-Bergmann, T., Cid Fernandes, R. & Schmitt, H. R. 1998, ApJ, 501, 94

Tran, H. D. 1995, ApJ, 440, 597

An Asymmetric Relativistic Model for FRII Radio Sources

T. G. Arshakian[1] and M. S. Longair

Cavendish Laboratory, Madingley Road, Cambridge, CB3 0HE

Abstract. An asymmetric relativistic model for FRII radio sources is described which takes account of both relativistic effects and intrinsic/environmental asymmetries to explain the observed structural asymmetry of their radio lobes. A key feature of the model is jet-sidedness, which can now be determined for about 80% of the FRII sources in the 3CRR complete sample. It is shown that a simple asymmetric relativistic model can account for a wide range of observational data, and that the relativistic and intrinsic asymmetry effects are of comparable importance. The mean translational speed of the lobes is $\bar{v}_{\text{lobe}} = (0.12 \pm 0.04)\,c$. The results are in agreement with an orientation-based unified scheme in which the critical angle separating the radio galaxies from the radio quasars is about 50°.

1. The problems with symmetric relativistic models

In the simplest kinematic model of FRII radio sources (Ryle & Longair 1967), the two components move out symmetrically from the central active galactic nucleus at the same velocity. Structural asymmetries of the core–hotspot angular distances are attributed to the difference in light travel time from each hotspot to the observer. Many studies have been made of the probability distribution of the velocities of the radio source components from the observed distribution of the ratio of core–hotspot distances (Longair & Riley 1979; Katgert-Meikelijn et al. 1980; Banhatti 1980; Best et al. 1995). The mean velocity of advance of the hotspots was found to be $\geq 0.2c$, with a considerable spread about the mean velocity, values greater than $0.4c$ being found.

The simple model makes a number of testable predictions about the structure of the sources. For example, the lobe approaching the observer should be longer than the receding lobe. Many of these sources are now known to contain one-sided radio jets emanating from the nucleus and so, if the cause of this asymmetry is relativistic beaming, the lobe on the jet side should be longer than the counterjet lobe. Saikia (1984) showed that, in about half of the 36 quasars in his sample, the jet was on the shorter side. The same effect was also present in the smaller quasar samples selected by Bridle (1994) and Scheuer (1995). The latest jet detections in high-resolution VLA observations indicate that there is no tendency for the brighter, or only, jet to lie in the longer lobe for radio galax-

---

[1]On leave from Byurakan Astrophysical Observatory, Byurakan 378433, Armenia.

ies (Hardcastle et al. 1997). The statistics of 103 FRII radio sources in the 3CRR complete sample (Laing, Riley & Longair 1983) show that, in about one third of the sources (28 radio galaxies and 8 radio quasars), the jet is on the short side. And so, *the approaching side is not always the longer side.* A further problem arises if the mean space velocity, estimated in a symmetric model to be $\bar{v}_{sm} = (0.27 \pm 0.16)\,c$, is compared with that determined from spectral ageing arguments, $\bar{v}_{sa} = (0.13 \pm 0.08)\,c$ (Alexander & Leahy 1987; Liu et al. 1992). This discrepancy suggests that the velocities of lobes may have been overestimated.

It is not at all unexpected that there should be problems with the simple symmetric model. For example, McCarthy et al. (1991) concluded that the one-sidedness of the optical emission-line regions in the vicinity of the radio galaxy provides evidence that *environmental effects contribute to the structural asymmetries of the radio sources.* As noted by Best et al. (1995), it is probable that both relativistic and intrinsic asymmetries contribute to the observed properties of the FRII sources.

2. An asymmetric relativistic model

To make quantitative estimates of the relative contributions of the relativistic and intrinsic/environmental effects, we describe an *asymmetric relativistic model*, in which intrinsically asymmetric jets advance through a clumpy asymmetric environment at an angle θ to the line of sight (Figure 1). The asymmetries associated with both the environment and the intrinsic properties of the jets can be described by assuming that the velocity of the lobe in the jet direction v_j and that in the counterjet direction v_{cj} are different. We define the structural asymmetry of radio lobes as the *fractional separation difference* $x \equiv (r_j - r_{cj})/(r_j + r_{cj})$, which is

$$x = \frac{v_j - v_{cj}}{v_j + v_{cj}} + \frac{2}{c}\frac{v_j v_{cj}}{v_j + v_{cj}} \cos\theta. \tag{1}$$

In the symmetric relativistic model, where $v_j = v_{cj}$, the value of x is always positive, but in the asymmetric relativistic model, negative values can be found if the first term on the right-hand side of (1) is negative and of greater magnitude than the second. By inspection of (1), we see that qualitatively we expect. (i) the radio axis of FRII sources with a jet on the short lobe side ($x \leq 0$, hereafter $-$FRII sources) must lie close to the plane of sky, whilst FRII sources with a jet on the long side ($x \geq 0$, hereafter $+$FRII) can be observed at all angles; (ii) large positive values of x are found if the lobes with the highest speeds expand in a direction close to the line of sight; small negative values of x are found when the lobes with the lowest speeds expand close to the plane of the sky; (iii) in the case of the symmetric relativistic model, all the sources would be $+$FRII source, whereas in the purely asymmetric model, with no relativistic effect, there would be equal numbers of $+$FRII and $-$FRII sources. Thus, we can define an *asymmetry parameter*

$$\varepsilon = 1 - 2\,\frac{\text{Number}(-\text{FRII})}{\text{Number}(+\text{FRII})}, \tag{2}$$

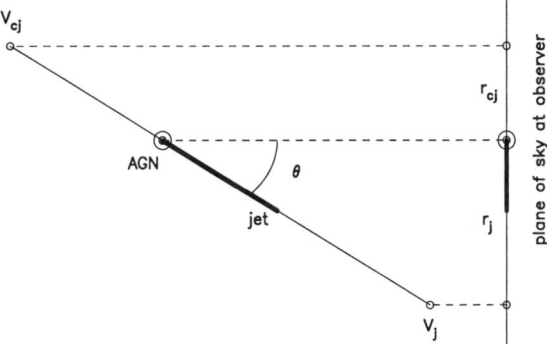

Figure 1. The asymmetric relativistic model of FRII radio sources.

which has the property that the relativistic effect is more important than the intrinsic/environmental effect if $0 < \varepsilon \leq 1$, and *vice versa*, if $-1 \leq \varepsilon < 0$. If $\varepsilon \sim 0$, the contribution of each effect is of comparable importance.

Furthemore, for relativistic sources we would expect that: (i) the mean linear size of $-$FRII sources should be greater than that of $+$FRII sources; (ii) there should be more two-sided jets in $-$FRII sources than in $+$FRII sources and more one-sided jets in $+$FRII sources than in $-$FRII sources; (iii) in orientation-based unified schemes, intrinsic asymmetries should be more significant for radio galaxies, while relativistic effects should be more important for quasars, that is, $\varepsilon_G < \varepsilon_Q$.

3. The properties of the FRII sources in the 3CRR sample

Of the radio sources in the complete 3CRR sample, jet-sidedness can be determined reasonably unambiguously for 103 FRII sources, values of x being found for the 71 FRII radio galaxies and 32 FRII radio quasars (for details of the procedures for estimating jet-sidedness, see Arshakian and Longair (1999)). The *asymmetry parameter* estimated for the joint sample has mean value $\varepsilon_{G+Q} = -0.07 \pm 0.02$, showing that relativistic effects and intrinsic/environmental asymmetry are of roughly equal importance in determining the structural asymmetries of the radio sources. The intrinsic asymmetry is more important for the radio galaxies, for which $\varepsilon_G = -0.3$, while relativistic effects are clearly more important for quasars for which $\varepsilon_Q = 0.33$. These results are in accord with the expectations of the asymmetric relativistic model.

The statistics of *one- and two-sided jets* are also consistent with the expectation of the model, namely that the percentage of two-sided jets is greater for $-$FRII (22%) than for $+$FRII radio galaxies (19%) and similarly for $-$FRII (50%) and $+$FRII radio quasars (13%). Since all the sources in the sample have jets, the reverse is true for one-sided jets: there are fewer one-sided jets in $-$FRII as compared with $+$FRII sources for both radio galaxies and quasars.

The *mean linear sizes* of 19 low luminosity ($P_{178} < 10^{27.7}\,\mathrm{WHz}^{-1}$) and 81 high luminosity ($P_{178} > 10^{27.7}\,\mathrm{WHz}^{-1}$) sources have also been studied. The expectation that the $-$FRII sources should be larger than the $+$FRII sources is found to be the case for the high-luminosity sources (Table 1). No such effect is seen for the low-luminosity sources. Furthermore, the differences in the mean linear sizes of −FRII and +FRII radio galaxies and radio quasars can be explained, if the latter are on average observed at smaller angles to the line of sight than the radio galaxies.

Table 1. The mean projected linear sizes of the sources in the sample with $P_{178} > 10^{27.7}\,\text{WHz}^{-1}$. The numbers of sources are in brackets.

FRII type RG/RQ	Mean linear size (kpc)	
	−FRII	+FRII
RG+RQ	290 (27)	220 (54)
RG	290 (19)	280 (30)
RQ	280 (8)	140 (24)

4. Expansion speeds and the unified scheme

In order to quantify the relative importance of intrinsic/environmental asymmetries as opposed to relativistic asymmetries, we can write the velocities v_j and v_{cj} as $v_j = v_0 + v_d$ and $v_{cj} = v_0 + v_{cd}$, where v_0 is the average velocity of lobes and v_d and v_{cd} are caused by intrinsic/environmental asymmetries on the jet and counterjet sides respectively. For the sake of illustration, we assume that the distribution function of v_0 is of gaussian form and that the dispersion in the intrinsic/environmental velocity dispersion is another independent gaussian. It is then straightforward to work out the expected distributions of x, selecting mean speeds and velocity dispersions to agree with the observed value $\varepsilon_{G+Q} = -0.07$. For the joint sample of radio galaxies and quasars, we find that the mean expansion speed of the lobes and its standard deviation are $\bar{v}_{\text{lobe}} \sim (0.12 \pm 0.043)\,c$, which is similar to the results of spectral ageing analyses, $(0.13 \pm 0.08)\,c$, significantly less than the mean speed estimated from the symmetric model. As noted above, the mean speeds are overestimated in the symmetric model because structural asymmetries are attributed entirely to the light travel times, whereas our improved analysis shows that, in fact, both relativistic effects and intrinsic asymmetry contribute equally to the structural asymmetry.

In orientation-based unified schemes for radio galaxies and quasars (Barthel 1989), these are the same class of object but viewed at different angles to the line of sight. A critical angle $\theta_c \sim 45°$ separates radio galaxies from the radio quasars. By modelling the observed distribution function of apparent velocities for different critical angles, the predicted asymmetry parameter for both radio galaxies and quasars can be estimated (Table 2). The observed values are $\varepsilon_G = -0.3$ and $\varepsilon_Q = 0.33$. Excellent agreement between the predicted and observed asymmetry parameters is found for $\theta_c \sim 50°$ as can be seen from Table 2. We conclude that the asymmetric relativistic model is in satisfactory agreement with an orientation-based unified scheme with $\theta_c \sim 50°$.

Table 2. Predicted asymmetry parameters for radio galaxies and quasars according to an orientation-based unified scheme for the asymmetric relativistic model.

θ_c^o	ε_G	ε_Q	θ_c^o	ε_G	ε_Q
10	−0.08	0.45	50	−0.27	0.33
20	−0.11	0.39	60	−0.35	0.27
30	−0.12	0.36	70	−0.46	0.17
40	−0.16	0.32	80	−0.81	0.06

5. Conclusions

- The proposed asymmetric relativistic model of FRII radio sources can take account of both relativistic and intrinsic/environmental asymmetry effects. The predictions of the model are in agreement with observational data, and the observational data indicate that both effects are of comparable importance.

- The mean expansion speed of lobes is about $(0.12 \pm 0.043)\,c$, a value consistent with values found from synchrotron ageing arguments.

- The data for radio galaxies and quasars are consistent with orientation-based unification schemes in which the critical angle separating the two types of object is $\theta_c \approx 50°$.

Acknowledgments. We are very grateful to Julia Riley for helpful comments and kindly supplying up-to-date information about the 3CRR sample, and to Andrew Blain for his critical reading of drafts of this paper.

References

Alexander, P., Leahy J. P., 1987, MNRAS, 225, 1
Arshakian, T., Longair, M.S., 1999. MNRAS, (in preparation)
Banhatti, D. G., 1980, A&A, 84, 112
Barthel, P. D., 1989, ApJ, 336, 606
Best, P. N., Bailer, D. M., Longair, M. S., Riley, J. M., 1995, MNRAS, 275, 1171
Bridle, A. H., et al., 1994, AJ, 108, 766
Hardcastle, M. J., et al., 1997, MNRAS, 288, 859
Katgert-Meikelijn, J., Levi, C., Padrielli, L., 1980, A&AS, 40, 91
Laing, R. A., Riley, J. M., Longair, M. S., 1983, MNRAS, 204, 151
Longair, M. S., Riley, J. M., 1979, MNRAS, 188, 625
Lui, R., Pooley, G., Riley, J. M., 1992, MNRAS, 257, 545
McCarthy, P. J., van Breguel, W. J. M., Kapahi, V. K., 1991, ApJ, 371, 478
Ryle, M., Longair, M. S., 1967, MNRAS, 136, 123
Saikia, D. J., 1984, MNRAS, 209, 525
Scheuer, P. A. G., 1995, MNRAS, 277, 331

Luminosity Correlation of the X-Ray Selected Radio-Loud AGNs

Q. Yuan[1], J. Wu[2], K. Huang[1]

[1] *Department of Physics, Nanjing Normal University, Nanjing 210097, China*

[2] *Department of Astronomy, Beijing Normal University, Beijing 100875, China*

Abstract.
This paper presents a test of the luminosity correlation of the X-ray selected radio-loud Active Galactic Nuclei (AGNs), based on a large sample constructed by combining our cross-identification of southern sky sources with the radio-loud sources in the northern hemisphere given by Brinkmann et al. (1995). All sources were detected both by the ROSAT All-Sky Survey and the radio surveys at 4.85 GHz. The broad band energy distribution confirms the presence of strong correlations between luminosities in the radio, optical, and X-ray bands which differ for quasars, seyferts, BL Lacs, and radio galaxies. The tight correlations between spectral indices α_{ox} and monochromatic luminosities at 5500 Å and 4.85 GHz are also shown.

1. Introduction

The luminosity correlation of the Active Galactic Nuclei (AGNs) has been an appealing subject. It is widely appreciated that the correlations of the AGNs and active galaxies among luminosities emitted from radio, optical and X-ray energy bands can give insight into the relevant physical processes in AGNs and serve as a qualitative test for existing central engine models of AGNs. The subgroup of X-ray selected radio-loud AGNs is particularly useful in understanding the unification schemes (Barthel 1989; Padovani & Urry 1992) and cosmological and evolutionary effects for AGNs, since the radio-loud AGNs with strong X-ray emission are the most luminous objects with large look back times in the universe.

After the launch of the ROSAT satellite, the correlations of the ROSAT detected sources with existing compilations of radio catalogues were extensively investigated. Brinkmann et al. (1995) studied the bulk properties of previously optically identified AGNs which were detected by the ROSAT All-Sky Survey (hereafter RASS) (Voges 1992), by using the source list from the Green Bank 4.85 GHz Survey (refered to as GB6) (Condon et al. 1989). However, all radio galaxies and AGNs are limited to the northern hemisphere, and the sample seems to be not large enough in size enough to claim the luminosity correlations.

This paper presents a test of the luminosity correlations of the ROSAT selected radio-loud AGNs. The AGN sample is constructed by combining our cross-identification list (Yuan et al. 1998) with the sources given by Brinkmann et al. (1995). Based on the larger sample, the luminosity correlations previously proposed can be verified in more details, and the implications relevant to AGN unification schemes can be also observed.

2. The Sample

The Parkes-MIT-NRAO (PMN) 4.85 GHz Radio Survey was conducted by Griffith & Wright (1993; 1991), using the Parkes 64m radio telescope with the NRAO multibeam receiver, with the same sensitivity as that of the GB6 survey. The PMN survey covers the southern hemisphere ($-87.^{\circ}5 < \delta < +10°$) with a limiting flux of about 35 mJy at 4.85 GHz frequency, unlike the GB6 survey (also at the same frequency) covering a region of $0° < \delta < +75°$. A cross-identification of the southern-sky objects with the RASS and PMN surveys yields a list of 642 sources. Merely 311 (52%) coincidences were previously optically identified as extragalactic objects, using the NASA/IPAC Extragalactic Database (NED). A significant fraction of them are radio galaxies (129) and AGNs (183) (including 119 QSOs, 39 Seyferts, and 25 BL Lacs). The fluxes emitted from radio, optical, and X-ray bands and the other important parameters, such as redshift (or radial velocity) and power law index in soft X-ray band, are compiled in Yuan et al. (1998), in which the statistical properties of our source list can be also found.

A similar list of the northern-sky objects was given by Brinkmann et al. (1995). The combination of these two source lists generates a large sample, containing 367 quasars, 86 seyferts, 80 BL Lacertae objects, and 282 radio galaxies, which allows an investigation on broad band energy distribution and other possible correlations among the spectral indices and luminosities for various types of AGNs.

3. Luminosity Correlation

Our sample includes a large number of various types of AGNs, which allows an investigation of the luminosity correlations for AGNs. The derivation of luminosity from the observed flux is given by Schmidt & Green (1986).

The correlation between the luminosities of X-ray and radio bands which has been emphasized in many previous studies strongly suggests a similar origin of the radiation in all radio sources. Figure 1b confirms the correlation between the monochromatic X-ray luminosity at 2 keV and radio luminosity at 4.85 GHz. The galaxies populate the low luminosity region and quasars locate the high luminosity end. The regression analysis for subsamples of galaxies and quasars shows that both of the linear correlation coefficients are significant at very high confidence level. The hypothesis that the slopes of the linear regressions are the same for galaxies and quasars can be ruled out with nearly 100% confidence. Compared with the previous studies by Fabbiano et al. (1984), our correlation is somewhat similar to that of the flat spectrum objects, hinting that radio-loud objects at high frequency band (4.85 GHz) seem to be enriched with flat radio spectrum objects. The quasar population seems to have a good proportionality between X-ray and optical emission, indicating a common radiation mechanism.

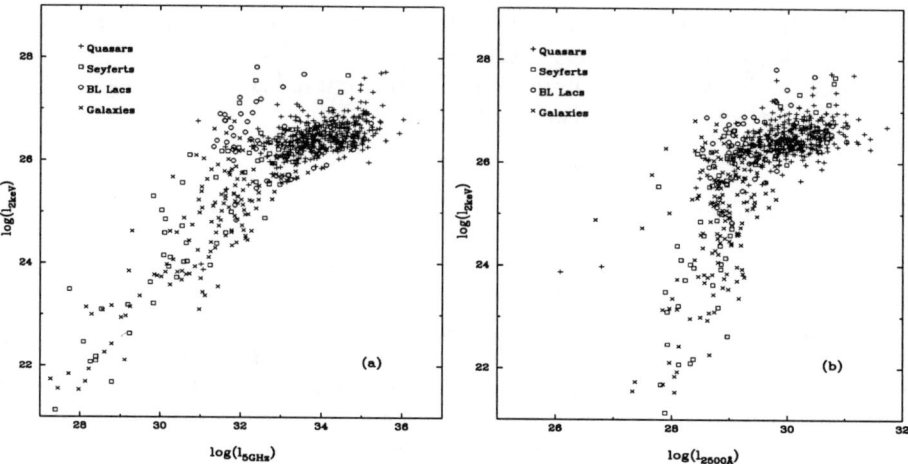

Figure 1. Monochromatic X-ray luminosity at 2 keV as a function of the monochromatic luminosites at 4.85 GHz (left panel) and at 2500 Å(right panel) (in erg s^{-1} Hz^{-1}), for all types of objects in our sample.

Figure 1b shows the monochromatic X-ray luminosity at 2 keV as a function of the optical luminosity at 2500 Å. However, it can be seen that the distribution of galaxies and Seyferts are completely different from that of quasars. Quasars have a typical monochromatic X-ray luminosity of 10^{27} erg s^{-1} Hz^{-1}, and cover a relatively wide region of optical luminosity. The slopes of the linear regressions for galaxies and quasars are significantly diverse.

Note that it is still uncertain about the extent of selection effects in observation and biases in measurements.

4. Correlations Between Spectral Index α_{ox} And Luminosities

Based mainly on geometrical and orientation arguments, unification schemes have been proposed for more than a decade to explain the connection of AGN properties. It is of great interest to analyze the probable correlations between the optical-to-X-ray spectral index α_{ox} and the monochromatic luminosities, since the spectral index α_{ox} may be sensitive to the structures and physical conditions of the outer regions, spanning from the accretion disk to the broad emission-line region (BELR), of different types of AGNs. The two-point spectral index α_{ox} is defined as: $\alpha_{ox} = -log(S_{\nu_x}/S_{\nu_{opt}})/log(\nu_x/\nu_{opt})$. The diagram of α_{ox} versus monochromatic luminosities might provide some clues to physical conditions of the BELR.

Figure 2 gives a global picture of correlations between the spectral index α_{ox} and luminosities emitted at 4.85 GHz (left panel), 5500 Å(middle panel), and 2 keV (right panel). It is interesting that radio galaxies adn Seyferts have a decreasing tendency systematically with increasing luminosities, which is completely different from quasars. A good linear correlation can be found for galaxies and Seyferts in Fig. 2c. As a sharp contrast, the quasars show a tight correlation between spectral indices α_{ox} and optical luminosities in Fig. 2b. Such a good correlation in Fig. 2b can be used to predict the X-ray luminosities of quasars according to their optical emissions. Similarly, Fig. 2c can be utilized to estimate the optical luminosities of radio galaxies and Seyferts with given X-ray emissions. Our results hint that the structure and physical conditions in outer regions are significantly different from galaxies to quasars.

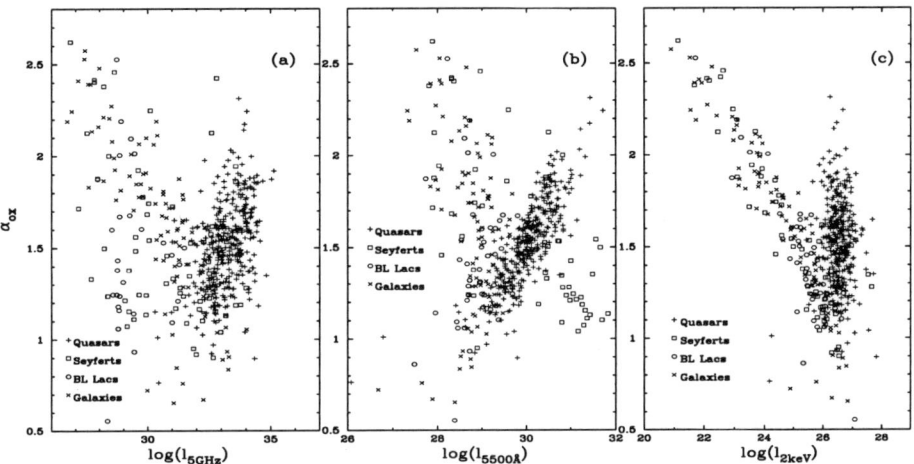

Figure 2. Plots of the spectral indices α_{ox} versus monochromatic luminosities at 4.85 GHz (left panel), 5500 Å (middle panel), and 2 keV (right panel), for all types of objects in our sample.

5. Discussion

To understand the intrinsic mechanisms and the mutual relations between the different categories of extragalactic objects, we constructed a large sample of X-ray selected radio-loud AGNs. This sample is inhomogenous and the selection is mainly a result of the X-ray flux limitation. The luminosity correlations given above strongly suggest that the radio, optical, and X-ray emission are essentially physically connected. Browne & Murphy (1987) presented a model of a 'canonical quasar' where the optical and X-ray emission both contain beamed and unbeamed contributions. The nuclear origin of the X-ray emission seems to be established (Fabbiano et al. 1984) even for radio galaxies, which indicates that the apparently different source types are intrinsically the same kind of objects, but seen under different geometric viewing conditions.

The impressive correlation between radio and X-ray luminosities, seen from Fig. 1a, confirms a similar origin for the radiation in all radio sources. However, it should be not easy for orientation-related unification schemes of AGNs to accommodate the fact that the slope of the correlation for radio galaxies and Seyferts is not so compatible with that for quasars.

We also find a tight correlation between spectral indices α_{ox} and optical luminosities for quasars, which is significantly different from the characteristics of the radio galaxies and Seyferts. According to the definition of spectral index α_{ox}, for quasars with a typical value of X-ray luminosity, the larger the optical luminosity is, the steeper the energy distribution between the optical and X-ray bands. The correlation might indicate that the core regions of quasars emit energy at X-ray energies of the same magnitude. On the other hand, the spectral indices α_{ox} of radio galaxies and Seyferts are found to be counter-correlated with X-ray luminosities. These findings do not seem to be the result of observational selection effects and measurement bias.

Acknowledgments. This research used the NASA/IPAC Extragalactic Database (NED) which is developed by the Jet Propulsion Laboratory, California Institute of Technology, under contract with the National Aeronautics and Space Administration. We also thank the ROSAT group for the release of All-Sky Survey database. This work was partly supported by the National Climbing Program and National Natural Science Fundation of China.

References

Barthel, P.D.: 1989, ApJ 336, 319

Brinkmann, W., Siebert, J., Reich, W., Fürst, E., Reich, P., Voges, W., Trümper, J., & Wielebinski, R.: 1995, A&AS 109, 147

Brown, I.W.A., Murphy, D.W.: 1987, MNRAS 226, 601

Condon, J.J., Broderick, J.J., & Seielstad, G.A.: 1989, AJ 97, 1064

Fabbiano, G., Miller, L., Trienchieri, G., Longair, M., Elvis, M.: 1984, ApJ 277, 115

Griffith, M.R., et al.: 1991, Proc. Astron. Soc. Australia, 9, 243

Griffith, M.R., Wright, A.E.: 1993, AJ 105, 1666

Padovani, P., Urry, M.C.: 1992, Physics of Active Galactic Nuclei. (eds. Duschl, W.J., Wagner, S.J.), Springer, Berlin, p.643

Schmidt, M. & Green, R.F.: 1986, ApJ 305, 68

Voges, W.: 1992, in Proc. of the ISY Conference "Space Science", ESA ISY-3, ESA Publications, 9

Yuan, Q., Wu, J., Yuan, W., Hu, F.: 1998, Publ. of Purple Mountain Obs., 17, No.1, p1

Black Hole Masses and Unification of Seyferts

R. Bachev, G. T. Petrov, L. Slavcheva and B. Mihov

Institute of Astronomy, Sofia, Bulgaria, e-mail: bachevr@astro.bas.bg

The most commonly invoked power source of Active Galactic Nuclei (AGN) is accretion of galactic gas (probably through a disk) onto a supermassive black hole in the center of the nucleus (Rees 1984). As is well known, a black hole is completely defined by its mass and angular momentum. The *unification scheme* of active galaxies assumes that two known *Seyfert types* (Sy1 and Sy2) are not intrinsically different, i. e. their black hole masses, accretion rates and the whole internal structures are identical (Antonucci 1993) and observed differences are due just to a different orientation to the observer of the axisymmetrical central structure (central engine, BLR and thick torus, shadowing broad lines from some directions).

To test this unification scheme we derived black hole masses for 143 Seyfert galaxies using an *emission-line (or dynamical)* method. This method is based on the assumption that the emitting gas is gravitationally bound, then the $FWHM$ of some line defines the Keplerian velocity (u) of emitting clouds, whose average distance (R) from the center could be approximately estimated from the line luminosity (L), gas density and covering factor. Ionisation equilibrium leads to: $L_{LINE}~R^2 J N n$, where J is the line emissivity, ω is the covering factor of the region and N column density. Velocity and distance of the gravitationally bound clouds determine the central mass (M): u FWHM $(GM/R)^{0.5}$. This central mass is probably dominated by the black hole mass (MBH???) or is proportional to it at least. The [OIII] 5007 narrow line is used (data from Whittle 1992, Nelson & Whittle 1995), as it is believed the narrow line regions are identical in all Seyfert types. Using that typically $n = 10^{4-5}$ cm^3, $N = 10^{20-21}$ cm -2, w = 0.1-0.01 and J[OIII] $= 1.2 \times 10^{-24}$ erg cm^{3}s^{-1} one can derive (Wandel & Mushotzky 1986):

$$\text{Log(MBH)} = 0.5\text{Log}(L[\text{OIII}]) + 2\text{Log}(u) - 17.7, \qquad (25)$$

where MBH is in Solar masses, L[OIII] - in erg s^{-1} and u in km s^{-1}. Most of obtained black hole masses are between 10^7 and 10^9 M$_\odot$. Mass distributions (see the figure) of all types do not show a statistical difference significant enough to separate nuclei into 2 groups based on this parameter, i. e. Seyfert types are indistinguishable through their black hole masses. This result indirectly supports the unified scheme.

Although the sample is not complete, it is interesting to note that Seyfert 2 galaxies with hidden Seyfert 1 nuclei (7 such "hidden Seyfert 1's" are found in the list of AGN used, see for details Antonucci & Miller 1985) possess the highest masses we derived - about 10^9 M$_\odot$. Thus, these objects fill the high mass tail of the Seyfert 2's distribution. By their masses, they could be well distinguished from both Seyfert 1 and Seyfert 2 galaxies, probably forming a separate subclass. Of course, another possible explanation is the narrow line clouds of these objects

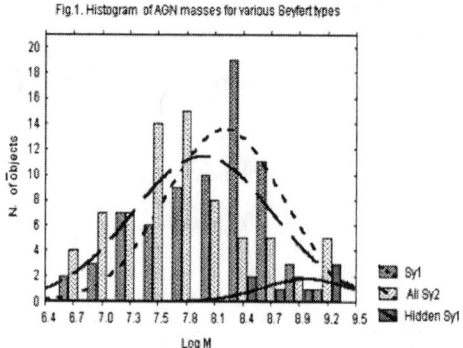

Figure 1. Distributions of black hole masses of Seyfert 1, 2 and "Hidden 1" types. Parameters of normal fits applied to the data are respectively 8.20.6, 8.00.7 and 9.00.4.

are nongravitationally accelerated to velocities quite above the Keplerian one, which may lead to significant overestimation of the mass of the central object. In this case, the emission-line method is inapplicable for the mass estimation. Radio luminosities of these hidden Seyfert 1's are unusually high as well (Moran & Halpern 1992). Anyway, it seems that these objects are not ordinary Seyfert galaxies and their place in the unified scheme could be quite special.

References

Antonucci R.R.J.,1993, ARA&A, [31], 473

Antonucci R.R.J.& J.S. Miller, 1985, Ap.J., [297], 621

Bachev R. & Petrov G. T., 1998, to be published in C. R. l'Acad. Bulg. Sci., [7/8].

Nelson C. & M. Whittle, 1995, Ap.J.S.S., [99], No.1

Moran E. C. & Halpern J. P, in "The Nature Of Compact Objects in AGN", eds. Robinson A., Terlevich R, Cambridge, 1992

Rees M, 1984, ARA&A,[22],471

Wandel A. & R.F. Mushotzky, 1986, Ap.J., [306], L61

Whittle M., 1992, Ap.J.S.S., [79], No.1

Bardeen-Petterson Effect and Broad HI Profiles of Seyferts

Rumen Bachev

Institute of Astronomy, BAS, 72 Tsar. Shose, 1784 Sofia, Bulgaria.
Email: bachevr@astro.bas.bg

A thin accretion disk could be not only the energy source of AGN but also the matter producing broad emission lines of Seyfert 1 type nuclei (Dumont & Collin-Souffrin, 1990 B). A possible mechanism for this is reprocessing of central hard X-ray radiation by the outer (at 10^{2-5} RG, RG is the *Schwarzschild radius*), low-temperature regions of the disk. This mechanism is effective enough especially if the disk is a *non-planar structure* (*a warped or twisted disk*), when the outer parts could be directly seen from the centre. An accretion disk around a *Kerr* black hole could be twisted if the angular momentum of the accreting gas is initially not aligned with the rotation axis of the hole. Due to the differential *Lense-Thirring precession* of orbits around a Kerr black hole, a viscous disk is a steady but non-planar structure. This is the well-known *Bardeen-Petterson effect* (Bardeen & Petterson, 1975). Near the hole, at distances R<RBP, where RBP is the *Bardeen-Petterson radius*, the flow is aligned with the equatorial plane of the black hole, while at larger distances it is tilted to its initial orbital plane (Fig. 1).

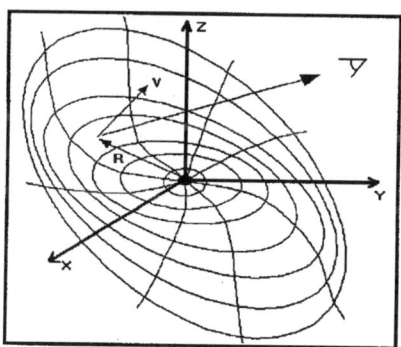

Figure 1. The illumination of a non-planar disk. Each gas volume reemits a part of the ionising continuum if direct illumination is possible. To a distant observer, the intensity and frequency shift of this emission depend on the velocity (**V**), the orientation and intercepted amount of hard radiation as well as on the reprocessing properties of the volume. The twisting structure of the disk is described by two Eulerian angles - slowly varying functions of ***R*** and the initial inclination angle. The solution, describing disk twisting, derived by Hatchett et al. (1981) is used here.

If it exists, an accretion disk in AGN should be non-planar in most cases as there is no reason to suppose that the angular momentum of accreting matter and the black hole spin axis are always coplanar. We studied numerically the profiles of the broad low-ionisation lines (H_β) produced by an irradiated twisted disk.

Reprocessing properties of the disk layers are taken into account following closely Collin-Souffrin and Dumont (1990). The central hard X-ray source is proposed to be point-like, at zero height above the disk plane. Since the twisted disk is *anisotropically* irradiated and just a part of the disk emits towards the observer's direction (Fig. 2), the profiles of the broad lines are generally *non-symmetric* and *frequency-shifted* with respect to the systemic velocity (Fig. 3). If the disk is transparent (a case not considered here), emission from the lower disk surface may come and profiles would be double-peaked and almost symmetric.

Because such frequency displaced, asymmetric profiles are seldom observed (several objects with similar profiles are known, for instance - 3C 227 and Mkn 668, Eracleous & Halpern, 1994), one may conclude that the bulk of the line emission of AGN arises from some more or less spherical or conical structure (a system of clouds or star atmospheres, jet, etc.) but not from a disk. In this case, the presence of an illuminated twisted accretion disk will cause only line asymmetry or a frequency displaced secondary peak. The absence of such asymmetry or displaced peaks may probably be an indication for the presence of non-rotating black holes in the centres of active galaxies when the disks are planar and the disk emission symmetric and double-peaked (Dumont & Collin-Souffrin, 1990 B). Asymmetrical double-peaked profiles could be observed from a warped disk if the ionising source is located far above the disk plane or a hot diffuse medium reflecting hard radiation back towards the disk is present (models not considered here). In this case the central parts of the disk will be illuminated isotropically and will produce the main double-peaked profile.

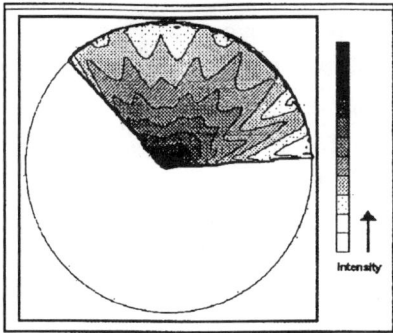

Figure 2. A schematic map of the H_β radiation flux from a self-irradiated warped disk. Standard thin Shakura-Sunyaev α-disk model is accepted. The initial inclination angle of the flow is 10°. Relative units are used to represent the emission-line flux from the disk. Black hole mass is 10^8 $M_\odot$, accretion rate 0.1 $M_\odot$ yr^{-1}. Dimensionless spin parameter $a=1$, viscosity parameter $\alpha=0.1$. Bardeen-Petterson radius (the radial distance where the infall time scale is comparable to the precession time scale) is about 300 R_G. Only the inner part ($R < 100$ R_{BP}) of the disk is presented here.

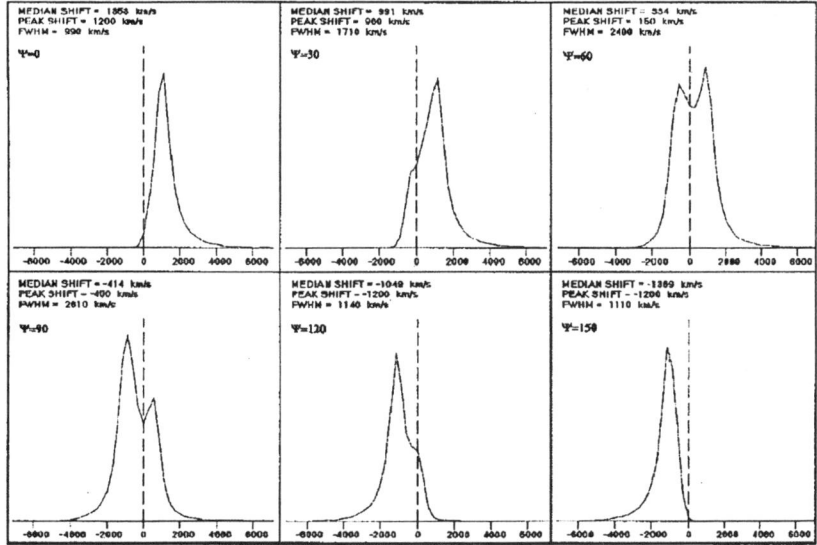

Figure 3. Line profiles of H_β emission from the disk on Fig. 2 and different orientation angles. The observer is tilted at an angle 30° with respect to the black hole equatorial plane. The azimuthal angle of the observer (Ψ) along the disk plane is respectively 0, 30, ..., 150°. The disk is extended up to 100 R_{BP}.

References

Bardeen J. M., Petterson J. A., 1975, ApJ, [195], L65
Collin-Souffrin S., Dumont A. M., 1990, A&A, [229], 292
Dumont A. M., Collin-Souffrin S., 1990, A&A, [229], 302, A
Dumont A. M., Collin-Souffrin S., 1990, A&A, [229], 313, B
Eracleous M, Halpern J. P.,1994, ApJS, [90], 1
Hatchett S. P., Begelman M.C. & Sarazin C.L., 1981, ApJ. [247], 677

Estimating the Number of Broad-Line Region Clouds

M. Dietrich

Landessternwarte Heidelberg, Königstuhl, D-69117 Heidelberg, Germany

The observed emission-line spectrum of active galactic nuclei is consistent with cold dense gas photoionized by a central continuum source (cf. Ferland & Persson 1989). The estimate of the filling factor of the line-emitting region yields f$\sim 10^{-6}$ only (cf. Osterbrock 1993) and motivated the picture of numerous clouds moving around a central black hole. Early estimates of the number of BLR clouds indicated a lower limit of 10^4 to 10^5 individual clouds (Capriotti et al. 1981; Atwood et al. 1982). Recently, Arav et al. (1998) estimated the number to be at least of the order of 10^7 discrete emitters. But the number and the nature of these clouds are still unknown.

A finite number of clouds will introduce a characteristic microstructure in the line profile which can be separated from the random observational noise. Hence, we observed 3C 273 with high spectral resolution ($\Delta v \sim 9$ km s^{-1}) with the NTT at La Silla, ESO. Proper parts of the outer wings were selected taking into account instrumental effects, presence of narrow emission lines, and contamination by atmospheric features. The line profiles were prepared following the method suggested by Tonry & Davis (1979). The resulting residuals of the Hα and Hβ emission lines were analysed using the interpolating cross-correlation method (cf. Gaskell & Peterson 1987). The significance of a cross-correlation peak was compared with the 3-σ threshold.

We studied simplified models in order to analyse the observed line profiles. We computed models with cloud widths of FWHM = 15 km s^{-1} to 100 km s^{-1}. The cloud ensembles ranged from 10^4 to 10^8 discrete identical emitters. The distribution of the ensemble was given by clouds moving on randomly orientated Keplerian orbits around a massive black hole ($M_{bh} = 10^8\, M_\odot$). The distance of the clouds to the central energy source was restricted to $r_{in} = 0.1$ light days and $r_{out} = 100$ light days. Photon noise was added to the residuals to simulate S/N ratios comparable to the observations. The simulated profiles were treated in the same way as the real observations.

The blue wing of the Hα line profile of 3C 273 is shown together with simulations of cloud ensembles with $N_{cl} = 10^8$ (Fig. 1). The cumulative profile of clouds with FWHM = 15 km s^{-1} still shows too much structure. The profiles become smoother for significantly larger width (FWHM = 100 km s^{-1}) of the clouds.

The ICCF of the real observation is displayed together with the corresponding ICCFs of the model residuals. No significant peak is visible in the ICCF calculated with the observed Hα and Hβ line profiles of 3C 273 (Fig. 1). Even for large cloud numbers ($N_{cl} = 10^8$) a significant ICCF peak can be expected for clouds with FWHM $\simeq 15$ km s^{-1}. The situation changes if the line width of an

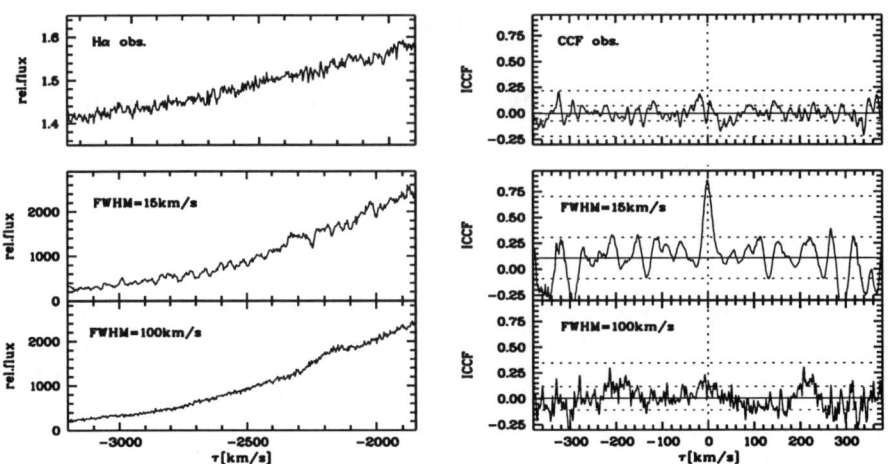

Figure 1. The observed Hα line profile of 3C 273 (top left side) and for cloud ensembles ($N_{cl}=10^8$) with FWHM = 15 km s^{-1} (middle) and FWHM = 100 km s^{-1} (bottom). The corresponding ICCFs are displayed in the three panels at the right side. The dashed lines represent the 1-σ and 3-σ thresholds.

individual cloud is significantly larger than 15 km s^{-1}. In the case of 10^8 clouds with FWHM = 100 km s^{-1} no significant ICCF peak is detectable.

This result indicates that the number of BLR clouds is larger than $N_{cl} = 10^8$ assuming a FWHM of 15 km s^{-1} for an individual cloud, i.e. the profile width is caused entirely by thermal broadening ($T_e \simeq 2\,10^4$ K). But the number of clouds might be of the order of 10^8 if the width of a discrete emitter is of the order of 100 km s^{-1} due to optical depth effects (Davidson & Netzer 1979) or electron scattering (e.g. Bottorff et al. 1997). However, the large number of clouds provides evidence that bloated star models can be ruled out if only red giant stars are taken as sources for the BLR clouds.

Acknowledgments. This work has been supported by DFG grant SFB 328.

References

Arav, N., et al. 1998, MNRAS, 297, 990
Atwood, B., Baldwin, J.A., & Carswell, R.F. 1982, ApJ, 257, 559
Bottorff, M., et al. 1997, ApJ, 479, 200
Capriotti, E., Foltz, C., & Byard, P. 1981, ApJ, 245, 386
Davidson, K. & Netzer, H. 1979, in Rev. Mod. Phys., 51, 737
Ferland, G.J. & Persson, S.E. 1989, ApJ, 347, 656
Gaskell, C.M. & Peterson, B.M. 1987, ApJS, 65, 1
Osterbrock, D.E. 1993, ApJ, 404, 551
Tonry, J. & Davis, M. 1979 AJ, 84, 1511

Diffusion Mechanism of Radiation of a Charged Particle on the Randomly Spaced Dust Grains in the X-rays

Zh. S. Gevorkian

Institute of Radiophysics and Electronics, Ashtarak-2, 378410, Armenia

V. V. Hambaryan and A. A. Akopian

Byurakan Astrophysical Observatory, Byurakan, 378433, Armenia

1. Diffusion radiation in the X-ray region

The theory of diffusion radiation of a charged particle on the fluctuations of the dielectric constant developed by Gevorkian can be explained as follows:

A charge moving in a medium creates an electromagnetic field (pseudophoton) which is scattered on the fluctuations of the dielectric constant (here, dust particles) and converted into radiation. In the wavelength region ($l \ll \lambda \ll L$) (l is the mean free path of the photon in the medium, L is the characteristic size of the system) the main mechanism of the radiation is the diffusion of the pseudophoton (Gevorkian & Atayan 1990, Gevorkian 1992, Gevorkian 1993).

For relativistic energies of charged particles ($\gamma > 10^3$) and characteristic sizes of dust particles ($a \sim 10^{-5} - 10^{-6} cm$), the characteristic frequency of diffusion radiation ($\omega_0 \sim \frac{c\gamma}{a}$) lies in the X-ray region. In this case, the intensity of diffusion radiation of one charged particle is given by (Gevorkian et al. 1998):

$$I^D(\omega) = 6\pi e^2 \frac{l_{in}(\omega)}{l^2(\omega)} \left(2\ln\frac{c\gamma}{a\omega} - 1\right) \quad (26)$$

where l_{in} is the inelastic mean free path:

$$l_{in} = \frac{1}{n_H \left(\sigma_H + \frac{\sum n_i \sigma_i}{n_i}\right)} \quad (27)$$

where n_H is the concentration of H atoms, σ_H is the photoionization cross section of H atoms, n_i and σ_i are concentration and cross section of the i-th element.

Thus the frequency dependence of l_{in} and I^D is determined by the frequency dependence of the photoeffect cross section. The elastic scattering mean free path is equal to:

$$l(\omega) = \frac{9\omega^2 c^2}{4\pi n a^4 \omega_p^4} \quad (28)$$

where n is the number density of dust particles, ω_p is plasma frequency.

The contribution of single scattering radiation is equal to (Gevorkian et al. 1998):

$$I^0(\omega) \approx 2\pi e^2 \frac{1}{l(\omega)} \left(2\ln \frac{c\gamma}{a\omega} - 1\right). \qquad (29)$$

Consequently, radiation emitted when charged particles pass through a dusty cloud could be much greater (by order of l_{in}/l) if one takes into account the diffusion term of radiation. We have estimated that in the range $2-10keV$, l_{in}/l can be at least $10-100$. In dense dusty clouds l_{in}/l may reach up to 10^3.

Using photoionization cross sections calculated by Brown & Gould (1970) we have obtained the theoretical spectrum of diffusion radiation. The spectral index for the range $2-10keV$ is equal to -0.8. In the general case, the expected indices lie in the range from -0.5 to -1.3 (Gevorkian et al. 1998).

2. AGNs as diffusion X - ray radiation sources?

A review of observational data obtained in the range $1-10keV$ let us select AGNs as possible sources of X-ray diffusion radiation. The following observational evidence supports our selection:

- Extended X-ray emission (up to $1Kpc$ in size) has been found around the nuclei of AGNs.

- In the nuclei of AGNs are all the necessary conditions (dust and relativistic electrons) for creating diffusion X-ray radiation. Most AGNs are a power source of IR and/or radio emission. There are some correlations between IR/radio and X-ray radiation. There is also a tight correlation between radio and X-ray morphologies.

- The spectral indices in $2-10keV$ of an overwhelming majority of AGNs lie between -0.4 and -1.1 with the mean ~ -0.8.

- In the expected range of diffusion radiation ($\sim 1-10keV$) observed spectra of AGNs often differ from the spectra of neighboring wavebands.

- The estimates (Gevorkian et al. 1998) of the X-ray luminosities of AGNs in the framework of a simple physical model are consistent with observational data.

References

Brown, R.L., & Gould,R.J., 1970, Phys.Rev.,D1, 2252
Gevorkian, Zh.S. 1992, Phys.Lett., A162, 187
Gevorkian, Zh.S. 1993, Radiofizika, 36, 36
Gevorkian, Zh.S. & Atayan, S.R. 1990, Phys.Lett., A144, 273
Gevorkian, ZH.S.& Atayan, S.R. 1990, Sov.Phys. JETP 71 (5), 862
Gevorkian, Zh.S., Hambarian,V.V., & Akopian, A.A., 1998, Astrofizika, 41, 443

General Discussion of Accretion Disks

Gurzadyan, Vahagn G.

Department of Theoretical Physics, Yerevan Physics Institute, Yerevan, 375036, Armenia. e-mail: gurzadya@lxz.yerphil.am

Even 25 years after the Shakura-Sunyaev seminal paper on the α-disk, we cannot claim that we have a reliable theory of accretion disks in galactic nuclei. Why? Because the problem is extremely complicated, it is essentially nonlinear and contains a number of parameters (i.e. is many-dimensional). The key point is whether it is possible to determine the magneto-hydrodynamical viscosity self-consistently, i.e. as a function of parameters of the disk - the temperature, matter and radiation densities, magnetic field, radius, etc., both in the radiation dominated and matter dominated regions. Another class of fundamental problems concerns the stability of the disk; Krolik mentioned only one instability - in the radiation dominated region, but there are many other types of instabilities which are quite sensitive to the physical conditions in the disk, for example, to the anisotropy of the ion pressure in the outer regions and possible electron-positron pair production near the inner edge of the disk. The other problems include those of the radiative transfer within the disk in various conditions, Comptonization of the outgoing radiation, radiation reflections by the desk, etc. Therefore it is not suprising that one can 'explain' almost whatever he wants - spectra, variability, jets, wind, etc., by proper fit of the 'free' (which are never free) parameters and ignoring the instabilities and so on.

Thus, both the apparent coincidence and disagreement (as claimed here by Sulentic) of certain observational data with the α-disk predictions, are equally not sufficient to conclude the existence or absence of accretion disks in those objects. Moreover, even the disk-type structures directly observed in certain galactic nuclei (they are extremely impressive!) cannot be interpreted uniquely as the accretion disks we are speaking about. Not every Keplerian disk has to be an accretion disk.

There is no hope that a complete self-consistent theory of turbulent accretion disks can be created in the near future. Thus, I see the following way for further development of the theory. Any nonlinear many-dimensional dynamical system (in particular case, given via a system of differential equations) due to unstable (chaotic) regions of the phase space can have both - properties very sensitive to the control parameters and boundary conditions, and those robust to the latter. Therefore it will be of particular importance to look for corresponding formulations of the disk problem and the search for their robust properties. Only those robust properties can make sense to be compared with the observations. An example of such a mathematically correct formulation of the problem can be the one reported by Fridman at this meeting.

To be brief, among many other problems, I mention only the following inevitable source of accreting matter which strangely was less discussed during present meeting - the tidal disruption of stars within the Roche lobe of the massive centre. The crucial point here is that, the character of accretion depends not only on the mass of the centre but also on the dynamical parameters of the surrounding stellar system (Nature, 280, 214, 1979). Given the present possibility of accurate determination of the parameters of the stellar systems surrounding the AGN engines, this effect has to be considered quite seriously.

Thus, it seems there is still an essential way to go both from the present toy disk models to more complex theories, as well as in the cautious and balanced interpretation of the ever increasing flow of excellent observations, in order to understand properly the basics of the AGN mystery.

Compact Nuclei of Galaxies and Sources of Their Energy

L.Sh. Grigoryan

Institute of Applied Problems in Physics, Yerevan, 375014, Armenia

G.S. Sahakian

Yerevan State University, Yerevan, 375049, Armenia

A model of compact nuclei of galaxies as spherically-symmetric star clusters is proposed. A concept of the equation of state for star clusters in statistical equilibrium is introduced (galactic nuclei are systems in statistical equilibrium if their age is of the order of the age of the Universe). It is shown that a statistically equilibrium star cluster is described by the equation of state of a polytrope $P = a\rho^3$, and with its help the main parameters of compact nuclei of galaxies are calculated. The formula $M = 2.524GR^5/a$ for mass M and radius R of the cluster is derived.

From the law of equipartition of internal energy among the degrees of freedom of stars (including the rotational degrees of freedom), it is proved that the main constituents of compact galactic nuclei are neutron stars and white dwarfs. The ordinary stars will be destroyed because of relatively fast rotation. The angular velocities of neutron stars in the nucleus of our Galaxy corresponding to the condition of statistical equilibrium are approximately the same as those observed for pulsars. Hence the nucleus of the Galaxy may be a possible source of pulsars. It is shown that 1.5% of stars can evaporate from the nucleus of our Galaxy. The number of pulsars among such stars is of the order of $3 \cdot 10^6$. The problem of energy sources of galactic nuclei is also discussed. It is shown that in the framework of the model under consideration the compact nuclei of galaxies are high power sources of hard γ-radiation ($L \approx 10^{48} N_8 \mu_{30}^2 (\Omega/50)^3 erg/s$, where μ is the magnetic moment, Ω - the angular velocity of neutron star rotation and N - the number of stars in the galactic nucleus) due to the curvature radiation from ultrarelativistic electron fluxes traversing along the channels of open magnetic lines of pulsars. The X-ray and ultraviolet radiation of galactic nuclei are due to the synchrotron radiation from these same electron fluxes in the magnetic field of galactic nuclei ($L \approx 10^{42} - 10^{44} erg/s$). The optical (ranges of visible and infrared radiation) and the radio-frequency radiation are caused by the bremsstrahlung from the electrons of interstellar space ($L \approx 6 \cdot 10^{46} N_8^2 (5/R_{pc})^3 erg/s$, where R is the radius of the galactic nucleus).

Magnetic Field Strengths of GPS Radio Sources

V. G. Panajyan

Byurakan Astrophysical Observatory, Armenia.

GHz peaked spectrum radio sources (GPS) are believed to be a subclass of compact steep spectrum radio sources (CSS) with high frequency spectral indices $\alpha < -0.5$ ($S \sim \nu^\alpha$), linear sizes of pc to kpc scale and turnover spectra near $1GHz$. Due to the work of many radioastronomers during the past two decades many properties of CSS and GPS radio sources at present are known (O'Dea,C.P. et al.1998, and references therein).

The origin of the low frequency turnover of the radio continuum spectra is disputable: there are mainly two ways to explain the low frequency turnover of the radio sources - 1. free-free absorption of the radiation and 2. synchrotron self absorption. For the latter case an equation was derived by V. Slish (Slish 1963) and by I.P.Williams (Williams 1963), connecting angular sizes of the radio sources with turnover frequency ν_{to}, turnover flux density S_{to}, redshift z and magnetic field strength B. This equation was used by many authors to estimate magnetic field strengths B of radio sources (for example, Artyukh et al. 1994a, 1994b). Recently I have completed a sample of GPS radio sources of intermediate flux densities (Panajyan 1988) containing 30 GPS radio sources. Unfortunately the data on the linear sizes of all radio sources of this sample and redshifts of most of them are unknown, therefore it is not possible to estimate their magnetic field strengths. In this paper I estimate magnetic field strengths of radio sources of the complete sample of strong GPS radio sources of Stanghellini and coworkers (Stanghellini et al. 1988) assuming that the low frequency turnover of the spectra is due to synchrotron self absorption. I have used the modified version of the synchrotron self absorption equation used by M.J.A. Oort (Oort 1988). From the equations for angular sizes and linear sizes from that paper for magnetic field strength B one can write

$$B = ((1+z)^2 \nu_{to}^{1.25} D / 0.85 z (1+z/2) S_{to}^{0.5})^4, \qquad (1)$$

(the case, when $H = 50 km/sMpc, q_0 = 0$), where B is in $10^{-4}G$, S_{to} is in mJy, ν_{to} is in GHz and D is in pc. The results of calculations are as follows. The values of magnetic field strengths of the complete sample of GPS radio sources spread in a broad region: from 10^3 to $10^{-5}G$, the median value of this sample is equal to $6G$, only for 40% of radio sources $B < 1G$. The values $B > 1G$ are too large to be real. Note that Athreya and coworkers got $1G$ for the steep spectrum radio cores' magnetic field strengths (Athreya 1997).

It is known that the accuracy of estimation of magnetic field strengths using equation mentioned above is not high; the errors range up to two orders. The largest contribution in the error is due to the linear sizes of the radio sources: the uncertainty is proportional to the third power of the linear sizes. Therefore

Magnetic Field Strengths of GPS Radio Sources

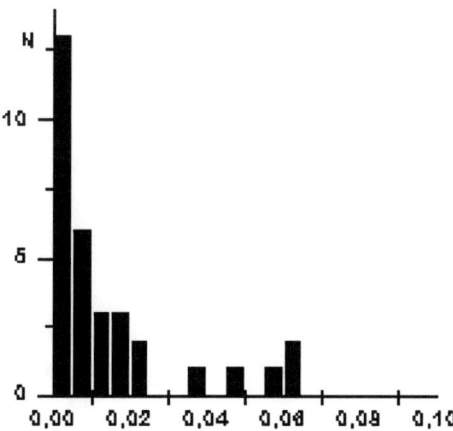

Figure 1. The histogram of the ratio $B^{0.25}/D$.

we calculated the ratio $B^{0.25}/D$ from the equation used above

$$B^{0.25}/D = (1+z)^2 \nu_{to}^{1.25}/z(1+z/2)S_{to}^{0.5} \qquad (2)$$

This ratio shows $B^{0.25}$ for angular size of $1pc$. But the magnetic field strength's connection with linear sizes is not simple, because linear sizes of the complete sample of GPS radio sources statistically are connected to the turnover frequency as follows (O'Dea C.P. 1997):

$$\nu_{to} \sim D^{-0.65} \qquad (3)$$

Fig. 1 shows the histogram of the $B^{0.25}/D$. One can see concentration of the values of the ratio $B^{0.25}/D$ around the median 0.0073.

References

Arthyukh, V. S., Hovhannissian, M. A., & Tyul'bashev, S. A. 1994, Pisma v AZh. 20, 178

Arthyukh, V.S., Hovhannissian, M. A., & Tyul'bashev, S. A. 1994, Pisma v AZh. 20, 258

Athreya, R. M., Kapahi, V. K., McCarthy, P.J., & van Breugel, W. 1997, MNRAS, 289, 525

O'Dea, C. P.,1998, PASP, 110, 493

O'Dea, C. P., & Baum, C. A. 1997, AJ, 113, 148

Oort, M. J. A. 1988, A&A, 192, 42

Panajyan, V. G. 1998, Astrofizika, 41, 377

Slish, V. 1963, Nature, 199, 682

Stanghellini, C., O'Dea C. P., Dallacasa, D., Baum, S. A., Fanti, R., & Fanti, C. 1998, A&AS, 131, 303

Williams, I. P. 1963, Nature, 200, 56

Quark-Hadron hybrid stars as remnants of Supernovae explosions

Poghosyan,Gevorg S.[1]

Yerevan State University, Alex Manoogian st.1, Armenia 375025

It is expected that at explosion of Supernovae the mechanism of producing a shock wave is based on subatomic interactions, and the remnant of the supernovae explosion can be a hybrid quark-hadron star. Since the temperature in the centre of collapsing stars reaches of order $T = 6 - 8 \cdot 10^{10} K$ and density $\rho = 10^{12} - 10^{13} \frac{g}{cm^3}$, electrons from the top of the Fermi sea can be captured and convert protons into neutrons via $e^- + p \rightarrow n + \nu_e$. The capture of electrons results in a neutronization burst (V.S.Imshennik 1988). Core collapse of the progenitor star becomes essentially a free fall with a time scale $t \sim \frac{1}{\sqrt{G\rho}} \sim 50ms$. When the central density of the core reaches supernuclear densities the repulsive QCD forces becomes essential. This can bring about manifestations of quark-hadron phase transitions (A.Dar 1997). After the explosion, from the remaining matter is probably formed a hybrid star.

Coming from the afore-mentioned, it is necessary to calculate stable stellar configurations with equations of state (EoS) that take into account quark-hadron phase transitions.I studied and obtained some results for the structure and stability of quark-hadron hybrid stars on the basis of Dynamic, confining (DC) (D.Blaschke 1998) model EoS and Bag model EoS (Ch.Kettner 1995) for quark phase and for hadronic phase. Relativistic nuclear matter EoS including pions and rho mesons was pioneered primarily by J.D.Walecka (J.Walecka 1974). For phase transition construction and mixed phase description, a method is used with more than one conserved charge (N.K.Glendenning 1992).

Figures show differences between hybrid EoS models which arise from mixed phase. The quarks became apparent in Bag model at 2.4 nuclear saturation density and in DC model at 3.2 nuclear density. The second figure shows, that the stable hybrid stellar configurations can exist with masses 2.5-2.6 M_{sun} with matter described by Bag-Walecka model and 2.8 $M_\odot$ by DC-Walecka model. Changes and improvement of EoSes used for quark phase can result in decreasing of masses.

---

[1]post-graduate student at Department of Theoretical Physics, e-mail: gevorg@darss.mpg.uni-rostock.de

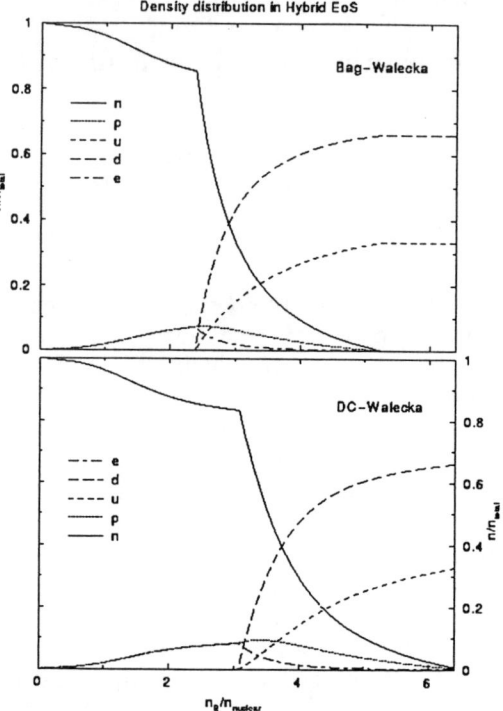

Figure 1. Relative densities depends on baryonic density: n-neutron, p- proton, u-up quark, d-down quark, e- electron

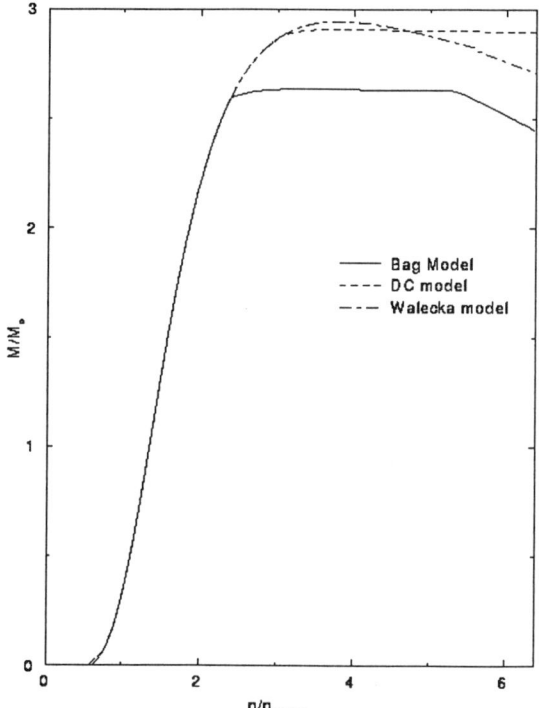

Figure 2. Mass curve of stellar configurations calculated by different EoS. Bag model- by Bag-Walecka hybrid EoS, DC- by dynamic confining-Walecka hybrid EoS, Walecka- pure hadronic EoS

References

D.Blaschke, C.D.Roberts, S.Schmidt, Phys.Lett.B, 425 (1998) 232

Arnon Dar,*"Supernovae 1987A-Ten years after"*, hep-ph/9707501,

N.K.Glendenning, Phys.Rev.D, 46, 1992,1275

V.S.Imshennik, D.K.Nodëzhin, Uspekhi Fizicheskikh Nauk, 1988 v.150, No. 4, 561

Ch.Kettner, F.Weber, N.K.Glendenning, etc Phys.Rev.D 51 (1995), 1440

Walecka, J.D., Ann.Phys., 83,491 (1974)

A New Numerical Method for Investigation of Evolution of Stars and Active Galactic Nuclei

Avetis Abel Sadoyan, Hovik Grigoryan and Gevorg Pogosyan

Yerevan State University, Yerevan, Armenia

We consider the problem of evolution for spherically symmetric configurations in the framework of Einstein theory. There are two independent functions that describe the "inner state" evolution of a configuration. In order to have an easy transformation to the Newtonian limit, where the problem of evolution is investigated, let us take as the basic functions density $\rho(t,r)$ and velocity $w(t,r)$ where r is the radius and t is time in the framework of the observer at the centre of the star. We write the equations of hydrodynamics in Einstein theory in a Schwarzschild coordinate system:

The equation of continuity

$$\rho_{,t} = -\frac{e^{\Phi-\Psi}}{1-w^2 c^2}\left[\rho_{,r} w \left(1-c^2\right) + (\rho+P)\left(w_{,r} + w\left(\frac{2}{r} - \frac{4\pi(\rho+P)r^2}{r-2m}\right)\right)\right]$$

and the Newton-Euler equation

$$w(t,r)_{,t} = \frac{e^{\Phi-\Psi}}{1-w^2 c^2}\left[\frac{-\rho_{,r} c^2}{(\rho+P)\gamma^4} - w(1-c^2)w_{,r} + [(\frac{2}{r}+\chi)w^2 c^2 - \xi]\gamma^{-2}\right]$$

where

$$\chi = \frac{m - 4\pi\rho r^3}{r(r-2m)},$$

$$\xi = \frac{m + 4\pi P r^3}{r(r-2m)},$$

and $w = V e^{\Phi(t,r)-\Psi(t,r)}$ is the velocity of matter in the comoving frame, while V is the derivative of radius with respect to time, $\gamma = (1-w^2)^{\frac{1}{2}}$ is the coefficient of relativity; it is one when the velocities are very small compared with the speed of light. The derivatives of a function with respect to time t and radius r are denoted as $f_{,t}$ and $f_{,r}$ respectively. Here we use a unit system, where the gravitational constant and the speed of light are equal to one. P is pressure on the sphere (remember that we consider spherical symmetry), c is the speed of sound in matter, $e^{2\Phi(t,r)} = g_{00}$, $e^{-2\Psi(t,r)} = g_{11}$, are metric coefficients, $m(r,t)$ is the accumulated mass inside the radius r.

These equations come to Oppenheimer - Volkoff equations when one takes $w = 0$ (Oppenheimer & Volkoff 1938).

The equation of state is assumed to be given in the form $P = P(\rho)$ for degenerate stellar matter, and no other limitations are used.

As we see, there are two additional functions $m(t,r)$ and $\Phi(t,r) - \Psi(t,r)$. We use two of the remaining 4 equations from the set of Einstein's equations and obtain

$$m(t,r) = \int 4\pi T_0^0 r^2 dr = \int 4\pi r^2 \frac{\rho + Pw^2}{1 - w^2} dr,$$

$$\frac{\partial \Phi(t,r)}{\partial r} - \frac{\partial \Psi(t,r)}{\partial r} = \frac{4\pi(P-\rho)r^2}{r - 2m} + \frac{2m}{r(r-2m)}$$

An algorithm and numerical code are developed for the solution of this system of equations and the time evolution of stellar bodies is investigated (Grigoryan, Sadoyan, 1994) for model examples.

The numerical integration is based on the method of characteristics, a more accurate and reliable method than the method of finite differences or spectral methods. The main advantage of this method is that there is no need to use sophisticated approaches such as calculations "against the wind" or introducing artificial viscosity, because it contains itself defence from such types of difficulties; the only weak point is the catastrophic increase of errors as have all other methods. We have developed some subroutines for special cases, for example for slow rotation of spherically symmetric superdense bodies in the frame of Einstein theory (A. Sadoyan, 1998) in $\Omega(t,r)$ approximation (Ω is the angular velocity of rotation). This approximation corresponds to spherical rotation with Coriolis forces taken into account .

Also a subroutine for investigation of shocks waves is established that can trace the origin, propagation and decease of shocks. We expect shocks in stellar evolution processes when other methods couldn't handle the situation.

Now we are looking for observational data concerning evolution of celestial bodies for the method developed and we hope that this new method can help to understand dynamical processes in Active Galactic Nuclei.

References

Oppenheimer, J. R., & Volkoff, G. M. 1938, Phys.Rev, 55, 374

Grigoryan, H., Sadoyan, A. 1994, Astrofizika, 37, 671

Sadoyan, A. 1997, in Abstracts of Plenary Lectures and Contributed Papers at General Relativity 15 meeting, Pune.

Can the Sonic Radius in ADAF Be Large?

F. Yuan

Astronomy Department, Nanjing University, Nanjing 210093, P.R.China

K.L. Huang

Physics Department, Nanjing Normal University, Nanjing 210097, P.R.China

Ju-fu Lu

Center for Astrophysics, University of Science and Technology of China, Hefei 230026, P.R.China

Abstract. We obtain the global transonic solutions by solving numerically a set of equations describing advection dominated accretion flows around black holes. We find that the sonic points can locate themselves at almost any radius, independent of the viscosity parameter α and the polytropic index γ.

Recently many authors have worked upon the global solutions for advection dominated accretion onto black holes(Chakrabarti 1996a, hereafter C96; Chen, Abramowicz & Lasota 1997, hereafter CAL; Narayan, Kato & Honma, hereafter NKH; Peitz & Appl 1997, hereafter PA; Igumenshchev et al. 1998, hereafter IAN). They obtain very different results especially on the important aspect such as the locations of the sonic points. C96 find that the sonic point can locate itself at almost any radius for polytropic index $\gamma = 4/3$ while the sonic radius R_s is small for $\gamma = 5/3$. Both CAL and NKH find that the sonic radius R_s can only be as large as several gravitational radii R_g. Although IAN think R_s can vary in a small range, its value is also within several R_g. PA's results are that R_s is small as the viscous parameter α is low but it becomes very large as α is high.

However, all the above results have discrepancies with that in inviscid adiabatic accretion flows which have been widely studied in the past years(see Chakrabarti 1996b for reviews). In that case R_s can be almost any value , regardless of γ and α. The ADAF equations are exactly the same as that in adiabatic flows in the limit of weak viscosity. So we expect the value of R_s should be similar in the two cases, i.e. R_s can be almost any value in ADAF. In fact, it should be so since the value of R_s should be determined by the outer boundary conditions which varies in different realistic systems.

Based upon the above considerations, we study the global solutions for advection dominated accretion flows. Equations are adopted from NKH but we don't select any boundary and don't assume any boundary conditions such as standard thin disk solutions since ADAF solutions may terminate at cer-

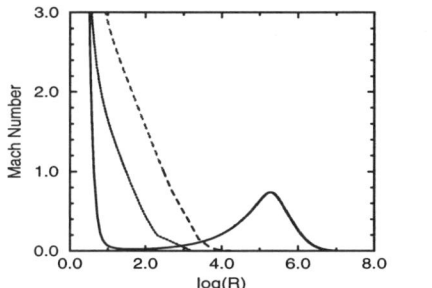 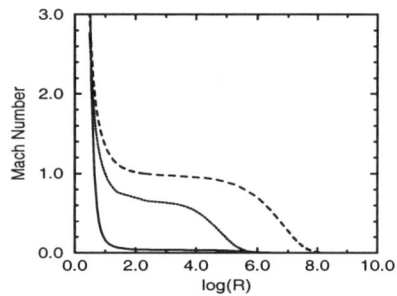

Figure 1. Left: the variations of Mach number with radius R for solutions corresponding to $\alpha = 10^{-3}, \gamma = 1.5$ and $R_s = 4.3, j = 3.696084$(solid); $R_s = 30, j = 1.75$(dotted) and $R_s = 300, j = 1.52$(dashed), respectively. Right: same as the left plot, but for $\gamma = 5/3$, and $R_s = 4.4, j = 3.4721$(solid), $R_s = 10, j = 2.033$(dotted), and $R_s = 200, j = 1.54$(dashed), respectively. $G = c = M_{\rm BH} = 1$ units are used.

tain radius, and we can't assure there exist solutions under arbitrarily selected boundary conditions. We use the method presented in C96 in which two free parameters R_s and j(the angular momentum per unit mass accreted by the central black hole) are selected and the equations are directly integrated from the sonic point inward to the horizon and outward. We find that all the solutions are terminated at certain limiting radius. When we integrate outward to this radius we find that the local sound speed suddenly begins to increase steeply with R which we think is unphysical. The value of the limiting radius is a function of α, R_s and j. What is more noticeable is that we obtain well-behaved global solutions under almost any values of parameter R_s, regardless of α and γ. We show some examples with $\alpha = 10^{-3}, \gamma = 1.5$(left plot) and $5/3$(right plot) in the figure. So our results are consistent with those in the case of inviscid accretion. This confirm our expectation.

Acknowledgments. This project was granted financial support from Pan-Deng Foundation.

References

Chakrabarti, S.K. 1996a, ApJ, 464, 664(C96)
Chakrabarti, S.K. 1996b, Physics Report, 266, No 5&6, 229
Chen, X., Abramowicz, M. A., & Lasota, J.-P. 1997, ApJ, 476, 61(CAL)
Igumenshchev, I. V., Abramowicz, M. A., & Novikov, I. D. 1998, MNRAS, 298, 1093(IAN)
Narayan, R., Kato, S. & Honma, F. 1997, ApJ, 476, 49(NKH)
Peitz, J. & Appl, S. 1997, MNRAS, 286, 681(PA)

M87 Solves the Problem of AGN

S. G. Iskudarian

Byurakan Astrophysical Observatory, Byurakan, 378433, Armenia

Abstract. The example of M87 shows that AGN are the most powerful units of protogalactic matter, which has built the large scale Universe during parts of the first second of the Big Bang. For fulfilment of such a mission it had to possess properties of elementary particles.

The active nucleus of M87 is the active nucleus of Our Supergalaxy (Iskudarian 1993). M87 itself is that "globular cluster" in the center of the Supergalaxy, a similar image of which is observed in the centers of Our Galaxy and the Andromeda nebulae in small scales (Waldrop 1985, Iskudarian 1995). M87 is the nearest cD galaxy (Suchkov 1988), it is the nearest "void" (Iskudarian 1995) with its immediate environment. Recent investigations showed that voids are not voids at all (Peterson et al. 1991). Their blue centers are surrounded by faint elliptical galaxies. In the case of M87 globular clusters rich population is in the role of superdwarf ellipticals, which does not contradict the reality.

M87 gave rise to a closed looplike superstring in Our Supergalaxy (Iskudarian) in the form of the rich group of galaxies N94 (Huchra & Geller 1983), which is connected with the other rich group of galaxies N106 physically and, it seems, was ejected from the entrails of the last one (in a protogalactic state certainly). Looking on Fig. 1a, b of Iskudarian (1993) one can see the connection of those two groups very well. Such a picture could form only during parts of the first second of the Big Bang, with the high energies released at the high temperatures. In such conditions takes place rather unusual phenomena in the micro world. For example, the sudden birth of a new particle or the decay of a nonstable particle in a heavy shower of other particles. There can meet particles also which combine in themselves characters of two quite different particles, so-called "mad" unity (Iskudarian 1997, Davies 1985). Observations show the existence of analogous processes in the macro world - in the world of galaxies. One may think with great conviction that those properties of elementary particles were kept in protogalactic matter as its "genes" (Iskudarian 1997). And AGN are those agents which carried the huge energies of the Big Bang to the remote corners of the observed Universe, drawing with its treks the large scale structure of the Universe. Later on, being centers of supergalaxies, and so the other potential centers of activity, they have built and build at present new structures in the form of new generations of stellar populations. The composite structure of the Virgo cluster is the best evidence of these last thoughts (Bingelli et al. 1987).

Undoubtedly, the thoughts expressed here need further study, investigation, and discussion. Nevertheless, it is clear that activity of AGN in its outward displays is diverse and many-coloured, but in the basis, there is the general regularity which acts in the micro and macro worlds of the Universe, ensuring by this way the unity of the Universe. It is the origin, formation, and ejection of the first type of stellar population from the entrails of the second one (Iskudarian 1993, 1996).

References

Bingelli,B., Tammann, G.A., & Sandage, A.R. 1987, AJ, 94, 251

Davies, P. 1985, Superforce, New York

Huchra, J.P., & Geller, M.J. 1983, ApJS, 52, 61

Iskudarian, S.G. 1993,"Is M87 the active nucleus of Our Supergalaxy?", International workshop on "Galaxy Clusters and Large Scale Structures in the Universe" Sesto Pusteria (Bolzano, Italy)

Iskudarian, S.G. 1995, "Similar structures-similar formations of large and small scales", International workshop on "Observational Cosmology: from Galaxies to Galaxy Systems", Sesto Pusteria (Bolzano,Italy)

Iskudarian,S.G. 1995, "The nearest "void?", International workshop on "Observational Cosmology: from
Galaxies to Galaxy Systems", Sesto Pusteria (Bolzano,Italy)

Iskudarian, S.G. 1996, "The Unity of the Universe", International workshop on "Hubble deep field", Baltimore, USA

Iskudarian, S.G. 1997,"The Universe of micro and macro scales", Meeting Astro-4, Moskow

Iskudarian,S.G. 1998, "An example of closed looplike superstring", (unpublished work)

Peterson, B.A., D'Odorico,S., Tarenghi M., & Wampler, E.J. 1991, The Messenger, N64, 1

Suchkov, A.A. 1988, "Familiar and Mysterious Galaxies", "Science", Moskow

Waldrop, M.M. 1993, Science, 230, N4722, 158

IV. AGN RELATED PHENOMENA

Alexei Fridman, Leonid Matveenko, Alexander Boyarchuk

Jack Sulentic, Halton Arp, Vahagn Gurzadyan

Activity in Interacting and Binary Galaxies

Yervant Terzian

Department of Astronomy and National Astronomy and Ionosphere Center, Cornell University, Ithaca, NY, 14853

Abstract.
This is a short review of recent studies and reports on interacting galaxies and binary galaxies. Early studies were primarily made at optical wavelengths, but recent ones have concentrated on observing HI at $\lambda 21$ cm. Application of the Virial Theorem to these observations indicates M/L values of the binary systems of $\lesssim 50$, suggesting the existence of moderate halos around spiral galaxies.

1. Introduction

If gravity is the dominant force at large scales in the Universe, as we believe, then we hope to learn about the large scale structure of the universe, including its mass content and distribution. Such advances are crucial for establishing a credible theory of cosmic evolution.

For more than fifty years one of the outstanding problems of cosmology has been the "missing mass" in the universe (Zwicky 1937). If the physics we have come to understand is universal, then the evidence for dark matter cannot be disputed. On the scale of galaxies, and even clusters of galaxies the need for dark matter is very well established (Ashman 1992). Much less is known about the dark matter needed to support a closed model of the Big Bang universe and this matter, if it exists, is probably nonbaryonic (Bahcall, Lubin and Dorman 1995).

In order to investigate the dynamical masses of galaxies, including their halos, studies of gravitationally bound binary galaxies have been made as discussed in the following sections. Evidence for interacting and merging galaxies has recently been discussed by Schweizer (1996) and Hibbard and van Gorkom (1996), and the dynamics of galaxy interactions by Barnes (1996). HI observations of binary galaxies have been discussed by Nordgren, et al. (1999) and references therein.

2. Interacting Galaxies

The most prominent signature of galaxy interactions is their neutral hydrogen distribution and kinematics (Sancisi, 1998). HI $\lambda 21$ cm observations of binary galaxies or small groups, normally show tidal interactions in the form of HI tails and bridges even when the optical images seem to be undisturbed (Rand, 1994; Nordgren, et. al. 1998). The extended HI features sometimes have a very large cross section and contribute to the halo material surrounding galaxies. Such extended halos could be the source of the absorption lines seen in the spectra of distant quasars. Sancisi (1998) has argued that major interactions between galaxies of similar masses cause large tidal effects and many lead to mergers and eventually to the formation of elliptical galaxies. Minor interactions are normally those between a normal galaxy and a small dwarf companion, and this leads into gas accretion in the galactic disk and may cause sudden star formation in starbursts. Tidal material falling back into the galaxies is clearly seen in the system NGC7252 mapped in HI with the VLA by Hibbard, et. al. (1994).

From a study of interacting galaxies Sancisi (1998) has concluded that at least 25 percent of field galaxies have undergone some kind of tidal interaction, and a larger percentage of galaxies may have been through one or more mergers.

Two very comprehensive reviews have been prepared in the Saas-Fee Advanced Course 26, in 1996, and appear in a volume *Galaxies: Interactions and Induced Star Formation*. François Schweizer wrote about "Observational Evidence for Interactions and Mergers", and Joshua E. Barnes wrote about "Dynamics of Galaxy Interactions." In the 1950s Fritz Zwicky began the examination of interacting galaxies and concluded that the extended filaments seen from close galaxy pairs must be due to tidal gravitational interactions (Zwicky 1956, 1959). Vorontsov-Velyaminov (1959) also noted the many peculiar, distorted images of galaxies in binary galaxy systems. Soon after, Arp (1966) produced his "Atlas of Peculiar Galaxies" thus documenting many examples of galaxy interactions. Then numerical modeling showed that tidal interactions can indeed produce extended galaxy filaments and bridges, of which the most famous is the galaxy pair NGC4038/4039, known as "The Antennae", successfully modeled by Toomre and Toomre (1972).

The numerical simulations of interacting galaxies have progressed from gravitating two-point masses, each with a disk of particles, to today's N-body simulations with $N \sim 10^6$ (Schweizer 1996). Tidal drag, fast encounters, dynamical friction, orbit decay, relaxation, and final mergers have now been incorporated in the sophisticated models of galaxy interactions as discussed by Barnes (1996).

A likely outcome of merging spiral galaxies could be the formation of elliptical galaxies, as simulations of merging remnants resemble ellipticals (Toomre and Toomre 1972). Indeed, Hibbard and van Gorkom (1996) have presented a systematic observational study of HI at $\lambda 21$ cm of a sample of galaxy systems involved in progressive stages of merging. The observations were performed using the VLA and included the systems Arp 295, NGC 4676, NGC 520, NGC 3921, and NGC 7252. Hibbard and van Gorkom conclude that as the merger rearranges the light profiles of the progenitor disk galaxies they evolve into elliptical galaxies. Interacting galaxies often show signs of vigorous star formation, and the Infrared Astronomical Satellite IRAS in 1983 made numerous observations of starbursts in merging galaxies (Sanders 1990).

3. Binary Galaxies

From statistical studies of stellar motions the mass to luminosity ratio M/L ($M_\odot=1$, $L_\odot=1$) in the vicinity of the sun is just a few, and for the entire galaxy could be as high as 20. The flat rotation curves of galaxies indeed indicate M/L values of the order of tens. Small groups of galaxies show M/L values reaching ~100 from the dispersion velocities of their members, and large galaxy clusters show values of a few hundred. In order to have a closed universe one would require an M/L value of about 1500. This missing mass problem has been with us since Zwicky pointed it out in 1937, and remains completely unresolved.

In order to investigate the M/L values in galaxies, including their halos, Page (1952) observed binary galaxies and applied the Virial Theorem to determine if significant dark matter was attached to the luminous galaxies. This method looks simple but has several limitations.

First we must make sure that we are observing real gravitating galaxy pairs and not detached "optical" pairs. In all cases we unfortunately can only measure the projected separation of the two galaxies, and we have no information on the orbital eccentricity and inclination. Hence we need to observe large samples of galaxy pairs to obtain statistical results. The one parameter that we can determine with great accuracy today is the systematic radial velocity of each galaxy in a pair and hence their velocity difference ΔV.

Observations of galaxy pairs at optical wavelengths were done by Page (1952), Arp (1966), Karachentsev (1972) and Turner (1976a, b) among a few others, with ΔV accuracies of ~30 to 50 km/sec. Radio HI observations began with Peterson (Peterson, 1979a, b; Peterson and Terzian 1979) who used large samples of spiral galaxies and who adopted criteria to confine the sample, such as the relative magnitude of the galaxy pair should not exceed 2.5 magnitudes, and an isolation criterion to insure that there were no other nearby galaxies gravitationally distrubing the system. Later Schneider, et.al. (1986) used a more expanded sample and more refined isolation criteria.

In 1993 more detailed studies were performed by carefully selecting the galaxy pairs from galaxy redshift catalogues, thus performing a three dimensional isolation analysis (Chengalur, et.al 1993, 1994, 1995, 1996). In this manner wide pairs as well as close pairs were identified and accurately observed. The wide pairs were observed with the Arecibo radio telescope (beamwidth 3 arc min), and the Parkes radio telescope (beamwidth 14 arc min). The close pairs were observed with the VLA and the Australia Array (angular resolution ~30 arc sec). These studies continued with even more conservative selection criteria, such as identifying pairs in low or medium galaxy density regions (Nordgren, et.al. 1997a, 1997b, 1998). In total 106 wide pairs were observed and 25 close pairs were mapped with synthesis arrays. The velocity accuracy of the radio observations was about 5 km/sec.

The new surveys were statistically complete samples that included only pairs in low and medium galaxy density regions. The results have shown that the distribution of ΔV peaks at zero and decreases smoothly with increasing ΔV, as expected from a random orientation of galaxy pairs (Fig 1. from Nordgren et.al. 1996). The HI maps of close pairs all show tidal interactions in the form of tails, bridges and common envelopes.

For wide pairs, which are isolated, the median ΔV ~30 km/sec and pairs with projected separation as large as 1.0 Mpc have a median ΔV ~50 km/sec and are probably bound pairs, since the random galaxy velocities are of the order of ≥ 200 km/sec. Nordgren, et.al. (1999) argue that as the separation between pairs decreases from ~1.0 Mpc the median ΔV should increase until dynamical friction from dark matter halos forces ΔV to decrease. The data, however, show no such increase in ΔV with decreasing separations, hence they conclude that large dark halos, $\sim$500 pc, exist. These could contribute to the HI line absorption seen in the spectra of quasars as mentioned earlier.

These modern data give a best fit that results into an overall M/L $\sim$ 50, with a mass per spiral galaxy of $\sim 4.5 \times 10^{11}$ $M_\odot$ (Nordgren, et.al. 1999). This value is a few times larger than that derived from rotation curves and strongly suggests moderate dark matter halos. It is to be noted that the matter needed to close the universe does not appear to be distributed in the vicinity of the galaxies.

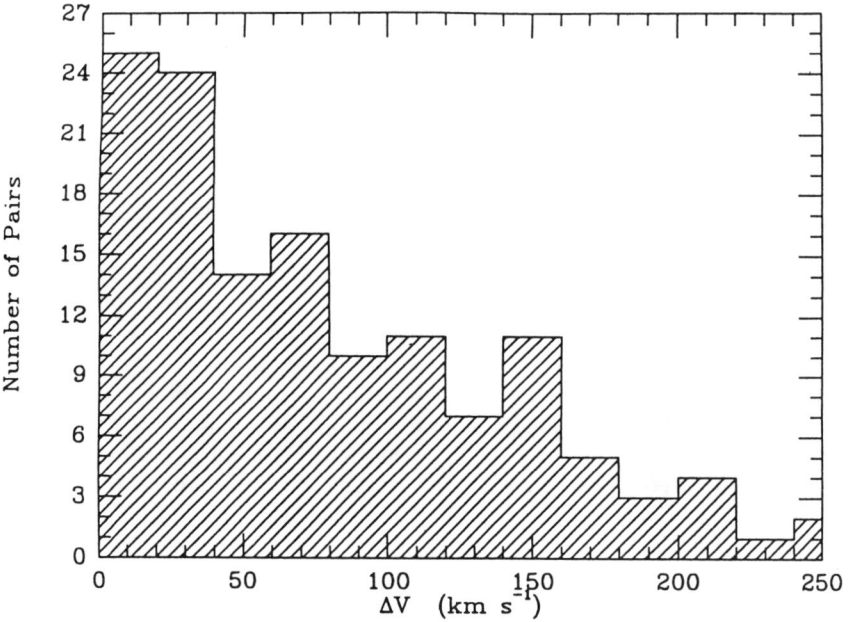

Figure 1. The distribution of velocity differences in galaxy pairs from the sample by Chengular et.al., and Nordgren et.al. cited in the text.

4. Velocity Quantization

In 1976 Tifft (1976) argued that the distribution of redshifts of galaxies shows periodic or quantized intervals separated by $\sim$72 km/sec. Significant literature has been devoted to this potentionally cosmologically very important assertion, and discussion can be found in Tifft and Cocke (1989), Tifft (1996), Napier and Guthrie (1996) and Nordgren et.al. (1996). This last reference provides observational data on the velocity difference between pairs of galaxies. The data is arguably the most accurate set of velocities with errors $\lesssim$5 km/sec, and includes ~130 isolated pairs. These observations show that the velocity difference of the pair sample decreases monotonically from zero and there is no indication of any redshift periodically in this sample (shown in Figure 1.).

A detailed analysis for searching for data clustering and periodicities has been given by Newman et.al. (1989, 1994) and by Newman and Terzian (1996). These authors critically discuss the approximate statistical methods used in concluding periodic effects, particularly when the data samples are very limited. Nevertheless, any credible velocity quantization effects, if real, would completely revolutionize our concepts of cosmology.

5. Conclusions

Although very significant new observational and theoretical studies have emerged on merging, interacting, and binary galaxies during the last few decades, much research work remains ahead. In particular detailed velocity structure of close and merging pairs is essential to understand the last phases of the mergers. Studies of small galaxy groups are also important to investigate the mutual gravitational perturbations that occur in these systems.

Acknowledgments. The author wishes to thank his close collaborators on the topics discussed above, who include J. Chengalur, M. Haynes, W. Newman, T. Nordgren and E. E. Salpeter. This work was partiallly supported by the National Astronomy and Ionosphere Center which is operated by Cornell University under a cooperative agreement with the National Science Foundation.

References

Arp, H., 1966, ApJS, 14, 1.
Ashman, K. M., 1992, PASP, 104, 1109.
Bahcall, N. A., Lubin, L. M., Dorman, V. 1995, ApJ, 447, 81.
Barnes, J. E., 1996, *Galaxies: Interactions and Induced Star Formation*, Saas-Fee 26.
Chengalur, J. N., Salpeter, E. E., Terzian, Y., 1993, ApJ, 419, 30.
Chengalur, J. N., Salpeter, E. E., Terzian, Y., 1994, AJ, 107, 1984.
Chengalur, J. N., Salpeter, E. E., Terzian, Y., 1995, AJ, 110, 167.
Chengalur, J. N., Salpeter, E. E., Terzian, Y., 1996, ApJ, 461, 546.
Hibbard, J. E., Guhathakurta, P., van Gorkom, J. H., Schweizer, F., 1994, AJ, 107, 67
Hibbard, J. E., van Gorkom, J. H., 1996, AJ, 111, 655.
Karachentsev, I., 1972, Com. Sp. Aps. Obs. USSR, 7, 1.
Napier, W. M., Guthrie, B. N. G., 1996, Ap&SS, 244, 111.
Newman, W. I., Haynes, M. P., Terzian, Y., 1989, ApJ, 344, 111.
Newman, W. I., Haynes, M. P., Terzian, Y., 1994, ApJ, 431, 147.
Newman, W. I., Terzian, Y., 1996, Ap&SS, 244, 127.
Nordgren, T. E., Terzian, Y., Salpeter, E. E., 1996, Ap&SS, 244, 65.

Nordgren, T. E., Chengalur, J. N., Salpeter, E. E., Terzian, Y., 1997a, AJ, 114, 77.

Nordgren, T. E., Chengalur, J. N., Salpeter, E. E., Terzian, Y., 1997b, AJ, 114, 913.

Nordgren, T. E., Chengalur, J. N., Salpeter, E. E., Terzian, Y., 1998, ApJS, 115, 43.

Nordgren, T. E., Chengalur, J. N., Salpeter, E. E., Terzian, Y., 1999, ApJ, (in press).

Page, T., 1952, ApJ, 116, 63.

Peterson, S. D., 1979a. ApJS, 40, 527.

Peterson, S. D., 1979b. ApJ, 232, 20.

Peterson, S. D., Terzian, Y., 1979, Jour. Roy. Ast. Soc. Can., 73, 215.

Rand, R. J., 1994, AA, 285, 833.

Sancisi, R. 1998. *Galaxy Interactions at Low and High Redshift*, eds. D. Sanders and J. Barnes (Kluwer, Dordrecht).

Sanders, D. B., 1990. *Dynamics and Interactions of Galaxies*, ed. R. Wielen (Springer Verlag, Berlin), 459.

Schneider, S., Helou, G., Salpeter, E. E., Terzian, Y., 1986, AJ, 92, 742.

Schweizer, F., 1996, *Galaxies: Interactions and Induced Star Formation*, Saas-Fee, 26.

Tifft, W. G., 1976, ApJ, 206, 38.

Tifft, W. G., Cocke, W. J., 1989, ApJ, 336, 128.

Tifft, W. G., 1996, Ap&SS, 244, 29.

Toomre, A. and Toomre, J., 1972, ApJ, 178, 623.

Turner, E. L., 1976a, ApJ, 208, 20.

Turner, E. L., 1976b, 208, 304.

Vorontsov-Velyaminov, B. A., 1959, *Atlas and Catalogue of Interacting Galaxies*, (Sternberg Institute, Moscow State University, Moscow).

Zwicky, F., 1937, ApJ, 86, 217.

Zwicky, F. 1956, Ergebnisse d. exakten Naturw., 29, 344.

Zwicky, F. 1959, Handbuch d. Physik, 53, 373.

Active Galactic Nuclei and Related Phenomena
IAU Symposium, Vol. 194, 1999
Yervant Terzian, Daniel Weedman, Edward Khachikian, eds.

Redshifts of New Galaxies

Halton Arp

Max-Planck-Institut für Astrophysik, Karl-Schwarzschild-Str. 1, 85740 Garching, Germany

Abstract.
Observations increasingly demonstrate the spatial association of high redshift objects with larger, low redshift galaxies. These companion objects show a continuous range of physical properties - from very compact, high redshift quasars, through smaller active galaxies and finally to only slightly smaller companion galaxies of slightly higher redshift. The shift in energy distribution from high to low makes it clear that are seeing an empirical evolution from newly created to older, more normal galaxies.

In order to account for the evolution of intrinsic redshift we must conclude that matter is initially born with low mass particles whose mass increase with time (age). This requires a physics which is non-local (Machian) and which is therefore more applicable to the cosmos than the Big Bang extrapolation of local physics. Ambartsumian's "superfluid" foresaw some of the properties of the new, low particle mass, protogalactic plasma which is required, demonstrating again the age-old lesson that open minded observation is much more powerful than theoretical assumptions.

Since the ejected plasma, which preferentially emerges along the minor axis of the parent galaxy, develops into an entire galaxy, accretion disks cannot supply sufficient material. New matter must be created within a "white hole" rather than bouncing old matter off a "black hole".

1. Introduction

Evidence for the association of high redshift objects with low redshift galaxies emerged in 1966 with completion of the systematic study in the Atlas of Peculiar Galaxies. For the most recent summary of the evidence to date see Arp (1998b). In the present report, however, we concentrate on the most recent observations and those which summarize best the empirical properties which characterize the associations. These latest discoveries reinforce a picture of empirical evolution which progresses from newly born, high redshift protogalaxies (quasars) to older, more normal, low redshift galaxies.

2. The Quasars around NGC5985

Fig. 1 shows one of the most exact alignments of quasars and galaxies known. Attention was drawn to this region when it was discovered that a very blue galaxy in the second Byurakan Survey (Markarian et al. 1986) had a quasar of redshift $z = .81$ only 2.4 arcsec from its nucleus (Reimers and Hagen 1998). Even multiplying by 3×10^4 galaxies of this apparent magnitude or brighter in their survey they estimated only a chance proximity of 10^{-3}. (Nevertheless they took this as proof that it was a chance projection! Also it was not referenced that G.Burbidge, in 1996 in the same Journal, had published extensive list of other quasars improbably close to low redshift galaxies).

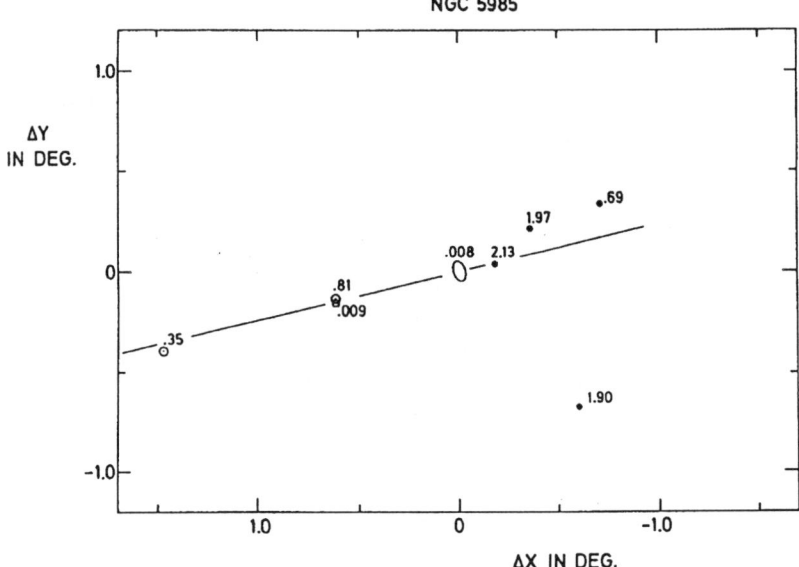

Figure 1. All catalogued active galaxies and quasars within the pictured area are plotted. Redshifts are labeled. The dwarf spiral 2.4 arcsec from the $z = .81$ quasar has $z = .009$ which marks it as a companion of the Seyfert NGC5985 at $z = .008$. The line represents the position of the minor axis of NGC5985.

But the galaxy was a dwarfish spiral, showing no active nucleus from which the quasar could have emerged. Proceeding on the by now overwhelming evidence that Seyfert galaxies eject quasars (Radecke 1997; Arp 1997a) I looked in the vicinity for a Seyfert. There was NGC5985, only 36.9 arcmin away on the sky! The chance of finding a Seyfert as bright as $V = 11.0$ mag. this close to the $z = .81$ quasar was less than 10^{-3}.

The next obvious question was: Were there other quasars in the field? Fig. 1 shows that there are 6 catalogued quasars discovered in a uniform search of the area in the Second Byurakan Survey. (P. Veron, in this Symposium, reports 66 - 88 % completeness for this survey - probably typical for quasar surveys). But five of these six quasars fall on a line through the Seyfert, with the dwarf galaxy along the same line. What is the probability that three or four objects would accidentally align within a degree or so of a straight line through the Seyfert? Conservatively this can be computed to be of the order of 10^{-4} to 10^{-5}. But the most astonishing result of all is that if one looks up the position of the minor axis in NGC5985 it turns out, as shown in Figs. 1 and 2, to be a line that looks as though it were drawn through the positions of the objects!

Just a simple visual evaluation of Fig. 1 would lead to the conclusion that the configuration was physically significant. A combined numerical probability of the configuration gives a chance of around 10^{-9} to 10^{-10} of being accidental (Arp 1998). Nevertheless several peer reviewers recommended against publication on the grounds that the accidental probability was "greater" than this. But, of course, several dozens of cases of anomalous associations had been reported since 1966 with chance probabilities running from 10^{-4} to 10^{-5}. What is the combined probability of all these previous cases? And what is the motivation to claim each new case is "a posteriori"?

3. The Companion Galaxies Around NGC5985

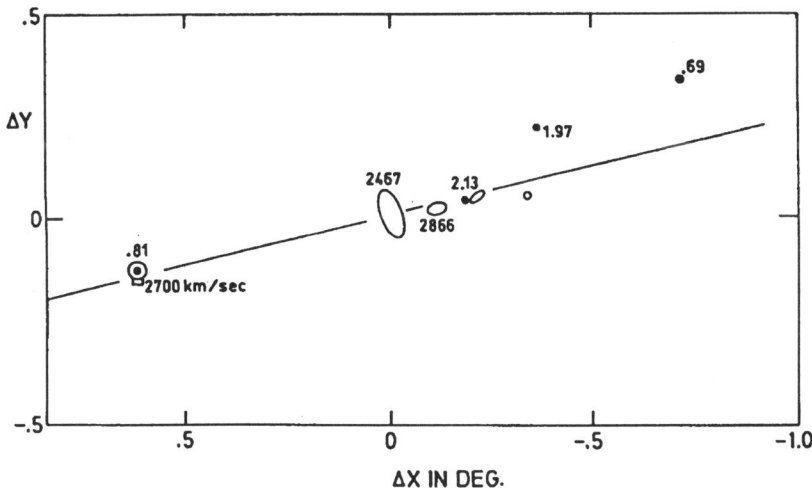

Figure 2. The central portion of Fig. 1 is here enlarged and all NGC galaxies are additionally plotted. Three additional companions are seen to lie along the WNW minor axis. The redshifts of the galaxies are now labeled in km/sec and it is seen that the two companions whose redshifts are known have slightly, but significantly, greater redshifts.

We have seen that the quasars are aligned along the minor axis of the central Seyfert. Sections that follow summarize that generally both quasars and companion galaxies are aligned along the minor axes. Is this true of the NGC5985 family? In Fig. 2 I have enlarged the central regions of Fig. 1 and plotted all the bright, NGC galaxies in the region. It is apparent that these NGC galaxies, which are fainter than NGC5985 as companions should be, are almost exactly aligned along the WNW extension of the minor axis! Taken together with the dwarf companion on the other side of the minor axis, this leaves no doubt that in the case of NGC5985 the companion galaxies and quasars are aligned exactly along the same minor axis. This then constitutes another proof that these objects of variously higher redshift have some physical relation to the low redshift,central galaxy.

There is confirming evidence in the measured redshifts of these companions. As Fig. 2 shows, the one to the ESE is about +230 km/sec with respect to NGC5985 and the one to the WNW is about +400 km/sec. These redshifts are close enough to that of the parent galaxy to conventionally confirm their status as physical companions. But at the same time they exhibit the systematic excess redshift of younger generation galaxies in groups (See Fig. 1 of preceding paper in this volume and Arp 1997). In the following sections we will interpret this excess redshift as indicating that they have more recently evolved from quasars. Since they are older than the quasars, they have had time to fall back from apgalacticon to within a few diameters of the parent galaxy.

4. The Empirical Model of Ejection and Evolution

As Fig. 3 shows, we can now combine the data from the last 32 years of study of physical groups of extragalactic objects. What results is a sequence of quasars emerging from a large galaxy along its minor axis, increasing in luminosity and decreasing in redshift as they move outward. When they reach a maximum distance of about 500 kpc they have started to turn into compact, active galaxies. As they age further into more normal galaxies they may fall back toward the parent galaxy. In that case they fall back along the minor axis because they emerged with little or no angular momentum component. *They move on plunging orbits.*

Ambartsumian (1958) noted that companion galaxies seemed to be ejected from larger galaxies. Arp (1967;1968) showed evidence that quasars were also ejected from galaxies. The astronomer most knowledgeable about disk galaxies, Erik Holmberg (1969), showed companion galaxies aligned along minor axes. He concluded they must be formed from gas ejected from the nucleus. Burbidge and Burbidge (1997) showed gas ejected along the minor axis of the strong Seyfert, NGC4258. Arp (1997b) showed a number of pairs of X-ray quasars ejected closely along the minor axis of Seyferts (e.g. NGC4235, NGC2639).

These observational results are now summarized in Fig. 4. It is seen that the quasars preferentially come out in a cone angle of about ±20°. The companion galaxies are preferentially confined within about ±35°. This difference is in the expected direction because as the quasars reach their maximum extension and slow down, they are vulnerable to perturbation by objects at that distance and hence will fall in again along slightly deviated orbits. Also since the companions

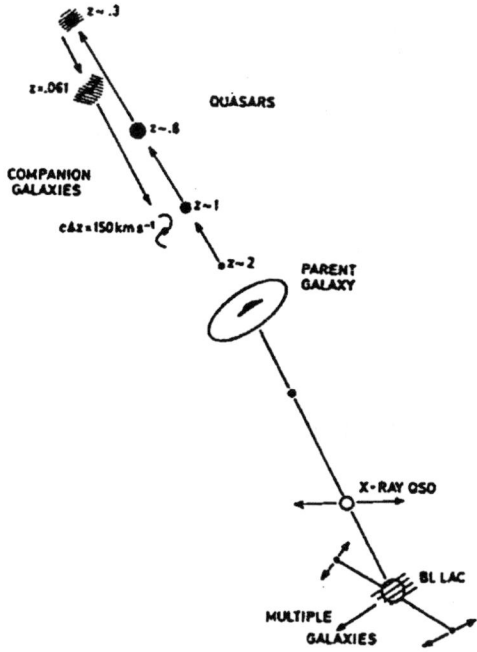

Figure 3. Schematic representation of quasars and companion galaxies found associated with central galaxies from 1966 to present. The progression of characteristics is empirical but is also required by the variable mass theory of Narlikar and Arp (1993).

are older, the more recent ejection axes of quasars in some cases have moved because of precession of the galaxy or the spin axis of the nucleus.

5. Verification of the Model

The above model combines many cases, each one of which may only contain a few elements. But it is now possible to show in Fig. 5 the active Seyfert NGC3516. This galaxy is exceptional because it confirms the essentials of the model in a single case. The objects are drawn from a complete sample of bright X-ray objects within about 22 arc min of the Seyfert (Chu et al. 1998). Notice there are 5 quasars and a BL Lac-type object, the latter object representing a transition between a quasar and a more normal galaxy. The redshifts of these six objects are strictly ordered with the highest closest to the Seyfert and the lowest furthest away. They define very well the minor axis of the galaxy.

As a side note it is clear that each of the six redshifts fits very closely to the quantized redshift peaks observed in quasars in general. Those peaks are: .06,

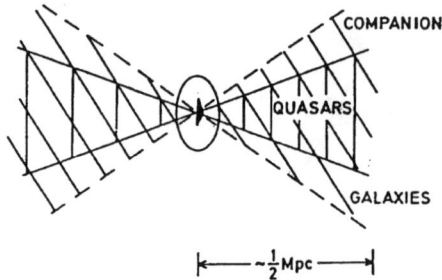

Figure 4. Schematic representation of distribution of companion galaxies along minor axes of disk galaxies (±35° from Holmberg 1969; Sulentic et al. 1978; Zaritsky et al. 1997). Quasars are observed ±20° from minor axis (Arp 1998a).

.30, .60, .96, 1.41, 1.96 (Arp et al. 1990). If these redshifts represented Doppler velocities, of course, quantization would be destroyed by random orientations to the line of sight. So quantization is direct proof that extragalactic redshifts are nonvelocity. The intrinsic redshifts are indicated to evolve in discrete steps as the quasars evolve into galaxies. The quantization of redshifts in smaller steps in the companion galaxies further supports their being end products of this evolution.

The NGC5985 case shown in Fig. 2 confirms very well the later stages of smaller redshift where the evolved companion galaxies have fallen back toward the parent galaxy. The very exact alignment of quasars and companion galaxies must mean that the minor axis of NGC5985 has stayed unusually well fixed in space over the time required to evolve from a high redshift quasar to a reasonably normal galaxy, or about 7×10^9 years (Arp 1991). Another case of a very narrow alignment of companion galaxies can be seen around our Local Group center, M31 (Arp 1998a).

The length of the NGC5985 line and the brightness of its components suggests that the group is closer than most Seyferts so far investigated. In particular the distribution of quasars is much closer to NGC3516 which appears to be more distant. It is also clear, however, from the measured ellipticities of the images, that the minor axis of NGC3516 is tipped much closer to our line of sight than in NGC5985. That accounts for some foreshortening of the NGC3516 line as well as an apparently greater spread off the line relative to the length of the line.

6. Statistics

From 1966 onward statistical associations of high redshift quasars with low redshift galaxies were reported (Arp 1987). Significances ranged from 16 to greater than 7.5 sigma. We will give only the latest example here as shown in Fig. 6. There it is shown that a nearly complete sample of Seyfert galaxies have a conspicuous excess of bright X-ray point sources within about 50 arcmin radius

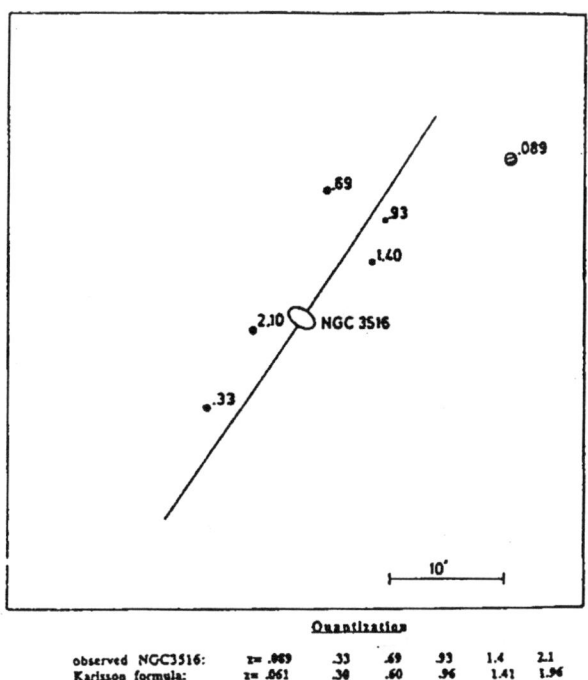

Figure 5. All bright X-ray objects around the very active Seyfert galaxy NGC3516 which have had their redshifts measured by Chu et al. (1998). Redshifts are written to the upper right of each quasar and quasar like object.

(Radecke 1997; Arp 1997a). Since these X-ray sources are essentially all quasar or quasar-like one can see at a glance that there are a large number of quasars physically associated with nearby Seyfert galaxies.

7. Theory

What would give an intrinsic (non-velocity) redshift to a galaxy - high when it is first born - and decreasing as it ages? The answer was supplied by Narlikar in 1977. It rests on the fact that the standard solution of the Einstein field equations as made by Friedmann in 1922 made one key assumption. It assumed that the elementary particles which constitute matter remain forever constant in time. A more general solution, as Narlikar showed, had the mass of particles increasing with time. This would furnish a series of answers to paradoxes which currently falsify the standard Big Bang solution:

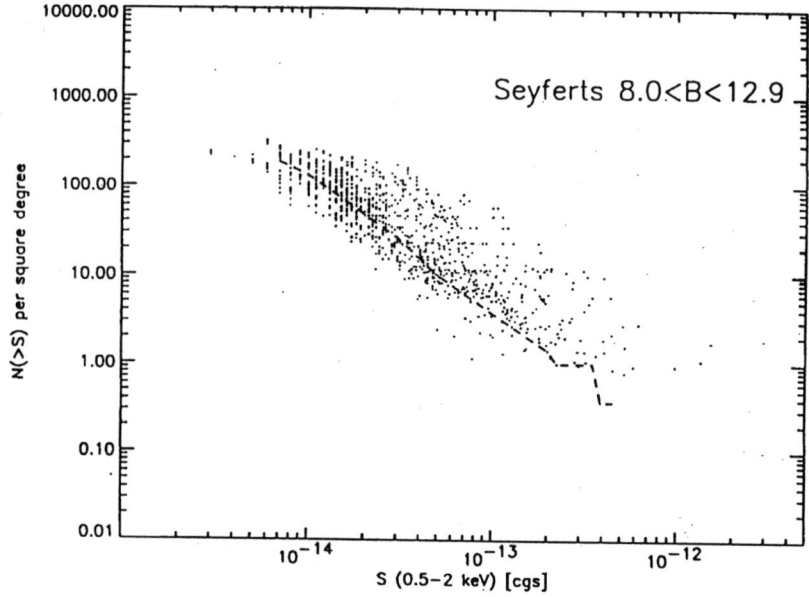

Figure 6. Cumulative number of X-ray sources brighter than strength (S) around a nearly complete sample of bright Seyfert galaxies. Dashed line represents counts in non-galaxy control fields.

1) Episodic creation near zero mass, giving initial outflow at signal speed, c.
2) Condensation into proto galaxies.
3) An intrinsic redshift initially high and decreasing with time.
4) Evolution into normal galaxies moving with small velocities.
5) Possibility of understanding quantized redshifts.
6) Transformation to local physics using local time scales.

The details of this theory are explained in recent publications (Arp 1998b). But here it should be emphasized that the observational disproofs of the conventional, singular creation theory can no longer be discarded with the claim that there is no viable alternative theory. The empirical conclusions of Ambartsumian and many others who followed, have now been supported by numerous amplifying observations and a unifying physical theory has been adduced.

It is clear from the experience of the last 40 years that influential astronomers will accept neither the observations which falsify the current beliefs nor the theory which enables these observations to be understood. Therefore it is of the utmost importance for each individual researcher to examine and judge the facts for themselves. The usefulness of each person's career labors and their usefulness to science will depend on their choosing the correct fundamental assumptions.

References

Ambartsumian, V.A. 1958, Onzieme Conseil de Physique Solvay, ed. R. Stoops, Bruxelles.
Arp, H. 1967 Ap.J. 148, 321.
Arp, H. 1968, Astrofizika 4, 59.
Arp, H. 1987, "Quasars, Redshifts and Controversies" Interstellar Media, Berkeley
Arp, H. 1991, Apeiron Nos. 9-10, 18.
Arp, H. 1997, Astrophysics and Space Sciences 250, 163.
Arp, H. 1997a, Astron. Astrophys. 319, 33.
Arp, H. 1997b, Astron. Astrophys. 328, L17.
Arp, H. 1998a, Ap.J. 496, 661.
Arp, H. 1998b, "Seeing Red: Redshift, Cosmology and Acad. Sci." Apeiron, Montreal
Arp, H., Bi, H., Chu Y., Zhu, X. 1990, Astron. Astrophys. 239, 33.
Burbidge, G.R., Burbidge, E.M. 1997, Ap.J. 477, L13.
Chu, Y., Wei, J. Hu, J., Zhu, X. and Arp, H. 1998, Ap.J. 500, 596.
Holmberg, E. 1969, Arkiv of Astron., Band 5, 305.
Markarian B.E., Stepanyan D.A., Erastova L.K. 1986 Astrophysics 25, 51
Narlikar, J.V. and Arp, H. 1993, Ap.J. 405, 51.
Radecke, H.-D. 1997, Astron. Astrophys. 319, 18.
Reimers D., Hagen H.-J. 1998, Astron. Astrophys. 329, L25.
Sulentic, J.W., Arp, H. and Di Tullio, G.A. 1978, Ap.J. 220, 47.
Zaritsky, D., Smith, R., Frenk, C.S. and White, S.D.M. 1997, Ap.J. 478, L53.

Active Galactic Nuclei and Related Phenomena
IAU Symposium, Vol. 194, 1999
Yervant Terzian, Daniel Weedman, Edward Khachikian, eds.

AGN Variability

Toshihiro Kawaguchi and Shin Mineshige

Department of Astronomy, Kyoto University, Sakyo-ku, Kyoto 606-8502, Japan

Abstract. A number of monitoring observations of continuum emission from Active Galactic Nuclei (AGNs) have been made in optical–X-ray bands. The results obtained so far show (i) random up and down on timescales longer than decades, (ii) no typical timescales of variability on shorter timescales and (iii) decreasing amplitudes as timescales become shorter. The second feature indicates that any successful model must produce a wide variety of shot-amplitudes and -durations over a few orders in their light curves. In this sense, we conclude that the disk instability model is favored over the starburst model, since fluctuations on days are hard to produce by the latter model.

Inter-band correlations and time lags also impose great constraints on models. Thus, constructing wavelength and time dependent models remains as future work.

1. Introduction

Emission from Active Galactic Nuclei (AGNs) shows rapid and apparently random variability over wide wavelength ranges from radio to X ray or γ ray (for a review, see Ulrich, Maraschi & Urry 1997). In spite of numerous intensive, multi-wavelength monitoring projects, there still remain major questions; what causes the variability? Which parts of nuclei are emitting in various energy bands? One of the goals of variability studies is to identify and characterize the physical processes responsible for the observed variability.

Most AGN light curves exhibit neither apparent periodicity nor a typical timescale. In fact, the AGN light curves fluctuate over wide timescales (e.g. Fahlman & Ulrych 1975, for decades' light curve; Peterson et al. 1994, for years; Clavel et al. 1991, for months; Korista et al. 1995, for weeks; Edelson et al. 1996, for days). Figure 1 shows schematic light curves of AGNs on various timescales, from decades to days/hours. We point out the following three features;

1. The top panel, light curves on decades, is qualitatively different from the lower three. It seems to just fluctuate almost randomly, up and down.

2. There is no typical timescale in variability over less than years (see the lower three light curves). Flux variations over years to days/hours resemble each other; each light curve shows one or two bumps and small fluctuations superposed on them. In other words, the variations are self-similar or fractal.

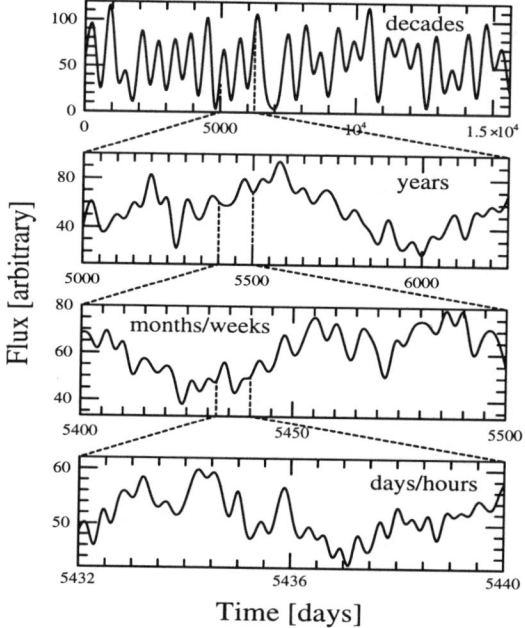

Figure 1. Schematic light curves of AGNs on different timescales.

3. Amplitudes of the variations over less than years decrease as the timescales become shorter.

To understand these features above in a statistical way, we often use the power-spectral density (PSD) and the structure function (SF). Observed PSD generally shows a power-law decline which is proportional to $f^{-\alpha}$. That corresponds to the fact that most AGN light curves show no typical timescale and the amplitudes of the variations decrease as the timescales become shorter. The power-law index (α) ranges from 1.5 to 2.0 (e.g. Leighly & O'Brien 1997; Hayashida et al. 1998). This index includes valuable information and is useful to discriminate among possible theoretical models. Generally, SF also has a power-law index, β, which is related to α by $\alpha = 1 + 2\beta$ (§2.1). On the other hand, when we analyze long light curves, we find one critical frequency (f_{break}) under which PSD flattens and one timescale (τ_{var}) over which SF flattens. The frequency and timescale are related to each other [$\tau_{\text{var}} \approx (2\pi f_{\text{break}})^{-1}$]. We often call τ_{var} a (maximum) variability timescale, and typically it is around 2 - 3 years (e.g. Hook et al. 1994).

Then, the main question in this paper arises;

- **What do the variability timescale (τ_{var}) and the power-law index of PSD (α) or that of SF (β) tell us?**

In Section 2, we focus on the question above, presenting our time series analysis of the light curves. There, we introduce the two possible models for

AGN optical variability: the disk instability model and the starburst model, and show how we can interpret the observed fluctuation properties based on the two models. Other observed properties, such as wavelength-dependence of the AGN variability, are discussed in Section 3. Finally, we summarize the conclusion and address issues remaining as future work.

2. Comparison of Models

In this section, we analyze light curves obtained by the observation, disk instability (DI) model (Mineshige et al. 1994), and starburst (SB) model (Aretxaga et al. 1997). The main issues are the variability timescale and power-law index of SF of AGN light curves (Kawaguchi et al. 1998).

2.1. Observed Fluctuation Properties

Light curves ideal for our analysis are long-term observational data over years, which have high sampling rates and good photometric accuracy. According to these criteria, we chose the optical light curve of the double quasar 0957+561 monitored by Kundić et al. (1997).

One might think that the flux variation of this macro-lensed quasar may be largely affected by microlensing events. Fortunately, however, most of the variability in this quasar is intrinsic (see Figure 4 of Kundić et al. 1997). Then, we adopt these light curves for testing models of intrinsic variability of AGNs.

Instead of PSD, we use a structure function $[V(\tau)]$ analysis, which is almost equivalent to PSD analysis but suitable for gapped data, as is often the case in AGN light curves. In short, SF expresses a curve of growth of variability with time-lag, and when a time series of magnitude $[m(t_i), i = 1, 2, \ldots; t_i < t_j]$ is given, it is defined as

$$V(\tau) \equiv \frac{1}{N(\tau)} \sum_{i<j} [m(t_i) - m(t_j)]^2, \qquad (1)$$

where summation is made over all pairs in which $t_j - t_i = \tau$, and $N(\tau)$ denotes a number of such pairs.

Usually, SF shows a power-law portion at smaller time-lags,

$$[V(\tau)]^{1/2} \propto \tau^\beta. \qquad (2)$$

This index β also contains important information concerning the variability mechanism, like a power-law index of PSD. In fact, their indices are related to each other by $\alpha = 1 + 2\beta$ in the limit of infinite and continuous data points.

We analyzed these light curves of Q0957+561 and showed that the power-law index β is about 0.35 (Kawaguchi et al. 1998), which is consistent with the value known for α (1.5 - 2.0). Although we can not see the turn-off in the SF, which corresponds to the variability timescale, we expect to find it around 2 - 3 years, as observed in other quasars (Hook et al. 1994), if we obtain a longer light curve.

AGN Variability

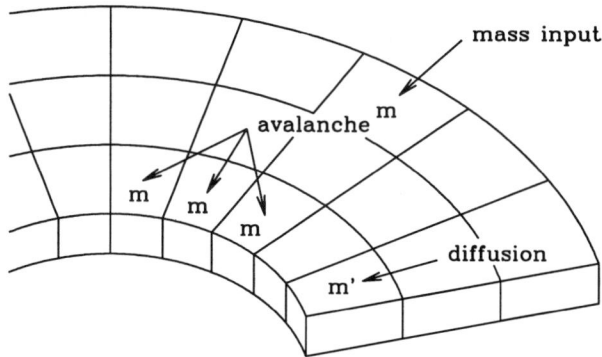

Figure 2. A schematic view of our cellular automaton model.

2.2. Disk-Instability Model

Procedure and Results

The first model we are discussing is the disk instability model, whose original idea was proposed by Bak et al. (1988) as a sand-pile model to explain fluctuation properties of complex systems, such as earthquakes. They considered random input of sand grains onto a sand pile, and imposed a certain rule; when a slope at some site exceeds the critical value, an avalanche occurs, and the slope decreases below the critical value. Then, they found that such a system evolves to and stays at a self-organized critical (SOC) state, in which any size of avalanche flows can occur, thus producing $1/f$-like fluctuations.

In fact, this SOC model has been modified to the case of black-hole accretion flows and applied to X-ray variability of Cyg X-1 (Mineshige et al. 1994; Takeuchi et al. 1995). We get great success in reproducing the basic shapes of PSD and the smooth shot-size distribution.

Here, we propose that an accretion disk (or flow) of AGN also stays at an SOC state, and calculate flux variations expected by the model. There are two key assumptions. The first is that the disk is locally unstable. Probably, the local instability is of magnetic origin, leading to magnetic reconnection (Matsumoto et al. 1998). The second assumption is that each site of the disk interacts with each other via avalanche.

The procedure of our cellular-automaton simulation is as follows (Figure 2).

(i) First, we divide the disk into numerous cells. Then, we assume each cell behaves as a reservoir, which is quiescent until the mass density exceeds the critical value. In a word, even when mass input is steady, output will be episodic.

(ii) We put one mass particle (m) at the outermost ring, which represents a mass supply to the disk.

(iii) Then, for unstable cells, where mass density exceeds a critical value, we set an avalanche flow. In other words, we move three mass ($3m$) particles into inner adjacent cells.

(iv) Aside from such critical behavior, we take account of viscous diffusion (m' at each annulus). In general, effect of the diffusion is much less than avalanche flow.

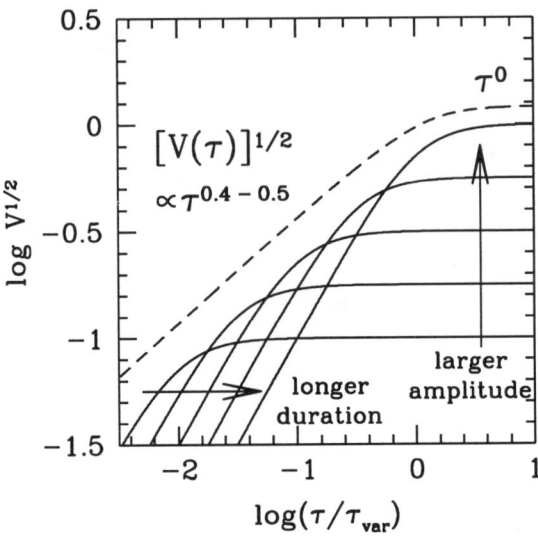

Figure 3. Typical structure function expected by the DI model (dashed line) and structure functions of individual flares (solid lines).

(v) We repeat the above procedures and we can draw the resultant light curve, assuming that the radiative energy is proportional to the potential energy released.

We calculated models for some parameter sets, finding that the resultant SFs exhibit power-law indices (β) of 0.4 - 0.5, which are close to the observed value ($\sim$ 0.35). Here, we should remark on the robustness of the cellular-automaton rule. The results are quite insensitive to the detailed rules, such as the number of cells, number of fallen particles, and prescription of the critical mass etc. The only influential parameters are the ratio of the diffusion flow over avalanche flow, m'/m, and the size of the disk.

What Determines the Power-Law Index and Variability Timescale?

Figure 3 illustrates what determines the power-law index and variability timescale in the DI model. The dashed line represents a typical SF; a gradual ($\beta \sim 0.4$ - 0.5) power-law increase at smaller time lags and a plateau over a variability timescale (τ_{var}). The solid lines show SFs of individual flares. In general, smooth time-symmetric flares produce steeper SFs than $\tau^{0.5}$, as displayed by those solid lines. Thus, in order to explain the observed gradual slope, 0.35, there must be numerous shots with a wide variety of shot-amplitude and -duration. Moreover, smaller-amplitude and shorter-duration flares should occur more frequently than larger and longer ones.

Thus, the index does not sensitively depend on the shot profile. Instead, the distribution of shot-amplitude and -duration is more important. Therefore, the index will not be largely changed even when the light curves of individual flares are modified.

According to this picture, the variability timescale is determined by the duration of the largest flare. If the fluctuating part of the disk is advection-dominated, that duration will be of order of the free-fall timescale from the outer rim. Then, variability timescales of several hundred days roughly correspond to the size of 100 Schwarzschild radii (r_g) for a black hole mass of $10^8 M_\odot$;

$$\tau_{\text{var}} \approx \frac{r}{v_r} = 160 \left(\frac{v_r}{0.1 v_{\text{ff}}}\right)^{-1} \left(\frac{r}{10^2 r_g}\right)^{3/2} \left(\frac{M}{10^8 M_\odot}\right) \text{ day}, \quad (3)$$

where v_r and v_{ff} are radial velocity and free fall velocity, respectively.

2.3. Starburst Model

Procedure and Results

As an alternative model for radio-quiet AGNs, the starburst model has been investigated by Prof. Terlevich and his collaborators (e.g. Terlevich et al. 1992). According to the starburst model, optical variability is explained as superposition of supernova (SN) explosions. In the superposition, the released energy and duration of each SN explosion are varied within a factor of 2 in a Gaussian way around the mean values given a priori. As the supernova rate increases, resultant light curves are less variable in magnitude. In other words, luminous AGNs must be less variable, which roughly agrees with the observed trend (Cristiani et al. 1996; see also Paltani & Courvoisier 1997).

We followed Aretxaga et al. (1997) and calculated light curves, finding that the power-law indices β expected from the model are around 0.7 - 0.9, which is significantly larger than the observed value, 0.35. Although the quasar we used here for the comparison is a radio-loud object and this may not be a fair test for the starburst model, we assume that optical variability of radio-loud objects is not very different from that of radio-quiet ones.

What Determines the Power-Law Index and Variability Timescale?

Then, why are the results so different between the two models? In both models, something like shots or flare-like events are superposed almost randomly in the light curves. The critical point is that in the SB model there will not be large variation in the shot-amplitude and -duration over many orders of magnitude; timescales and released energies of each explosion can vary only by some factor. This is the substantial difference from the disk instability model.

As a result, both the timescale and power-law index of SF are determined by a typical light curve of SN explosions, as shown in Figure 4.

To produce the gradual index in the SF, we need much shorter events; for instance, SN explosions decaying over days. This is probably not the case in the actual situation. It is possible that a thermal instability within SN remnants may change the typical light curve to fluctuate more over short timescales (Cid Fernandes et al. 1996), so that the resultant power-law index will be consistent with the observed one.

3. Other Issues

We have focused on the AGN light curve itself, with structure function analysis. In this section, we will discuss several related issues, especially the wavelength dependency of the variability.

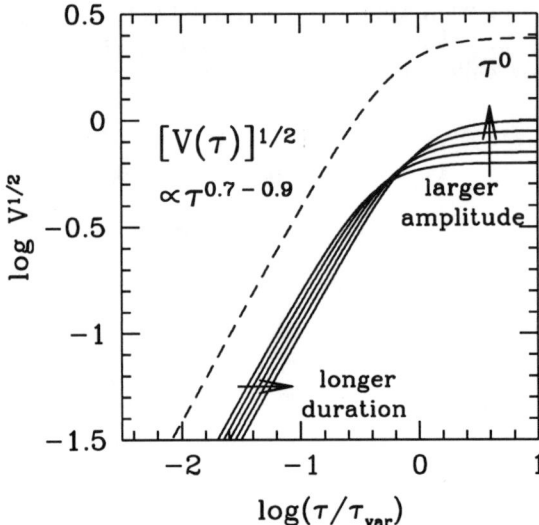

Figure 4. Typical structure function expected by the SB model (dashed line) and structure functions of individual explosions (solid lines).

First, there is a clear anti-correlation between fractional amplitudes and observed wavelength (e.g. Di Clemente et al. 1996). In other words, flux at shorter wavelength varies with larger amplitudes, compared to that at longer wavelength. It is also known that optical/UV continuum becomes "harder" as it gets brighter (Maoz et al. 1993; Peterson et al. 1994; Trèvese 1998). These properties should contain important clues to the understanding of the physics and site of variability generation. However in dealing with the spectral variability, we must be careful about the mixture of numerous emission lines and Balmer continuum, and also contaminations of other components.

Next, the flux variations at different wave-bands are almost simultaneous (e.g. Edelson et al. 1996), but recently it is reported that there seem to exist some lags by several days or less than a day; (i) Optical variations tend to lag behind UV variations by ~ 1 day (Peterson et al. 1998; Courvoisier 1998) (ii) Flux variations of NGC 7469 in UV band sometimes lead X-ray ones by ~ 4 days and sometimes they are simultaneous (Nandra et al. 1998).

In tentative pictures for central engines of AGNs, such as a cold accretion disk with hot corona or cold clouds within a hot corona, we had expected that X-ray variations lead Opt./UV variations, if we could detect such small lags. The new results of Nandra et al. (1998) are challenging to the tentative pictures. Thus, it is a good time to construct a wavelength-dependent model of AGN variability including an anti-correlation of wavelength – variation amplitude and spectral variability.

4. Summary

(1) As far as the power-law index in structure functions is concerned, the disk instability model is favored, since this model can produce a wide variety of shot-amplitudes and -durations, as found in observed light curves.

(2) Observed variability timescales and power-law indices are the key properties to testing models.

(3) Constructing a wavelength and time dependent model of AGNs remains as future work.

Acknowledgments. One of the authors (TK) is grateful to Prof. Edward Khachikian for the invitation to this stimulating conference. He also thanks the Yamada Science Foundation for financial support.

References

Aretxaga, I., Cid Fernandes, R., & Terlevich R. J. 1997, MNRAS, 286, 271
Bak, P., Tang C., & Wiesenfeld K. 1988, Phys.Rev.A, 38, 364
Cid Fernandes, R. et al. 1996, MNRAS, 283, 419
Clavel, J. C. et al. 1991, ApJ, 366, 64
Courvoisier T. J.-L. 1998, these proceedings
Cristiani, S. et al. 1996, A&A, 306, 395
Di Clemente A. et al. 1996, ApJ, 463, 466
Edelson, R. A. et al. 1996, ApJ, 470, 364
Fahlman, G. G., & Ulrych T. J. 1975, ApJ, 201, 277
Hayashida, K. et al. 1998, ApJ, 500, 642
Hook, I. M. et al. MNRAS, 268, 305
Kawaguchi, T., Mineshige S., Umemura M., & Turner E. L. 1998, ApJ, 504, 671
Korista, K. T. et al. 1995, ApJS, 97, 285
Kundić, T. et al. 1997, ApJ, 482, 75
Leighly, K. M., & O'Brien, P. T. 1997, ApJ, 481, L15
Maoz D. et al. 1993, ApJ, 404, 576
Matsumoto, R., & Shibata, K. 1998, in preparation
Mineshige, S., Takeuchi, M., & Nishimori, H. 1994, ApJ, 435, L125
Nandra K. et al. 1998, ApJ, 505, 594
Paltani S., & Courvoisier T. J.-L. 1997, A&A, 323, 717
Peterson, B. M. et al. 1994, ApJ, 425, 622
Peterson, B. M. et al. 1998, PASP, 110, 660
Takeuchi, M., Mineshige, S., & Negoro, H. 1995, PASJ, 47, 617
Terlevich, R. et al. MNRAS, 255, 713
Trèvese, D. 1998, these proceedings
Ulrich, M.-H., Maraschi L., & Urry, C. M. 1997, ARA&A, 35, 445

Supernova Types, Star Formation and AGN

Massimo Turatto and Enrico Cappellaro

Osservatorio Astronomico di Padova, vicolo dell'Osservatorio 5, 35122 Padova, Italy

Artashes R. Petrosian

Byurakan Astrophysical Observatory, Aragatsotn Province, Armenia

Abstract. New values of the frequencies of SNe are presented and discussed in relation to their use as SF indicators. The rate of core–collapse SNe is correlated to the colors and the FIR excesses of the parent galaxies in the sense that galaxies with blue colors and strong infrared excess have higher occurrence of type II and Ib/c SNe than other galaxies. This is in agreement with the expectation that they contain a higher fraction of massive stars. Instead no correlation is present for SNIa. The SN frequency does not correlate with the galaxy activity probably because searches are unable to discover SNe in the nuclear regions of galaxies.

A number of SNe with spectra similar to those of AGN exist. Their characteristic features are explained with explosion of SNe in dense environments, reminding of cSNR's invoked in the starburst model for AGNs. Some recent, peculiar SNe seem linked to GRB's opening the possibility that at least some GRB's arise from this kind of stellar explosion.

1. Introduction

The common property of all kinds of SNe, independent of the explosion mechanism, is the amount of energy (of order 10^{51} ergs) emitted as kinetic and luminous energy. After that each SN type has distinctive characteristics.

Type Ia SNe are discovered in all kind of galaxies and show a high degree of homogeneity. They have been explained by thermonuclear explosions of accreting WD's, hence associated with Population I. Recent findings have shown that some SNIa reached magnitudes at maximum considerably different from the average and challenged the standard scenario.

Type II and Ib/c SNe result instead from the explosions of massive stars (M> 8 $M_\odot$) via core collapse. The optical displays are governed mainly by the configurations of the precursor stars at the moment of explosion and by the amount of radioactive material. If the stars retain the H envelopes, the SN shows Balmer lines and is classified as type II. On the other side, if the H envelope (and, in cases, even the He layer) has been lost by the progenitor via wind or binary interaction, the SN is classified as SNIb (or SNIc). The discovery of few but significant cases of transitions from early SNII to late SNIb/c spectra, has firmly established a physical link between these classes of objects, hence confirmed the association of SNIb/c with young progenitors.

For their descent from different progenitors, SNe have been used in the past as probes of stellar populations in galaxies. In particular, SNII and Ib/c have been used as indicators of star formation (e.g. Petrosian & Turatto 1990, 1995). Clearly differently from other SFR indicators such as FIR luminosity, intensity of recombination lines, etc., SNe cannot be used to test individual galaxies but can give the SFR in a given sample.

In addition to these topics, we will discuss the properties of a particular subclass of type II SNe, the SNIIn, which for their overall spectral similarity to AGN are of special interest to the subject of this conference. Finally, the recent association of some SN to GRB's will be briefly addressed.

2. The Supernova and SF rates

Given a sample of N_{gal} galaxies, the rate of SNe per unit time and luminosity is given by

$$\frac{N_{SN}}{\sum_1^{N_{gal}} t_i L_i k_i}, \qquad (4)$$

where N_{SN} is the number of SNe discovered in the sample, t_i and L_i are the surveillance time and the luminosity of the i-th galaxy, and k_i is the correction term, relative to that galaxy, which takes into account different search biases.

Currently, the efforts to improve the accuracy of the observational rates are focused in increasing the statistics, i.e. N_{SN}, and in determining appropriate corrections for the selection effects (cfr. Cappellaro et al. 1997 and references therein).

By grouping the databases of five SN searches (updated versions of those described in Cappellaro et al. 1997) we have compiled a sample of about $\simeq 10^4$ RC3 galaxies in which 137 SNe have been discovered, the largest SN sample ever used for such purposes. The results are summarized in Table 1. Note that the rates scale as $(H/75)^2$. The reported errors take into account the statistics of SNe and the uncertainty on the input parameters and the selection effects.

Table 1. The rate of Supernovae in a sample of about 10^4 RC3 galaxies and 137 SNe.

galaxy	n. of SNe			rate [SNu]			
type	Ia	Ib	II	Ia	Ib	II	All
E-S0	22.0			0.18 ± 0.06	< 0.01	< 0.02	0.18 ± 0.06
S0a-Sb	18.5	5.5	16.0	0.18 ± 0.07	0.11 ± 0.06	0.42 ± 0.18	0.72 ± 0.14
Scd-Sd	22.4	7.2	31.5	0.21 ± 0.08	0.14 ± 0.07	0.86 ± 0.36	1.21 ± 0.46
Other	6.8	2.2	5.0	0.40 ± 0.19	0.22 ± 0.17	0.65 ± 0.37	1.26 ± 0.40
All	69.6	14.9	52.5	0.20 ± 0.06	0.08 ± 0.04	0.40 ± 0.15	0.68 ± 0.11

in SNu = 1 SN $(100yr)^{-1}$ $(10^{10} L_\odot^B)^{-1}$

Somewhat different from older determinations (e.g. Cappellaro et al. 1993) the rate of SNIa does not increase significantly from ellipticals to late spirals. This is due both to a new determination of the correction factors (Cappellaro et al. 1997) and to new galaxy data (here we adopted the RC3 updated version from CDS, 1995). Instead, the rate of core–collapse SNe is rapidly increasing from zero in ellipticals to about 0.5 SNu in early spirals and close to 1 SNu in late spirals. This resembles the average increase of SFR along the Hubble sequence (Kennicutt 1998).

The correlation of SN rates with other SFR indicators has been controversial. For instance, Kennicutt (1984) found that the rate of SNII was strongly correlated to the total massive star formation rate in galaxies as measured by their total H_α emission. Instead, on a small sample of galaxies, Richter & Rosa (1989) suggested that the star formation determined from the UV luminosities was only slightly correlated to the SN productivity. Statistical analyses of the star forming Markarian galaxies and of AGN host galaxies have not found enhanced SN rates but only a tendency of SNe to explode closer to the nuclear regions (Turatto et al. 1989; Petrosian & Turatto 1990).

We have computed again the rates of SNe in AGN host galaxies by extracting from our general sample the active galaxies according to Veron–Cetty & Veron (1998). Though only 3 % of the galaxies of the sample are active, they hosted 12% of SNe. Yet, the rate expressed in SNu is not different from that in normal galaxies. This apparent contradiction is due to the fact that both the average control time and luminosity of the AGN of the sample are higher than normal galaxies. In other words, AGN host galaxies were surveyed better and they are bigger than other galaxies of our sample. A similar conclusion, i.e. that the SN frequency in star forming galaxies is not enhanced, was reached by Richmond et al. (1998) from a dedicated SN search in a sample of 142 starburst galaxies. We note however that SN rates are not able to monitor SNe in the central regions of the galaxies because of increased extinction.

The correlation of SN rate with SF indicators is possibly hidden by normalization factors. Actually, the fact that the rate of core–collapse SNe in SNu(B) increases towards late spirals indicates that the blue luminosity is not an optimal indicator of current SF (e.g. Sage and Solomon 1989), otherwise one should expect a constant rate per unit luminosity.

Alternatively one may use L_{FIR}, which is often used as a direct measure of the SFR. Only 30% of our galaxy sample have FIR data. The results, reported in Tab. 2, show that the rate of core–collapse SNe normalized to FIR luminosity increases along the Hubble sequence. This means also that L_{FIR} is not a universal measure of SFR.

A possible explanation is that the FIR luminosity of the galaxies is the sum of two components, one taking place in the circumnuclear regions and one in the disks. The two components are decoupled with the circumnuclear component strongly correlated to the nuclear SF. Instead the disk component is contaminated by the IR emission of the dust heated by evolved stellar populations, which varies significantly with the Hubble type. On the other hand, as said before, because of strong extinction optical SN searches cannot discover objects in the nuclear regions and the observed SN rates test SFR only in the disks. The SN rates in units of FIR luminosity shows the same dependence as a function of the Hubble types as those expressed in SNu(B) (cfr. Tab. 2).

More reliable to this purpose can be the FIR excess, L_{FIR}/L_B, though also this SF indicator has been questioned in the past because both the FIR and B

Table 2. The rate of Supernovae in units of FIR luminosity.

galaxy type	rate [SNu$_{FIR}$]		
	Ia	II+Ib/c	All
E-S0	1.8 ± 0.8		1.8 ± 0.8
S0a-Sb	0.6 ± 0.2	2.0 ± 0.5	2.7 ± 0.5
Sbc-Sd	0.6 ± 0.1	3.5 ± 0.6	4.1 ± 0.6
All	0.7 ± 0.1	2.5 ± 0.3	3.2 ± 0.3

1 SNu$_{FIR}$ = 1 SN $(100yr)^{-1}$ $(10^{10} L_\odot^{FIR})^{-1}$

luminosity somehow correlate with the SFR (e.g. Kennicutt 1998). In Tab. 3 the SN rates in SNu(B) for three groups of galaxies with different FIR excess, i.e. those with strong FIR excess, those with low excess and those with no FIR flux, are shown. A trend is evident for type II+Ib/c SNe with higher rates in galaxies with stronger FIR excess. Such effect is not seen for SNIa. One may conclude that the FIR excess is a better SFR indicator than L_{FIR}.

Table 3. SN rates in SNu(B) for galaxies with different FIR excess

galaxy type	not detected by IRAS		$L_{FIR}/L_B < 0.35$		$L_{FIR}/L_B > 0.35$	
	Ia	II+Ib/c	Ia	II+Ib/c	Ia	II+Ib/c
E-S0	0.2 ± 0.1		0.4 ± 0.2		< 0.5	
S0a-Sb	0.2 ± 0.1	0.3 ± 0.2	0.2 ± 0.1	0.5 ± 0.2	0.3 ± 0.1	1.1 ± 0.4
Scd-Sd	0.3 ± 0.1	0.7 ± 0.3	0.2 ± 0.1	1.0 ± 0.2	0.2 ± 0.1	1.2 ± 0.3
All	0.2 ± 0.1	0.2 ± 0.1	0.2 ± 0.1	0.9 ± 0.1	0.3 ± 0.1	1.1 ± 0.2

in SNu(B)

As a further check we have estimated the SN rates in galaxies with different colors extending the analysis carried out by Tammann (1974). An increase of the SN rate of type II and Ib/c SNe is visible in all galaxy bins while that of SNIa is insensitive to the colors of the host galaxies. This is expected since bluer galaxies are expected to host more young stars than red ones.

Table 4. SN rates for galaxies of different (B-V) colors

galaxy type	$<(B-V)_T^0>$	rates [SNu]		$<(B-V)_T^0>$	rates [SNu]	
		Ia	II+Ib		Ia	II+Ib
E-S0	0.86	0.3 ± 0.1		0.95	0.2 ± 0.1	
S0a-Sb	0.60	0.2 ± 0.1	0.6 ± 0.2	0.80	0.2 ± 0.1	0.5 ± 0.2
Sbc-Sd	0.45	0.1 ± 0.1	1.5 ± 0.3	0.62	0.3 ± 0.1	0.9 ± 0.2
All	0.56	0.2 ± 0.1	1.0 ± 0.2	0.88	0.2 ± 0.1	0.1 ± 0.1

in SNu(B)

We conclude, therefore, that the rates of core–collapse SNe correlate with the FIR excess and the galaxy colors once the dependence on the Hubble type is removed. This finding confirms that the SN rates can be used as probes of SF. Conversely, Tables 3 and 4 confirm that the parent population of type II and Ib/c SNe is constituted by young massive stars.

2.1. Supernovae and galaxy activity

The source of the energy in active galaxies is, according to the standard model, the accretion of material onto a supermassive nuclear black hole. As an alternative, Terlevich and collaborators have developed in the last decades the starburst model, which does not require the presence of a black hole to explain the AGN's observed properties (Terlevich et al. 1992, 1995). The model explain the time variability, X-ray and radio emission, BLR properties and NLR ionization by means of *compact Supernova Remnants* (cSNR), i.e. ordinary SNRs evolving in a dense (10^6 cm^{-3}) circumstellar medium which are able to radiate most of their kinetic energy in a few years.

Without entering the dispute of how good the starburst model is in explaining the AGN properties as compared with the standard model, there is no doubt that a number of SNe with spectra impressively similar to those of AGN exist. Such *Seyfert 1 impostors*, as named by Filippenko (1989), or SNIIn, as most commonly called by supernovists, are indeed normal SNe exploding in dense environments.

The best example of such a SNe is 1988Z (Stathakis & Sadler 1991, Turatto et al. 1993). Contrary to normal SNII, the spectrum of SN 1988Z does not show the P–Cygni profiles but only relatively narrow emission with complex profiles. Shortly after the explosion the most intense component of Hα was 2200 km s^{-1} wide (FWHM) and less intense, but clearly visible, were a broad component (FWHM$\sim$ 15000 km s^{-1}) and an unresolved one (FWHM < 700 km s^{-1}). A number of unresolved faint lines of very high ionization (e.g. [FeX], [FeVII], [FeVI]) were also present. The line intensity ratios of SN 1988Z indicate densities of the order of 10^6 to 10^7 cm^{-3} and this was confirmed by the non-detection of the [NII] 6548-6583A lines.

In addition to the peculiar spectral features, the temporal evolution of this object was unusually slow. After the maximum at about M= -18 the SN declined only 3 mag in 3 yr in the R band while normal SNe fade the same amount in few months. Indeed, despite the relatively large distance, SN 1988Z is still reachable nowadays, 10 years after the discovery.

Several supernovae with narrow emission lines have been grouped in the SNIIn subclass (Schlegel 1990). Though it is not yet clear if all such objects undergo similar physical processes, it is clear that at least a fraction of them share also other features of SN 1988Z, i.e. relatively narrow lines, very slow evolution, radio and X-ray emission and in some cases, weak unresolved lines of very high ionization.

Remarkable is the spectral similarity of SN 1988Z with SN 1995N (Turatto et al. 1998) and, to a lesser extent, with SN 1995G (Pastorello 1998). Figure 1 compares the spectra of these three SNe to that of a Sy 1.5 galaxy.

The observations of these SNe can be interpreted by a model in which the bulk of emission is due to the interaction of the ejecta with a dense CSM (Chugai & Danziger 1994). The broad emissions arise at the contact discontinuity between the high–velocity shocked ejecta and the shocked circumstellar wind, while the lines of intermediate widths come from shocked clumps embedded in the rarefied medium. The model also explains well the radio and X–ray emission from the shocked material and the narrow coronal lines, since temperatures as high as 10^6 K are reached.

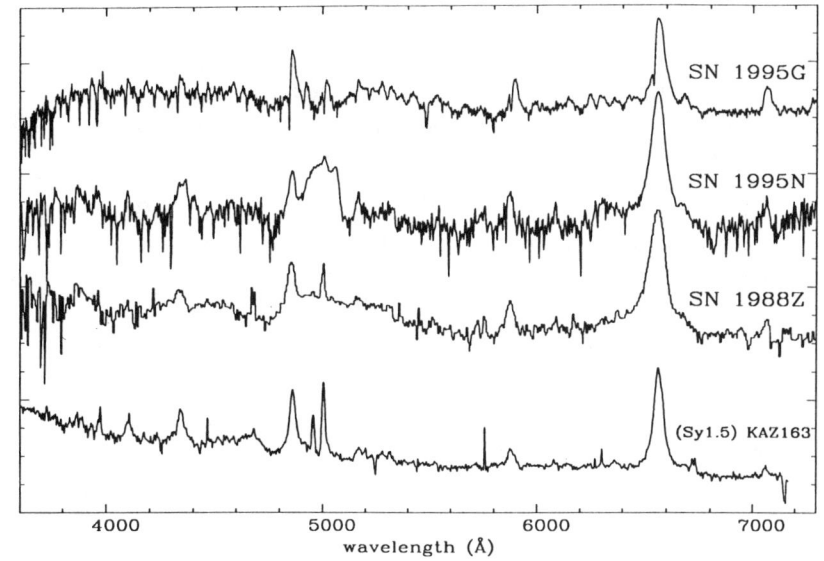

Figure 1. Comparison of the spectra of three SNIIn, SNe 1988Z, 1995G and 1995N at about 2 years past maximum, with a Seyfert 1.5 spectrum.

The existence of SNe with spectra similar to AGN, which are interpreted by the interaction of the ejecta with dense circumstellar matter, renews interest in the starburst model.

3. γ–ray activity

Gamma Ray Bursts are one of the mysteries of current astronomy. A recent breakthrough was obtained thanks to the Italian–Dutch Beppo–SAX satellite which allowed the optical identifications of a number of GRB's. It turned out that some of them are related to some sort of activity in distant galaxies. One outstanding exception was GRB 980425 associated with the relatively nearby (z=0.0085) SN 1998bw. This SN was discovered within the GRB error box soon after the detection (Galama et al. 1998). The a *posterior* probability to find a SN inside the error box of WFC is only 10^{-4}

The spectrum appeared unusual while showing very broad features typical of SNe. The light curve was also unprecedented. Extensive multifrequency observations, still in progress, are showing that this is the most powerful radio SN ever, and that relativistic outflow is required to explain the spectral features. If the SN and the GRB are indeed associated, i.e. if we place the GRB at the distance of the SN parent galaxy, the peak luminosity and total budget of the γ–ray event are orders of magnitude fainter than other GRB's.

The properties of SN 1998bw have been modeled by the collapse into a black hole of a C+O star with mass of 12–15 $M_\odot$. A core like this might originate from a progenitor with initial mass of about 40 $M_\odot$ stripped of its H and He either via stellar wind or binary interaction. The explosion energy required is huge for a SN, of the order of 10^{52} ergs, and the mass of ^{56}Ni produced is 0.6–0.8 $M_\odot$ (Iwamoto et al. 1998). High energy photons are emitted via the synchrotron mechanism when, after the shock breakout, external layers are highly accelerated to produce a relativistic shock with the circumstellar material. It must be noted however that, although this was a relatively weak GRB, the γ–rays predicted by such a model are not sufficient to explain the observations unless the emission was collimated (Iwamoto et al. 1998).

The discovery of this SN–GRB event has triggered the search for angular and temporal correlations of other SNe with detected GRB's. The results are contradictory (Wang & Wheeler 1998, Kippen et al. 1998). The straight association of normal SNe with GRB's seems extremely weak, while it seems possible that faint GRB's are associated with peculiar SNe.

Woosley et al. (1998) have pointed out among others the interesting association of SN 1997cy with GRB 970514, which are compatible both in epoch and location. This SN has been extensively observed by us at ESO and the preliminary reduction of the data shows that both the spectra and the light curve share several of the characteristics of SN 1988Z, namely narrow emission lines with broad wings (Fig. 2) and the slow light curve. The study of this SN is at present in progress but there is evidence that such event was also particularly energetic: with $M_V(max) > -20$ this would be the brightest SN ever.

The discovery and the study of cases like these will finally prove the association of GRB's with peculiar SNe and define the role of SNe in powering these extraordinary events.

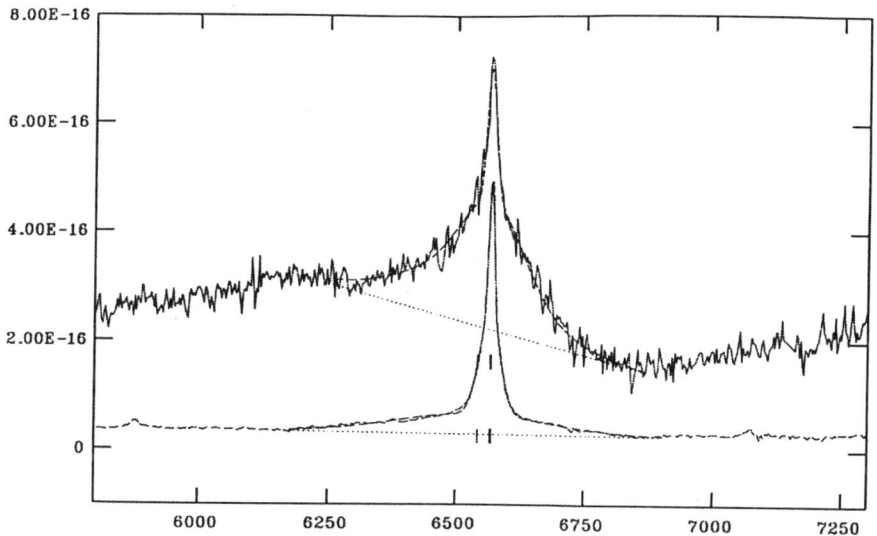

Figure 2. Comparison of the H_α emission of SNe 1988Z (bottom) and 1997cy (top) in spectra taken short after maximum. Relatively narrow lines with broad wings are evident in both cases.

Acknowledgments. We are grateful to Andrea Pastorello for the preparation of the figures.

References

Cappellaro, E. Turatto, Benetti, S., M. Tsvetkov, D.Yu. Bartunov, Makarova, I.N., 1993, A&A, 273, 383

Cappellaro, E. Turatto, M. Tsvetkov, D.Yu. Bartunov, O.S. Pollas, C. Evans, R. Hamuy, M., 1997, A&A, 322, 431

Chugai, N., Danziger, I.J., 1994, MNRAS, 268, 173

Filippenko, A., 1989, AJ, 97, 726

Galama, T. J. Vreeswijk, P. M. Van Paradijs, et al., 1998, Nature, 395, 670

Iwamoto, K. Mazzali, P. A. Nomoto, et al., 1998, Nature, 395, 672

Kennicutt, R.C., 1984, AJ, 277, 361

Kennicutt, R.C., 1998, preprint astro-ph 9807187

Kippen, R. M. Briggs, M. S. Kommers, J.M., et al., 1998, ApJ, 506, L27

Pastrorello, A., 1998, Tesi di laurea, Univarsità di Padova

Petrosian, A.R., Turatto, M., 1990, A&A239, 63

Petrosian, A.R., Turatto, M., 1995, A&A297, 49

Richmond, M.W., Filippenko, A., Galisky, J., 1998, PASP, 110, 553

Sage, L.J., Solomon, P.M., 1989, ApJ, 344, 204

Stathakis, R.A., Sadler, E.M., 1991, MNRAS, 250, 786

Tammann, G.A., 1974, Supernova and Supernova Remnants, ed. C.B. Cosmovici, Dordrecht, Reidel, p.155

Terlevich, R.J., Tenorio-Tagle, G., Franco, J., Melnick, J., 1992, MNRAS, 255, 713

Terlevich, R.J., Tenorio-Tagle, G., Franco, J., Rozyczka, M., Melnick, J., 1995, MNRAS, 272, 192

Turatto, M., Cappellaro, E. Danziger, I. J. Benetti, S. Gouiffes, C. Della Valle, M., 1993, MNRAS, 262, 128

Turatto, M., Benetti, S., Cappellaro, E., Danziger, I. J. Mazzali, P., 1998, *SN 1987A: ten Years after*, eds. Phillips, M.M., Suntzeff, N.B., Fifth ESO/CTIO/LCO Workshop, in press

Veron-Cetty, M.P., Veron, P., 1998, Catalogue of Quasars and Active Galaxy Nuclei, ESO Scientific Report 18.

Woosley, S.E., Eastman, R.G., Schmidt, B.P., 1998, ApJ, in press

Active Galactic Nuclei and Related Phenomena
IAU Symposium, Vol. 194, 1999
Yervant Terzian, Daniel Weedman, Edward Khachikian, eds.

The Gravitational Lens System B1030+074. Discovery and Follow-up.

E. Xanthopoulos, I. W. A. Browne, L. J. King, N. J. Jackson, D. R. Marlow, P. N. Wilkinson

University of Manchester, NRAL Jodrell Bank, Macclesfield, Cheshire SK11 9DL, England

L. V. E. Koopmans

Kapteyn Astronomical Institute, P. O. Box 800, 9700 AV Groningen, The Netherlands

A. R. Patnaik and R. W. Porcas

Max-Planck-Institut für Radioastromomie, Auf dem Hügel 69, D 53121, Bonn, Germany

Abstract. We report the discovery of a new double image gravitational lens system B1030+074 which was found during the Jodrell Bank - VLA Astrometric Survey (JVAS). We have collected extensive radio data on the system using the VLA, MERLIN, the EVN and the VLBA as well as HST WFPC2 and NICMOS observations. The lensed images are separated by 1.56 arcseconds and their flux density ratio at centimetric wavelengths is approximately 14:1 although the ratio is slightly frequency dependent and the images appear to be time variable. The HST pictures show both the lensed images and the lensing galaxy close to the weaker image. The lensing galaxy has substructure which could be either part of the galaxy or a companion object. We have modeled B1030+074 using a Singular Isothermal Ellipsoid that yielded a time delay of $156/h_{50}$ days. This lens is likely to be suitable for the measurement of the Hubble constant.

1. Introduction and observations

The Jodrell-Bank VLA Astrometric Survey (JVAS) is a survey of flat-spectrum radio sources one of whose purposes is to search for gravitational lens systems (Patnaik et al. 1992; Browne et al. 1998, Wilkinson et al. 1998). We report here the discovery of such a double system, B1030+074.

The discovery map of B1030+074 is presented in Figure 1. It shows two distinct compact components with a separation of 1.56 arcsec, with the fainter image, B, at a PA of 142° relative to image A. Table 1 presents all the radio observations. In all the radio maps the two components are unresolved. Only the VLBA 5 GHz data with 3 mas resolution were able to resolve at least one of the components, A, which shows a jet-like extension to the North-East (PA of 65 degrees) and 20 mas in length.

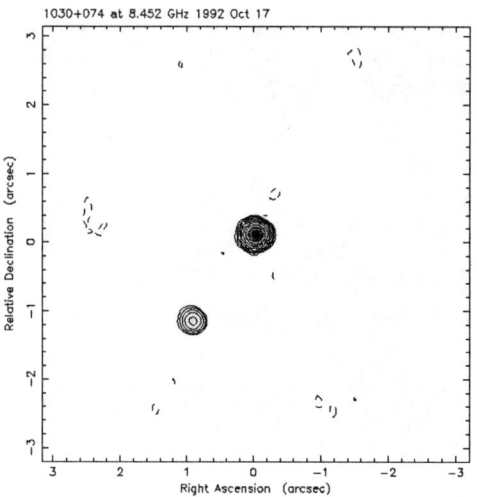

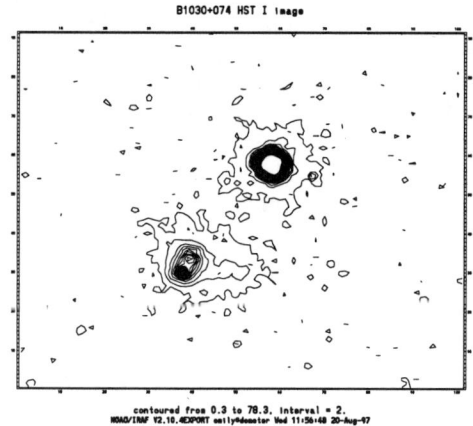

Figure 1. Top: VLA 8.4 GHz 1992 discovery map restored with a 200 × 200 mas beam. The contours are 0.00037 Jy per beam × (-2, 2, 4, 8, 16, 32, 64, 128, 256, 512), and the peak brightness is 0.189 Jy per beam. Bottom: The HST WFPC2 I image. North is up and East to the left. The lowest contour level is 2σ of the sky background value and consecutive contours differ by a factor of 2 in intensity.

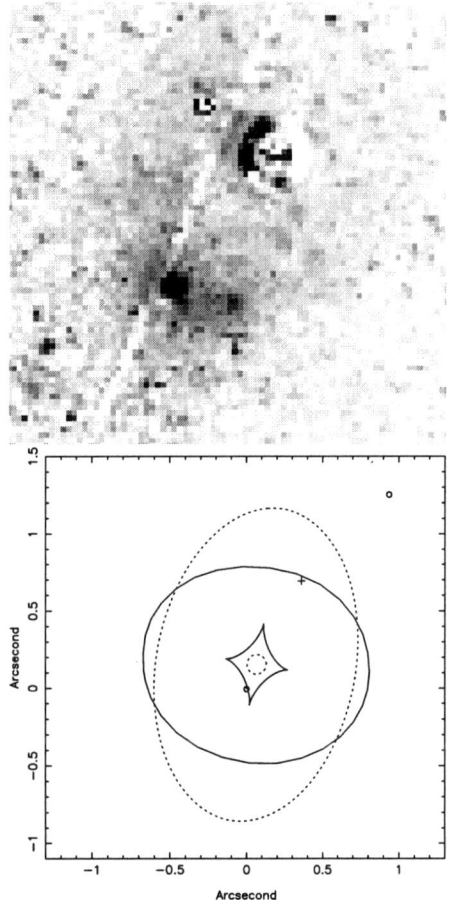

Figure 2. Top: HST NICMOS H band image after the two point objects A and B have been subtracted. The image is 3.69 arcsec on a side. The very bright and close to saturation A image makes a good PSF subtraction difficult. Bottom: The critical (dashed) and caustic (solid) structure of the "best" model of B1030+074. The circles indicate the observed image positions, the cross the inferred source position.

Table 1. Radio observations of B1030+074. Absolute amplitude errors are estimated to be 5%; the flux density ratios are accurate to ≈1%.

Telesc.	Obs. date	Frequ. (GHz)	Resol. (arcsec)	Flux A (mJy)	Flux B (mJy)	Flux (A/B)
EVN	1994 05 15	1.7	0.015	147	8.1	18.1
EVN	1994 11 18	5	0.005	173	10.9	15.9
MERLIN	1993 09 27	1.7	0.150	186	9.8	18.8
MERLIN	1996 12 27	5	0.050	326	27.3	12.0
VLBA	1995 11 12	5	0.003	248	19.1	13.0
VLA	1992 10 17	8.4	0.240	202	16.0	12.6
VLA	1994 02 22	8.4	0.240	197	12.9	15.2
VLA	1994 02 22	15	0.140	208	14.8	14.0
VLA	1995 12 20	15	0.140	295	24.4	12.1
VLA	1994 02 22	22	0.080	184	15.3	12.0
VLA	1995 12 19	22	0.080	219	12.2	18.0

HST images were obtained of B1030+074 using WFPC2 in two filters F555W and F814W. The contour plot of the I band image is shown in Fig. 1. The contour plots reveal compact optical objects corresponding to both radio components, together with a galaxy between them and very close to the B component. These data leave no doubt that B1030+074 is a gravitational lens system. The lensing galaxy appears not to have a simple smooth light distribution nor is it symmetric about the galaxy core. The overall extension of 1.035 arcsec is in a position angle nearly perpendicular to the image separation (6.1 kpc at the redshift 0.599 of the galaxy assuming $H_0 = 75$ km sec^{-1} Mpc^{-1} and $q_0 = 0$). The redshift of the source is 1.535. We computed and subtracted scaled PSFs from the direct CCD images. More detail of the lensing galaxy is then revealed. The profile information seems to support the spiral galaxy interpretation for the lensing galaxy since an $r^{1/4}$ law does not fit the surface brightness profile of the galaxy.

NICMOS observations were taken with Camera 1 through the F160W filter. In Fig. 2 we show the NICMOS 1.6μm picture. The peak of the galaxy light lies on the line joining the two images but there is also a secondary emission feature to the SE of the main part of the galaxy. The H-band data indicate that the colour of this feature is similar to that of the rest of the lensing galaxy, possibly suggesting that it is not a spiral arm but either part of the main galaxy or a companion object. Table 2 presents the flux and positions of the two components of the lens system and the lensing galaxy from all three HST band images. The colours of the lensed images are consistent with the lensed object being a quasar or BL Lac object. The optical/infrared flux density ratios of the images are considerably higher than any of those measured in the radio. It is tempting to attribute such differences to extinction but the fact that the H-band ratio is even larger than V and I ratios argues against this. We suggest that the differences are best explained as arising from microlensing.

Table 2. Optical and F160W image photometry and positions. The galaxy photometry refers to the light within the Einstein radius. Plate scales are 45.5 mas/pixel for the 555-nm and 814-nm images and 43 mas/pixel for NICMOS. Errors in the radio positions are 1 mas or less.

Object	Image	Flux density (555 nm)	Flux density (814 nm) 10^{-20}W m^{-2}nm^{-1}	Flux density (1.6 μm)
B1030+074	A	26.94±0.10	27.37±0.07	47.6±0.4
	B	0.9±0.2	1.17±0.17	1.40±0.09
	GAL	6?	7.9±0.7	10.8±0.5

	Offset from A (RA,δ,err) / mas			
555 nm	814 nm	1.6 μm	Radio	
930,−1256,4	931,−1247,4	918,−1244,5	935, −1258	
882,−1155,10	878,−1143,10	864,−1162,5		

B1030+074 as a lensed system: There is no doubt that B1030+074 is a gravitational lens system. We have modeled B1030+074 using a Singular Isothermal Ellipsoid (SIE) mass distribution (Figure 2). To investigate the stability of this model, we performed 10,000 Monte-Carlo simulations, by adding Gaussian distributed errors to the image positions (0.3 mas), galaxy position (4 mas) and flux density ratio (20%) (all 1σ errors). The predicted time delay is around $156/h_{50}$ days with an error of only a few percent.

Acknowledgments. This research was supported by European Commission, TMR Programme, Research Network Contract ERBFMRXCT96-0034 "CERES".

References

Browne, I. W. A., Patnaik, A. R., Wilkinson, P. N., Wrobel, J. M. 1998, MNRAS, 293, 257
Patnaik, A. R., Browne, I. W. A., Wilkinson, P. N., Wrobel, J. M. 1992, MNRAS, 254, 655
Wilkinson, P. N., et al. 1998, submitted

Emily Xanthopoulos

Active Galactic Nuclei and Related Phenomena
IAU Symposium, Vol. 194, 1999
Yervant Terzian, Daniel Weedman, Edward Khachikian, eds.

An Analysis of 900 Optical Rotation Curves

D. F. Roscoe

School of Mathematics, Sheffield University, Sheffield, S3 7RH, UK.

Abstract. Persic & Salucci (1995; hereafter, PS) reduced raw H_α data from a large sample of southern sky spirals obtained by Mathewson, Ford & Buchhorn (1992), to present 900 H_α rotation curves. It is found that a power-law, $V_{rot} = A R^\alpha$, imposes extremely detailed correlations between the kinematic and the luminous properties of the galaxies through the model parameters (A, α); this law can be inverted to give a form of super Tulley-Fisher relationship for the determination of absolute magnitudes in terms of velocity information. Furthermore, there is very strong evidence suggesting the existence of discrete kinematic states for rotation curves (RCs hereafter) - which ultimately implies the luminosity evolution of galaxies to be similarly constrained to occupy discrete states.

1. Introduction

This analysis is based on the hypothesis that RCs in optical discs (away from the dynamical effects of the bulge) can be reasonably described by the law $V = A R^\alpha$, where $R > R_{min}$ and R_{min} represents an estimate of the transition radius from bulge-dominated to disc-dominated dynamics. This latter estimate arises from the following process: if the *innermost* data point on any given RC is classified (by the regression software) as statistically 'unusual' in relation to the remaining data on the RC, then it is assumed to be part of the bulge-dominated dynamical regime, and so is deleted. The process is applied iteratively until an innermost point is defined which is *not* classified as statistically unusual.

$V = A R^\alpha$ is then shown to provide an extremely detailed statistical resolution of kinematic data in terms of luminosity data. It is subsequently found that ln A appears to favour certain discrete values; given the detailed correlation between kinematic properties and luminosity properties, the final conclusion is that it is the luminosity evolution of galaxies which is ultimately constrained to occupy discrete states, implying a large-scale cosmic coherence phenomenon.

2. A Necessary Condition For The Power-Law Hypothesis

Given $V = A R^\alpha$ then, excepting for inevitable noise, a plot of any RC in the $(\log R, \log V)$ plane would lie on a straight line. Since RCs differ, for the 900 RCs, we would obtain 900 different straight-line plots in the $(\log R, \log V)$ plane. Since the mean plot of 900 different plots of this type must also be a straight line, it follows that a necessary condition for optical RC data to be described by $V = A R^\alpha$ is that the mean plot of all the RCs in the $(\log R, \log V)$ plane must be a (statistically) straight line. So, the strategy of the initial analysis is described as follows:

- Reduce all the data in the combined sample of 900 RCs to a uniform linear scale based on Tulley-Fisher (TF hereafter) distances; Superimpose all the data of the combined sample into a single data set; Divide the data set into bins of (convenient) width $0.057\,kpc$; Form the average of the data in each bin and plot it.

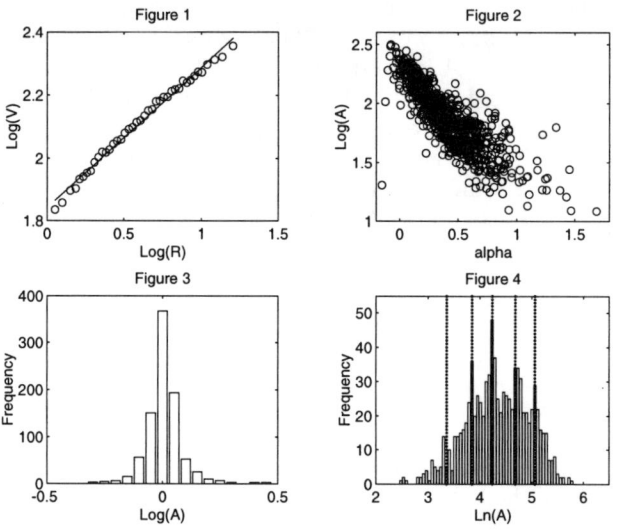

Figure 1. Mean Of 900 Individual Rotation Curves

The results, plotted in Figure 1, provides solid support for the idea that $V = A R^\alpha$ is, at the very least, a good working approximation to H_α RC data, and provides the necessary rationale for a detailed investigation of the power-law hypothesis.

Given $V = A R^\alpha$, then $\log V = \log A + \alpha \log R$. The parameters $(\alpha, \log A)$ are estimated by regressing $\log V$ on $\log R$ for each of 900 RCs and the results, plotted in Figure 2, demonstrate an extremely strong $(\alpha, \log A)$ correlation. A detailed analysis of the correlation, described in the following, shows it strongly orders the 900 galaxies according to their luminosity properties.

3. The Regression Model And A Testing Of It

A detailed analysis shows the best model of Figure 2 to be $\log A = \log V_0 - \alpha \log R_0$ so that

$$\frac{V}{V_0} = \left(\frac{R}{R_0}\right)^\alpha, \quad \text{where} \tag{5}$$

$$\log V_0 = -0.584 - 0.133\,M - 0.000243\,S,$$
$$\log R_0 = -3.291 - 0.208\,M - 0.00292\,S,$$

and S is surface brightness in solar luminosities per square parsec. This model accounts for over 90% of the variation in Figure 2.

As a final test of the model, for each galaxy we computed (R_0, V_0) according to (5), formed the scaled profile $(R/R_0, V/V_0)$ and regressed $\log(V/V_0)$ on $\log(R/R_0)$. Given (5), then the resulting regression constant should be *statistically* zero. The distribution of the 900 regression constants is shown in Figure 3, for which a 95% confidence interval for the mean value is computed as $(-0.003, +0.007)$. Thus, for all practical purposes, the regression constant can be considered as statistically zero, thereby confirming the quality of (5).

4. A Super-TF Relationship

Evaluating (5) at $R = R_{opt}$ and inverting, gives $M = f(V_{opt}, R_{opt}, S, \alpha)$. Since (a): S is a function M and R_{opt}; (b): the radial scale of any given galaxy is ultimately determined as a function of its angular scale, its apparent magnitude and its absolute magnitude via the relation $m - M = 5\log d - 5$, where d is object distance; (c): α is independent of any linear scale change, and is therefore determined entirely by velocity information; then $M = f(V_{opt}, R_{opt}, S, \alpha)$ can be seen as an implicitly defined non-linear equation for the determination of M given total rotational velocity information. It can therefore be seen as a form of super-TF relation for the determination of absolute magnitudes.

5. The Discrete Dynamics Hypothesis

The power-law hypothesis was originally tried against a subsample of the 21 RCs of late-type spirals given in the classical paper of Rubin, Ford & Thonnard (1980; hereafter RFT); this trial took the form of performing a linear regression of $\ln V_{rot}$ on $\ln R$ for that subset of twelve RCs which manifested purely *monotonic* behaviour, and then recording the regression constants. The results of this preliminary exercise, quoted to two significant figures, are condensed in Table 1. The salient feature of this table is that, after taking into account the numerical rounding process, $\ln A$ only takes values which lie within ± 0.15 of an integer or half-integer.

Table 1: $V = A R^\alpha$

Galaxy	Ln(A)	Galaxy	Ln(A)	Galaxy	Ln(A)	Galaxy	Ln(A)
N3672	3.6	U3691	3.6	N3495	4.0	N4605	4.0
I0467	4.1	N0701	4.1	N1035	4.1	N4062	4.5
N2742	4.5	N4682	4.5	N7541	4.6	N4321	4.9

The likelihood of this result is determined as follows: the probability that a single number, chosen at random from the real line, will lie within ± 0.15 of an integer or half-integer is exactly 0.6. Consequently, the probability that every one of a sample of twelve numbers, chosen on the basis of an independent prior criterion, lying within ± 0.15 of an integer or half-integer is $0.6^{12} \approx 0.002$. Thus, the RFT data suggests that $\ln A$ might be periodic with period 0.5 and zero phase.

Considering details, we noted how RFT's use of H = 50 km/sec/Mpc to set linear scales (against today's consensus that H > 50 km/sec/Mpc by a large margin) implies these linear scales to be too big by a similar margin. Consequently, to make PS data (with its TF-based linear scales) comparable to the RFT data, it was necessary to scale the PS data in some suitable way. This problem was addressed by noting that, if for example, H = 70 km/sec/Mpc was the correct value of H, then RFT linear scales would be too great by a factor of 1.4. Given the actual uncertainty in H, this led to the formulation of the specific hypothesis that, when $V = A (R/R_0)^\alpha$ is fitted to the PS data, there exists a scaling value, $1/R_0$, somewhere in the range $1.0 \leq 1/R_0 \leq 2.0$ which is such that $\ln A$ is distributed with period 0.5 and zero phase.

6. The Search and Results

The analysis introduced eleven experimental scalings, $1/R_0 = 1.0, 1.1, ...2.0$ and, at each of these, the specific question posed was *how many $\ln A$ values lie within ± 0.15 of either integer or half-integer values?* The results are condensed into Table 2 for two partitions of the data; a 'Hit' is defined to be when a particular $\ln A$ lies within ± 0.15 of an integer or half-integer.

Table 2				
Galaxy Types	Optimal R_0	Sample Size	Number Of Hits	Single Trial Probability
0..5	1/1.5	485	304	0.123
6..9	1/1.7	415	292	0.731×10^{-5}

The final column gives the probability of the recorded result occurring by chance alone *assuming the optimal scaling was chosen correctly without searching*. Given that, in practice, a search over a range of eleven (adjacent) experimental scalings was executed, then the actual probabilities are at least an order of magnitude less significant and, as is clear from the table, the only significant result occurred for late-type spirals.

The actual probability of obtaining the result of Table 2 for late-type spirals when a search over eleven adjacent experimental scalings was used were assessed using Monte-Carlo simulations based on synthetic data generated from the data of Figure 2. This probability was estimated at 6.2×10^{-5} over 2×10^6 simulations.

7. All Spirals Together

The analysis only found a positive result for late-type spirals. However, further investigation showed that α-values for early-type spirals were significantly noisier than for late-type spirals, and this lack of fidelity was sufficient to degrade any possible result for early-type spirals.

The problem was avoided in the following way: for the late-type spirals, it was found that there existed a direct 1:1 correspondence between the integer/half-integer peaks in the $R_0 = 1/1.7$ rescaled $\ln A$ data, and a set of identifiable peaks in the TF-scaled ln A data. But now it was found that these peaks were exactly mirrored by corresponding peaks in the TF-scaled ln A data for *early*-type spirals. Figure 4 shows the distribution of ln A for all 900 RCs together; the dotted lines define the positions of the ln A peaks in the *late*-type spiral data which correspond exactly to the 3.0, 3.5, 4.0, 4.5, 5.0 and 5.5 peaks in the $R_0 = 1/1.7$ rescaled data. The effect of combining all the data into Figure 4 has been to dramatically amplify these peaks well above the background noise and, given the hypothesis originally raised on RFT data, the overall probability of this result has been conservatively estimated at about 10^{-7}.

8. Conclusions

The first part of the analysis has shown that, once the effect of the bulge on disc-dynamics has been accounted for then, at the very least, the power-law model must be considered as an extremely good approximation for the kinematical behaviour of an idealized disc (that is, one without the irregularities that inevitably occur in real discs).

The second part of the analysis tested an hypothesis, raised initially on RFT data, that disc-kinematics have a preference for certain discrete states; this hypothesis has been confirmed at an extremely high level of confidence. Given the correlation between the kinematic and luminosity properties of spiral galaxies, this translates into the hypothesis that spiral galaxies exist in discrete luminosity states.

Finally, if (5) is inverted, and applied at the optical radius, then it gives $M = f(V_{opt}, R_{opt}, \alpha)$. Given that R_{opt} is defined as a function of M, m and the angular size of an object, this latter equation can be seen as a super-TF relationship for the determination of absolute magnitudes given V_{opt}.

References

Mathewson, D.S., Ford, V.L., Buchhorn, M. 1992 A Southern Sky Survey of the Peculiar Velocities of 1355 Spiral Galaxies *Astrophys J. Supp.* **81** 413-659.

Persic, M., Salucci, P., 1995 Rotation Curves of 967 Spiral Galaxies *Astrophys. J. Supp.* **99** 501-541.

Rubin, V.C., Ford, W.K., Thonnard, N. 1980 Rotational Properties of 21 Sc Galaxies With A Large Range Of Luminosities And Radii From NGC 4605 (R=4 kpc) To UGC 2885 (R=122 kpc) *Astrophys. J.* **238** 471-487.

Variable Sources in Active Galactic Nuclei

V. A. Hagen-Thorn, S. G. Marchenko, V. A. Yakovleva, A. V. Hagen-Thorn

St.-Petersburg State University, Bibliotechnaya Pl.2, St.-Petersburg 198904, Russia, vaht@aispbu.spb.su

1. Introduction

The problem of the activity of extragalactic objects is one of the main problems of current astrophysics. There are many manifestations of activity but the photometric and polarimetric variability in IR, optical, UV and X-ray regions is probably one of the most important. The sources responsible for the variability are placed in the nearest neighbourhood of a central engine. Clarifying the nature of these sources might give a tool to the solution of the problem of nuclear activity in general.

Several components (host galaxy, accretion disk, etc) contribute to the observed radiation. Because in the photometric and polarimetric observations their radiation is fixed in common, extracting the radiation of the variable source is not a simple task. In various spectral regions the contributions of those components are different. However in all regions the radiation of active sources may be extracted by the most reliable way on the basis of variability studies.

For clearing up the nature of the variable sources, it is of first rate importance to know their spectral energy distributions. These distributions may be found from the analysis of the multicolour data on photometric variability.

Another important feature of the active sources is polarization of their radiation. The analysis of the data on polarimetric variability may give information about their polarization properties.

It is well known that the highest activity among the extragalactic objects is observed in blazars. Therefore, just the blazars are investigated more extensively with the purpose to find the properties of variable sources though the other active extragalactic objects (for instance, Seyfert galaxies) are suitable for such analysis also.

At present the instantaneous spectra of active objects in a wide frequency range (from radio to X-rays or even γ-rays) are constructed as a result of cooperative work. These spectra are obtained for different brightness levels of the objects, and some correlations between various *observed* parameters are studied. But the success of such an approach for studying the properties of *variable sources* is limited and some conclusions about their nature are not strictly based.

The comparison of variations in different regions shows that variable sources seen in these regions are connected with each other because the variations (maybe with the exception of X-rays) are correlated (sometimes with time lag). But to confirm the identity of these sources it is necessary to compare the spectral and polarization properties of the sources acting in various spectral regions.

2. The technique of extracting variable sources from the observations of photometric and polarimetric variability

Let us suppose that the variability within some time interval is due to a single variable source and we have multicolour photometric and polarimetric data distributed within this interval. As was shown by Choloniewski [1] for photometric data and by the author [2] for polarimetric ones, if the variability is caused only by flux variations but the relative spectral energy distribution (for photometry) and relative Stokes parameters (for polarimetry) of the variable source are unchanged, then in the flux space $\Phi_1, \ldots, \Phi_n$ (for photometry, here n is the number of spectral bands used) or in the space of absolute Stokes parameters I, Q, U (for polarimetry) the observational points must lie on straight line. The direction of this line give the flux ratios for the *variable source*, i.e. its relative spectral energy distribution (for photometry) or relative Stokes parameters for the source (for polarimetry).

Thus if observational points in fact lie on the stright line within the observational errors (of course we may verify this considering two-dimensional pictures) one can conclude that the model of a single variable source with unchanged spectral energy distribution (for photometry) or with unchanged relative Stokes parameters (for polarimetry) is valid. Determination of direction (the slope) of the straight line gives the relative spectral energy distribution of the variable source (for photometry) or its polarization parameters (for polarimetry).

This technique has two advantages. (1) The information about the variable source may be obtained *without* the knowledge of its contribution to the total flux (for photometry) or to the observed absolute Stokes parameters (for polarimetry). (2) For the construction of the relative spectral energy distribution of the variable source in a wide spectral range it is not necessary to have *simultaneous observations in all wavelengths*. For instance, the variable source may be extracted in JR and optical separately and then these spectra may be connected by the comparison of the simultaneous data in only two spectral bands (say, K and B).

The limitation of this technique is obvious: it is applicable in the case when within a given time range the single variable source determines the behaviour of the AGN and the variability is due to variations in its flux level only. But often this is the case. The details of using this technique may be found in our paper [3].

3. Results

The technique of extracting variable sources described above was used by the author and his colleagues many times (it was used also by Winkler et al. [4]). The most detailed investigations made by us are for blazars 3C 345 [5-7], OJ 287 [3,8-9] and BL Lac [10,11]. In these works one can find many examples confirming the constancy of polarization parameters and spectral energy distributions of variable sources responsible for photometric and polarization behaviour of the blazars.

Here we give one example concerning the Sy 2 galaxy NGC 1275 (Fig. 1). The data are from Lyuty [12,13]. The time range of observations is several years. We see that the points lie on straight lines quite well. This means that the spectral shape of the variable source was unchanged. The shifted points belong to one outburst of 100 days duration (J.D. 2441218 - 321). Arrows show

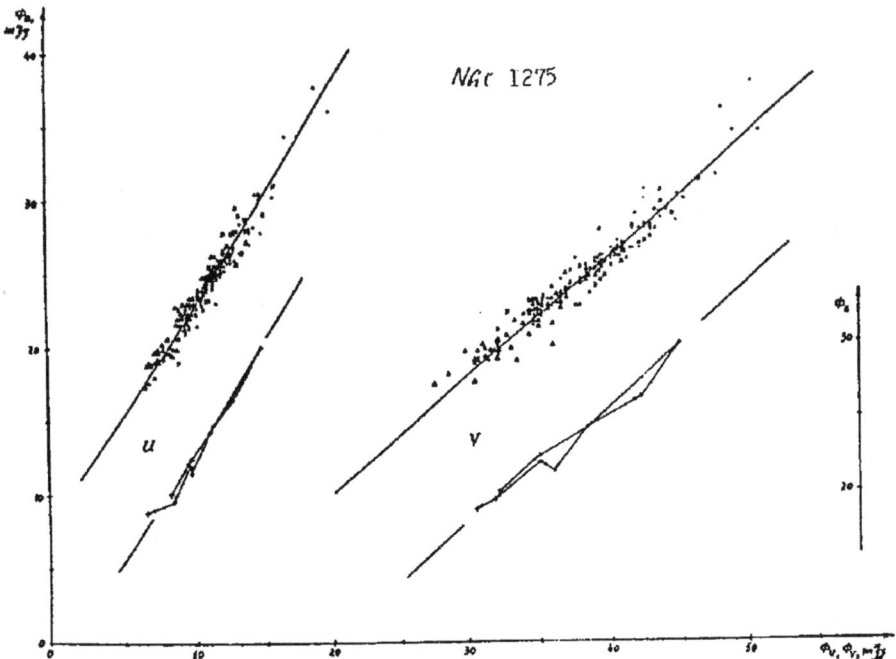

Figure 1. Flux-flux diagram for the galaxy NGC 1275, the shifted points belong to one outburst of 100 days duration.

its temporal development. The points move exactly along the straight line both at flux rise and flux decrease. This is a very important result because we must reject any variability mechanism resulting in changes of spectral shape of the variable source. In particular, if the radiation is of synchrotron nature (see below) we don't see an influence of synchrotron losses in this spectral range. One can often find in the literature that observed reddening of the spectrum in decreasing part of the outburst is due to synchrotron losses in the active point source. This is not the case.

Usually the sources in JR and optical are extracted separately and then the spectrum in the whole spectral interval is built by comparison of K and B fluxes. As a rule in the composite spectrum neither shift nor break exist. This means that the *same* source acts in both regions.

The relative spectral energy distributions obtained for variable sources are found to be well represented by the spectrum of a homogeneous synchrotron source with or without high-frequency cutoff [6-7,9]. In the first case the estimation of the critical frequency $v_c = 1.608 \times 10^{13} H E_{max}^2$ may be found with high precision.

If there are no temporal changes of the relative spectral energy distribution in the region of the high-frequency cutoff, the only reason for flux variability is variation in the number of relativistic electrons in the source.

On the other hand, some flattening of the spectrum in JR region (but not in the optical) at the very top of the outburst was found for 03 287 [9]. The most probable explanation of this fact is synchrotron self-absorption.

The results of extracting the sources of polarized radiation at different variability time scales are given in Table 1. The polarization degree found for individual sources may be as high as 50%. This confirms their synchrotron nature.

Table 1. The sources of polarized radiation

OJ287	($\Theta_{pref} = 82°$)		
Intraday:	$40\% < P < 50\%$		March 15, 1972
Interday:	$P = 43\%$	$\Theta_0 = 101°$	J.D. 2441803-808
BL Lac	($\Theta_{pref} = 20°$)		
Interday:	$P = 27\%$	$\Theta_0 = 73°$	J.D. 2443017-022
	$P = 56\%$	$\Theta_0 = 27°$	J.D. 2443786-789
Long-term:	$P = 23\%$	$\Theta = 3°$	1972
3C 345			
Long-term:	$P = 53\%$	$\Theta = 15°$	Feb.-July, 1983

4. Conclusions

The main results of our investigations may be formulated as follows.

a) In many cases the photometric behaviour of AGN at different time scales and in different spectral regions may be explained by the existence of a single variable source which has variable flux but unchanged spectral energy distribution. In particular, this concerns the behaviour in the flares. As a rule the spectral shape in optical-UV region is the same from the very beginning to the end of each event.

b) The distributions are well represented by the spectrum of a homogeneous synchrotron source with or without a high-frequency cutoff. In some cases at the very top of the light curve in outburst the synchrotron self-absorption may exist.

c) The spectral shape constancy excludes all variability mechanisms resulting in a change of spectral energy distributions (for instance, fading because of synchrotron losses). Probably, the variability within each event is due to variation in the number of relativistic electrons in the source.

d) Polarization behaviour is determined by a single variable source very rarely; but if it is the case the polarization degree for the source may be as high as 50%. This may be considered as evidence of its synchrotron nature.

References

Choloniewski J., 1981, Acta Astron., 31,293-311

Hagen-Thorn V.A., 1981, Trudy Astron. Oba. Leningrad Univ., 36,20-26 (in Russian)
Hagen-Thorn V.A., 1997, Astronomy Letters, 23, 23-29
Winkler H., Glass I.S., van Wyk F. et al., 1992, MNRAS, 257, 659
Hagen-Thorn V.A., Mikolaichuk O.V., 1988, Afz, 29, 604-610
Hagen-Thorn V.A. & Yakovleva V.A., 1994, MNRAS, 269, 1069-1076
Hagen-Thorn V.A., Marchenko S.G., Takalo L.O., Sillanpää A., 1996, A&A, 306, 23-26
Hagen-Thorn V.A., 1980, A&SS, 73, 263-277
Hagen-Thorn V.A., Yakovieva V.A., Takalo L.O., Sillanpää A.,1994, A&A, 290, 1-6
Hagen-Thorn V.A., Marchenko S.G., Yakovleva V.A., 1985, Afz., 22, 1-6
Hagen-Thorn V.A., Marchenko S.G., Yakovleva V.A., 1986, Afz., 25, 634-640
Lyuty V.M., 1972, Astron. Zh., 49, 930-942
Lyuty V.M., 1977, Astron. Zh., 54, 1153-1167

Active Galactic Nuclei and Related Phenomena
IAU Symposium, Vol. 194, 1999
Yervant Terzian, Daniel Weedman, Edward Khachikian, eds.

A UV Flare at the Center of the Elliptical Galaxy NGC 4552

Lucio M. Buson[1], Francesco Bertola[2], David Burstein[3], Michele Cappellari[2], Sperello di Serego Alighieri[4], Laura Greggio[5,6] & Alvio Renzini[7]

Abstract. A self-consistent analysis of near-UV, HST/FOC images of the elliptical galaxy NGC 4552 is used to show that its central spike has brightened by a factor ~ 4.5 between 1991 and 1993, and has decreased its luminosity by a factor ~ 2.0 between 1993 and 1996. A strong UV continuum over the energy distribution of the underlying galaxy is concurrently revealed shortward of $\lambda \sim 3200$ Å by our FOS spectra extending from the near-UV to red wavelengths. Nuclear emission-line profiles of both permitted *and* forbidden lines are best modelled with a combination of broad and narrow components, with FWHM of ~ 3000 km s^{-1} and ~ 700 km s^{-1}, respectively. Current diagnostics based on the emission line intensity ratios definitely places the spike among AGNs, just at the border between Seyferts and LINERs. This evidence argues for the variable central spike being produced by a modest accretion event onto a central massive black hole (BH), with the accreted material having possibly being stripped from a star in a close fly-by with the BH. In this regard, one has to look at NGC 4552 as *the faintest known* AGN.

1. The Serendipitous Discovery

With the intent of studying the stellar populations of early-type galaxies, in 1993 we obtained FOC images in several ultraviolet bands of the central regions of a few nearby ellipticals galaxies, including NGC 4552. The latter is a rather typical giant elliptical in the Virgo cluster whose absolute magnitude ($M_B = -20.2$, adopting a distance of 15.3 Mpc), structure, stellar population and metal content appear quite normal. At the time of our observation a point-like source with a photometric profile indistinguishable from the PSF of the aberrated HST

---

[1] Osservatorio di Capodimonte, Napoli, Italy

[2] Dipartimento di Astronomia, Università di Padova, Padova, Italy

[3] Department of Physics & Astronomy, Arizona State University, Tempe, AZ, USA

[4] Osservatorio di Arcetri, Firenze, Italy

[5] Dipartimento di Astronomia, Università di Bologna, Bologna, Italy

[6] Sternwarte der Universität München, München, Germany

[7] European Southern Observatory, Garching bei München, Germany

was present at its center (Bertola et al. 1995). In principle this is not surprising, as Maoz et al. (1996) show that ~ 10% of nearby galaxy cores sampled in the ultraviolet with the FOC, do contain a unresolved point source in their centers. FOS spectra reveal, in turn, that UV-bright sources of this kind can be simply sub-parsec size star clusters (as in the case of NGC 2681; Cappellari et al. 1998a) or a centrally located starburst (Koraktar, this meeting).

The case of NGC 4552 turned out to be quite odd, however. By comparing our UV observation with an identical, pre-existing FOC image of this galaxy secured in 1991, we soon realized that we had detected its central spike just at a time it had increased its luminosity by a factor $\sim 7 \pm 1.5$ between the two epochs – as roughly estimated from aberrated data. The estimated luminosity of the UV-bright source in 1993 was $\sim 10^6$ $L_\odot$ (Renzini et al. 1995). To get further clues to this phenomenon we obtained additional HST observations, including both UV/optical spectroscopy with FOS and FOC UV imaging, in 1996. These last observations confirm the variability of the central source (in the sense it is presently fading). So, though providing quite a loose sampling of the phenomenon, the existing FOC photometry strongly suggests that the spike which became visible at the center of the Virgo Elliptical NGC 4552 *is a single, protracted UV flare caught in mid-action* (see Fig. 1). This conclusion was made possible by fitting to the observed images a properly PSF-convolved galaxy model, consisting of an underlying constant nuker-law profile and a central, variable excess representing the spike. Such a procedure takes into account all needed correction factors for sensitivity differences and possible non-linearity effects of each FOC image, thus providing a self-consistent recalibration of the whole set of data (see Cappellari et al. 1998b for details).

With a bit of handsight it is quite clear one could have traced several hints of secondary events in the galaxy history as well as of BH-related nuclear activity well before our detection of the time-variable spike. Major signatures include extended Hα emission from the galaxy center out to ~ 2.5 kpc – likely the result of a recent accretion of a small gas rich satellite – (e.g. Macchetto et al. 1996), together with inner dust patches first recognized by van Dokkum & Franx (1995) on early WFPC images, which turned out to form a circumnuclear, dusty ring-like structure in more detailed WFPC2 optical frames. Our own subtraction of more recent archive WFPC2 images does show this inner dusty feature in its full shape, i.e. as a complete ring encircling the central bright spike (Fig. 2).

2. The UV-Dominated Continuum Spectrum of the Spike

The availability of FOS, subarcsec UV/optical spectra of the flaring central spike proved to be an unprecedented source of physical insight onto the nature of this phenomenon, too. In Fig. 3 we compare the merged FOS spectra of the central $0\rlap.{''}21 \times 0\rlap.{''}21$ region of NGC 4552 with a composite UV/optical spectrum meant to represent the underlying spectrum of the inner giant elliptical. Having normalized at the V band, we see that this composite spectrum is a good match in overall spectral energy distribution for the FOS spectrum, with two notable exceptions. Unlike the underlying spectrum the spike shows strong emission lines and, what is more, a remarkable continuum UV excess shortward of λ 3200 Å. As far as the *intrinsic* UV SED of the spike is concerned, both the galaxy-subtracted

A UV Flare at the Center of NGC 4552

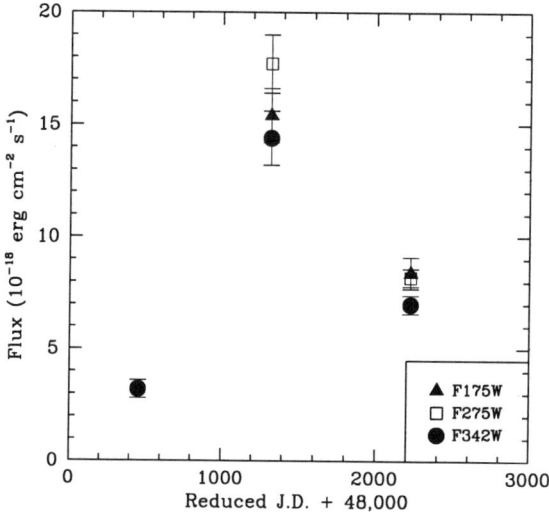

Figure 1. The light curve of the spike in the F175W, F275W and F342W passbands. The three epochs are July 1991, November 1993 and May 1996, respectively.

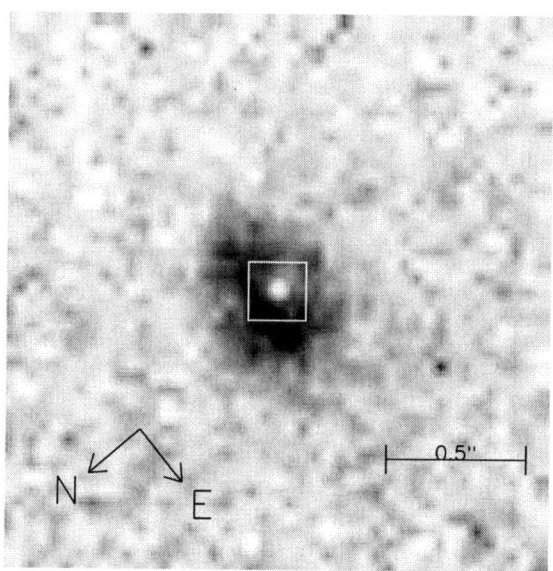

Figure 2. The $V - I$ (F555W-F814W) map of the central regions of NGC 4552. The darkest color correspond to $V - I \simeq 1.44$ mag, while the background is at the $V - I \simeq 1.34$ mag level. The central square represents the size of the $0\rlap{.}''2 \times 0\rlap{.}''2$ FOS aperture used for the spectroscopic observations. The central spike is clearly visible as a bright spot.

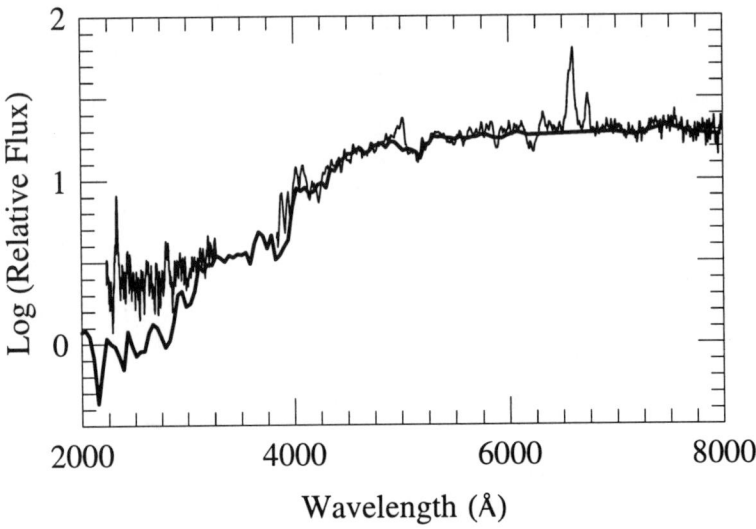

Figure 3. The overall 1996 FOS central spectrum of NGC 4552 within the 0″.2×0″.2 (thin line) superimposed to a scaled combination of the IUE 10″×20″ aperture spectrum of NGC 4552 and a ground-based optical spectrum of the virtually identical giant elliptical NGC 4649. The spectra have been normalized to the visual region.

FOS SED and FOC fluxes in contiguous UV bands indicate a temperature of $T \sim 15,000$ K for the spike in 1996 (if a thermal origin for the UV flux is assumed).

3. The Uncommon Emission-Line Spectrum of the Spike

The continuum-subtracted emission-line FOS spectra have been modeled as a whole by applying the Levenberg-Marquardt algorithm to fit a non-linear function. The resulting emission line fits permit us to draw many (some of which quite unusual) conclusions: [i] satisfactory fits of the emission lines identified in nuclear FOS spectra can be obtained only by resorting to a combination of broad and narrow components for *both* permitted *and* forbidden lines. This result is at variance with the behavior of classical AGNs – where the broad component is usually present *only* in the permitted lines [ii] the shape of the Hα+[NII] complex has changed from the May 1996 spectrum to the January 1997 spectrum. In the context of our procedure, a satisfactory fit can be *formally* achieved only allowing for a shift to the blue of ~ 230 km s^{-1} of the whole (narrow + broad) Hα line in the latter spectrum [iii] the ratios of the *narrow* emission line components of the spike spectrum place this feature definitely within the AGN region, just at the borderline between Seyferts and LINERs. Therefore, the spike in NGC 4552 can be either classified as a very high excitation LINER or a very low excitation Seyfert.

4. Towards a Self-consistent Interpretation

The phenomenology of the spike in NGC 4552 is consistent with a scenario in which a central, UV-bright flare is caused by the tidal stripping of a star in a close flyby with a central supermassive BH. Other interpretations (a central supernova, a central starburst, ...) were already considered unlikely. Circumstantial evidence in favor of the tidal stripping option comes now also from the noticed shift in the broad Hα emission between May 1996 and January 1997. Tidal stripping/disruption is indeed predicted to give rise to an *elliptical* accretion disk, the precession of which results in sizable Hα line profile variations. The stripping-induced flare is predicted to be very bright ($\sim 10^{10}$ L$_\odot$) for several years, much brighter than our observed flare, however. This indicates that if the flare in NGC 4552 was indeed caused by a tidal stripping in a BH-star flyby, it led to only *partial* stripping. As far as the broad forbidden lines of the NGC 4552 nucleus are concerned, taking into account we are dealing with a very low luminosity AGN, one can expect the *forbidden lines* to originate from the *low-density* disk itself, rather than from the usual distinct region farther away. If the whole emission line radiation does originate from a (thin) disk, then the gradient in rotational velocity should be responsible for the actual line profile, with the *broadest part* of it originating in the inner regions. Adopting the above scenario, the combined availability of high-resolution imaging and sub-arcsec spectroscopy allows us to put constraints to the mass of the central BH. Our rough estimates, leading to a BH mass between 3×10^8 to 2×10^9 M$_\odot$, appear consistent with the mass of $M \simeq 4 - 6 \times 10^8$ M$_\odot$ of the supermassive central BH in NGC 4552, obtained by Magorrian et al. (1998), using simple *isotropic* dynamical models based on HST photometry and ground based kinematics.

References

Bertola, F., Cappellari, M., Burstein, D., Greggio, L., Renzini, A., di Serego Alighieri, S. 1995, Stellar Populations, ed. G. Gilmore & P. van der Kruit (Dordrecht: Kluwer), p. 445

Cappellari, M., Bertola, F., Burstein, D., Buson, L.M., Greggio, L. & Renzini, A.: UV Spikes in Bulge-Dominated Galaxies, 1998a, in: "STScI Workshop: How and When Do Bulges Form and Evolve?", C.M. Carollo, H.C. Ferguson & R.F.G. Wyse (eds.), ASP Conf. Series, ASP, San Francisco

Cappellari, M., Renzini, A., Greggio, L., di Serego Alighieri, S., Buson, L.M., Burstein, D. & Bertola, F. 1998b, ApJ, submitted

Macchetto, F., Pastoriza, M., Caon, N., Sparks, W.B., Giavalisco, M., Bender, R., & Capaccioli, M. 1996, A&AS, 120, 463

Magorrian, J., Tremaine, S., Richstone, D., Bender, R., Bower, G., Dressler, A., Faber, S.M., Gebhardt, K., et al. 1998, AJ, 115, 2285

Maoz, D., Filippenko, A.V., Ho, L.C., Macchetto, F.D., Rix, H.-W., & Schneider, D.P. 1996, ApJS, 107, 215

Renzini, A., Greggio, L., Di Serego Alighieri, S., Cappellari, M., Burstein, D., & Bertola, F. 1995, Nature, 378, 39

Van Dokkum, P.G. & Franx, M. 1995, AJ, 110, 2027

Active Galactic Nuclei and Related Phenomena
IAU Symposium, Vol. 194, 1999
Yervant Terzian, Daniel Weedman, Edward Khachikian, eds.

Emission-line Galaxies in Shahbazian Compact Groups

H.Tiersch [1,2] and D.Stoll[1]

Sternwarte Koenigsleiten, München, Germany

S.Neizvestny

Special Astrophysical Observatory, Nizhny Arkhys, Russia

A.S.Amirkhanian and A.G. Egikian

Byurakan Astrophysical Observatory, Byurakan, 378433, Armenia

Based on the CCD photometric and spectroscopic observations taken with 1.23 m and 2.2 m telescopes at the German-Spanish Astronomical Center in Calar-Alto (Almeria, Spain) as well as with the 1.54 m telescope at the European Southern Observatory (La Silla, Chile) some emission-line galaxies (ELGs) surprisingly have been found in Shahbazian compact groups of galaxies (SHCGs) by our research group.

The majority of galaxies in the SHCGs usually have an absorption spectra typical of ellipticals and S0s. But some of these objects turn out to be the ELGs including two broad-line AGNs (of classical Seyfert 1 type) and the narrow emission-line galaxies. Their redshift makes these galaxies a physical member of the host group.

Unlike well studied Hickson compact groups (HCGs) (see e.g. Hickson 1993; Hickson 1997) not many SHCGs of galaxies have been investigated photometrically and spectroscopically till quite recently (Tiersch et al. 1998 and references therein). For this reason a program has been started in the University of Potsdam, Potsdam Astrophysikalisches Institut in cooperation with other observatories to obtain photometric and spectroscopic properties of the galaxies in SHCGs (Tiersch et al. 1994; Oleak et al. 1995; Tiersch et al. 1995).

The main goal of our program is to study the physical nature of these groups, their past and future evolutionary histories. With this purpose it is necessary to know the morphology of member galaxies, the photometric data, i. e. B, V, R apparent magnitudes, radial velocities, velocity dispersion, mass-to-luminosity ratio, crossing times, dynamics, chemical abundance, metallicity as well as the evolution effects.

Three-colour (B, V, R) photometric and spectroscopic observations have been carried out at Calar Alto and La Silla Observatories by Tiersch and Stoll. The data reduction procedure used the MIDAS software package at Potsdam Astrophysikalisches Institut.

---

[1] Visiting astronomer at the German-Spanish Astronomical Center, Calar Alto, Spain

[2] Visiting astronomer at the European Southern Observatory, La Silla, Chile

The emission-line spectra are indicative of galaxies with active galactic nuclei (AGNs), star-forming galaxies which have had large bursts of star formation and normal late-type galaxies.

In the present study a total of 100 galaxies in 15 SHCGs has been spectroscopically investigated based on observations at Calar Alto. Eleven objects have emission-line spectra. One of them is a background AGN with the high - excitation ([OIII]/$H_\beta \geq 3$) spectrum. This is galaxy number 7 in group Sh 354 (Sh 354/7) - our first emission object with $z = 0.1804 \pm 0.0001$. The mean redshift of the group has been found to be $z = 0.0707$.

A total of 90 galaxies in 20 SHCGs has been spectroscopically observed at La Silla. Twelve objects exhibit emission lines. Two of them are foreground AGNs.

This sample is not at all complete. In order to complete, one must study all 377 groups in Shahbazian original lists (Shahbazian 1973; Baier and Tiersch 1979 and references therein).

According to these preliminary data about 10% of member galaxies in SHCGs are ELGs. This can be compared with the results obtained by Dressler, Thompson and Shectman (1985) for rich clusters of galaxies. Their statistics of emission - line galaxies are based on the spectra of 1268 galaxies in the fields of 14 rich clusters. About 7% of member galaxies in nearby ($z \sim 0.04$) clusters have emission-line spectra.

A typical Shahbazian compact group is a relatively isolated association of five to fifteen galaxies within a small sky area of a few arcminutes across (Shahbazian 1973). The resulting spectra of the member galaxies usually show characteristics of K-type stars with absorption features of the H_β, MgIb, NaID and H_α lines, which were used for the determination of the redshift. Reproduction of the absorption spectrum of the galaxy Sh 344/4 is shown in Figure 1. It is representative of an E type galaxy. The radial velocities were determined for five galaxies in the group Sh 344 from eight in the original list. All except one are group members with absorption spectra. One object is a foreground galaxy.

The first emission-line object in SHCGs was discovered by Spanish astronomers del Olmo and Moles (1991) at Calar Alto Observatory using the 3.5 m telescope. It is galaxy number 4 in group Sh 278. Sh 278/4 is the only active object found by them among a total of 49 galaxies in 11 SHCGs spectroscopically investigated. This low luminosity active nucleus hosted in an early-type galaxy has a remarkable resemblance with QSO or QSO-like objects.

Our spectrum of Sh 278/4 (see Fig.1) taken with the 1.54 m telescope at La Silla includes the H_α region which is absent in the spectrum optained at Calar Alto. The spectrum of the galaxy shows typical characteristics of broad-line AGNs, i.e. broad hydrogen lines H_γ, H_β, H_α with a high velocity FWHM of $\sim 7000\ km/s$. There are also the narrow forbidden lines of [OIII] 4959,5007, [NII] 6548,6584, [SII] 6717,6731 and weaker line of [OII] 3727. H_α is contaminated by [NII] 6548,6584. We classified Sh 278/4 as a Seyfert 1 type galaxy.

The second active object found in our research is the galaxy Sh 355/4 (Tiersch et al. 1996). It is a classical Seyfert 1 type galaxy with very broad emission Balmer lines H_β and H_α (see Fig. 1). The FWHM of 250 A is determined by the undisturbed H_β line and corresponds to a very high velocity of about 15000 km/s. The spectrum also presents the strong forbidden lines N1, N2

[OIII] and stellar absorption features MgIb and NaID which are absent in the spectrum of Sh 278/4.

All other emission-line objects are narrow emission-line galaxies, among those Sh 307/7, Sh 331/1 and Sh 331/4 (see Fig.1). In their spectrum the permitted hydrogen lines are narrow as well as the forbidden ones. There are two member ELGs (numbers 1 and 4) in group Sh 331. Note the strongest H_α in the spectrum of Sh 302/7 and Sh 331/1.

The ELGs discovered by our group in these and other SHCGs now investigated will be the subject of a later more detailed spectrophotometric study. It is very important to know what is the population of emission-line galaxies found in SHCGs. What is their nature ? Which is their activity class ?

Extensive spectroscopic studies of distant clusters of galaxies performed by Dressler, Gunn and collaborators (1992) over one decade showed that in distant clusters with redshifts between 0.37 and 0.55 22% of all galaxies exhibit emission lines and about 13% reveal post-starburst features, while in nearby ones there are only 7% and 0.3%, respectively. This seems to confirm the "Butcher-Oemler effect" which is the first observational evidence for the coeval evolution of galaxies (Butcher and Oemler 1984). Although the Butcher - Oemler effect was regarded with a lot of skepticism 15 years ago, it is now well established that galaxies in high redshift ($z \sim 0.4$) rich clusters show a population of galaxies that are significantly bluer than their lower redshift counterparts (Dressler et al. 1994; Fabricant et al. 1994; Thimm et al. 1994). The fraction of blue galaxies increases from 5% at $z < 0.1$ to 20% at $z = 0.4$ (Dressler et al. 1985) and to 80% at $z = 0.9$ (Rakos and Schombert 1995).

The Butcher-Oemler effect is also observed in radio selected groups of galaxies (Zirbel 1994). At high redshifts ($z \sim 0.4$) rich groups (of average density comparable to Abell richness class 0) have bluer colours. The blue galaxy population in these high redshift rich groups can be divided into two classes: the bright blue galaxies ($M_v < -21.5$) that have disappeared today and the fainter blue galaxies whose evolution is more gradual. The bright blue galaxies may be identified with the "classical" Butcher-Oemler galaxies, post-starbursting ellipticals with strong Balmer lines (named E+A galaxies) discovered by Dressler and Gunn (1983, 1992), while the fainter blue galaxies may be late type galaxies (Zirbel 1994). Recent Hubble Space Telescope data showed that many of these "bluer" galaxies reveal a disturbed and peculiar morphology (Dressler et al. 1994).

On the other hand, now it is an observational fact that at low redshifts ($z < 0.4$), i. e. in the local Universe the QSOs and Butcher-Oemler galaxies have disappeared in the rich clusters of galaxies. Why? At redshifts 0.4 to 0.6 radio-loud QSOs are found in clusters of moderate to low richness (Ellingson et al. 1991). At lower redshifts the QSOs usually live in small groups of galaxies with a high density (Hutchings et al. 1993).

The comparison of nearby and distant galaxy groups' emission-line population probably is the key in understanding the formation and strong evolution of their member galaxies.

Acknowledgments. This work is supported by a grant of the "Deutsche Forschungsgemeinschaft".

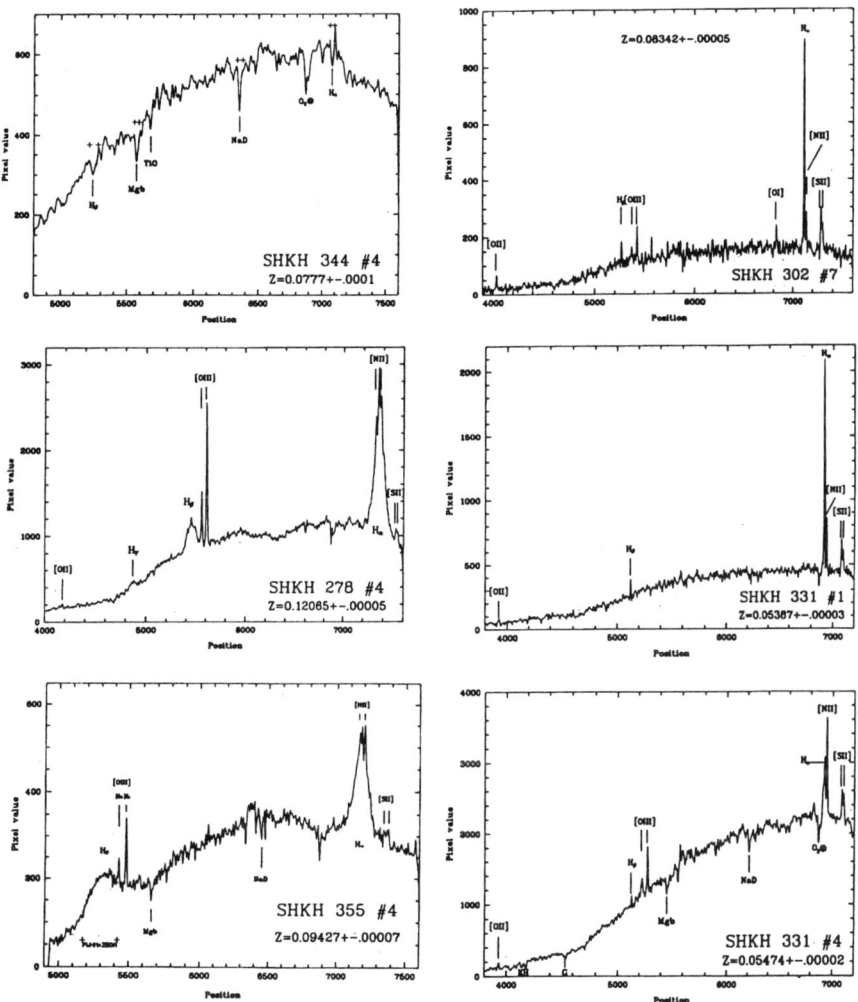

Figure 1. The absorption spectrum of Sh 344/4 and the spectra of some ELGs in SHCGs. All except one were taken with the 1.5 m telescope at La Silla. The spectrum of Seyfert 1 galaxy Sh 355/4 was obtained using the 2.2 m telescope at Calar Alto. Original dispersion is 2.9 Å per pixel, integration 3600s.

References

Baier, F.W., & Tiersch, H. 1979, Astrofizika, 15, 33
Butcher, H., & Oemler, A. 1984, ApJ, 285, 426
del Olmo, A., & Moles, M. 1991, A&A, 245, 27
Dressler, A., & Gunn, J.E. 1983, ApJ, 270, 7
Dressler, A., Thompson, I.B., & Shectman, S.A. 1985,ApJ., 288, 481
Dressler, A., Gunn, J.E., & Schneider, D.P. 1985, ApJ, 294, 70
Dressler, A., & Gunn, J.E. 1992, ApJS, 78, 1
Dressler, A.,Oemler, A., Sparks, W.B., & Lucas, R.A. 1994, ApJ, 435, L23
Ellingson, E., Green, R.F., & Yee, H.K.C. 1991, ApJ, 378, 476
Fabricant, D.G., Bautz, M.W., & McClintock, J.E. 1994, AJ, 107, 8
Hickson, P. 1993, Astrophys.Lett.Commun., 29, Nos.1-3
Hickson, P. 1997, Ann. Rev. Astron. Astrophys., 35, 357
Hutchings, J.B., Crampton, D., & Persram, D. 1993, DAO preprint
Oleak, H., Stoll, D., Tiersch, H., & MacGillivray, H.T. 1995, AJ, 109, 1485
Racos, K., & Schombert, J. 1995, ApJ, 439, 47
Shahbazian, R.K. 1973, Astrofizika, 9, 495
Thimm, G.J., Roser, H.-J., Hippelein, H., & Meisenheimer, K. 1994, A&A, 285, 785
Tiersch, H., Oleak, H., Stoll, D., & Böhringer, H. 1994, in Astronomy from Wide-Field Imaging, IAU Symp.161, ed. H.T.MacGillivray et al. (Dordrecht, Kluwer), 623
Tiersch, H., Oleak, H., Stoll, D., Neizvestny, S., Amirkhanian, A.S., & Egikian, A.G. 1995, Astrofizika, 38, 688
Tiersch, H., Oleak, H., Stoll, D., Amirkhanian, A.S., & Neizvestny, S. 1996, Astron. Soc. Pac. Conf. Ser., 98, 523
Tiersch, H., Stoll, D., Neizvestny, S., Amirkhanian, A.S., & Egikian, A.G. 1998, in press
Zirbel, E.L. 1994, STSI preprint No. 856

Morphology and Photometry of the UV-Excess Galaxy Pair NGC 7770/7771

G. B. Ali and A. M. I. Osman

Astronomy Department, National Research Institute of Astronomy and Geophysics 1142 Helwan, Cairo, Egypt E-mail:gamal@frcu.eun.eg

Abstract. Morphology of the nuclear region of the component NGC 7770 indicates that it has a complicated structure in the B-band, while the V-band shows a double nuclear structure. The brightness and color of the nuclear condensations are given. Detailed photometric analysis of this component including luminosity profiles along major and minor scans as well as position angle and ellipticity curves is obtained. Integration and decomposition of the luminosity profiles are also performed. The color distribution and integrated color are also studied taking into account that the blue color together with the unusual color distribution are characteristics relevant to UV-excess galaxies. The integrated photometric parameters are also given.

The state of interaction between the components of the pair is also investigated. The peculiar features appearing in the isophotal maps, the truncation of the luminosity profile along the side of interaction and the unusual appearance of the structural profiles are discussed and considered as strong evidence for the real state of interaction between the components of the pair. A lower limit of the total orbital mass of the system is found to be $M_t=1.36\times 10^{10} M_\odot$, and the total mass-to-luminosity ratio is f=0.63 $f_\odot$.

1. Introduction

This pair forms together with a third galaxy, NGC 7769, a group of triple galaxies described by Karachentsev (1987) as an isolated triple system. The components of the pair form a double system with about 1.'2 separation, and embedded in a common envelop at B$\geq$ 24.5 and V$\geq$ 23.5 mag.arcsec^{-2}. The outer envelopes enclosing the pair are clearly irregular and shredded, so according to the Karachentsev scheme of interaction signs (Karachentsev 1990) this pair could be described as a pair of shredded atmosphere with interactive class ATM(sh). According to the 21-cm observations made by Sulentic and Arp (1983), the HI profiles of NGC 7770 and NGC 7771 are classified as SA class (asymmetric profile with a single peak) and C class (complicated profile) respectively; the classes which are an indication for undergoing violent interaction. The components of the system are also included in the 3rd Kazarian list for galaxies with UV-excess, (Kazarian and Kazarian 1980). Detailed UBV photometry for this system was made by Tamazyan (1984) who showed that the nuclei of these galaxies are blue objects with various degrees of ultraviolet continuum. Assuming that the components of the pair are in a circular orbit relative to each other, a rough estimate of the lower limit of the total orbital mass of the system is calculated from the relation $M_t = 32 r_p (\Delta v)^2/(3\pi G)$ (Karachentsev 1990), where r_p is the projected separation, Δv is the velocity difference between the components and G is the universal constant of gravity. The total orbital mass of the system is found to be $M_t = 1.36 \times 10^{10} M_\odot$ while the total mass of the system NGC 7769/7771 was estimated by Karachentsev (1987) to be $2.74 \times 10^{11} M_\odot$. Detailed photographic B and V surface photometry is performed only for the small companion; NGC 7770; since the other component has been studied by Osman (1986). The basic parameters of the components of the pair are collected from RC3, and listed in Table 1.

Table 1. Basic data for NGC 7770 and NGC 7771, (RC3).

Parameter	NGC 7770	NGC 7771
Coordinates α (1950)	23h 48m 49.9s	23h 49m 52.3s
δ (1950)	19° 49' 13"	19° 50' 08"
Morph. Type	S0/a, T=0	SBa, T=1
Magnitude BT	14.40 0.30	13.08 0.05
BT	14.16	12.49
mB	14.38	12.94
mFIR		9.92
Colors $(B-V)_T$	0.59 0.06	0.83 0.05
$(B-V)_{T0}$	0.51	0.69
$(U-B)_T$	0.34	0.09
$(U-B)_{T0}$		0.26
HI Index HI		2.04
Mean HRV, Vopt	4264 52 km/sec	4298 30
Position angle		68°

2. Observations

Three photographic plates in each of the B and V bands have been obtained at the Newtonian focus (22.″53/mm) of the 188 cm telescope, Kottamia Observatory (Egypt). The photographic calibration of the plates was performed using the Kodak calibrated step wedge on each plate (Osman 1985). The plates were scaned and measured employing the PDS microdensitometer in RGO, UK.

3. The Component NGC 7770

NGC 7770 (=Kaz 347 =UGC 12813 =MCG +03-60-0034) is a small close companion of NGC 7771 (Kaz 348) and is displaced about 1.′2 southwest from it. It appears to be a compact galaxy with the main body of elliptical shape (0.′18 × 0.′14 at B=22.5 mag.arcsec^{-2}) embedded in a faint envelop extended nearly perpendicular to the main body. Many authors have confused the morphological type of this galaxy. Pettit (1954) and de Vaucouleurs and de Vaucouleurs (1964) have classified it as SBb and Sb respectively. But de Vaucouleurs et al. (1976) have classified it as S0/a, i.e. a transition stage between late lenticular and early spirals. Egiazaryan (1983) described it as a very bright compact galaxy measuring about $12'' \times 16''$, with complicated and composite spiral structure. Tamazyan (1984) has showed that it does not exhibit a developed spiral structure and classified it as S0 galaxy. NGC 7770 is a galaxy with strong UV continuum of early type and spectroscopically classified as d2 by Kazarian and Kazarian (1980). Spectral observations made by Kazarian and Kazarian (1989) showed that the spectra of this galaxy cover both long and short wavelengths including emission lines of [SII] $\lambda\lambda 6731/17$, [NII] $\lambda 6584$, [OIII] $\lambda\lambda 5007, 4959$, $H_\beta \lambda 4861$, and [OII] $\lambda 3727$.

3.1. Morphology of the Nuclear Region

Although the appearance of this galaxy is uniform, it has a complicated nuclear region. The B and V isophotal maps show that the structure of the nuclear region is more complicated in the B-band than in the V-band. It shows multi condensations of different sizes, with the major one off-centered and shifted towards the southwest direction. This major condensation appears to be composed of many small condensations. On the other hand, the V-isophotes show a double nuclear structure. This complicated structure of the nuclear region is also confirmed in the intensity profiles, in which there are a number of peaks representing the condensations noticed along the adopted major (p.a.$\simeq 90°$) and minor axes. The condensation noticed in the B-band along the major axis has average brightness of 20.22 mag.arcsec^{-2} with relatively blue color B-V=0.09.

3.2. Luminosity Profiles

The B and V luminosity profiles along major and minor axes have the major axis taken at p.a.$\simeq 90°$. The luminosity profiles appear to possess a general smooth structure except for the complicated structure of the nuclear region and the presence of some humps in the outer region, especially along both sides of the major axis and the south side of the minor one. These outer humps occur at relatively high level, and we expect represent traces of unresolved spiral structure. The N-side (side of interaction) of the minor profile is clearly truncated at $0.'36$ due to the presence of the other component, NGC 7771.

3.3. Equivalent Luminosity Profiles, Integration and Decomposition

The B and V equivalent luminosity profiles are followed to logI=-1.2 at r =$0.'69$ and up to logI=-1.1 at r =$0.'36$ in the B and V bands respectively. The truncation of the northern side prevent us from reaching fainter levels. However, the profiles, within r =$0.'36$, are integrated and decomposed.

Integration and total magnitude The integrated B and V photographic magnitudes within r =$0.'36$ are: m_T=14.^{m}51 and m_V=14.^{m}05 respectively, with the color index (B- V)=0.^{m}46. The total B magnitude is m_{BT}=14.^{m}25 which is about 0.^{m}15 brighter than that reported in RC3, while the total color, (B-V)$_T$=0.^{m}44 is about 0.^{m}15 bluer.

Profile Decomposition The B and V luminosity profiles are decomposed into spheroidal bulge and flat disk components following Kormendy (1977). The decomposition parameters (effective intensity, effective radii, and luminosity of

the bulge and disk components) as well as bulge-to-disk ratio are obtained in
the B and V bands and given numerically in Table 4.

The bulge-to-disk ratio is found to be 2.77 and 1.07 in the B and V bands
respectively. The Freeman's parameters are: $B_c = 23.48 \pm 0.04$ mag.arcsec^{-2} and
$\alpha = 7.96 \pm 0.56$ kpc.

3.4. Color Distribution and Integrated Color

The integrated color along equivalent radius is found to be of abnormal distribution in comparison with normal spirals. The color index within the nuclear region (r =0.'16) is: $(B-V)_{nuc} = 0.^m38$, while that within r =0.'36 is $(B-V) = 0.^m46$ i.e. the color of the nuclear region is bluer than that of the outer region. The blue color of the nuclear region relative to that of the outer region, and the blue total color index $(B-V)_T = 0.^m44$ together with the color index $(U-B) = 0.^m13$ (Tamazyan 1984) are relevant properties of the UV-excess galaxies. Hence these results confirm the classification of this galaxy as UV-excess one. The blue color of the outer region of this galaxy probably confirms the presence of spiral structure mentioned by Egiazaryan (1983).

3.5. Position Angle and Ellipticity Curves

The isophotal parameters of the best fit ellipses are calculated. The curves show clearly that this galaxy has two main parts of different behavior, the inner part represents the main body and the outer part represents the outer envelope in which the main body is embedded. The outer part is clearly affected by the presence of the other component NGC 7771. The position angle within a=0.'182 is found to be $62.°72 \pm 5.°01$, while that of the outer envelope for a > 0.'23 is $-7.°12 \pm 4.°79$ which means that the outer envelope is oriented more or less perpendicular to the main body of the galaxy. This could be explained as the outer envelope is subject to a strong twist due to the presence of the other component. The ellipticity increases from 0.54 for the 2nd isophote until it reaches a maximum value of 0.96 for the 7th isophote; then it decreases in the outer three isophotes.

4. Discussion and Conclusion

The state of interaction between the components of the pair is confirmed by the presence of the peculiar features noticed in the distortion of the isophotes and in the unusual appearance of the structural profiles (ellipticity and position angle profiles). A lower limit of the total orbital mass of the system is $1.36 \times 10^{10} M_\odot$, with total mass-to-luminosity ratios $f = 0.63$ $f_\odot$. The complex structure of the nuclear region of the small component NGC 7770 is confirmed in the B-band, while the V-band shows double nuclear structure. A faint disk component is recognized from the decomposition of the equivalent luminosity profile.

References

de Vaucouleurs, G., de Vaucouleurs, A. 1964, Reference Catalogue of Bright Galaxies, Texas University, Austin de Vaucouleurs, G., de Vaucouleurs, A., and Corwin, H. G. 1976, Second Reference Catalogue of Bright Galaxies. (RC2) (Austin: University of Texas Press)

Egiazaryan, A. A. 1983, Astrofiz. 19, 631

Karachentsev, I. D. 1987, in Binary Galaxies, Moscow, Nauka

Karachentsev, I. D. 1990, in IAU Colloquium 124, Paired and Interacting Galaxies, ed. J. W. Sulentic, W. C. Keel, and C. M. Telesco, NASA PC-3098, p. 3

Kazarian, M. A., and Kazarian, E. S. 1980, Astrofiz., 16, 17

Kazarian, M. A., and Kazarian, E. S. 1989, Astrofiz., 30, 575

Kormendy, J. 1977, ApJ, 217, 406

Osman, A. M. I. 1986, Ap&SS, 124, 345

Pettit, E. 1954, ApJ, 120, 415

Sulentic, J. W. and Arp, H. 1983, AJ, 88, 489

Tamazyan, V. S. 1984, Astrofiz. 20, 43

The Orientation of Extragalactic Radiosources Relative to the Optical Axes of their Host Galaxies

R. Andreasyan[1]

Byurakan Astrophysical Observatory, Byurakan, 378433, Armenia

H. Sol

DARC, Observatoire de Paris, 92195 Meudon Cedex, France

Abstract. The relative orientation of radio and optical axes of radiogalaxies has been analysed for a sample of about 300 sources. Radio major axes have a tendency to align with optical minor axes of the host galaxies for the more elongated radiosources while they appear correlated with optical major axes for the less elongated radiosources.

1. Samples of radiosources and classification

Many attempts have been done to analyse the relative orientation of radio and optical images of low-redshift elliptical radio galaxies (Palimaka et al. 1979, Bottistini et al. 1980, Guthrie 1980, Valtonen 1983, Andreasyan 1984, Sansom et al. 1987) However, to our knowledge, no statistical trends were found up to now, except within the alternative classification of radiosources introduced by Andreasyan (1984). Following this approach, radiosources can be of two types, (i) the elongated ones for which K, the ratio of major to minor radio axes, is larger than 2.5, and (ii) the stocky ones, for which K is smaller than 2.5. Here we compare this alternative classification with the standard Fanaroff-Riley (FR) classification and further analyse the trends previously found by Andreasyan for the relative orientation.

The present sample of about 300 sources comprises radio galaxies with published radio maps, and identified with elliptical galaxies mainly brighter than 17 magnitude. Values of the position angles of the radio and optical images and elongation of the radio structure have been collected from the literature, or determined on the maps. FRII type sources appear often elongated (62 of 81 cases) while FRI type sources are preferentially stocky (52 of 73 cases). However, the two classifications do not completely overlap.

---

[1]DARC, Observatoire de Paris, 92195 Meudon Cedex, France

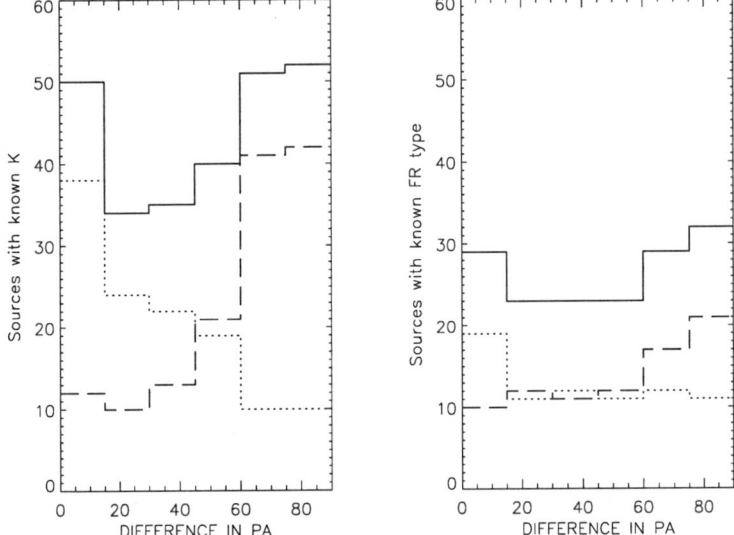

Figure 1. Histogrammes of apparent angle between major radio and optical axes of radiogalaxies, for classification by elongation on the left and for FR classification on the right. Full lines are for whole subsamples, dotted lines for stocky or FRI sources, dashed lines for elongated or FRII sources

2. Analysis of relative orientation of radio and optical axes

Histogrammes of the apparent angle between the major radio and optical axes of the radio galaxies are shown in Fig.1, for all the sources and for subsamples of stocky, elongated, FRI and FRII sources. There are two significant peaks, one at zero for the subsample of stocky sources and one at ninety degrees for the subsample of elongated sources. The same two peaks are still visible when using the FR classification, but they are much less significant. So there is a clear tendency to alignment (respectively orthogonality) of major radio axis and optical minor axis for elongated (respectively stocky) radiosources.

The finding of such a correlation between radio and optical axes orientation of radiogalaxies may be important for understanding the mechanism of formation and evolution of these objects.

References

Andreasyan R.R. 1984, Astrofizika, 21, 409
Bottistini P., et al, 1980, A&A, 85, 101
Guthrie B.N.G. 1980, Ap&SS, 70, 211
Palimaka J.J., et al, 1979, ApJ, 231, L7
Sansom A.E., et al, 1987, MNRAS, 229, 15
Valtonen M.J. 1983, Ap&SS, 90, 207

Rapid Variations in the Broad $H\beta$ Profile of the Radio Galaxy 3C 390.3

N. S. Asatrian, E. Ye. Khachikian

Byurakan Astrophysical Observatory, 378433 Byurakan, Armenia

P. Notni

Astrophysikalisches Institut Potsdam, An der Sternwarte 16, D-14482 Potsdam, Germany

1. Observations and data reduction

We report on rapid variations of the $H\beta$ profile shape in 3C 390.3 over a period of one hour and discuss the implications for the velocity field of the BLR.

Four spectra of the radio galaxy 3C 390.3 were obtained 1991 October 12 at the 6-m telescope using a TV scanner. The mean $FWHM$ of comparison lines is typically ~ 4 Å. The S/N ratio per resolution element in the continuum near the $H\beta$ line for individual spectra is 20 − 35. No correction for spectral sensitivity has been applied.

To illustrate the rapid variability of the shape of the broad line profiles we used the difference spectrum obtained by subtracting the mean of our last two spectra from the first one. Before subtraction, 1) an average continuum has been subtracted from each spectrum; 2) after this, the spectra have been internally calibrated to a common flux scale, using the total flux over two bands (λ_{obs}: 5030 − 5100 Å and 5160 − 5210 Å) in the $H\beta$ wings (cf. Eracleous & Halpern 1993).

2. Results

The difference spectrum is shown in Figure 1 together with the corresponding parent spectra. The difference $H\beta$ profile reveals small, but significant changes in the form of three narrow, positive and negative bumps located at the blue and red sides of the line (marked by arrows). The positions and the S/N ratios of the bumps are −3700, −2300 and 4700 km/s and 6.1, 4.6 and 3.7, respectively. These changes in the shape of $H\beta$ have occurred on a time scale of 1 hour. (The high residuals at the positions of the strong narrow lines in the difference spectrum are a consequence of the TV scanner saturation.)

Similar but long-term profile variation of the $H\beta$ profile of 3C 390.3, resembling our two spectra in Figure 1, has been reported by Osterbrock, Koski and Phillips (1976), Veilleux and Zheng (1991) and Dietrich and Kollatschny (1994). Comparison of our data with these observations provide evidence in favour of the reality of the changes observed by us which occured, however, on a time scale of only one hour.

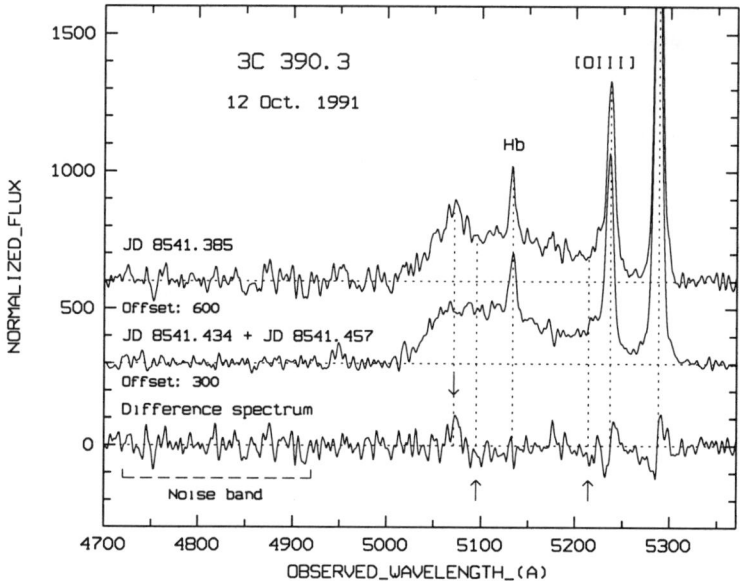

Figure 1. Spectra of 3C 390.3 obtained 1991 October 12 and their difference.

3. Discussion

The variations occur simultaneously at the blue and red sides of the $H\beta$ profile. This may indicate the response of a circularly rotating (accretion) disk to a light pulse from a central source. In this case the two negative features observed in the $H\beta$ difference profile at -2300 and $4700\,km/s$ ($\lambda_{obs} = 5095$ and 5214 Å) are formed in two opposite zones of gas close to the line of nodes. The apparent redward asymmetry in the positions of these bumps (i.e. the redshift of their midpoint) relative to the central line may be due to transverse Doppler and gravitational redshifts in the field of a massive central object (e.g., Netzer 1977). A possible counterpart of the observed blue positive residual (i.e. a red positive one) is not visible in the red wing of the $H\beta$ difference profile. This is possibly related to the superposition of the high residual of the strong $[OIII]$ line with an expected weaker, red positive bump.

The observed "doublet" of a positive and a negative residual appearent on the blue wing can be interpreted as the signature of an intensity peak moving in velocity space. Such small-amplitude profile variations, which take the form of narrow bumps drifting across the line profile, have been predicted by Antokhin and Bochkarev (1983) and Stella (1990).

Thus, the observed $H\beta$ profile variability properties appear to favour models of a relativistic circular disk in the BLR.

References

Antokhin, I. I., & Bochkarev, N. G. 1983, AZh, 60, 448

Dietrich, M., & Kollatschny, W. 1994, in Multi-Wavelength Continuum Emission of AGN, T. J.-L. Courvoisier & A. Blecha, Dordrecht: Kluwer, 444

Eracleous, M., & Halpern, J. P. 1993, ApJ, 409, 584

Netzer, H. 1977, MNRAS, 181, 89

Osterbrock, D. E., Koski, A. T., & Phillips, M. M. 1976, ApJ, 206, 898

Stella, L. 1990, Nature, 344, 747

Veilleux, S., & Zheng, W. 1991, ApJ, 377, 89

Active Galactic Nuclei and Related Phenomena
IAU Symposium, Vol. 194, 1999
Yervant Terzian, Daniel Weedman, Edward Khachikian, eds.

Rapid Profile Variations in the Broad Hα Line of the Seyfert Galaxy Markarian 6

N. S. Asatrian, E. Ye. Khachikian

Byurakan Astrophysical Observatory, 378433 Byurakan, Armenia

P. Notni

Astrophysikalisches Institut Potsdam, An der Sternwarte 16, D-14482 Potsdam, Germany

1. Observations and data reduction

We report on rapid variations of the Hα profile in Mark 6 over a period of about one hour and the implications for the velocity field of the BLR.

Two telescopes were used — the 6-m telescope of the SAO, equipped with a TV scanner, and the 2.6-m telescope of the BAO, using an image tube. A total of 6 $H\alpha$ spectra were recorded for Mark 6 during two nights at the two telescopes. The spectral resolution is typically $\sim 4\,\text{\AA}$ for scans and $\sim 6\,\text{\AA}$ for photographic spectra. The S/N ratio in the continuum for individual spectra is 10 – 15. No correction for spectral sensitivity has been applied. Individual spectra were continuum subtracted and calibrated scaling the total flux over two bands (λ_{obs}: 6500 – 6600 Å and 6740 – 6800 Å) in the $H\alpha$ wings.

2. Results

The $H\alpha$ difference profiles shown in Figures 1 and 2 display small but significant changes occuring on a time scale of ~ 1 hour. These changes take the form of narrow bumps (marked by arrows). The S/N ratios for the bumps are 3.6 – 9.0. (The high residuals at the positions of narrow $H\alpha$ and $[NII]$ lines in the difference spectra are consequences of guiding and centering errors.)

Interestingly, in the literature there are four $H\beta$ spectra of Mark 6 obtained at two different observatories in the same night 5 hours apart (Khachikian et al. 1982, Chuvaev 1991). These spectra show clear differences in the short wavelength shoulder of the broad profile during this time interval. This provides further evidence in favour of the reality of the variablity in Hα observed by us.

These rapid variations in the emission line profiles of Mark 6 are very similar to those observed in 3C 390.3 (see our other poster in this volume) and favour models of a relativistic inhomogeneous circular disk in the BLR.

References

Chuvaev, K. K. 1991, Izv. Krimsk. Astrofiz. Obs., 83, 194
Khachikian, E. Ye., Popov, V. N., & Egiazarian, A. A. 1982, Astrofizika, 18, 541

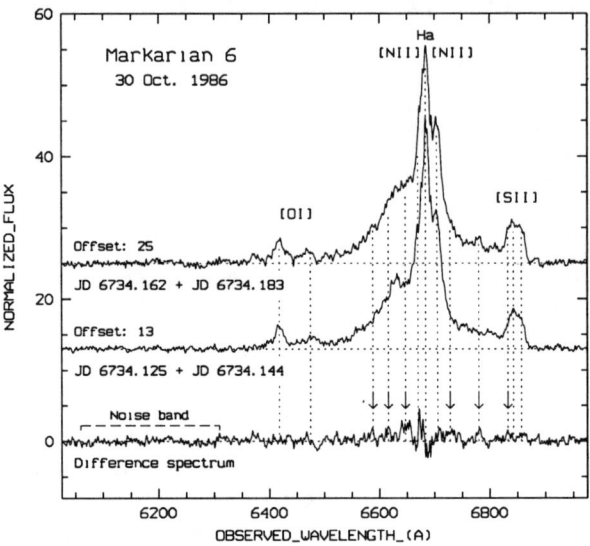

Figure 1. Image tube spectra of Mark 6 obtained 1986 October 30 and their difference.

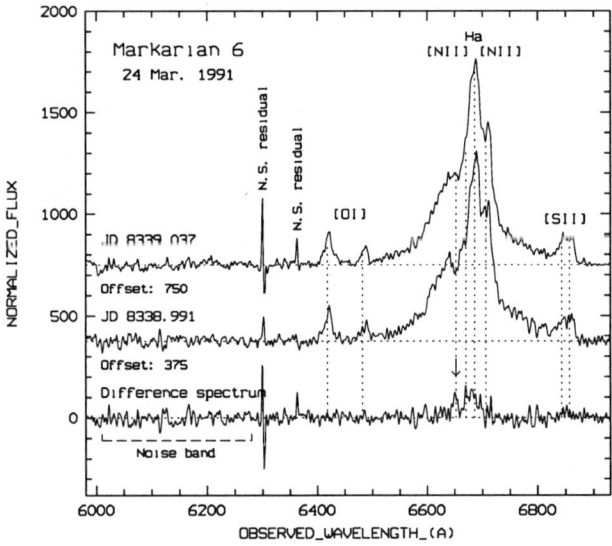

Figure 2. TV scanner spectra of Mark 6 obtained 1991 March 24 and their difference.

Multiwave Monitoring of the Nuclei of Seyfert Galaxies and Quasars at Several Telescopes of FSU and European Countries in the Frame of an INTAS Program.

N.G.Bochkarev

Sternberg Astronomical Institute Universitetskij Prosp., 13, 119899 Moscow, Russia boch@astronomy.msk su

A.I.Shapovalova, A.N.Burenkov, V.V.Vlasyuk

Special Astrophysical Observatory of RAS 357147 Nizhnij Arkhyz, Karachaevo-Cherkessia, Russia

Abstract. Several groups of astronomers in FSU (Former Soviet Union) and European countries have started, in 1998, a cooperative spectral and broad-band optical and UV-monitoring of the nuclei of selected Seyfert galaxies and quasars (10 objects). The monitoring is to be continued until Oct. 30, 2000. The spectral optical monitoring with CCD-cameras in the spectral range $\lambda\lambda 4000$-$8000 Å$ is being carried out at two observatories: SAO RAS, Russia (6-m and 1-m telescopes) and at Kazakhstan (70-cm telescope). The resolution is about 3-10$Å$, the S/N ratio is $\approx$50-100. The broad-band optical monitoring with BVRI filters (with a 1% accuracy at the R-band) is being done at 5 observatories. NLR and ENLR in monitored galaxies will also be investigated.

1. Introduction

A program of AGN spectral monitoring based on theoretical analyses (Blandford and McKee, 1982; Bochkarev and Antokhin, 1982; Antokhin and Bochkarev, 1983) was initiated at the 6-m telescope observatory (SAO RAS). The first test observations were carried through in 1983; regular observations started in 1986 continue up to now (see e.g. Bochkarev et al., 1990, 1991). Observations with one single telescope, however, are not sufficient for determination of AGN physical characteristics. Therefore we joined the global "AGN Watch" program (see, e.g., Peterson et al., 1991) at its starting time in 1988 and continue to work in its frame. To increase Russian/the Former Soviet Union (FSU) input to the "AGN Watch" we are starting now, in collaboration with colleagues from West-European countries, a FSU-European program supported by INTAS .

2. Claimed objectives

The aim of this program is to determine the fundamental characteristics of the physical structure and the kinematics of the broad (BLR) and narrow (NLR and ENLR) emission line regions in a representative sample of Seyfert galaxies.

Specific objectives are to:
1) determine the BLR structure by application of cross-corelation methods;
2) determine the long-term stability of the BLR;
3) determine the physical state of the BLR by modeling the observed spectra;
4) determine the spatial structure, the physical state and the ionization state of the NLR and ENLR;
5) compare the line and continuum properties of Seyfert galaxies of different types covering a large range in nuclear luminosity.

3. Spectral and broadband monitoring

Nuclear spectra of the selected galaxies will be taken with three telescopes using CCD: with the 6-m and 1-m telescopes of SAO RAS ; with the 70cm telescope in Kazakhstan; spectral range $\lambda\lambda 4000\text{-}8000 \text{Å}\text{Å}$, spectral resolution 3-10Å and signal-to-noise ratio of 50-100 in the continuum.

Photometric observations with an accuracy of about 1% in R filter will be carried out at 5 observatories: SAO RAS with CCD in BVR filters on 1-m and 60-cm telescopes; Landessternwarte, Heidelberg, with the 70-cm telescope in BVRI filters; Crimea (Ukraine) with the 60-cm telescope equipped with an electro-photometer and a simple CCD-camera in UBVR filters; Lesniki (near Kiev, Ukraine), with a 70-cm or 50-cm telescope with an electro-photometer in UBVR filters, after equipping it with an ST-6 CCD-camera; Maidanak Mountain Observatory (Republic of Uzbekistan), with a 60-cm telescope with an electro-photometer in the UBVR filters.

4. Spectrophotometry of the NLR

Spectra of the NLR and ENLR will be obtained at the prime focus of the 6-m telescope of SAO RAS using the multi-pupil field spectrograph (MPFS) + CCD in the range $\lambda\lambda 4000\text{-}8000 \text{Å}\text{Å}$, with a resolution of about 4-6Å, and a signal-to-noise ratio of 50-100 in the nuclear region and 10 in the off- nuclear regions. Blocks of 16x16 micro-lenses will be used in the spectrograph MPFS, and it will be possible to register simultaneously 256 spectra of the galaxy over a region of 16x16 arcsec.

5. Data analysis

Time lags of prominent emission lines will be determined by cross- correlation methods probably by two independent teams. We plan to produce images in lines and continuum, maps of velocity fields, relative intensities, line FWHI distribution etc. The archival spectra from the 6-meter telescope, from Kazakhstan, and from Crimea obtained in 1976-1996 for NGC 4151, 3516, 7469, will be used to study the long-term evolution of the BLR in these galaxies. Photometric data for the same period will be used also. The data obtained will be interpreted in the framework of photoionization models. Interpretation and theoretical researches will be done by five teams: in France, GB, Germany, Kazakhstan, and Russia.

This research was supported by INTAS-grant 96-328 and by the Russian Basic Research Foundation through grant 97-02-17625.

References

Antokhin, I.I., Bochkarev, N.G., 1983, Astron.Zh., v.60, p.448.
Blandford, R.D., McKee, C.F., 1982, Astrophys.J., v.255, p.419.
Bochkarev, N.G., Antokhin, I.I., 1982, Astron.Tsirc., No. 1228, p.1
Bochkarev, N.G., Shapovalova, A.I., Zhekov, S.A., 1990, Astron.J., v.100, p.1799.
Bochkarev, N.G., Shapovalova, A.I., Zhekov, S.A., 1991, Astron.J., v.102, p.1278.
Peterson, B.M., et al., 1991, Astrophys.J., v. 368, p.119.

Active Galactic Nuclei and Related Phenomena
IAU Symposium, Vol. 194, 1999
Yervant Terzian, Daniel Weedman, Edward Khachikian, eds.

Variability of the broad emission H_β profile in 3C390.3

A.N. Burenkov, A.I. Shapovalova, V.V. Vlasyuk

Special Astrophysical Observatory of Rus.Ac.Sci., Nyzhnij Arkhyz, Russia

and N.G. Bochkarev

Sternberg Astronomical Institute, Moscow, Russia, 119899

1. Introduction

The object 3C390.3 is a well known broad-line radio galaxy (z=0.056), whose very broad emission lines are variable in time (see Veilleux and Zheng 1991, and Dietrich at al. 1998 for references). It is known as the prototypical source of broad double-peaked H_β and H_α emission lines. In this paper we present the preliminary results of spectral observations of 3C390.3 in 1995 – 1998 on 6-m and 1-m telescopes of the Special Astrophysical Observatory of RAS.

2. Observation and data reduction

The spectra were obtained with the long-slit spectrograph + CCD of the 6-m telescope ($\approx$ 80%) and with the long-slit spectrograph + CCD of 1-m telescope ($\approx$ 20%). The processing of the spectra was made using software developed in SAO RAS by Vlasyuk (1993). The spectra were scaled to constant flux of [OIII] emission lines. The linear approximation of continuum in a spectrum was made via two areas ($\lambda 4930 \pm 5 \text{Å}$ and $\lambda 5313 \pm 5 \text{Å}$). The profile of the emission H_β was fitted by compilation of 3 Gaussians (broad blue, broad red and narrow components). Fig.1 presents results of measurement of the spectra.

3. Results

We studed the variability of the broad H_β profile during the last 3 years: from April 1995 to April 1998.

    1. The variations of the H_β and broad red component fluxes followed the continuum (Icont) during the period of observations, excluding dates dJD before 10000 and dJD after 10800. The possible delay $\approx$ 200 days of the emission lines variability to continuum was present (Fig.1, top).

    2. The local continuum varied quasi-periodically with possible period $\approx$ 800-1000 days (Fig.1, top).

    3. The relation of broad blue component flux to red was almost constant $\approx$ 0.25 until dJD 10500, increased rapidly up to 2.5 at dJD 10600 and later turned back. For testing this result we averaged spectra inside dJD 10599 – 10701,

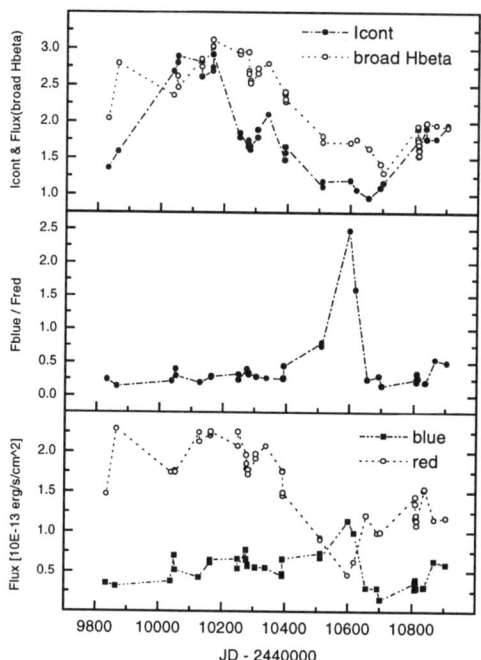

Figure 1. Top: Flux density in continuum (Icont) is measured within the wavelength range $\lambda\lambda 5100 - 5150\text{Å}$ and the full broad H_β flux. Middle: Relation of blue/red components of broad H_β. Bottom: Fluxes of blue and red components of broad H_β. Icont is shown in $10^{-15} erg\ s^{-1}\ cm^{-2}\ \text{Å}^{-1}$ and fluxes are present in $10^{-13} erg\ s^{-1}\ cm^{-2}$ (for the [OIII] 4959+5007ÅÅflux = $1.7 \cdot 10^{-13} erg\ s^{-1}\ cm^{-2}$)

repeated the same analyses as for individual spectra and confirmed the strong decrease of the broad red component at this time.

Acknowledgments. This work was supported by INTAS grant 96-328 and by Russian Basic Research Foundation grants 94-02-4885 and 97-02-17625.

References

Dietrich M. et al., 1998, ApJS, 115, 185.
Veilleux S. and Zheng W., 1991, ApJ, 377, 89.
Vlasyuk V.V., 1993, Astrofiz. issled. (Izv. SAO RAS), 36, 107.

Active Galactic Nuclei and Related Phenomena
IAU Symposium, Vol. 194, 1999
Yervant Terzian, Daniel Weedman, Edward Khachikian, eds.

Flare in the Hydrogen Line Region of the NGC 3227 Nucleus on January 12-15, 1997

I.Pronik[1], L.Metik[1], N.Merkulova[1]

Crimean Astrophysical Observatory, Crimea, Ukraine

Introduction. The nucleus of NGC 3227 was classified as a Sy2 type before 1974, and a Sy1 type after 1974. The Seyfert type of the nucleus changed corresponding to its brightness: U brightness of the nucleus increased in 1974: $\Delta U \sim 1^m$. Maximum brightness of the nucleus was observed in 1975 – 1977.

Observations. 57 spectrograms in the spectral region 3700–7300Å were obtained with the 6-m telescope on January 12-15, 1977 during maximum of the nucleus brightness. Spectral resolution was ~ 8Å. Seeing was $(1-3)''$.

Equivalent widths (W_λ) and profiles of the emission lines H_δ, H_γ, H_β, [OIII]$\lambda\lambda$4959,5007ÅÅ, H_α+[NII], and [SII]$\lambda\lambda$6717,6731ÅÅ were averaged by series. Duration of the series of observation was $\sim$25 min.

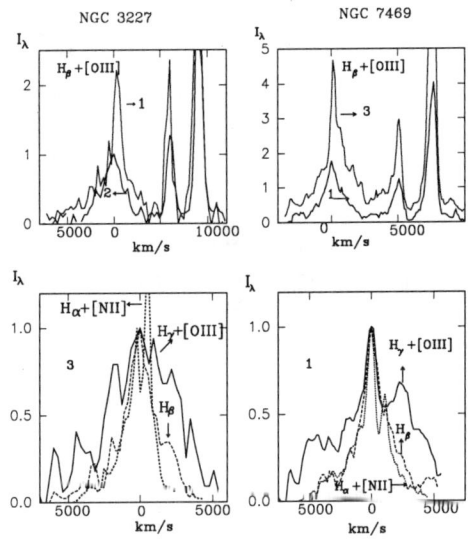

Figure 1.   Night-to-night variations in line profiles (see text).

Equivalent widths of lines showed asynchronous variations. This fact leads us to suspect, that there were night–to–night variations of intensities of emission lines with upper limit $\sim$ 1.5 times.

Profiles of Balmer lines H_γ, H_β and H_α in the spectrum of the NGC 3227 nucleus showed remarkable similarity in characteristics of night–to–night varia-

---

[1]Isaac Newton Institute of Chile, Crimean Branch

tions with the same lines in the spectrum of the NGC 7469 nucleus (Pronik et al., 1997).

Comparison of results can be seen in the Figure, where figures on plots signify the ordinal numbers of nights from the beginning of the event. Profiles on the top panel, where I_λ are given in continuum units show that:

1. Central narrow peaks of the H_β line are double in both cases.
2. The ratio I_b/I_r of the narrow H_β line smoothly decreases from $I_b/I_r \geq 1$ to $I_b/I_r \leq 1$ when W_β is increased (i.e. continuum of the nucleus is decreased).

On bottom panel I_λ of profiles are given in units of peak brightness of Balmer lines. These profiles show that:

3. Strong brightening of H_γ broad wings were observed with lag 1-2 days from the beginning of the brightening of the blue narrow H_β component.

Discussion. We suppose that night–to–night variability of profiles of Balmer lines in spectra of NGC 3227 and NGC 7469 nuclei are connected with short time flares in emission line regions of these nuclei. Two variable components of a narrow H_β line can reflect the existence of two variable streams in both nuclei. In that case, radial velocities of the streams emitting H_β light are about –250 km/s and –550 km/s compare to the recession velocities for the galaxies NGC 7469 and NGC 3227 correspondingly.

The observational indication of the beginning of a short–time flare appears to be a relative brightening of the blue narrow peak of the H_β line and an increasing of continuum flux of the nucleus. Variations in streams approaching observers can be interpreted as an ejection from the nucleus with radial velocities of about 200–500 km/s. This ejection influences the gaseous regions emitting broad Balmer lines: brightening of narrow blue H_β component was accompanied by an increase with 1 day lag of the intensity of the broad H_γ wings with the radial velocities equal to $\pm$ 6500 km/s. Such high intensity of broad wings was not observed in the H_β and the H_α profiles. One can argue that the emission of broad H_γ wings has an inverse Balmer decrement. Comparison with theoretical models permits us to suppose that this gas could be opaque, hot and inhomogeneous in physical conditions of plasma (T_e = 25 000 K, $n_e = 10^{12} - 10^{14}$ cm^{-3}), and it is ionized and excited mainly by a collisional process.

Almost simultaneous variability of the broad H_γ line and the narrow component of the H_β line during the flare shows that regions emitting broad lines and narrow lines are overlapped. These regions are not more than several light days ($\sim 4.5 \cdot 10^{15}$ cm) in dimension. One can speculate that regions of flares are excited by shocks acting inside jets.

The flare in NGC 7469 was observed during the minimum brightness of its nucleus (Pronik et al., 1997). The flare in NGC 3227 was observed during the maximum brightness of its nucleus. Therefore short time flares did not connect with the general brightness of nuclei of galaxies. One can suspect that there are two independent sources of nuclear activity. One of them is connected with the general brightness of the nucleus, and another one is not.

References

Pronik, I., Metik, L., & Merkulova, N. 1997, A&A, 318, 721

Active Galactic Nuclei and Related Phenomena
IAU Symposium, Vol. 194, 1999
Yervant Terzian, Daniel Weedman, Edward Khachikian, eds.

Strong Radio Outbursts in Six Active Galactic Nuclei in 1997–1998

Y.Y. Kovalev
*Astro Space Center of the Lebedev Physical Institute,
Profsoyuznaya 84/32, Moscow, 117810 Russia*

Abstract. The beginnings of strong radio flares in six AGNs were detected and their evolution was studied using 1–22 GHz five epoch instantaneous spectral observations in 1997–1998. The phenomena started at the highest frequency and moved to the lower frequency in a regular fashion. Such behavior of the flares in AGNs can be explained by synchrotron emission from a compact relativistic jet implying different physical models which are discussed. For different types of AGNs the same behavior of flares is revealed, in favor of the same basic physical model for BL Lacs and quasars. A birth of new VLBI components in compact jets associated with these outbursts is predicted.

1. Observations

Observations were made at 31, 13, 7.6, 3.9, 2.7, and 1.4 cm, simultaneously for all frequencies, with the RATAN–600 in the framework of a monitoring program (see Kovalev, 1998). The beginnings of strong radio flares in some AGNs were detected in 1997. We selected six of them for an analysis: quasars 0906+01 (z=1.02), 1622−25 (z=0.79), 1958−17 (z=0.65), 2121+05 (z=1.94), BL Lacertae object 0235+16 (z=0.94), quasar or compact galaxy 0007+10 (z=0.09).

2. Radio flares evolution

Comparing the behavior of flares with the modeled synchrotron radiation of a blob of relativistic particles evolving in a radial magnetic field (see e.g. Kovalev & Mikhailutsa, 1980; Kovalev & Larionov, 1994), one can see a very good agreement between them. In this model the outbursts and their evolution are explained by an increasing and a decreasing of the density of relativistic particles, emitted from the core region to the jet.

Considering the flare evolution in respect to the model of a shock wave passing through a conical adiabatically expanding jet (Marsher & Gear, 1985), we have analyzed the different stages of flares. During a growth the synchrotron self-absorption turnover is going to lower frequencies and the turnover flux is increasing (a Compton phase). During a plateau phase the turnover frequency decreases while the turnover flux remains approximately constant (a synchrotron phase). A fading of a flare is well consistent with a final adiabatic expansion phase, both the turnover frequency and the flux decreasing in time.

The similar behavior of flares is revealed for all objects, which are displayed on figure 1 (note that the strength of flares and the velocities of evolution are different). This favors the hypothesis of the same basic physical model for BL Lacs and quasars. The change of the polarization of a VLBI core at the beginning of these flares could happen (as was mentioned for BL Lacs by Gabuzda et al., 1994). We predict a birth of new VLBI components in compact jets associated with these outbursts. These components may appear in the VLBI maps in a few years (depending on the source) after the beginning of a flare, if a compact jet will be resolved from a VLBI core.

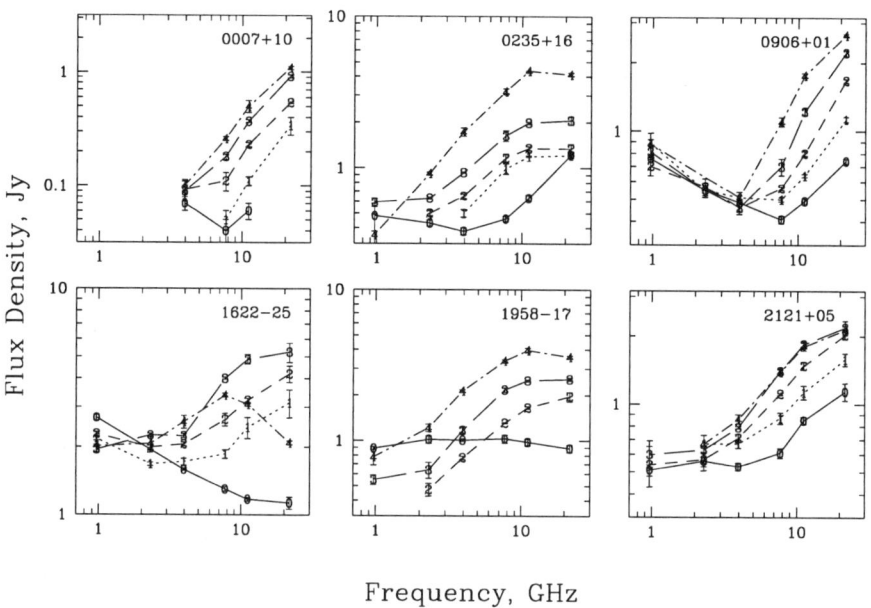

Figure 1. The evolution of the instantaneous total flux density spectra for six AGNs. '0', '1', '2', '3', '4' denote points at March, June, September, December, 1997, and April, 1998, respectively.

Acknowledgments. The author is thankful to the LOC of the IAU Symposium 194 and INTAS for support of his stay in Byurakan during the conference and to the ISSEP for support by the PhD student grant. This research was supported in part by the Russian State Program "Astronomy".

References

Gabuzda, D.C., et al. 1994, ApJ, 435, 140
Kovalev, Yu.A., & Mikhailutsa, V.P. 1980, Soviet Ast., 24, 400
Kovalev, Yu.A. 1998, Bull. SAO, 44, 50
Kovalev, Y.Y., & Larionov, G.M. 1994, ALett., 20, 3
Marsher, A.P., & Gear, W.K. 1985, ApJ, 298, 114

Active Galactic Nuclei and Related Phenomena
IAU Symposium, Vol. 194, 1999
Yervant Terzian, Daniel Weedman, Edward Khachikian, eds.

Long Term X-ray Variability of Galactic Nuclei

J. Cunniffe & E.J.A. Meurs

Dunsink Observatory, Castleknock, Dublin 15, Republic of Ireland.

Abstract. We present some results on the long term variability of the X-ray emission from optically catalogued galaxies, especially in view of their nuclear properties. For this investigation we use data from the ROSAT PSPC archive. Some future directions are also discussed.

1. Introduction

The ROSAT All Sky Survey (RASS) and subsequent pointed observations with the PSPC instrument have provided a database with a sufficient timebase to examine the long term variability of the X-ray emission from galactic nuclei. Many of the detected sources are recognised AGN but some will be galaxies which are not expected to have large X-ray brightness variations. NGC 5905, a proposed nuclear starburst galaxy (Bade, Komossa & Dahlem 1996), was detected during the RASS in July 1990 but over subsequent observations its count rate dropped by a factor of approximately 100 taking it to the limit of detectability. This preliminary investigation aims to determine how many other galaxies which are not classified as AGN exhibit similar variations. Eventually, we hope to find candidates for variation in normal nuclei as has been suggested by the case of NGC 5905.

2. Data

An optical sample of galaxies was chosen from the CfA Redshift Catalogue (1995) (see Huchra et al. 1992) for those objects with type $T \leq 20$ and $M_B < 13.21$ (the catalogue claims to be complete to this magnitude.)

These criteria produce 1774 objects which were compared with X-ray source detections from the Pointed Phase observations with the ROSAT PSPC instrument given in the WGA (Rev.1) Catalogue (White et al. 1994). Hardness ratios were defined as follows:

$$HR1 = \frac{A - (B + C)}{A + B + C}, \quad HR2 = \frac{(A + B) - C}{A + B + C},$$

where A=PSPC channels 11 to 39, B=40 to 84, and C=86 to 200.

3. Results

Those optical positions matching to within 30" of the X-ray error circle (quoted by WGA as 13" within 20' of PSPC axis, and 50" outside 20') are accepted as matches at this stage. This yields 237 objects with X-ray detections. Four of these objects are an alternative optical identification for the X-ray source and are thus rejected, leaving:

- 162 X-ray sources detected once (90 detected in a single pointing, 72 sources detected only once from multiple pointings.)

- 71 X-ray sources detected more than once.

At this point only the 71 sources with more than one X-ray detection will be discussed further as no variability information is directly available from the single detections. 3 sources are excluded as they contain X-ray detections which are too widely spread to be due to a single point source and a clear discrimination is not straightforward. The remaining objects are classified as active/non-active based on their inclusion/non-inclusion respectively as a Seyfert/QSO/BL Lac in the compilation of Véron-Cetty & Véron (1998). Of these 68 objects, 30 are classified as active and 38 as non-active.

4. Discussion

The levels of variability of the PSPC count rate for all the sources here are relatively limited (certainly compared with the factor of 100 mentioned in the Introduction) with the highest factor being ~ 7.5 (active) and ~ 3 (non-active). 89% of non-active type galaxies have variabilities less than a factor of 2, compared to 67% of active types. A greater level of variability may be seen when the 72 (out of 162) objects detected once from multiple pointings are analysed and upper limits calculated.

For non-active galaxies, the hardness ratios show little significant change with variation in count rate. Active galaxies, on the other hand, appear to show an overall trend towards greater softness at the lower count rate. However, this trend is absent in the few high variability cases in the sample of active types with little change in hardness ratio shown.

References

Bade,N., Komossa,S. & Dahlem,M. 1996, A&A, 309, L35

Huchra,J., Geller,M., Clemens,C., Tokarz,S. & Michel,A. 1992, Bull. Inf. C.D.S. 41, 31

Véron-Cetty & Véron 1998, ESO Scientific Report 18

White,N.E., Giommi,P., Angelini,L. 1994, BAAS, 185 #41.11

Active Galactic Nuclei and Related Phenomena
IAU Symposium, Vol. 194, 1999
Yervant Terzian, Daniel Weedman, Edward Khachikian, eds.

Searching for Low-Mass Supermassive Black Holes

Michele Cappellari[1], Francesco Bertola[1], Enrico M. Corsini[1],
José G. Funes, S.J.[1], Alessandro Pizzella[2] & Juan C. Vega Beltrán[3]

[1] *Dipartimento di Astronomia, Università di Padova, Italy*

[2] *European Southern Observatory, Santiago, Chile*

[3] *Osservatorio Astronomico di Padova, TNG, Italy*

It has become generally accepted that most or possibly all ellipticals and bulges of spirals harbor supermassive black holes in their center (see Ho 1998 for a recent review).

We have constructed a model for a thin gaseous disk orbiting in the combined potential of a central point-like mass embedded in a diffuse component mimicking the stellar disk. We have used this model to reproduce the peculiar bidimensional shape of the emission lines in a sample of early type disk galaxies, observed with ground based telescopes (e.g. Fig. 1). In this way it has been possible to deduce the presence of nuclear mass concentrations of the order of 1×10^9 $M_\odot$ in four objects (Bertola et al. 1998).

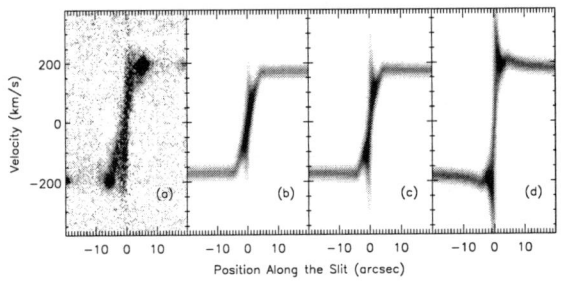

Figure 1. *(a):* The Hα emission line, observed along the major axis of NGC 2179, after subtraction of the stellar continuum. *(b), (c)* and *(d):* Models of the NGC 2179 Hα line [shown in the same scale of panel *(a)*] obtained with different point-like central masses. *(b):* $M_\bullet = 2 \times 10^8$ $M_\odot$. *(c):* $M_\bullet = 1 \times 10^9$ $M_\odot$ corresponding to our best-fit model. *(d):* $M_\bullet = 5 \times 10^9$ $M_\odot$.

In some of the observed galaxies (e.g. NGC 5064) the rotation curve does not show any peculiar bidimensional shape, but rises almost linearly from $v = 0$ at $r = 0$ up to an almost constant value at larger radii (Fig. 2a). By applying our modeling technique, taking into account the observational parameters of our ground based observations, we have estimated that in these cases the central mass, if present, has to be lower than 5×10^7 $M_\odot$ (Fig. 2b).

In Fig. 3 we present a *simulation* of an HST spectrum obtained with STIS (grating G750M with the 52"×0.2" slit) for the same gaseous disk modeled in

Fig. 2b. It is apparent that the much higher spatial resolution of HST would still allow us to detect the central perturbation due to the 5×10^7 M$_\odot$ black hole adopted for this model.

We think that disk galaxies like NGC 5064 are good candidates to harbor low-mass supermassive black holes. This example shows that the combination of ground based observations, with properly equipped telescopes, and follow-up HST observations could be an efficient way to explore the low end of the black-hole mass function.

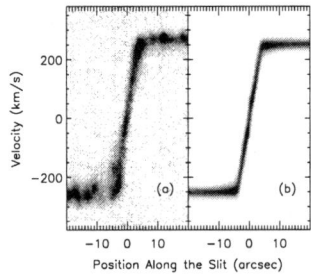

Figure 2. (a): The Hα emission line, observed with the ESO 3.6-m telescope equipped with CASPEC echelle spectrograph, along the major axis of NGC 5064, after subtraction of the stellar continuum. (b): Model of the NGC 5064 Hα line obtained with the highest point-like central mass which can be added without significantly disturbing the general shape of the line ($M_\bullet = 5 \times 10^7$ M$_\odot$).

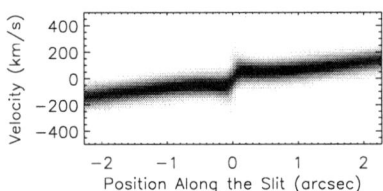

Figure 3. *Simulation* of an HST spectroscopic observation with STIS (grating G750M with the 52"×0.2" slit) for the same gaseous disk modeled in Fig. 2b. The reciprocal dispersion Δv is $\sim 8\times$ smaller (better) in the case of the ground based observation of Fig. 2b, but the spatial resolution is $\sim 24\times$ better for the STIS observation. Note that only the very inner region of the profile is shown in this Figure.

References

Bertola, F., Cappellari, M., Funes, J.G., Corsini, E.M., Pizzella, A., & Vega Beltrán, J.C. 1998, ApJL, in press [astro-ph/9807075]

Ho, L.C. 1998, in Observational Evidence for Black Hole in the Universe, ed. S.K. Chakrabati (Dodrecht: Kluver Academic Publishers), in press [astro-ph/9803307]

Seyfert Activity in Ring Galaxies due to Galactic Interactions

Tapan K. Chatterjee

Facultad de Ciencias, Fisico-Matematicas, Universidad A. Puebla, Apartado Postal 1316, Puebla, Mexico. e-mail: tchat@mailexcite.com

Observations indicate that ring galaxies are more active than spirals (e.g., Ghigo et al., 1983 Appleton and Struck-Marcell, 1987). However, Seyfert activity is noted in only a few ring galaxies, e.g. NGC 985 (de Vaucouleurs & de Vaucouleurs 1975), NGC 1144 = Arp 118 (Huchra et al. 1982), ring galaxy in Sextans–optical counterpart of the IRAS source 09595-0755 (Wakamatus & Nishida). This is indicative of the fact that this type of activity requires favorable circumstances. In this context a previous work on the stellar episode of ring formation, Chatterjee (1984), is extended, including gas and studying the evolution of the ring structure, due to rebounds of the compact elliptical about the plane of the disk.

The spiral galaxy (of radius R) is modeled by an exponential disk of scale length $\alpha = 4/R$, witha (static) thickness (Chatterjee 1990a), and a spherical polytropic bulge (n = 0.3.4 equally weighted combination) containing 1/3 of the mass; about 20% of the mass of the disk contains gas particles. The disk is embedded in a static halo (of mass 5 times the optical parts) with an asymptotic dependence $M(r) \sim r$ (c.f., Allen & Martos 1986) as in one of the moddles of Chatterjee, 1990b. The elliptical is modeled identically as the bulge. The gravitational potential is softened with sofetning constants of $\epsilon = r_o/5$, $r_o/3$, $0.8r_o$, $r_o/4$ for the bulge of the spiral as well as the elliptical, stellar and gasous components of the disk, and for mutual gravitational interaction, respectively (r_0 being the radius containing 75% of the total mass of the spiral). The Lagrangian Equations (in polar cooriditates r and θ) determine tyhe relative orbit of the elliptical (Galaxy e) with respect to the spiral (Galaxy s),

$$dt = [2\mu^{-1}\{E - W(r) - L^2/2\mu r^2)\}]^{-0.5} dt, \text{ and } d\theta = [L/(\mu r^2)] dt,$$

where μ, L and E are the reduced mass, angular momentum and the orbital energy, respectively, and W(r) is the instantatneious potential energy of interaction between the galaxies. At small intervals of r and θ, the tital effects and the corresponding changes in velocities of the stars are determined, such that the instantaneous internal energy changes of the two galaxies $\Delta U_s(t)$, can be determined; the instantaneous relative velocity follows from,

$$V(t) = [2\mu^{-1}\{E_i - W(t) - \Delta U_e(t) - \Delta U_s(t)\}]^{-0.5},$$

E_i being the inital value of E. The galaxies are interchanged at each interval to determine the relevant quantities for each one. The three-body approach is adopted to study the motion of the stellar and gasous particles, (but taking the extended nature of the particles into account through gravity softening). Marginally bound conditions are used so that rebounds of the spherical intruder occur; such that during the collision episode the perturber does not espcape abruptly but oscillates about the plane of the target disk.

Results indicate that after the formation of the stellar ring sturcture, the same propagates outwards, and then inwardws; such that it can be considered as a density wave with damped oscillatory behavior. The gas inner to the stellar ring, and leading it, experiences a clumping and decelerating torque, that causes the same to lose angular momentum and fall inward; while the gas outer to the ring gains angular momentum and moves outwards in expanding orbits.Though the ring disappears in due course of time, the same is reformed due to rebounds of the ellipti8cal intruder; such that we have a feedback process. Basically a bi-symmetric potential develops due to the formation of the stellar ring; the same dissolves and redovelops, as the perturber oscillates about the plane of the disk, with growing amplitude. This process leads to a falttening of the surface density distribution and redistribution of the angular momentum of the gas in the disk. In this way the angular momentum exchange induces a turbulence in the gas content of the disk. This will render the ring galaxies more active than normal spirals, as is observed. The cumulative effects of the repeated oscillations of the intrudr is the accumulation of gas immediately surrounding the nucleus, in the form of an accretion disk. It is found that in the prescence of a deep potential well (corresponding, for example to that of a black hole), the gas will flow into the nucleus. As only several ring galaxies manifest Seyfert activity, this counde be indicative of the fact that Seyfert ring galaxies have compact central objects.

Acknowledgments. It is a pleasure to thank the organizers for an excellent hospitality

References

Allen, C., Martos, M.A., 1986, Rev. Mexicana. Astron, Astrof. 13, 137.

Appleton, P.N., Struck-Marcell, C., 1987, ApJ, 312, 566.

Chatterjee, T.K., 1984, Ap&SS, 106, 309.

Chatterjee, T.K., 1990a, Ap&SS, 163, 127.

Chatterjee, T.K., 1998b, IAU Symp. 124, "Paired and Interacting Galaxies" (NASA, USA).

de Vaucouleurs, G., de Vaucouleurs, Al., 1975, ApJ, 197, L1.

Ghigo, F.D., et al., 1983, AJ, 88, 1587

Huchra, J.P., Wyatt, W.F., Davis, M., 1982, AJ, 87, 1628.

Wakamatsu, K., Nishida, M.T., 1987, ApJ, 315, L23.

The Role of Neutral Hydrogen in the Evolution of Sprial and Irregular Galaxies

V.A. Ambartsumian[1] and A.L. Gyulbudaghian

Byurakan Astrophysical Observatory, Armenia, e-mail agyulb@bao.sci.am

The Hubble classification suggested for the spiral galaxies, Sa–Sb–Sc–Sm–Irr, was considered by many astronomers as an evolution succession. The version of transformation from early spirals to the late spirals and vice versa was considered. The existence of several parameters, monotonically increasing or decreasing along this succession, together with the parameters which do not depend on the structural type of the galaxies of this succession, is a confirmation of this succession to be a succession of stages of evolution. Let us consider in detail both types of parameters. About the full masses and luminosities of galaxies we can state, that except Irr and maybe Sm-galaxies there is not a clear expressed difference of mean values of these parameters for galaxies of different structural types. Irr galaxies are the systems with low masses and luminosities, but on the (M-L) diagram a part of them is obviously mixed with the late spirals. There is a strong change of the M/L ratio with changing of the structural type, though the early type galaxies on average apparently have the least values of that ratio. As monotonically changing parameters we can suggest the following three:

- Relative amount of HI (M(HI)/M(tot)).

- Relative amount of HII(M(HII)/M(tot).

- The ratio of mass of H_2 to HI ($M(H_2)/M(HI)$).

All these three parameters are changing monotonically along the succession Sa–Irr, moreover the third parameter is substantially decreasing (20 times), and the first and second are increasing correspondingly two times and several dozen times. These figures show, that along the succession Sa–Irr the quantity of H_2 is abruptly decreasing, and the quantity of HI, on the contrary, is increasing. We think that as time elapses, the transformation of H_2 into HI takes place, that is why we must take Sa–Irr as a succession of evolution. Now the opposite point of view is widespread: the galaxies with high relative amount of HI are considered as an early phase of evolution of spiral galaxies. It is also assumed that the clouds of two other types (H_2 and HII) must be formed from the HI clouds. We have here an analog of a long existing dilemma: the transformation of diffuse matter into the more dense bodies (nobody observed such a process) or the transformation of more dense objects into dilute matter, and the first concept is more widespread than the second, which has direct observational confirmations. The question of direction of evolution from the molecular clouds

---

[1]Deceased

to the clouds of HI is one of the components of a larger question about the direction of evolution of galaxies. Let us summarise the results of observations, presented in the literature.

- There is a clear increase of relative amount of HI along the succession Sa–Sb–Sc–Sm–Irr.

- Many irregular and some Sc galaxies have large HI envelopes.

- There are dense HI intergalactic clouds, some of them are 4×10^9 years old (in Leo) and have the masses of dwarf galaxies.

- The H_2 clouds, as well as HI clouds are concentrated towards the spiral arms (in our Galaxy, as well as in other galaxies), but the arrangement along the spiral arms in CO is more clear, than in HI.

- We have a case of transformation of molecular clouds into HI clouds within the galaxy M83 (under the influence of bright stars radiation).

- The investigation of radial systems of dark globules in our Galaxy shows, that a transformation of H_2 clouds into HI clouds takes place (with an intermediate stage of HII).

The first two points are possible to interprete from the point of view of transformation of H_2 into HI and vice versa (we have then for the evolution of galaxies correspondingly Sa–Sb–Sc–Sm–Irr or vice versa), but the other points are clearly in favour of the transformation of H_2 clouds into HI clouds. It means that the older galaxies have higher values of the ratio M(HI)/M(tot). We have the following observational data (from the literature), about clusters of galaxies.

- When we are approaching the center of a cluster, the deficit of HI in the spiral galaxies is increasing.

- The deficit of HI is higher in the cluster galaxies than at the field galaxies.

- The clusters having high X-ray radiation have also more galaxies with HI deficit.

Taking into account the preceding three points, we can make the following conclusion: the galaxies situated in the younger systems have also higher HI deficit. We can make three main conclusions from the whole paper:

- In our Galaxy, as well as in other spiral galaxies, a transformation of H_2 clouds into HI clouds takes place.

- The older the galaxy, the higher the percentage of HI quantity (in some cases almost the whole galaxy can consist of HI).

- The evolution path for the spiral galaxies passes from the early spirals to the irregular galaxies.

Are there Quasars at Non-Cosmological Distances?

Haik A. Harutyunian

Byurakan Astrophysical Observatory, Armenia.
E-mail: hhaik@bao.sci.am

In Dravskikh & Dravskikh (1996) the results of a statistical analysis of quasar luminosity as a function of redshift, if it is assumed to be cosmological, are given (see Fig.1). Three quasar samples were analyzed separately:

- quasars with absorption lines;

- quasars associated with galaxies or association-quasars not having absorption lines (from Burbidge et al. 1996);

- other quasars not included in the first two groups.

The first thing to attract attention is that the curves for these samples differ sharply from one another. The difference is that galaxies with absorption lines have a higher luminosity than those of the other two types at any redshift. Association quasars at small redshifts also show an increased luminosity compared to quasars of the third sample, although they lose this feature with increasing z. The authors state that these results have fairly high statistical significance with a confidence level of over 0.99.

Let us consider these results assuming that some fraction of quasars have a local origin and were ejected from the nuclei of nearby galaxies. We designate as v_L the average quasar ejection velocity. Then it will be positive if the ejection occurred into the half-space "behind the galaxy" relative to the observer and it will have a negative sign, if the ejection occurred between the galaxy and observer.

Furthermore, it is easy to see that the radiation from quasars ejected into the half-space "behind the galaxy", other conditions being equal, will be more subject to absorption from the generating galaxy (as well as in the gaseous shelf that accompanies such an ejection), than the radiation from quasars ejected with a negative velocity. Burbidge (1996) notes on the similar physical assumption that absorption lines should be observed in the spectra of about 50% of ejected quasars, while there should be no such lines in the spectra of the remaining galaxies.

From this physical picture the following conclusion could be drawn. To the value of the redshift of the quasars that exhibit absorption is added some positive component due to the ejection velocity. On the other hand, for quasars ejected toward the observer and therefore free of absorption in the generating galaxy the redshift is decreased due to the negative velocity. The observable redshift z_Q for these objects will then be determined by the equation

$$1 + z_Q = (1+z_c)(1+z_a)(1 \pm z_L) = 1 + z_c + z_a(1+z_c) \pm z_L(1+z_c)(1+z_a), \quad (6)$$

where z_c is the cosmolgical and z_a is the anomalous quasar redshift, and finally, z_L is the redshift due to ejection from the galaxy.

It thus becomes clear that if the observed redshift z_Q is treated as entirely cosmological, an error depending on the anomalous and local redshifts is thereby introduced. In this case the cosmological redshift differs from its true value by an amount

$$\Delta z = [z_a \pm z_L(1+z_a)](1+z_c), \qquad (7)$$

where the "+" sign pertains to quasars that are more subject to absorption and the "-" sign to association quasars.

It is seen that if the expression $z_a \pm z_L(1+z_a)$ is larger than zero, then the quasar's distance and luminosity are overestimated. Moreover, if at least one of the quantities z_a and z_L differs from zero, the luminosity of quasar with absorption is found to be larger than the true value. The overestimate for quasars with absorption is therefore more substantial than for association quasars. In converting to quasar luminosities, therefore, we find that quasars with absorption lines have a higher luminosity, on the average, than association quasars.

The model considered here is an extremely simplified one when all quasars are assumed to have been ejected from nearby galaxies and to have retained their initial anomalous redshift. Nevertheless, it would be more natural to assume that the anomalous redshift has some maximum value at the very onset of formation of a quasar, and subsequently it gradually decreases in the course of evolution and ultimately disappears. On the other hand naturally there should be quasars at cosmological distances which could be named "first generation" or "previous generation" quasars.

According to our current concepts, a quasar ultimately evolves to the stage of a galaxy, in the spectrum of which no anomalous redshift should be observed. This means that among quasars, including those ejected from galaxies, there are objects whose spectra are not distorted by an anomalous redshift. Such quasars obviously, with high probability, lie at large distances from the site of their formation and could not be included in the first two samples.

It is thus natural to assume that if one is calculating the luminosities of quasars that do not have absorption lines and are not in associations, then they will very likely include quasars whose anomalous redshift is close to zero. This means that their observed redshift consists mainly of two terms, which are the cosmological and the local redshifts, so we have

$$\Delta z = \pm z_L(1+z_c) \qquad (8)$$

and we no longer have to talk about the systematic overestimation of quasar redshift. We can therefore conclude that the overestimation of distances and luminosities is less in this case. It is then natural that the luminosities of these quasars calculated from their redshifts will be lower than for quasars associated in one way or another with neighboring galaxies. In other words, the redshifts of quasars with absorption lines and association quasars differ more from their cosmological values, on the average, than for other quasars.

Let us try to interpret now a feature at large redshifts mentioned at the very beginning. That is that at large redshifts the association quasars show a lower luminosity, on the average, than quasars of the third type. So, if we assume the

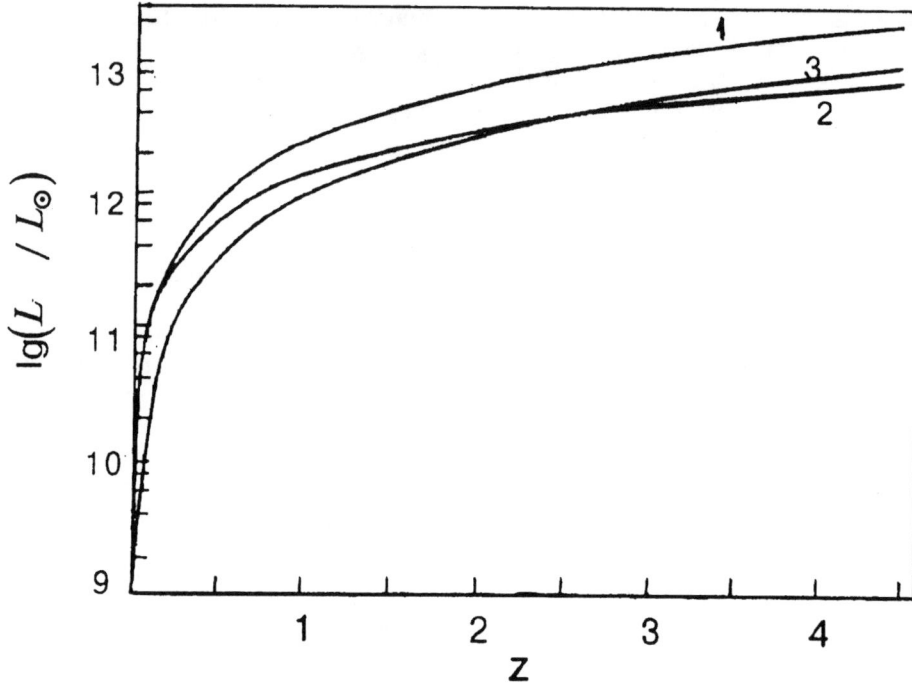

Figure 1. The distribution of average luminosities of quasars depending on z as given by Dravskikh & Dravskikh (1996).

existence of an anomalous component in quasar redshift, then this component should be manifested more often in quasars with large observed redshifts than in quasars with small redshifts. It is thus natural to expect the presence of an anomalous component in the spectrum of quasars in the third sample with large redshifts to be more likely than in the spectrum of quasars of the same sample but with small redshifts.

Thus it is understandable that if all three samples are under the same conditions from the standpoint of the presence of an anomalous redshift component, then the luminosities of quasars in the third sample should be closer to the average of the luminosities of the other two types. And in this case we can actually say that the luminosities of quasars in the first sample, which include quasars with "positive" velocities, are overestimated the most, that the luminosities of quasars in the second sample, having "negative" velocities, are overestimated the least, and that the luminosities of quasars in the last sample, for which averaging over all velocities occurs, should be overestimated less than the former and more than the latter. Just such a picture is observed.

So, the assumption about the existence of the quasars ejected from nearby galaxies with an anomalous redshift allows interpretation of some observational data unexplained till now. We believe this is an argument speaking in support of existence of "local" quasars. This is a very important fact in favour of Ambartsumian's concept. Briefly, it appears that some part of the matter existing in the nuclei of galaxies (ejecting as "local" quasars) has identical properties with the matter generated at the very beginning of the Universe ("first generation" quasars).

References

Dravskikh, A.F. & Dravskikh Z.V. 1996, Astron. Zh., 73, 19
Burbidge G., Hewitt A., Narlikar J.V., & Das Gupta P.1990, ApJS, 74, 675
Burbidge G., Astron. Astrophys., 1996, 309, 9

Haik Harutyunian

First Ranked Galaxy Morphology and Morphological Content in Groups of Galaxies

A. P. Mahtessian and V. H. Movsessian

Byurakan Astrophysical Observatory, Byurakan, 378433, Armenia

V. Ambartsumian (1956) has shown that the observable quantity of double and multiple galaxies much exceeds that expected, based on the assumption about dissociative balance. He has concluded that the members of double and multiple systems, as well as the clusters, could not arise independently of each other, and only then to be united in systems, by mutual capture. They should arise in common. Moreover, when in a group are large luminous central galaxies, the origin of the weak members of the group should be caused by activity of a nucleus of the central galaxy (Ambartsumian 1962). On the other hand, recently, there was widespread the opinion that the properties of central galaxies of groups and clusters are caused by interactions with environmental galaxies.

Mahtessian et al. (1995) has confirmed the earlier result (Mahtessian 1982, Tikhonov 1987), that there is not sufficiently strong influence on the morphology of galaxies from the environment.

Hickson et al. (1984) deduced that in compact groups in which the first-ranked member is a spiral galaxy, the relative number of spirals is higher than in the other groups.

The population of ellipticals in the group with a first-ranked elliptical member turns out to be significantly higher than in the other groups (Ramella et al. 1987, Wirth 1983).

Hickson et al. (1988) has shown that the distributions of morphological types of first-ranked galaxies and all galaxies in compact groups do not differ from each other. The relative quantity of spiral galaxies for all of the sample is 0.49, which almost coincides with the relative quantity of spiral galaxies among the first-ranked members of groups (0.48). Besides in these groups is found morphological concordance. As has appeared, in 20 of 58 quartets, all four galaxies have the same general type (early or late). The probability of such an event, counted from the assumption of independence of morphological types, is small (10^{-5}).

Mahtessian (1982) found that in groups with higher relative number of elliptical and lenticular galaxies among spiral members early spirals meet more often.

A study of double galaxies (Karachentsev, Karachentseva 1974, Noerdlinger 1979) determined, that the tendency of pairs have galaxies with close morphological types. Similar conclusion have come by Yamagata (1989), where 16930 pairs of nearest neighbor galaxies were investigated.

Dependence of morphological types of non-first-ranked and second-ranked galaxies from a morphological type of the first-ranked galaxy in loose groups (Mahtessian 1988, 1997, 1998) has been studied. Compared also are distributions of morphological types of first-ranked, second-ranked and ordinary (not first-ranked) galaxies.

There is a significant tendency that second-ranked and ordinary galaxies of groups have similar morphological types as the first-ranked member.

There are not significant differences between distributions of morphological types of first-ranked, second-ranked and ordinary galaxies of groups.

These results allow us to conclude, that the observable morphological types of galaxies can be a consequence of the initial conditions (for example, mass, its density, moment of rotation of protogroup) in the epoch of their formation.

Two occurrence following mechanisms of galaxies in groups are possible:

a) At first, the protogroup disintegrates to parts, hereafter from which are formed galaxies.

b) At first, the first-ranked galaxy of the group is formed, from which later on by means of ejection (or disintegration) are formed other members of the group.

In both cases is expected a correlation between morphological types of the members of groups (in particular between morphological types of first-ranked galaxies and other galaxies). And the prevailing morphological type in the group should "remind" of the initial conditions present in the protogroup at the moment of occurrence. For example, if rotational moment per unit mass of the protogroup is small, there can arise many elliptical galaxies.

References

Ambartsumian, V. A. 1956, Izv. Akad. nauk Arm. SSR, seria fiz.-mat. nauk, 9, 23

Ambartsumian, V. A. 1962, in Transaction of the IAU, vol. XIB, Academic Press, London, New York, 145

Hickson, P., Kindl, E., & Huchra, J. 1988, ApJ, 331, 64

Hickson, P., Ninkov, Z., Huchra, J., & Mamon, G. 1984, in Clusters and Groups of Galaxies, F. Mardirossian, G. Giuricin, & M. Mezzetti, Dordrect: Reidel, 367

Karachentsev, I. D., & Karachentseva, V. E. 1974, AZh, 51, 724

Mahtessian, A. P. 1982, Comm. of Byurakan obs., 53, 102

Mahtessian, A. P. 1988, Astrofizika, 28, 255

Mahtessian, A. P. 1997, Astrofizika, 40, 45

Mahtessian, A. P. 1998, Astrofizika, 41, 473

Mahtessian, A. P., Movsessian, V. H., Khachikian, E. Ye., & Tiersch H. 1995, Astron. Nachr., 316, 143

Noerdlinger, P. D. 1979, ApJ, 229, 877

Ramella, M., Giuricin, G., Mardirossian, F., & Mezzetti, M. 1987, A&A, 188, 1

Tikhonov, N. A. 1987, Comm. of Special astr. obs., 52, 51

Wirth, A. 1983, ApJ, 274, 541

Yamagata, T., Nogouchi, M., & Iye, M. 1989, ApJ, 338, 707

Activity Phenomena Observed at Radio Frequencies in Sprial Galaxies

V.H.Malumian

Byurakan Astrophysical Observatory, Byurakan, 378433, Armenia

Abstract. It has been shown that the stage of activity of spiral galaxies depends on both the distance between them and their nearest neighbours and the structure of their central parts.

In the end of the seventies on the basis of radio observations of many spiral galaxies it has been shown, that among galaxies which are the members of pairs of galaxies, especially of close and interacting pairs, radio sources with the level of radio emission above a fixed limit occur significantly more frequently than among single ones (Stocke 1978). Later on it has also been shown that spiral members of pairs of galaxies are 2 to 2.5 times stronger radio sources than single galaxies of the same absolute magnitude (Altschuler & Pantoja 1984; Malumian 1986, 1987a, 1989a, 1990). Interesting results were obtained on the basis of statistical investigations of the radio properties of spiral galaxies which are the members of loose groups of galaxies from the lists of Turner and Gott (Turner & Gott 1976) and Geller and Huchra (Geller & Huchra 1976). It turned out that if the members of groups were considered altogether irrespective of ranks among them the radio emission with the flux density above a fixed level occurs significantly more frequently than among single galaxies but noticably rarer than among the members of pairs of galaxies of the same absolute magnitude (Tovmassian & Shakhbazian 1981; Malumian 1987b). When the spiral members of the groups were considered separately by rank, it turned out that radio emission among the first ranked spiral members of groups was observed to be not rarer than among the members of pairs of galaxies (Malumian 1987b).The relative numbers of radio sources is the highest among the first ranked spiral galaxies. The detailed analysis of radio properties of isolated spiral galaxies and spiral members of pairs and groups indicates that radio luminosities of spiral galaxies are determined by the presence of close neighbours rather than by the space density of galaxies around them. The rate of occurence of the radio sources and their radio luminosities in all probability do not depend on whether are members of pairs or groups of galaxies or they are isolated but depend on the presence of the neighbour. The probability of radio emission for spiral galaxies, if other conditions are the same, depends on the separation between them and their nearest neighbours. The shorter this separation the higher the probability of radio emission (Malumian, 1996). Now it is clear why the spiral members of groups from the list of Geller and Huchra (1976) are the less powerful radio sources than the members of pairs from Karachentsev's catalogue (Karachentsev's catalogue 1987). The matter is that in the samples considered the separations between the members of the vast majority of pairs are much less than that between the members of groups and their nearest neighbours. For example, the number of pairs with projected separation between their members $R < 50 kpc$ is nearly 80% but among galaxies within groups the objects with projected separation between them and their nearest neighbours $R < 50 kpc$ are only 16%. In Malumian (1996) it was shown that members of groups with $R < 200 kpc$ are on average 1.5 times stronger radio sources than members with $R > 200 kpc$. The relation between radio luminosity and the structure of the central parts of spiral galaxies was ascertained. The structure of the central parts of galaxies is determined by the Byurakan classification. It was shown that galaxies with stellar, semistellar and split nuclei or with central condensations (marked 5, 4, 2s and 2 according to Byurakan classification) are 2 to 3 times more powerful radio emitters than galaxies without such signs (marked 3 and 1 according to Byurakan classification) (Malumian 1989). It was also ascertained that among galaxies of classes 2, 2s, 4 and 5 are dominant objects with flat spectra of radio emission, and vice versa among galaxies of classes 1 and 3 are dominant the objects having steep spectra of radio emission (Malumian 1983). From our investigations of spectra of radio emission of spiral galaxies it follows that among galaxies of the classes 2, 2s, 4 and 5 the compact, recently formed radio sources occur more frequently. They have flat spectra of radio emission which became steeper for a comparatively short time because of losses of energy mainly due to synchrotron emission by emitting relativistic electrons. On the contrary, among the galaxies of classes 1 and 3 there is a large excess of objects which have steep spectra of radio emission. Summarizing the aforementioned facts one can state that the stage of activity of spiral galaxies depends on both the distance between them and their nearest neighbours and the structure of their central regions. The shorter this separation the higher the stage of activity. The stage of activity is also higher in spiral galaxies with any peculiarities in their central parts.

References

Altschuler, D.R., & Pantoja, C.A. 1984, AJ, 89, 1531
Geller, M. & Huchra, J. 1976, ApJS, 32, 409
Karachentsev, I.D. 1987, Dvojnie Galaktiki (Double Galaxies): Moscow
Malumian, V.H. 1983, Astrofizika, 19, 251
Malumian, V.H. 1986, Astrofizika, 25, 19
Malumian, V.H. 1987a, IAU Symposium 121, p.24
Malumian, V.H. 1987b, Astrofizika, 26, 311
Malumian, V.H. 1989a, Astrofizika, 30, 223
Malumian, V.H. 1989b, Astrofizika, 31, 241
Malumian, V.H. 1990, Astrofizika, 32, 507
Malumian, V.H. 1996, Astron. Nachr., 317, 101
Stocke, J.T. 1978, AJ, 83, 348
Turner, E., & Gott, J.R. 1976, ApJS, 32, 409
Tovmassian, H.M., & Shakhbazian, E. 1981, Astrofizika, 17, 265

Simultaneous UBVRI Light Curves of the Seyfert Galaxy NGC 4151 During the Extraordinary Brightening in 1989-1996

N.Merkulova[1], L.Metik[1], I.Pronik[1]

Crimean Astrophysical Observatory, Crimea, Ukraine

Introduction. The nucleus of the Seyfert galaxy NGC 4151 is known to be variable in the optical on different timescales: from minutes to tens of years. A new cycle of activity of the nucleus is investigated from 1989.

Observations. 1500 measurements were obtained simultaneously in each of the UBVRI bands at Crimean Astrophysical Observatory with the 1.25-m telescope in an aperture of 20 arcsec during 96 nights between 1989 February 11 and 1996 June 14. The estimated accuracy of each measurement is no more than $0.^m01$.

UBVRI Light curves in each filter showed an increase of the nuclear brightness from 1990 February. Until 1996 June the unusual historically extraordinary long-term brightening was going on. Amplitudes and averaged gradients of the brightness variations during 7.3 years of our observations decreased from the U to I band. The amplitude amounts to $2.^m0$, $1.^m5$, $0.^m9$, $0.^m8$ and $0.^m7$ in UBVRI bands, respectively. Gradients of brightnesses were: 0.30; 0.23; 0.15; 0.13; 0.11 mag/year for UBVRI bands correspondingly.

There were intranight variations and flares with durations 10–147 days. Amplitudes for flares $A = F_{max}/F_{min}$ decreased from the U to I band. Maximum amplitudes for U and I bands were 3.4 and 1.3 correspondingly.

Gradients of the nuclear brightening during flares in all spectral bands are more than those for total observing time. The largest gradients were observed at the end of the observing period.

UBVRI energy distributions for flux excesses during the flares and during the whole period of observations showed a power-law form $F_\nu \sim \nu^\alpha$. Calculated spectral slopes for flares were $+0.18 \geq \alpha_{pl} \geq -1.83$, $\Delta\alpha_{pl} = 2.01$. Obtained data were discussed in the frame of the model of synchrotron nature of UBVRI variations of the NGC 4151 nucleus. The increasing of spectral indices with time suggests that the optical depth of synchrotron emission clouds increased from the beginning to the end of the nuclear brightening.

Structure Function analysis. Evolution of the process causing the variation of the nucleus in all spectral bands was investigated by the method of the structure function (SF). The slope of SF b=dlog(SF)/dlogΔt characterizes the nature of the process: b = 0 – flicker-noise, b = 1 – shot-noise. SF was calculated separately for four observational time intervals: I.89 – V.90; XI.90 – V.92; XII.92 – I.94 and I.95 – VI.96. The figure shows that:

---

[1]Isaac Newton Institute of Chile, Crimean Branch

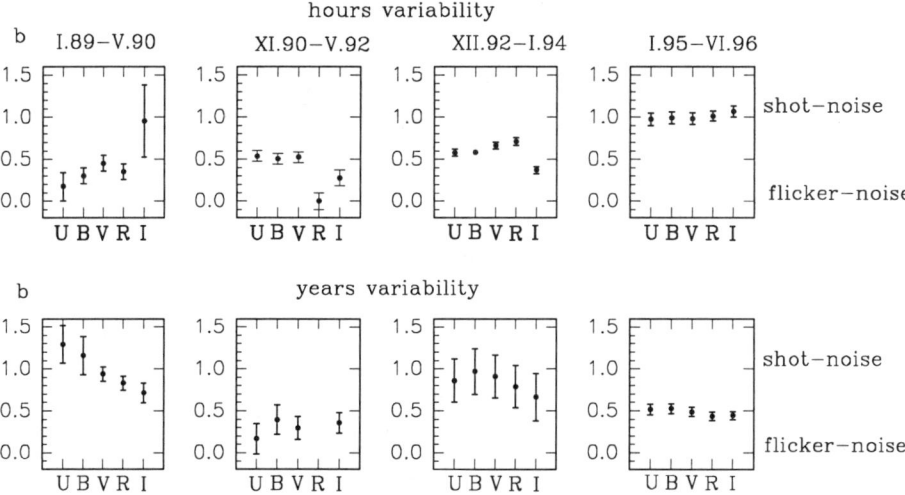

Figure 1. Variation of SF slope "b" with time. Top – for hours variability, bottom – for years variability.

1. Intranight and years UBVRI nucleus variations are caused by a process variable with time. The process varied between flicker-noise (b = 0) and shot-noise (b = 1).

2. The nuclear process evolved with time in different directions, intranight variations evolved from flicker to shot noise, and years variations vice versa.

Discussion. Difference in direction of the evolution of the process acting in the nucleus of NGC 4151 within nights and during 1989 – 1996 years revealed from SF slopes variations suggests that variation of the nucleus is connected with two different sources. One of them acts on a time scale less than one night but another one – on a time scale of years. The source of years time-scale variations at the beginning of the observations was caused by a strong shot–noise process: SF slope "b" for all UBVRI bands were near or more than 1. Emission in U band of this source is characterized by the highest amplitude of variations and highest slope "b". We suppose that increasing of the nuclear brightening on time scale of years was connected with the source of high ultraviolet emission. At the beginning of the nuclear brightening this source was strongly fueled. To the end of the nucleus brightening fueling of the process decreased; it evolved from shot–noise to flicker–noise.

Variations of the source acting on time-scale of hours at the beginning of the nuclear brightening was caused by a flicker-noise process, slope "b" of SF was near 0. But during the nuclear brightening SF slope "b" gradually increased. At the end of the nuclear brightening variations of this source were caused by a strong shot-noise process. This fact suggests that fueling of the source acting on time-scale of hours is increased with increasing nuclear brightening contrary to the source acting on time-scale of years.

Active Galactic Nuclei and Related Phenomena
IAU Symposium, Vol. 194, 1999
Yervant Terzian, Daniel Weedman, Edward Khachikian, eds.

The Anisotropy in the Distribution of the Major Axes of Elongated Shakhbazian CGCG

R.A. Vardanian

Byurakan Astrophysical Observatory, Byurakan, 378433, Armenia

In 1978 we have shown that Shakhbazian CGCG have an elongated form (Vardanian & Melik-Alaverdian 1978). Morever, they turned out to be of comet forms (1996), at the head of which basically are situated their bright members. This enables determination with high accuracy of the position angles θ of major axes of the elongated Shakhbazian CGCG. A closer examination of these angles shows that distribution of orientations of the major axes of groups is anisotropic and has two maximums (Vardanian 1996). (See Table, where θ is position angle, N is number of CGCG). These maximums are oriented in mutually perpendicular directions ($\theta = 45°$ and $\theta = 135°$).

No	θ	N	θ	N
1	0° - 10°	3	90° - 100°	6
2	10 - 20	6	100 - 110	11
3	20 - 30	11	110 - 120	15
4	30 - 40	14	120 - 130	13
5	40 - 50	27	130 - 140	25
6	50 - 60	16	140 - 150	14
7	60 - 70	15	150 - 160	10
8	70 - 80	8	160 - 170	9
9	80 - 90	3	170 - 180	4

References

Vardanian, R.A., & Melik-Alaverdian, Yu. K. 1978, Astrofizika, 14, 195
Vardanian, R. A. 1996, Astrofizika, 39, 585

Active Galactic Nuclei and Related Phenomena
IAU Symposium, Vol. 194, 1999
Yervant Terzian, Daniel Weedman, Edward Khachikian, eds.

NGC 3077: H_α Features and the Young Population

H. Abdel-Hamid[1] and P. Notni

Astrophysikalisches Institut Potsdam, D-14482 Potsdam, Germany

1. Introduction

NGC 3077, an Irr II galaxy in the M81/M82 group of galaxies at 3.63 Mpc, suffered a tidal encounter with M81 about 400 Myr ago. Tidal structures of HI in this group of galaxies can best be seen in a VLA image by Yun et al. (1994).

This close encounter, perhaps, led to an accumulation of gas in the centre of NGC 3077 and triggered star formation there.

2. The young stellar population

The optical image of NGC 3077 is governed by a central irregularly distributed blue population and dust superposed on a regular old red stellar background of a dwarf Elliptical. Several nearly starlike blue knots have been selected from the clumpy light distribution. Some of them may be young star clusters (as shown by a comparison with an HST image), others may be the brightest young stars. The youngest of the knots have ages between 1 - 5 Myr derived from their colours and magnitudes, only slightly depending on the interpretation (stars or clusters).

A population analysis method using only broad-band colours (U,B,V,R) was applied to study the stellar population in NGC 3077, see Abdel-Hamid 1998. The method enables us to decompose the observed light at each point in the galaxy into two populations. We derive the intensity distribution of both populations I and II together with their age and extinction distributions. The surface brightness distributions of the two populations, taken from isophotal ellipse fits, are fitted well by exponential laws. The age and the extinction distributions show a grand design circular structure centred approximately on the optical centre. The same structure is seen in polarisation measurements published by Scarrott et al. (1990) and in excess H_α radiation.

3. The H_α morphology

Large scale structures: Near the centre, H_α mimics the centrally symmetric distribution of the blue stellar population. A network of arcs, loops, filaments and knots is seen superposed onto this huge central HII-region, the contrast growing with distance. There is generally *no* obvious relation between the H_α features and distinct star forming regions. In the outskirts, large partial shells

---

[1]Present address: National Research Institute of Astronomy and Geophysics, Helwan, Egypt

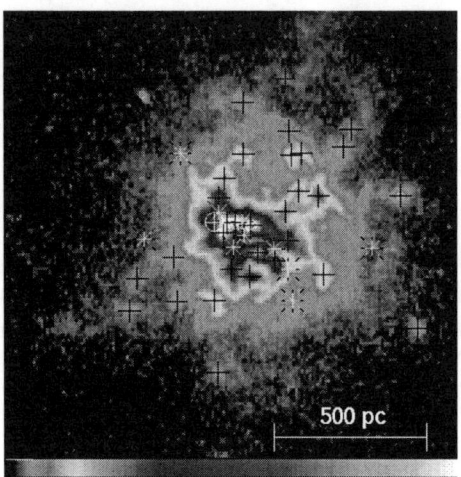

Figure 1. Ionized gas structure (H_α) in NGC 3077. Position of blue knots (*), HII regions (+) and the physical centre (⊕)

are distributed at both sides of the central region along an inclined axis (N–S) up to a distance of about 1.5 kpc . None of these structures has an obvious continuum source inside. A faint glow of diffuse emission extends to the NW side.

HII regions: We selected 36 comparatively compact condensations ("HII regions"). Most of them are *not* geometrically correlated with a blue knot (Figure 1); the distribution of H_α seems to be largely matter-dominated, knots and shells signify turbulence and local motions. The luminosity function of the knots is fitted by a power law with power index $\alpha = -1.3 \pm 0.2$. Surprisingly, this agrees well with that of dwarfs (Miller & Hodge 1994, Hodge et al. 1989) despite the fact, that we are discussing mostly knots in shell structures, whereas normally HII regions are associated directly with young star forming regions.

H_α-ring: Comparison of H_α, extinction and continuum polarisation (Scarrott et al.) shows that *enhanced* H_α often coincides with enhanced extinction and enhanced polarisation, all showing the nearly centrally symmetric grand design mentioned in Section 2. The polarisation pattern is roughly symmetric about the centre. This hints at an *illumination of the dust by a central light source* and *a scattered light origin also of part of H_α in these places*.

References

Abdel-Hamid, H.A. 1998, Dissertation, Uni. Potsdam, Wissensch.-verlag Berlin.
Hodge, P., Lee, M., & Kennicutt, R. 1989, PASP, 101, 32.
Miller, B.W., & Hodge, P. 1994, ApJ, 427, 656.
Scarrott, S.M., Rolph, C.D. & Semple, D.P. 1990, in IAU Symposium 140, 245.
Yun, M.S., Ho, P.T.P. & Lo, K.Y. 1994, Nature 372, 530.

Active Galactic Nuclei and Related Phenomena
IAU Symposium, Vol. 194, 1999
Yervant Terzian, Daniel Weedman, Edward Khachikian, eds.

A New Sample of Candidate GPS Radio Sources with Intermediate Flux Densities

Gabriel A. Ohanian

Byurakan Astrophysical Observatory, Byurakan, 378433, Armenia

A new sample of faint Gigahertz Peaked Spectrum (GPS) sources is presented here as an addition to our previous one (Ohanian,1995). Comparison of the results of investigation of these samples with those of previously studied brighter samples (e.g. O'Dea et al. 1991) will delineate the properties of GPS sources over a wide range of luminosity, redshift and rest frame peak frequency. This will give noticeable information relevant to their cosmological evolution and relationship with other classes of extragalactic radio sources.

As the basis for our selection the data of 4 catalogues have been used: Texas survey (UTRAO 365 MHz, Douglas et al. 1980), Arecibo 611 MHz survey for which reliable positions and flux densities were determined with the NRAO 300 ft transit telescope at 4755 MHz (Lawrence et al. 1983), Green Bank 1.4 GHz (White & Becker, 1992) and 4.85 GHz (GB6, Gregory et al. 1995).

The Texas 365 MHz survey $+18°$ strip and the Arecibo survey overlap from RA 0^h to 12^h and Dec. from $13°$ to $19°$. In this range there are 716 Arecibo sources, with $b^{II} > 10°$. Fifty six sources (8%) were selected with negative spectral indeces ($S \sim \nu^{-\alpha}$) in the range of 365- 611 MHz , and positive in the range 611- 4850 MHz. Since the interferometer used for the Texas survey underestimates flux densities of sources larger than $15''$ and does not detect emission on angular scales $> 2'$, we are excluding all matches in the Texas survey with angular sizes $> 15''$ and matches in GB6 survey which are marked as extended. Besides that, there are 25 Arecibo sources with flux densities $> 5\sigma$, having no matches both in Texas and GB6 surveys. If they are not spurious sources, they can be good candidates for a sharply peaked spectrum, a few examples of which have been reported by O'Dea et al. (1991). High sensitivity multifrequency observations are needed for these sources.

Our sample contains 56 radio sources. The upper limits of flux densities at 305 MHz of 24 sources are less than 250 mJy. The 365 MHz flux densities of the remaining 32 sources lie in the range 163-880 mJy with a median value of $S_{365} = 357 mJy$.

We looked for optical identifications on the POSS plates for the sources, which overlap on the Minnesota APS fields (in sum 45 a radio sources). From 45 sources 13 (29%) are identified with galaxies. One source is identified with radio quiet quasar and one with a radio loud quasar, 10 radio sources are identified with stellar objects, which are possible quasars (in the end 12 (27%) are possible quasars). The remaining 20 (44%) radio sources are EF. From 13 identified galaxies 3 are nearby and radio quiet galaxies ($z < 0.1, L_{rad} = 10^{38-40} erg\, sec^{-1}$). All the identified objects among the bright members of GPS sample (e.g. O'Dea et al. 1991) are distant ($z > 0.1$) and luminous. The only exception is Mrk 668, being one of the brightest Sy1 galaxies. From 9 comparatively bright galaxies

($m_r = 9.2^m - 17.6^m$) 6 (60%) have close companions and/or are members of clusters and groups of galaxies. These properties also are typical for bright members of GPS radio galaxies (see e.g. Stanghellini et al. 1993).

In addition to the above mentioned data, we have searched from available literature objects which are identified with GPS radio sources.

First, we have formed a sample of 18 nearby and radio quiet galaxies which are identified with GPS type radio sources or contain a core dominated GPS component. From 18 galaxies 5 are elliptical and the remainder are spirals. Fourteen galaxies from 18 are interacting or pairs of galaxies, and/or members of clusters and/or groups of galaxies. Mrk 231, IIIZw35, Arp 220, which are contained in this sample, are well known galaxies in which extensive bursts of star formation take place. The presence of infrared and HI emission from the remaining galaxies also suggests that in these galaxies are star formation processes.

Second, we have examined the spectra of fainter radio sources ($S_{1.4GHz} < 50mJy$) from the Leiden - Berkeley Deep Survey presented by Oort (1988). Radio sources of this survey are identified with faint ($m_F = 22^m - 24^m$) galaxies. It was found that 12 sources are GPS type objects. From 12 radio sources 8 are identified with blue galaxies. These galaxies which are called blue compact radiogalaxies have optically peculiar morphology. Their radio luminosity is a factor of 10 -100 higher than that of local spirals and is intermediate between that of Seyferts and radio quiet quasars. The blue colour of these galaxies suggests active star formation. From 12 objects 6 are pairs of galaxies or have close companions and/or lie in the direction of clusters of galaxies.

From the data obtained we can conclude that GPS radio sources appear in the inner regions of active galaxies of different Hubble types. The presence of GPS type sources in host galaxies suggests active star formation in them.

References

Douglas, J. N., Bash, F. N., Torrence, G. W., & Wolfe, C. 1980, The Univ. of Texas Publ. in Astronomy, 17

Gregory, P. C., Scott W. K., Douglas K., & Condon, J. J. 1995, The GB6 catalog of radio sources, Charolttesville, NRAO, 1-587

Lawrence, C. R., Bennet, C. L., Garrisa-Barreto, J. A., Greenfield, P. E., & Burke, B. E. 1983, ApJS, 51, 67

O'Dea, C. P., Baum, S. A., & Stanghellini, C. 1991 ApJ, 380, 66

Ohanian, G. A. 1995, Astrofizika, 38, 694

Oort, M. J. A. 1988, A&A, 192, 42

Stanghellini, C., O'Dea, C. P., & Baum, S. A. 1993, ApJS, 88, 834

White, R. L., & Becker, R. H. 1992, ApJS, 79, 331

Determination of the OB stellar population of IZw18 on the basis of its H_γ and H_δ absorption lines

A. Sinanyan

Byurakan Astrophysical Observatory, Byurakan, Armenia

D. Kunth

Institut d'Astrophysique de Paris, France

J. Lequeux

Observatoire de Meudon, France

G. Comte

Observatoire de Marseille, Marseille, France

A. Petrosian

Byurakan Astrophysical Observatory, Byurakan, Armenia

Abstract.
On the basis of new spectroscopic observations of the blue compact dwarf galaxy IZw18 in the narrow spectral range between 4000Å and 4500Å absorption components of H_γ and H_δ lines were discovered. Equivalent widths of H_γ and H_δ lines have been measured. From available data the OB population of IZw18 was analyzed.

1. Introduction

IZw18 is the prototype of an isolated low metallicity blue compact dwarf galaxy experiencing an intense episode of star formation (Miguel Mas-Hesse & Kunth; 1998). During the last few years more than 120 scientific papers have been addressed to this galaxy.

The morphology of IZw18 has been extensively studied. The existence of an older population underlying the present young stellar population was discussed by several authors (Loose & Thuan 1986).

The very existence of Balmer absorption lines in the spectrum of this galaxy is the marker of its massive stellar population. From new high dispersion deep spectral observations of the galaxy IZw18, absorption components for H_γ and H_δ lines were discovered. In this poster, the results are presented and the contribution of the OB stellar population was determined.

2. Observations, data reduction and results

The spectra of IZw18 were obtained at the OHP 193cm telescope. The equipment used was the CARELEC spectrograph. Spectra have been taken in the blue, between 4000Å and 4500Å with a dispersion of 33Å/mm. For the interpretation, published equivalent widths of OB stars have used (Cananzi et al.; 1993).

The data have been reduced using IRAF package. The wavelength calibration was done with a IRAF procedure relying on the argon spectrum. The one-dimensional spectra have been derived from the two-dimensional ones by averaging the CCD columns illuminated by the object. These one-dimensional spectra have been normalized by taking the continuum windows into account (Fig. 1). After removing of the emission components, equivalent widths of the Balmer H_γ and H_δ absorption lines were computed (Fig. 2).

The obtained equivalent widths are the following:

(1) $EW(H_\gamma) = 2.38\text{Å} \pm 0.1$

(2) $EW(H_\delta) = 2.51\text{Å} \pm 0.1$

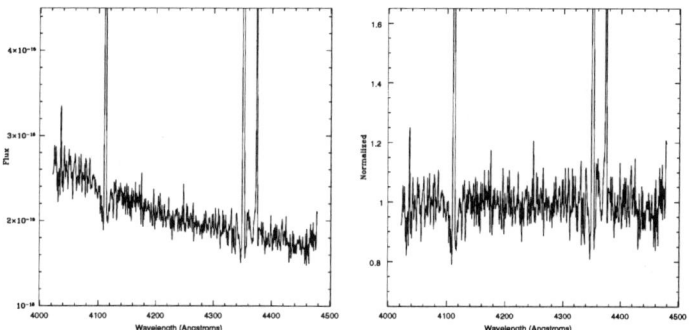

Figure 1. Left: The spectrum of IZw18 in the narrow spectral range between 4000Å and 4500Å Right: The normalized spectrum of IZw18 in the same spectral range.

3. Discussion

A galactic spectrum is the sum of the spectra of individual stars with different characteristics. Other factors, like star velocity dispersion for instance, also play a role in the final appearance of a galactic spectrum.

Synthetic equivalent widths have been calculated using the tabulated equivalent widths of H_γ and H_δ lines for the OB main sequence stars (Cananzi et al.; 1993)

The stars are combined by O type and B type. The average equivalent width of O and B type stars was calculated. For the O type stars the average equivalent width of H_γ and H_δ lines are 2.32Å and 2.48Å respectively, for the B type stars the average equivalent width of H_γ and H_δ lines are 6.84Å and

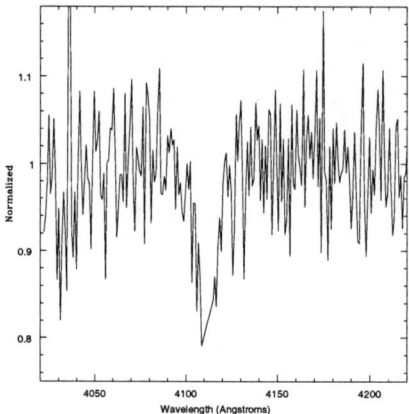

Figure 2. Smoothed spectrum. Absorption component of H_δ line of IZw18 after removing of emission line.

7.1Å respectively. Taking to account the calculated equivalent widths (1) and (2) from our observed spectra we can make up following equations:

(3) $2.32k_O + 6.84k_B = 2.38$, for H_γ line

(4) $2.48k_O + 7.1k_B = 2.51$, for H_δ line

Where k_O is a coefficient for O type stars and k_B is a coefficient for B type stars.

From (3) and (4) we obtain

$k_O \approx 0.55$,

$k_B \approx 0.16$

Now we can deduce the relative contribution of the OB stars as follows:

$$\alpha_O = \frac{k_O}{k_O + k_B} = \frac{0.55}{0.55 + 0.16} \approx 0.77,$$

$$\alpha_B = \frac{k_B}{k_O + k_B} = \frac{0.16}{0.55 + 0.16} \approx 0.23$$

Where α_O is the percentage of the O type stars (77%) and α_B is the percentage of the B type stars (23%).

Further work is in progress in a sample of 7 starburst galaxies.

References

Bica E., D Alloin 1986, A&A, 162, 21-31

Canazi K., R Augarde and J. Lequeux 1993 A&AS, 101, 599-619

Lequex J., M. Peimbert, J. F. Rayo, A. Serrano and S. Torres-Peimbert 1979, A&A, 80, 155-166

Loose H.-H., Thuan T. X. 1986, Star Forming Dwarf Galaxies, eds. D. Kunth, T. X. Thuan, J. T. T. Van, p. 73

Olofsson K. 1995, A&A, 293, 652-664

Olofsson K. 1995, A&AS, 111, 57-73

Petrosian A. R., J. Boulesteix, G. Comt, D. Kunth, E. LeCoarer, 1997, A&A, 318, 390

Zinnecker, Hans 1995 The Interplay between massive star formation, The ISM and Galaxy Evolution,Proceedings of the 11th IAP Meeting

V. CONFERENCE SUMMARY

Another Meal

Mary Kaiser and Byurakan employee

Active Galactic Nuclei and Related Phenomena
IAU Symposium, Vol. 194, 1999
Yervant Terzian, Daniel Weedman, Edward Khachikian, eds.

Future Directions in AGN Research

Julian H. Krolik

Physics and Astronomy Department, Johns Hopkins University, Baltimore MD 21218

Abstract.
A review is given of the principal successes in AGN research (black holes as the central engines, radiation mechanisms, population studies) and the most important open questions (creation of AGN, dynamics of the accretion flow, acceleration and collimation of relativistic jets). Some recent work that gives hope for progress towards answering these questions is also discussed.

1. Successes

Before considering where the field may be headed in the future, it would be worthwhile to review how far we have already come. Although many important questions remain unanswered, there have also been a number of notable successes.

1.1. Black Holes

Perhaps the single most important of these is the identification of accretion onto massive black holes as the fundamental energy source for AGN. The most powerful argument for this idea is that nothing else surpasses the possible energy production efficiency, $\eta \sim 0.1$ in rest-mass units, when matter accretes into a relativistically deep potential.

Moreover, even when such a high efficiency is realized, the accumulated mass must still be very large. By counting how many AGN we see as a function of the flux they deliver at Earth, we can "scoop up" a sample of the total photon density created by AGN over the lifetime of the Universe (Sołtan 1982). If we ascribe a mean fuel efficiency to the production of these photons, we may then estimate the mean density today of AGN "ash". Carrying out this exercise with modern data yields a mean remnant mass per bright galaxy

$$\langle M_{\rm rem} \rangle \simeq 1.6 \times 10^7 \left(\frac{F_{\rm bol}}{10 F_{\rm B}}\right) \left(\frac{h}{0.75}\right)^{-3} \left(\frac{\langle 1+z \rangle}{3}\right) \left(\frac{\eta}{0.1}\right)^{-1} \frac{M_\odot}{L_*\text{-galaxy}}, \quad (1)$$

where the ratio of bolometric flux to B-band flux is scaled to a factor of 10, the Hubble constant is $100h$ km s^{-1} Mpc^{-1}, and the mean redshift of AGN light is scaled to 2.

Several important conclusions follow from this estimate. First, the mean remnant mass is minimized if *every* galaxy once housed an AGN; if only select galaxies ever contained an AGN, the remnant mass per once-active galaxy must be even greater. Even with an efficiency close to the maximum conceivable, and with the assumption that the remnant mass is spread as evenly as possible over contemporary galaxies, the mass budget for AGN light production is interesting on a galactic scale. If the efficiency were substantially smaller, or if the remnants were found only in some galaxies, the mass per remnant would be even greater.

This theoretical argument for black holes as the prime movers of AGN has been strongly bolstered by recent observations. Fe Kα profiles with velocity widths several tenths of c have now been measured in many nearby AGN (Nandra et al. 1997). Clearly, at least mildly relativistic dynamics are required; orbits just outside a black hole are a very plausible location.

A second line of approach toward the identification of black hole central engines and AGN remnants has been to study the dynamics of the inner-most regions of galaxies, searching for the telltales of a Keplerian potential. Several different methods have all proved fruitful in different contexts: observing the kinematics of H_2O mega-maser spots (as reviewed, for example, by Willem Baan in this volume); using Hα emission to measure the rotation curves of orbiting gaseous disks (e.g. Harms et al. 1994); and studying how the stellar surface brightness and velocity dispersion behave as functions of radius from the galactic center (Magorrian et al. 1998). The specific results in terms of mass within a given radius vary from case to case, but discovering $\sim 10^8 M_\odot$ within ~ 10 pc of a galactic nucleus appears to be quite common.

1.2. Radiation Mechanisms

Major advances have also been made in the quest to work out the specific mechanisms by which different portions of the photon spectrum are made. Ordering by increasing frequency, we now have good evidence that

- Relativistic electrons generate the radio spectrum by synchrotron radiation. In blazars, the importance of synchrotron radiation extends to much higher frequencies, sometimes even into the X-ray band (see the reviews by Rita Sambruna and Marina Romanova in these proceedings).

- In non-blazar AGN, the infrared continuum can be safely attributed to thermal dust emission, although there is some uncertainty about how much of the energy can be attributed to the AGN proper, and how much to a nearby starburst region.

- The optical/ultraviolet continuum is very likely due to quasi-thermal emission from the surface of an accretion disk, although there are a great many detailed discrepancies between model predictions and observed spectra (Koratkar & Blaes 1999).

- Photoionization is almost certainly the culprit for both powering, and controlling the selection of, the rich spectrum of optical and ultraviolet emission lines generally seen in AGN. Although this conclusion had commanded general agreement long before, the monitoring experiments of the past few years have strongly confirmed this belief.

- Thermal Comptonization appears to provide an excellent explanation for the character of the hard X-ray spectrum (again excepting blazars).

- Non-thermal inverse Compton scattering is by far the most likely way to produce the high-energy photon continua of blazars, which often extend all the way up to GeV, and sometimes even to TeV energies (this subject is also discussed in the reviews by Sambruna and Romanova).

1.3. Population Studies

We also have a reasonable understanding of the character of AGN populations, and how they have evolved over the lifetime of the Universe. Both Pat Osmer and Vahé Petrosian reviewed for us the state of the art in determining how the luminosity function of AGN has changed from $z > 3$ to the present epoch. We have learned that extremely strong quasar activity was a distinctive mark of the period from $z \simeq 3$ to $z \simeq 1$, the time when, as we are now beginning to understand, the Universal star formation rate was at its peak (e.g., Madau, Pozzetti & Dickinson 1998).

The last decade has also seen a great improvement in our ability to recognize when two AGN are truly intrinsically different, and when they merely appear different due to anisotropic radiation and the accident of different viewing angles. Several different mechanisms are now known to create anisotropic appearance, and we can now, in many cases, trace their impact quite reliably.

Compact, flat-spectrum radio sources are almost always found in those AGN exhibiting dramatic variability, strong polarization, and powerful high-energy γ-ray emission. Objects such as these might appear to be quite distinct from extended, steep-spectrum radio sources, which are generally far less variable, much more weakly polarized, and undetectable above a few tens of keV. However, we now understand that these AGNs are in fact one and the same. They simply look very different because they contain jets of plasma moving at relativistic speed. The former phenomenology appears when the jet is moving almost straight at us; the latter is what we see when our line-of-sight is directed otherwise.

Similarly, there is now strong evidence for extremely optically thick belts of dusty gas occluding the majority of solid angle around the nuclei of both low-luminosity radio-quiet AGN (Seyfert galaxies) and high-luminosity radio-loud AGN (radio-loud quasars and radio galaxies). When our line of sight passes along the system axis, we are privileged to see the nucleus in all its glory, and we call it a type 1 Seyfert or a radio-loud quasar; when our line of sight is blocked by obscuring gas, we cannot easily detect the true nucleus, and rename the object a type 2 Seyfert or a radio galaxy.

2. Open Questions

That we have learned much does not negate the fact that much remains to be learned. We have made but little progress toward answering quite a number of the most fundamental questions about AGN that we might wish to ask.

2.1. Why do they exist?

Perhaps the most basic of these unanswered questions is why AGN should exist at all. Although it somehow seems "natural" that stars should form from the collapse of interstellar gas clouds, we could easily imagine a Universe without any AGN at all. If massive black holes ready to accrete are indeed the *sine qua non* of active galactic nuclei, this question may be rephrased as "What makes massive black holes?"

We still cannot say whether the origin of massive black holes may be found in the direct collapse of very large interstellar gas clouds, or in stellar collapse and subsequent growth. Whichever mechanism creates the original event horizon, we do know that their growth must be, in a well-defined sense, very "rapid". Because the maximum mass accumulation rate due to an efficiently radiating accretion flow (i.e., the Eddington accretion rate) scales in proportion to the central mass, black holes accreting in this fashion grow exponentially with a characteristic timescale called the "Salpeter time"

$$t_S \simeq 4 \times 10^7 \left(\frac{\eta}{0.1}\right) \left(\frac{L}{L_E}\right)^{-1} \text{ yr.} \qquad (2)$$

Note that to grow from, say, 10 $M_\odot$ to $\sim 10^8$ $M_\odot$ requires 16 *e*-foldings. On the other hand, the age of the Universe at the beginning of the quasar epoch ($z = 5$) was only

$$t_U \simeq (5\text{—}10) \times 10^8 \left(\frac{h}{0.75}\right)^{-1} \text{ yr,} \qquad (3)$$

where the range corresponds to a range of reasonable values for q_o. Sixteen *e*-foldings barely fit within the time available (perhaps that's why quasars may have first appeared around then).

A closely related question is why certain galaxies are active (at least at any one time), and why certain kinds of galaxies prefer certain kinds of nuclear activity. In the contemporary Universe, essentially all radio-quiet AGN are in disk galaxies (Adams 1977, Huchra & Burg 1992) while essentially all radio-loud AGN are in ellipticals (Martel et al. 1998). On the other hand, at slightly earlier times ($z \simeq 0.3$), there are indications that this clear division may break down (McLure et al. 1998).

We might likewise ask, "How long does activity last?" and "Once it ceases in a certain galaxy, does it recur?" The answers to these questions are bound up with the inquiry into what controls the fueling of active nuclei. The gravitational influence of even a very large central black hole cannot extend far enough into the host galaxy for it to be able to control its own fuel supply; the radius outside of which the deepening of the gravitational well due to the central black hole becomes negligible is

$$r_* \simeq 40 \left(\frac{M}{10^8}\right) \left(\frac{\sigma_*}{100 \text{ km s}^{-1}}\right)^{-2} \text{ pc,} \qquad (4)$$

where σ_* is the *rms* stellar orbital speed. There simply isn't enough fuel within that small a portion of a galaxy to supply what is needed, so it must be events farther out, connected with the ordinary life of the galaxy, that ultimately control how much fuel the nucleus is permitted to consume.

As mentioned earlier, the whole phenomenon of galactic nuclear activity reached its peak somewhere between $z \simeq 3$ and $z \simeq 1$. This fact immediately raises the suspicion that there is something about the youth of galaxies that encourages nuclear activity. Just what that might be, however, remains unknown. Many have speculated that this epoch was specially favorable to AGN ignition and fueling because it was a time when the ratio of gas mass to stellar mass in galaxies was relatively high. Because fluids are much more dissipative than collisionless systems like stars, it might therefore have been much easier to move matter inward then. On the other hand, others have favored the hypothesis that AGN activity was so much stronger then because the rate of major galaxy-galaxy encounters was very high. Even if the co-moving number density of galaxies hasn't changed since that time, the physical density was, of course, larger by a factor $(1+z)^3$, so the rate of encounters should have been that much greater. Moreover, many of the currently most fashionable theories about galaxy formation also posit that galaxies are assembled from small pieces, so that the *co-moving* number density of galaxies was also much larger then than now. We do not know which of these suggestions is more nearly correct, or whether some other effect was more important.

2.2. Accretion dynamics

A second area where we are truly far short of our goals is our understanding of accretion dynamics. That we do not genuinely understand accretion is immediately demonstrated by the gap between what we might predict for the output spectrum of AGN and reality.

According to the standard picture of accretion dynamics, the gas settles down into a geometrically and optically thick system. It should therefore radiate in a quasi-thermal fashion, with the temperature declining outward roughly as $r^{-3/4}$:

$$T_d \simeq 6.8 \times 10^5 \eta^{-1/4} L_{46}^{-1/4} \left(\frac{L}{L_E}\right)^{1/2} x^{-3/4} R_R^{1/4}(x) \text{ K}, \tag{5}$$

where L_{46} is the total luminosity scaled to 10^{46} erg s^{-1}; $x = rc^2/GM$, the radius in gravitational units; and R_R is a correction factor that includes both the outward flow of energy carried with the outward flow of angular momentum and general relativistic effects. R_R approaches unity at large x, and falls to zero at the innermost ring of the disk. This model would predict that AGN would radiate predominantly in the UV, with little power emerging at wavelengths either much longward of the optical band, or much shorter than a few tens of eV.

The reality is, of course, dramatically different. AGN radiate their power almost even-handedly in luminosity per logarithm of wavelength from the mid-infrared to at least hard X-rays, and sometimes beyond. There is often a modest local maximum in the UV, where accretion disk models predict that it should be found, but this maximum does not dominate the total luminosity anywhere near as thoroughly as the simple disk model would lead us to expect. Although, as mentioned in §1, the infrared continuum can be ascribed to reradiation of a primary continuum carried initially in higher energy photons, distant thermal reprocessing of ultraviolet photons cannot explain the strength of even the EUV continuum, much less hard X-rays or γ-rays.

There are also smaller scale problems. Virtually every detailed calculation of AGN accretion disk atmospheres has predicted some sort of Lyman edge feature, although effects such as oblique view of an accretion disk around a rapidly-spinning black hole can reduce the apparent strength of such features.

The failure to explain the very broad-band character of AGN emission may be interpreted in another way as a fundamental failure to understand why, in AGN accretion disks, a substantial fraction of the dissipated heat is concentrated onto a small fraction of the mass. Such a segregation of the dissipated energy is the only way to channel enough energy into "coronal" plasma capable of creating the observed EUV and X-ray continua.

In fact, there is an even deeper dynamical problem. Canonical disk models (in which the local stress is proportional to the total pressure) predict that the inner regions of AGN disks should be dominated by radiation pressure, and are therefore both thermally and viscously unstable. No satisfactory model has yet been found for these regions that is free of instability, although a number of speculative alternatives have been proposed.

2.3. Formation of relativistic jets

Yet another mystery about AGN on which we have made little headway is the nature of relativistic jets. That we see such jets is incontrovertible: from VLBI observations of superluminally-separating radio spots to the strength of high-energy γ-ray emission in blazars (a fact that requires relativistic outflow in order to reduce the apparent γ-γ opacity of the source plasma), we see unmistakable signatures of relativistic motion. The problem is that we have no answers to the most basic questions about them.

Although we can estimate global quantities such as total mechanical power and momentum flux, we do not know the nature of the plasma inside them: Is it an ordinary electron-ion plasma, or are most of the positive charges provided by positrons? Likewise, is most of the energy carried by matter or by Poynting flux?

We are also entirely ignorant about how they are accelerated and collimated. Most workers in the field believe that somehow they are expelled from the immediate vicinity of the central black hole by some combination of MHD effects and rotation, but there is no consensus beyond that vague statement.

Nor have we been able to identify clearly the energy source for jets. Some argue that their energy comes predominantly from the same accretion flow that powers everything else; others insist that the energy is drawn from the rotational energy of the black hole.

Finally, there is the basic mystery of why strong jets occur in only about 10% of all AGN, the radio-loud minority. This question is, of course, coupled to the one raised earlier about why (at least at the present epoch) radio-loud AGN (i.e., strong jet AGN) can be found only in elliptical galaxies, and never in disks.

3. Hopes for Progress

Having posed these questions, it is time to point out the directions in which we may hope for some progress toward answering them.

3.1. Existence

Several different lines of approach may soon yield clues to the creation of quasars. Major programs are being pursued to image the inner regions of AGN hosts, both near and far. Of course, the central technical problem to overcome is eliminating the bright point source that is the AGN itself. Starlight in the host can only be seen clearly after the contribution of the AGN is isolated and removed. Consequently, the best data come from those instruments with the highest angular resolution; in this case, that means the HST.

One way to further ensure complete removal of the AGN light is to look at AGN in which the nucleus is thoroughly obscured by opaque matter very close to the galactic center. Type 2 Seyfert galaxies are, of course, ideal for this purpose. HST observations (González Delgado et al. 1998, Storchi-Bergmann in these proceedings) have now shown that bright starbursts often occupy the inner few hundred parsecs of the hosts of type 2 Seyfert galaxies. One naturally wonders whether the existence of these starbursts has any causal relation to the creation of an active nucleus.

When looking at higher luminosity AGN, i.e. quasars, the number of known obscured nuclei is so small as to preclude making use of such a natural "coronagraph". Instead, one has no choice but to try to subtract out the point source light. Combining the results of Bahcall, Kirhakos & Schneider (1996) with those of McLure et al. (1998), we have learned that the hosts of quasars are an unusual lot, with many showing signs of disturbance.

Further pursuit of this program should bring further rewards, but genuine progress will require moving from the collection of examples to systematic statistical studies. Fortunately, the construction of such studies is about to become much easier, for several enormous surveys are about to burst onto the scene. Over the next five years, the Sloan Digital Sky Survey will image 1/4 of the entire sky, collecting spectra for 10^6 galaxies and 10^5 quasars. Meanwhile, the 2MASS (Two Micron All Sky Survey) is roughly halfway through covering the entire sky in the near-infrared. It will ultimately produce a catalog of $\sim 10^6$ galaxies brighter than $K \simeq 13.5$ mag. In a few years, the GALEX spacecraft will obtain UV images of 10^7 galaxies and 10^6 quasars, and UV spectra for 10^5 galaxies and 10^4 quasars. The FIRST (Faint Images of the Radio Sky at Twentyone centimeters) survey has produced VLA images of roughly 10% of the sky, with an ultimate goal of covering about 25%. Its ultimate catalog should contain $\sim 10^6$ entries.

These enormous databases should be useful in several ways. Clear sample definition will become much easier because they have well-defined, homogeneous selection rules. Because of their tremendous size, it should be possible to find statistically significant samples even after taking "cuts" in a variety of parameters. Such overwhelmingly large catalogs should also turn up examples of unusual and rare varieties, some of which may be especially instructive.

Yet another positive trend, remarked upon by numerous speakers at this Symposium (Malcolm Longair, Dave Sanders, and Gene Smith), is the convergence between galaxy formation and AGN studies. With our newfound ability to observe galaxies in the same redshift range where the quasar activity peak is located, we can now seriously begin to develop the connection between galaxy formation and nuclear activity that has so long been merely a matter of speculation. Real data about the character of inactive galaxies at $z \simeq 2$ may be extremely helpful in understanding what made so many other galaxies' nuclei active at that time.

3.2. Accretion dynamics

There is also a great deal of activity in the area of accretion dynamics. Not so long ago, state-of-the-art accretion disk atmosphere codes incorporated only Thomson and free-free opacity. Now they include: bound-free opacity for both H and He, with a non-LTE treatment of their ionization balance; opacity due to heavy element resonance lines and ionization edges; non-diffusive Comptonization; and explicit treatment of vertical force balance, including radiation pressure (Koratkar & Blaes 1999). These dramatic improvements in calculational quality should permit the confrontation between theoretical predictions and observations to become ever more pointed. Soon we should be able to say clearly whether the almost-universal absence of Lyman edge features truly poses a fundamental problem for conventional ideas about disk structure. Similarly, theorists should soon be able to make clear statements about when they would predict HeII edges, and about the character of UV polarization to be expected.

The last few years have also seen the emergence of a number of important new ideas about fundamental disk dynamics. The invention of the Balbus-Hawley mechanism (reviewed in Balbus & Hawley 1998), has finally given us a plausible physical picture of how angular momentum is transported in accretion disks. With this advance, we are now poised to learn how and where the accretion energy is dissipated into heat, so that it may eventually be radiated as photons. A description from first principles of where the dissipation occurs will remove a major systematic uncertainty from disk atmosphere models.

Another important new development in fundamental disk physics is a deeper understanding of disk equilibria and stability. Narayan & Yi (1995) emphasized the importance of the fact that low accretion rates may permit "advection-dominated accretion", i.e., a condition in which the gravitational potential energy of the accreting matter is dissipated into heat, but is advected into the black hole (or used to drive a wind) rather than being lost in photons. Others have stressed that high accretion rate disks are subject to such strong radiation pressure-driven instabilities that the conventional Shakura-Sunyaev equilibrium likely does not exist in Nature. One suggestion is that the result is a limit-cycle (Szuszkiewicz & Miller 1998); another is that the disk adopts a qualitatively different equilibrium, in which most of the matter is found in relatively high-density clumps, and the rest is more smoothly distributed. In this latter picture, the low density gas rises to a temperature so high that it cools by Compton scattering and produces the observed hard X-ray emission (Krolik 1998).

In fact, the interplay between X-rays produced by Comptonization and their reprocessing in the (presumably) adjacent accretion disk has opened up a whole new field of observational studies, coordinated monitoring of X-ray and ultraviolet monitoring (as reported, for example, in the talks by Toshihiro Kawaguchi and Thierry Courvoisier). By following the variations in both bands, we can reasonably hope to understand better the double feedback loop (ultraviolet photons from the disk upscattered to X-rays, X-rays partially absorbed by the disk and reprocessed into the ultraviolet) that connects the two systems.

Still more information on disk dynamics should be forthcoming as the quality of Fe Kα line profiles improves. Hitherto, these have been badly limited by the inability of ASCA data to clearly define the continuum shape (ASCA's sensitivity falls drastically above 6 keV). However, data should be available very soon from coordinated ASCA and RXTE observations, in which RXTE spectra define the continuum shape that can then be subtracted from the ASCA spectra in order to isolate the Kα profile. Moreover, the order of magnitude increase in sensitivity promised by XMM should permit both much higher S/N Fe Kα profiles in bright AGN, and the extension to high redshift of the entire field of study. Indeed, there may be some quasars whose redshifts bring the line within the range spanned by the high-resolution spectrometer on XMM; from those quasars (if they are bright enough), we may be able to obtain a much finer measure of the shape of the Kα line.

3.3. Jets

The complexities of jet acceleration and collimation physics demand numerical simulation (reviewed in this volume by Dick Lovelace). Fortunately, the rapid growth in computational power makes possible today calculations that even a few years ago were only pipedreams. Reasonable resolution 2-d MHD simulations in a general relativistic background are quite feasible today; 3-d should not be far in the future.

Moreover, advances in disk physics should contribute to new steps in understanding the origins of jets. The Balbus-Hawley mechanism doesn't only explain inter-ring torques in disks; it also explains (and predicts) the strength and structure of disk magnetic fields. A physical understanding of dissipation in disks should also be of great aid when we try to discover how disk heating can help launch jets.

On the observational side (as discussed by Thierry Courvoisier), monitoring campaigns have much to offer this subject, just as they do for disk-corona relations. Multiwavelength monitoring campaigns can trace the propagation of shocks down jets, constraining our ideas about how and where relativistic electrons are accelerated and cooled, as well as ideas about the source and transmission of the seed photons later inverse Compton scattered up to high energies.

References

Adams, T.F. 1977, ApJS 33, 19
Bahcall, J.N., Kirhakos, S. & Schneider, D.P. 1996, ApJ 457, 557
Balbus, S.A. & Hawley, J.F. 1998, Revs. Mod. Phys. 70, 1
González Delgado, R.M., Heckman, T., Leitherer, C., Meurer, G., Krolik, J., Wilson, A.S., Kinney, A. & Koratkar, A. 1998, ApJ505, 174
Huchra, J.P., and Burg, R. 1992, ApJ 393, 90
Koratkar, A. & Blaes, O.M. 1999, PASP in press
Krolik, J.H. 1998, ApJ 498, L13
Madau, P., Dickinson, M. & Pozzetti, L. 1998, ApJ 498, 106

Magorrian, J., Tremaine, S.D., Richstone, D., Bender, R., Bower, G., Dressler, A., Faber, S.M., Gebhardt, K., Green, R.F., Grillmair, C., Kormendy, J. & Lauer, T. 1998, AJ 115, 2285

Martel, A.R., Baum, S.A., Sparks, W.B., Wyckoff, E., Biretta, J.A., Golombek, D., Macchetto, F.D., de Koff, S., McCarthy, P.J. & Miley, G.K. 1998, ApJS in press

McLure, R.J., Dunlop, J.S., Kukula, M.J., Baum, S.A., O'Dea, C.P. & Hughes, D.H. 1998, preprint astro-ph/9809030

Nandra, K., George, I.M., Mushotzky, R.F., Turner, T.J. & Yaqoob, T. 1997, ApJ 477, 602

Narayan, R. & Yi, I. 1995, ApJ 452, 710

Sołtan, A. 1982, M.N.R.A.S. 200, 115

Szuszkiewicz, A. & Miller, J.C. 1998, MNRAS 298, 888

VI. REFLECTIONS OF V.A. AMBARTSUMIAN

V.A. Ambartsumian and Yervant Terzian
(1995, Byurakan)

Influence of V. A. Ambartsumian on the Development of Astronomy

Alexander Boyarchuk

Institute of Astronomy of the Russian Academy of Sciences, 48 Pyatnitskaya Str., 109017, Moscow, Russia. Electronic Mail: aboyar@inasan.rssi.ru

This is a great honour for me to present a report on the influence of the outstanding scientist academician V.A. Ambartsumian. I was well familiar with V.A. Ambartsumian, but unfortunately, I did not manage to work with him in scientific research. When, in 1948, I arrived in the Leningrad State University (LSU) V.A. Ambartsumian already had left LSU and worked in Armenia. My teacher in LSU, academician V.V.Sobolev, was the best student of V.A. Ambartsumian. Therefore, to some degree, I can consider myself as a scientific grandson of V.A. Ambartsumian. However, I very frequently met V.A. Ambartsumian at scientific conferences, discussed with him scientific problems and problems of organization of scientific research. Contacts with this outstanding, strong and talented person have rendered large influence on my formation as a scientist.

When I was asked to prepare a lecture on scientific activity of V.A. Ambartsumian, I reviewed his scientific articles and came to the conclusion that it is an excessive task for a half-hour or even one-hour report. Therefore, having consulted with my colleagues, I have decided to choose several results, which have exerted decisive influence on the development of astronomy and which characterize the versatile activity of V.A. Ambartsumian. Certainly it is difficult to make an unequivocal choice from the large scientific heritage of V.A. Ambartsumian, therefore my choice is rather subjective. Someone else would make his choice on other results. But I hope I will manage to show an outstanding role of V.A. Ambartsumian in the development of astronomy, as well as other sciences.

V.A. Ambartsumian lived a long scientific life and results achieved by him 70 years ago now may seem trivial, but in that time they were certainly new data frequently overthrowing existing views on those or other phenomena.

One should bear in mind the level of development of astronomy in those days when V.A. Ambartsumian began his research activity. Nuclear reactions were unknown as the energy source in stars, the chemical composition of stars was not determined. There were neither ultraviolet nor infrared observations. Radio astronomy and high-energy astronomy did not exist. Astronomers knew only positions of stars in the sky, their proper motions (for nearby stars) and radial velocities (for bright stars), as well as visual magnitudes and an approximate temperature scale.

Among early works in the field of study of non-stationary stars I would like to pay attention to the method of determination of temperatures of nuclei of planetary nebulae proposed by V.A. Ambartsumian.

Planetary nebulae are very interesting objects. The variety of their forms is amazing. Their spectra consist mainly of intensive emission lines drastically different from absorption spectra of stars.

Shortly before the work of V.A. Ambartsumian an opinion was stated, that the radiation from a planetary nebula is a re-emission of the radiation of the central star. Zanstra offered a method to determine temperatures of central stars, using intensities of emission lines of hydrogen measured with respect to the continuous spectrum of the star. The temperature, a very important characteristic of the star, appeared to be very high, close to 100000 K. The difficulty in applying the method by Zanstra was that it was necessary to measure very intense emission lines relative to a very weak continuum, where the star's radiation itself is strong. So, 24-year old V.A. Ambartsumian, in 1932, offered to determine temperatures from the ratio of the HeII line 4686 and the hydrogen line 4861, located close to each other in the spectrum and having comparable intensities [1]. He considered in detail the process of formation of the recombination spectrum of HeII. He showed that, with the exception of chemical composition, this ratio only depends on temperature.

V.A. Ambartsumian calculated and tabulated the function $F(T)$ which describes the dependance of the ratio of line intensities on stellar temperature. Having determined from observations the ratio of intensities, we immediately obtain from the table the temperature of the central star of the planetary nebula. It is a very simple and fast method of determining the temperature. A graceful method!

This work put a beginning to the whole cycle of works devoted to problems of radiation transfer in planetary nebulae and envelopes of non-stationary stars [2]. The results of these works were generalized in the book by V.A. Ambartsumian "Theoretical astrophysics" [3], the first book of this sort which has rendered huge influence on a whole generation of astronomers.

The cycle of works concerning the study of processes of scattering of light adjoins the listed works. The problem of scattering of radiation is paramount in astronomy, as all our knowledge of celestial objects is received from analysis of electromagnetic radiation frequently named by the short word light. Mostly between a source of radiation and detectors there is absorbing and scattering matter. We can include here the atmosphere of a star, circumstellar envelopes, interstellar medium, and terrestrial atmosphere. Their influences should be carefully taken into account to deduce correct data on the source of radiation, as well as characteristics of the scattering matter. For this reason the works of V.A. Ambartsumian are of great importance.

The theory of radiation transfer is rather complex and it comes down to the solution of a system of integro-differential equations. Its presentation for non-specialists, which a large audience inevitably consists of, is rather boring business, but the merits of V.A. Ambartsumian in this area are so great, that it is impossible to pass by in silence this topic. Therefore I shall dwell upon one point, which I consider to be one of the masterpieces of scientific ideas, the method of invariancy [4,5].

V.A. Ambartsumian attacked the problem completely from another side. He considered a case, when radiation S falls on a semi-infinite medium at an angle ξ and it is necessary to determine the intensity I of radiation scattered at an angle η. In real life there are many such cases. For example, a sea illuminated by the Sun.

V.A. Ambartsumian proposed to add a layer of very small optical thickness to the layer of infinite optical thickness. It is clear, that general characteristics of the whole layer do not change. But this small addition allowed him to consider how the radiation in a direction η arises. This is scattering in the layer $\Delta \tau$, reflection from the border A, scattering of the reflected radiation, and secondary reflection. Additional absorbtion in a layer $\Delta \tau$ was also taken into account. The appropriate equations were compiled and their solution resulted in a functional equation.

The functional equation can be rather simply solved by the successive approximation method. It was many a time and oft done later, and there exist rather detailed and exact tables of their values. Also, using this method V.A. Ambartsumian solved the problem of radiation transfer through a layer of restricted optical thickness, by taking away a small layer at one border of the medium and adding it to the other border. The problem was reduced to solution of two functional equations by the successive approximation method. The paper by V.A. Ambartsumian was published in the Reports of the Academy of Sciences of the USSR in 1943 [4]. The same solution was achieved by the great American scientist Chandrasekhar in 1947, who recognized the priority of V.A. Ambartsumian.

The method offered by V.A. Ambartsumian and named as the method of invariancy has found wide application not only in astronomy and not only in investigations of radiation transfer, but also in other areas of science and in studies of other phenomena, for example scattering of neutrons in nuclear reactors. This is surely one of the great achievements of V.A. Ambartsumian. He not only proposed the method but also obtained solutions for various configurations of scattering matter and different kinds of the scattering medium.

In the 1930s V.A. Ambartsumian published some outstanding papers on stellar dynamics, which gave him, a 28-year old scientist, world popularity. In those days stellar dynamics was one of most active directions of astronomy.

One of most interesting problem in those times, as well as now, is determination of the age of the universe. Just then, practically the only way to determine the age of the universe or one of its parts was a statistical investigation of the general characteristics of stars.

It was supposed that after formation of stars their ensemble eventually comes to statistical equilibrium as a result of gravitational interaction. The lowmass stars would have high spatial velocities, and vice versa. There should be a very small number of wide pairs, and so on.

Jeans calculated that the time for establishment of equilibium, that is the relaxation time of the system, amounts to 10^{13} years. He also found that, provided the distribution of binary stars over energy, i.e. over distance between components, is in equilibrium, the distribution of eccentricities of orbits should be proportional to ε^2, as is observed. Based on this point, Jeans concluded that the age of the universe is more than 10^{13} years. It agrees well with his hypothesis concerning the evolution of stars along the main sequence. V.A. Ambartsumian convincingly showed that the consideration of Jeans is wrong [6-9]. He showed that irrespective of the kind of distribution of binary stars over the sizes of their orbits, the distribution of binary stars over eccentricities mentioned above always should be fulfilled. Therefore, the analysis of eccentricities yields no answer for the question whether there is a statistical equilibrium. Thus, the conclusion by Jeans on the long time scale appeared to have no foundation. V.A. Ambartsumian analysed the distribution of binary stars over energy, that is the orbit size, the ratio of the numbers of wide and close pairs, the ratio of the number of wide pairs to the number of single stars, etc. He came to the conclusion that the ensemble of stars is still very far from equilibrium and its age is 10^{10} years (the short time scale). The modern data on the age of the Universe yields $(1.5-2.0)^{10}$ years. The accuracy is amazing. The discussion lasted about two years. There was an exchange of letters to the world-wide journals. V.A. Ambartsumian won. V.A. Ambartsumian acquired fame due to this discussion and since then his name was usually considered in relation with a foundation of modern astronomy.

It is pleasant to note that V.A. Ambartsumian was awarded the State Prize of Russia in 1996 for this cycle of works on stellar dynamics.

In the middle 40s V.A. Ambartsumian paid attention to the fact that the non-stationary stars such as T Tauri are located not homogeneously but in separate groups. Their sizes are appreciably larger than those of usual star clusters. V.A. Ambartsumian called them stellar associations (T-associations). The T Tau-type stars seem to be young by reason of many of their features. Therefore it was decided to analyze how the other young stars behave. The O-type stars also appeared to group in the sky, forming O-associations. The evidence that associations are young is given also by the presence in their structure of multiple stars of the Trapezium type, where the distances between components are close unlike usual multiple stars with a hierarchy of distances. For example, ξ U MA is a usual multiple system, θ Ori is a Trapezium. V.A. Ambartsumian showed that the time of disintegration of a Trapezium amounts to a few million years.

Astronomers all over the world began to actively investigate associations which were called aggregates by Western astronomers.

The main idea in investigation of associations is that the process of star formation occurs till now and stars are born in groups. A new direction of astronomy has appeared. Certainly, during the past 50 years, particularly due to appearance of infrared observations, the accents of the problem of star formation have been strongly shifted to researches of protostars, cocoons, and other objects. But we should remember that the study of associations was one of the basic sources of the science of star formation.

The ideas of V.A. Ambartsumian have also given rise to other important diretion of astronomy, to which the present astronomical symposium in Byurakan is devoted - investigation of active galactic nuclei. It seems to me that conferencies on this subject occur more often than on any other topic of astronomy. It means that this subject is of current interest.

The main idea put forward by V.A. Ambartsumian was as follows. Most galaxies can be roughly subdivided into elliptical (with few details), spiral (with details observed as spirals) and irregular (with unclear details). V.A. Ambartsumian noted that besides this subdivision the galaxies differ from each other by the brightness and size of the central nucleus. There are galaxies having no nucleus, but in others nuclei are very strong.

As the other parts of a galaxy rotate round the nucleus and the density of the nucleus is several orders higher than that of the other parts of a galaxy, V.A. Ambartsumian argued that the nucleus essentially influences the life of the galaxy as a whole and suggested that astronomers investigate the nature of this influence; many of them have studied the problem until now.

The recognition of the importance of the galactic nuclei problem raised by V.A. Ambartsumian was demonstrated by the fact that he was invited to present several reviews on the topic in the early 1960s [12,13], including the review lecture at the IAU General Assembly in Berkeley in 1961 [14], and also by the huge development of these investigations during the last 60 years. Let me not dwell more on this subject. There are here a lot of astronomers much more qualified about the topic than I am. Those who wish to learn more about galactic nuclei will do it during the symposium.

Let me finish a brief and subjective review of the huge scientific heritage of V.A. Ambartsumian.

I would like also to say some words about organizational activity of V.A. Ambartsumian in the field of astronomy.

V.A. Ambartsumian was the Vice-President of the Astronomical Council at that time when it was a powerful organization, the President of the IAU, the President of the Scientific Council of Scientific Unions. He created the Department of Astrophysics in Leningrad University and the Byurakan Astrophysical Observatory in Armenia. He was a Full Member of the USSR Academy of Sciences and a Member of the Academies of more than a dozen countries, including the main ones. Besides he occupied many responsible posts in governmental circles, i.e. he was a rather influential man. People came to him for help and advice and he never refused support if the business concerned the development of science. It can be surely said that no single serious measure in the field of astronomy has been made without the participation of V.A. Ambartsumian. The shining example was the construction of 2.6-m and 6-m telescopes, which are the largest ones in Russia.

It may safely be said that the influence of V.A. Ambartsumian on the development of astronomy, particularly in the USSR and Armenia, was very great. We were lucky, that such a brilliant person as V.A. Ambartsumian was with us.

References

1 Ambartsumian, V. A. 1932, Circ. Pulkovo Obs., 4
2 Ambartsumian, V. A. 1993, Izv. Pulkovo Obs., 114, 1
Ambartsumian, V. A. in The Theoretical Astrophysics. Moscow-Leningrad, 1939
Ambartsumian, V. A. 1943, Rept. USSR Acad. Sci., 38, 257
Ambartsumian, V. A. 1943, J. Experim. and Theoret. Phys., 13, 224
Ambartsumian, V. A. 1935, Observatory, 58, 152
Ambartsumian, V. A. 1936, Nature, 137, 537
Ambartsumian, V. A. 1937, Russian Astron. J., 14, 207
Ambartsumian, V. A. 1949, Russian Astron. J., 26, 3

Ambartsumian, V. A. 1954, Byurakan Obs., 15, 3
Ambartsumian, V. A. 1955, Observatory, 75, 72
Ambartsumian, V. A. 1966, IAU Trans., vol. XIIB, Academic Press, London, p.578
Ambartsumian, V. A. 1965, InterScience Publishers, London, p.1
Ambartsumian, V. A. 1962, IAU Trans., vol. XIB, Academic Press, London, p.145

Ambartsumian's Greatest Insight - The Origin Of Galaxies

Halton Arp

Max-Planck-Institut für Astrophysik, Karl-Schwarzschild-Str. 1, 85740 Garching, Germany

Abstract.

From simply looking at pictures of galaxies Ambartsumian realized that new galaxies were formed in ejections from old galaxies. In the ensuing 40 years, observations have supported in increasing detail his original insight. We can now empirically outline the development of compact objects emerging from the nuclei of active galaxies into young star forming galaxies and finally into aggregates of old stars.

The observations actually require galaxies to continually originate in a low particle mass plasma which has remarkably similar properties to the "superfluid" which Ambartsumian foresaw. He had the courage to present these conclusions to influential astronomers who still today reject any origin of galaxies other than in the Big Bang. In this most important subject of science, the nature of our universe, Ambartsumian's revolutionary insights are now increasingly vindicated by observation.

1. Looking at Galaxies

In order to communicate a feeling for how my appreciation of Ambartsumian's work grew over time, I wish to recount a series of seminal events roughly in chronological order:

In 1949 I came to Cal Tech. My first job as a graduate student was to pick guide stars for the Palomar Schmidt Sky Survey (PSSS). In the following years the Sky Survey was carried out. Famous astronomers such as Hubble, Baade, Zwicky, Minkowski and Abell noticed peculiar and unusual galaxies from time to time. But I felt that in order to understand their physical nature one had to classify their properties and particularly to study the possible causes of the continuity of change of these properties. Between 1961 and 1965 I finally got the opportunity to do systematic photography of the 333 most peculiar galaxies with the highest resolution telescope of that era, the 200-inch at Palomar. After assembling the Atlas of Peculiar Galaxies my conclusion was clear: Galaxies could eject material which resulted in the birth of smaller galaxies!

Then a shocking event occurred in my young researcher's life. I realized that I had forgotten that Ambartsumian had come to exactly the same conclusion about eight years earlier. What impressed me the most was that he had come to this conclusion by just looking at the Schmidt PSSS prints which had much less detail than my reflector plates. When I asked some of the older scientists, they told me that he had presented his conclusions at the prestigious Solvay conference in about 1957. They also related that this select group of the best known scientists in the world had either been completely baffled or laughed privately at these crazy ideas.

At that point I recognized that Ambartsumian would be one of the few people in the world who would really understand and appreciate the peculiar galaxies classified in my Atlas. Some of them were Ambartsumian's original objects now viewed with higher resolution. So I determined to travel to Armenia and present him with the first copy of the Atlas. I wrote to the Russian Embassy informing them that Ambartsumian had invited me and requested a visa. It was delivered to me as I was going up the steps to the airplane to fly to Europe. Only when I was on my out of Russia did I learn that I was in the disturbingly irregular position of being neither Intourist nor a guest of the Academy of Sciences. Ambartsumian's influence had apparently secured a singular visa for me at the last moment.

I remember walking up the steps of the Byurakan Observatory where Ambartsumian met me and took me to his office. There, together with the staff of the Observatory, we first looked at the birds nesting outside his window and then proceeded to go through the Atlas of Peculiar Galaxies, photograph by photograph, discussing and commenting on each one. Later walking in the garden with him he said to me, "Well you don't need to come all the way to Armenia to talk to me when you have so many famous astronomers in Pasadena to discuss these matters with." I was startled and quickly looked at him. He was smiling and I knew that this was a private joke that referred to the enormous schism that had developed between the majority of astronomers who were sure galaxies all were born at the same time in the beginning of the universe and that small minority who believed galaxy creation was continually going on.

2. The Theory Freezes

Four years earlier, in 1961, the International Astronomical Union (IAU) had met in Berkeley, California. Ambartsumian assumed Presidency of the Union at that time and was also present at IAU Symposium 15 at Santa Barbara California. That symposium brought together all the reigning experts in extragalactic astronomy. Among them were Sandage, de Vaucouleurs, Zwicky, Abell, Vorontsov-Velyaminov, Holmberg, Minkowski, Hoyle, Ryle and most prominent of all, Jan H. Oort, the retiring president, and the astronomer credited with the discovery of galactic rotation and the Oort cloud of comets.

Because Ambartsumian gave an invited discourse at the General Assembly which followed some days later in Berkeley, he gave only a minor talk in Santa Barbara. In his Berkeley talk the summary was exciting and masterful but many of the audience did not understand it. An example from his printed discourse, however is clear "...the jet joining the dwarf galaxy to the large one...there remains no doubt that the dwarf galaxy has been detached from the prinipal galaxy...the bridges and the filaments as resulting from the formation of two galaxies from one." Geoffrey Burbidge knew well the proposals of Ambartsumian, and in his summary talk in Santa Barbara, reviewed the evidence for young galaxies. The opinion of the conference was swept, however, by Jan Oort. It is not evident from his printed paper, but from his oral presentation and many comments to informal discussion groups it was clear that he stressed arguments supporting the "all galaxies are old" school of opinion. That turned out to be the defining event of the conference. Even to this day, it is the unshakable belief of official astronomy that no young (newly created) galaxies can exist.

The reason that this assumption cannot be questioned is that the Big Bang theory of the creation of the universe requires that all galaxies formed at the same time about 15 billion years ago. This was basically the argument that was used at the Santa Barbara conference in 1961 and it has somehow come to be regarded as proof that the many observations of young and forming galaxies are not what they seem. It is therefore extremely important for all astronomers, now and in the future, to realize that they must make a conscious choice in their basic assumptions in extragalactic astronomy. If they choose the hypothesis that galaxies are not being created today, and if this is not true, then their research efforts will be, for the most part, wasted and their entire cosmology embarassingly incorrect.

3. What do People Really Think?

In 1973 the IAU was held in Australia. Because that was so distant for Europeans, exceptionally a second section was convened in Krakow. Due to the fact that I returned from Australia via the Tata Institute in Bombay to see Narlikar, I was able to also attend the Polish part of the meeting. One moment of that latter meeting stands out vividly in my memory. Ambartsumian was chairing a session of contributed papers and I happened to be sitting next to Jan Oort. I knew Oort fairly well by then. Looking at Ambartsumian he leaned over and whispered in my ear:

"You know, Ambartsumian was right about absolutely everything."

I was rather stunned, and I have been thinking about that remark ever since. I finally have concluded that Oort must have wavered from time to time in his opinions in spite of his having such strong control over the adherence of the community to the orthodox doctrine. It also serves to highlight the paradox which Ambartsumian represented to the established paradigm. It was unanimously agreed that Ambartsumian was a great astronomer. At the same time his statements about the most important fundamentals in astronomy were not believed. This is even more true today. Is this another example of people thinking they believe one thing and then acting in a completely opposite manner? Is there an insincerity or inconsistency in the higher levels of astronomical theory?

Ironically, as the evidence has grown in favor of continual creation of galaxies, the data has been increasingly citicized and rejected from main stream journals. In all honesty, of course, there has arisen a complication in that some researchers, including myself, feel that when matter is created it has high intrinsic redshift. Then, as it ages and evolves, the redshift falls to more normal values. This enables quasars to be low luminosity, nearby proto-galaxies and to evolve into second generation companion galaxies with slightly higher redshifts as observed. Up until about the middle 1980's I know Ambartsumian did not believe in the existence of non-velocity redshifts. I have no reason to believe he changed his mind after that. But it is true that what kept the establishment from believing the evidence for non-velocity redshifts was fear for the expanding universe hypothesis. That same fear kept them from believing Ambartsumian's evidence for new galaxies. Put another way, Ambartsumian's arguments had already contradicted the Big Bang so it was no longer necessary to save it by rejecting the existence of non-Doppler redshifts. But there was another amazing link between quasars as high redshift proto galaxies and Ambartsumian's picture of the creation of new galaxies.

4. Ambartsumian's "Superfluid"

It was clear that in order to "detach" a new galaxy from an old galaxy that it would have to emerge from the very small active nucleus and then expand into a normal appearing galaxy with apparently normal stars. Ambartsumian apparently had the physical intuition to realize that this would be impossible to do with an ordinary hot plasma. Therefore he proposed that the galaxies emitted a "superfluid". In conventional plasma this would imply a superdense state. It was the obvious difficulties with this high density which, I presume, prevented him from going farther with the model.

Having to account for the high intrinsic redshifts of the new galaxies, however, cast a startling new light on the nature of the emerging material. The only way matter could be understood to emit redshifted photons which was not due to recession velocity was if the constituent particles of the matter were of low inertial mass. Narlikar in 1977 showed that a *general* solution of the General Relativity field equations required just that - particle masses changing with age since creation and hence intrinsic redshifts initially high and decreasing with time. But with particle masses low and interaction cross sections high, this was a more or less perfect description of a fluid. Moreover it could emerge as a very small object with nearly zero mass from the small active nucleus. As time passed it would then develop into a more normal mass plasma.

More details of this galaxy forming process are considered in my paper in the this symposium. But here it is appropriate to emphasize that more than 30 years ago, from just the small telescope observations which were available, Ambartsumian foresaw the essential physical processes which would be required to account for the most fundamental entities in cosmology - the galaxies. Still today the vast majority of astronomers have not been able to reconcile the observations with a simple, understandable picture of how the universe works. Ambartsumian was, in my opinion, a Copernican man in Ptolemaic times. With such huge changes in concept about the most fundamental aspects of nature the paradigm takes a very long time to shift. From the common sense of Aristarchus and Eratothsenes there was the interregnum of Ptolemy for 1800 years before the clarity of Copernicus. I would hope, with the light of Ambartsumian shining ahead, that today's astronomers would more quickly relinquish their concentration on epicycles required to shore up a theory of everything created instantaneously out of nothing and follow the observational path to a more profound understanding of how the universe really works.

The following picture shows a galaxy with conspicuous jet activity which was originally noted by Ambartsumian. (Photograph taken by Arp with the 200-inch Palomar telescope).

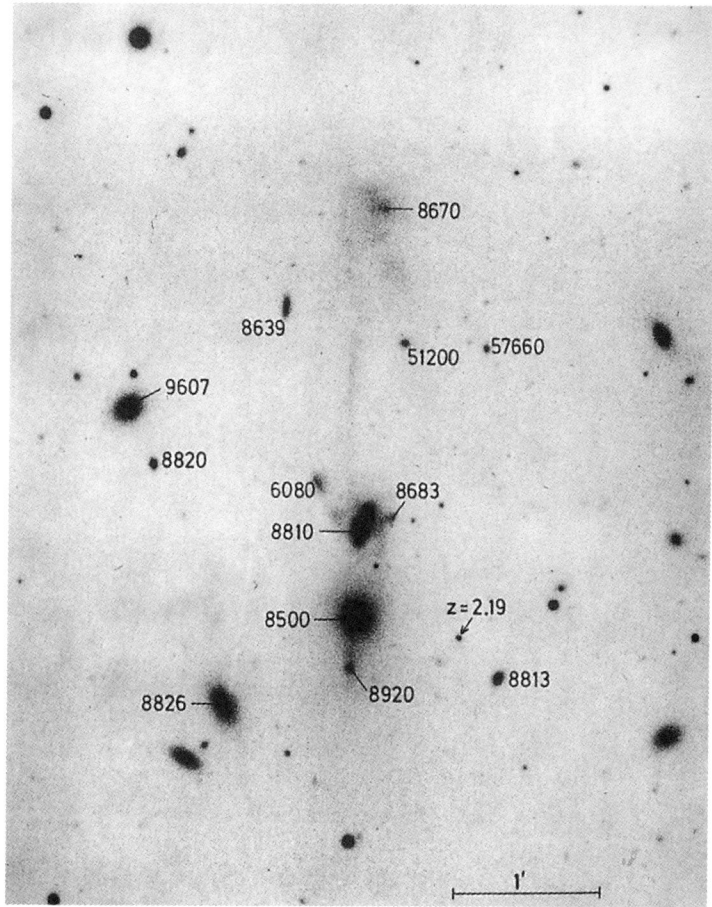

Figure 1. "Ambartsumian's knot" is shown being ejected directly southward from the giant E galaxy NGC3561B (also known as Arp Atlas No. 105). This small companion as well as 12 out of 13 of the surrounding companion galaxies have higher redshifts than the parent galaxy. In this single picture one can view the evolution from the young, high redshift quasar ($z = 2.19$) through to the oldest second generation galaxies of only a few hundred km/sec higher redshift.

INTERNATIONAL ASTRONOMICAL UNION
SYMPOSIA
Published by the Astronomical Society of the Pacific

Vol. No. 190	NEW VIEWS OF THE MAGELLANIC CLOUDS eds. Y.-H. Chu, N. Suntzeff, J. Hesser, and D. Bohlender ISBN:
Vol. No. 191*	ASYMPTOTIC GIANT BRANCH STARS eds. T. Le Bertre, A. Lebre, and C. Waelkens ISBN: 1-886733-90-2
Vol. No. 192*	THE STELLAR CONTENT OF LOCAL GROUP GALAXIES eds. P. Whitelock and R. Cannon ISBN: 1-886733-82-1
Vol. No. 193*	WOLF-RAYET PHENOMENA IN MASSIVE STARS AND STARBURST GALAXIES eds. K. A. van der Hucht, G. Koenigsberger, and P. R. J. Eenens ISBN: 1-58381-004-8
Vol. No. 194*	ACTIVITY IN GALAXIES AND RELATED PHENOMENA eds. Y. Terzian, E. Khachikian, and D. Weedman ISBN: 1-59391-008-0

Please note volumes with an asterisk (*) and ISBN are already printed as of August 1999

Complete lists of proceedings of past IAU Meetings are maintained at the
IAU Web site at the URL: http://www.iau.org/publicat.html

Volumes 32 - 189 in the IAU Symposia Series may be ordered from
Kluwer Academic Publishers
P. O. Box 117
NL 3300 AA Dordrecht
The Netherlands

EDITORIAL/PUBLISHING OFFICE:
Managing Editor
PO Box 24463
211 - KMB
Brigham Young University
Provo UT 84602-4463
USA

(801) 378-2298   Phone
(801) 378-2265   Fax
pasp@astro.byu.edu   E-mail

CATALOG/BOOK ORDERS:
IAU Publications
390 Ashton Avenue
San Francisco CA 94112-1722
USA

(415) 337-1100   Phone
(414) 337-5205   Fax
catalog@aspsky.org   E-mail

OHIO UNIVERSITY LIBRARY
Please return this book as soon as you have finished with it. In order to avoid a fine it must be returned by the latest date stamped be-